# ELEMENTS OF HEAT TRANSFER

# ELEMENTS OF HEAT TRANSFER

**Ethirajan Rathakrishnan**

CRC Press is an imprint of the
Taylor & Francis Group, an **informa** business

CRC Press
Taylor & Francis Group
6000 Broken Sound Parkway NW, Suite 300
Boca Raton, FL 33487-2742



Version Date: 20120227

International Standard Book Number: 978-1-4398-7891-0 (Hardback)



**Library of Congress Cataloging-in-Publication Data**

Rathakrishnan, E.
Elements of heat transfer / Ethirajan Rathakrishnan.
p. cm.
Includes bibliographical references and index.
ISBN 978-1-4398-7891-0 (hardcover : alk. paper)
1. Heat--Transmission. I. Title.

QC320.R375 2012
536'.2--dc23 2012000318

**Visit the Taylor & Francis Web site at**
**http://www.taylorandfrancis.com**

**and the CRC Press Web site at**
**http://www.crcpress.com**

# Dedication

This book is dedicated to my parents,

Mr. Thammanur Shunmugam Ethirajan

and

Mrs. Aandaal Ethirajan

Ethirajan Rathakrishnan

# Contents

# Preface

This book is developed to serve as a text for a course on heat transfer at the introductory level for an undergraduate course and for an advanced course at the graduate level. The basic aim of this book is to make a complete text covering both the basic and applied aspects of heat transfer. The philosophy followed in this book is that the subject of heat transfer is covered combining the theoretical analysis, physical features and the application aspects.

The principles of heat transfer are covered as the subject is treated at the undergraduate level. Beginning with introducing the elementary ideas of the heat transfer process which results in change of flow parameters, such as pressure and temperature, and developing the methods to achieve the desired state, are presented in depth, together with their limitations and the features influencing them.

Basic concepts and definitions, which form the basis for learning the science of heat transfer, are discussed precisely to the point. Basic principles of thermodynamics are discussed in the first chapter to make the readers comfortable with heat transfer principles, processes and application aspects covered in the rest of the book. The treatment of the material is such that it emphasizes both the theory and application simultaneously.

The entire spectrum of heat transfer is presented, with necessary explanations on every aspect, introducing the subject in a simple and effective manner, in Chapter 2. One-dimensional, steady-state conduction is presented in the subsequent Chapter 3. All aspects of the conduction process, and the associated theory and application are discussed with appropriate examples. The simplifying assumptions rendering conduction through a plane wall, cylindrical and spherical shells are stated explicitly to gain an insight into the theory, application and its limitation.

Unsteady heat conduction is discussed in Chapter 4. Beginning with the transient conduction through simple geometry of infinite plane wall, the subject is taken to an advanced level of transient conduction through multidimensional systems. At every stage, the parameters influencing the heat transfer process and their grouping, leading to different dimensionless parameters and the physical significance of those dimensionless numbers, are systematically presented. Suitable examples involving these parameters are included to assimilate and enhance the understanding of the theory studied.

Convective heat transfer is discussed in Chapter 5. This is the theory involving a significant amount of flow physics; a thorough treatment of the science of flow physics is presented along with a large number of solved examples to make the reader comfortable with this involved process of heat transfer.

Chapter 6 on radiation heat transfer deals with the theory and application aspects of the mechanism of thermal radiation. Chapter 7 on mass transfer presents all the important aspects of mass diffusion process. Chapter 8 on boiling and condensation deals with the different regimes of boiling and condensation and the calculation of heat flux associated with these regimes.

The chapter on heat exchanger, Chapter 9, deals with the classification, performance and design aspects of heat exchangers. The popular methods used for calculating the performance of heat exchangers are discussed in detail, along with their merits and demerits.

The material covered in this book is designed so that any beginner can easily follow it. The topics covered are broad-based, starting from the basic principle and progressing towards the physical aspects which govern the heat transfer process. The book is organized in a logical manner and the topics are discussed in a systematic way.

The student, or reader, is assumed to have the background on the basics of fluid mechanics and thermodynamics. Advanced undergraduate students should be able to handle the subject material comfortably. Sufficient details have been included so that the text can be used for self-study. Thus, the book can be useful for scientists and engineers working in the field of thermal science in industries and research laboratories.

My sincere thanks to my undergraduate and graduate students at the Indian Institute of Technology Kanpur, who are directly and indirectly responsible for the development of this book. My special thanks to my doctoral students, Mrinal Kaushik and Arun Kumar, for checking the initial version of the manuscript.

I thank Shashank Khurana, doctoral student, Graduate School of Frontier Sciences, the University of Tokyo, Kashiwa Campus, Japan, and Yasumasa Watanabe, doctoral student, Department of Aeronautics and Astronautics, the University of Tokyo, Hongo Campus, Japan, for their valuable help in checking the manuscript of the book and its solutions manual.

I would like to place on record my sincere thanks to my friend Professor Kojiro Suzuki, Department of Advanced Energy, the Graduate School of Frontier Sciences, the University of Tokyo, Kashiwa, Japan, for extending his full-hearted help in the finalization of this book, during my stay in his lab, in the year 2011, as Visiting Professor.

For instructors only, a companion Solutions Manual is available from Taylor & Francis, CRC Press, that contains typed solutions to all the end-of-chapter problems. I am grateful for the financial support extended by the Continuing Education Centre of the Indian Institute of Technology Kanpur, for the preparation of the manuscript.

**Ethirajan Rathakrishnan**

# The Author

**Ethirajan Rathakrishnan** is Professor of Aerospace Engineering at the Indian Institute of Technology Kanpur, India. He is well-known internationally for his research in the area of high-speed jets. The limit for the passive control of jets, called *Rathakrishnan Limit*, is his contribution to the field of jet research, and the concept of *breathing blunt nose (BBN)*, which reduces the positive pressure at the nose and increases the low-pressure at the base simultaneously, is his contribution to drag reduction at hypersonic speeds. He has published a large number of research articles in many reputed international journals. He is Fellow of many professional societies, including the Royal Aeronautical Society. Professor Rathakrishnan serves as the Editor-In-Chief of the *International Review of the Aerospace Engineering* (IREASE) journal. He has authored eight other books: *Gas Dynamics*, 3rd ed. (PHI Learning, New Delhi, 2010); *Fundamentals of Engineering Thermodynamics*, 2nd ed. (PHI Learning, New Delhi, 2005); *Fluid Mechanics: An Introduction*, 2nd ed. (PHI Learning, New Delhi, 2007); *Gas Tables*, 2nd ed. (Universities Press, Hydrabad, India, 2004); *Instrumentation, Measurements, and Experiments in Fluids* (CRC Press, Taylor & Francis Group, Boca Raton, FL, USA, 2007); *Theory of Compressible Flows* (Maruzen Co., Ltd. Tokyo, Japan, 2008); *Gas Dynamics Work Book* (Praise Worthy Prize, Napoli, Italy, 2010); and *Applied Gas Dynamics* (John Wiley, New Jersey, USA, 2010).

# Chapter 1

# Basic Concepts and Definitions

## 1.1 Introduction

*Heat transfer* is the science of energy transfer due to a temperature difference. We know that thermodynamics deals with energy balance in a variety of physical situations. In other words, thermodynamics deals with the amount of heat transfer as a system undergoes a process from one equilibrium state to another. Heat transfer, on the other hand, deals with the rate at which heat is transferred as well as the temperature distribution within the system as a function of time. Basically thermodynamics deals with systems in equilibrium. It may be used to predict the amount of thermal energy required to change a system from one thermal equilibrium state to another. But it cannot predict how fast this change from one equilibrium state to another will take place, because the system is not in equilibrium during the process. For example, consider the cooling of a hot metallic bar placed in a water bath. Thermodynamics may be used to predict the final equilibrium temperature of the metallic bar-water combination. But thermodynamics cannot answer the question, *how long* the process would take to reach this equilibrium or what would be the temperature of the bar after a certain time interval before the equilibrium is attained? Whereas, the science of heat transfer can answer these questions. That is, heat transfer can predict the temperature of both the metal bar and the water as a function of time. In other words, unlike thermodynamics, heat transfer can answer the transient energy transfer questions such as the following:

- Can heat be supplied to a system without employing high temperature difference?
- How long would it take to supply a certain amount of heat energy to the

system?
• How much heat energy is transferred between two specified instances of time during a process?
• What sort of temperature distribution exists in the system?
• How large an area is necessary to transfer the desired heat energy?

Heat transfer finds application in all processes involving energy transfer. In this introductory text we will discuss heat transfer briefly, highlighting the basic principles of the subject.

Basically there are three modes of heat transfer: conduction, convection and radiation.

• *Conduction* is an energy transfer process from more energetic particles of a substance to the adjacent, less energetic ones as a result of the interaction between the particles.
• *Convection* is the mode of heat transfer between a solid surface and the adjacent liquid or gas that is in motion.
• *Radiation* is a heat transfer mode in which the energy is emitted by matter in the form of electromagnetic waves (or photons) as a result of the changes in the electronic configurations of the atoms or molecules, dictated by their temperature.

### 1.1.1 Driving Potential

We know that, in an electric circuit the current flow through a wire depends upon the potential difference across the two ends of the wire, that is,

$$\text{Current flow} = \frac{\text{Potential difference}}{\text{Resistance}}$$

Analogous to this, we can state that in a heat transfer process, the heat flow is given by

$$\text{Heat flow} = \frac{\text{Thermal potential difference}}{\text{Thermal resistance}}$$

From the above observations, we may define in a simple, but general, manner that, "*heat transfer is the process of energy transfer due to temperature difference.*" Whenever there is a temperature difference in a medium or between media, heat transfer occurs.

Examine the heat transfer processes illustrated in Figure 1.1. From the process shown in Figure 1.1, we can infer that,

- Heat transfer due to a temperature gradient in a stationary solid or fluid medium is referred to as *conduction.*

- Heat transfer between a solid surface and a moving fluid when they are at different temperatures is referred to as *convection.*

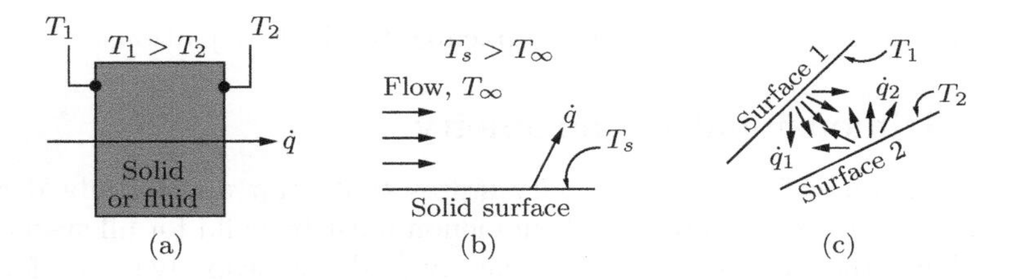


**Figure 1.1**
Conduction, convection, and radiation heat transfer modes: (a) conduction through a solid or a stationary fluid, (b) convection from a solid surface to a moving fluid, (c) net radiation exchange between two surfaces.

**Table 1.1** Common system of units

| Quantity | Unit | SI | CGS | FPS | MKS |
|---|---|---|---|---|---|
| Mass | kilogram | kg | g | lb | kg |
| Length | meter | m | cm | ft | m |
| Time | second | s | s | s | s |
| Temperature | kelvin | K | °C | °F | °C |

- The third mode of heat transfer shown in Figure 1.1(c) is termed *thermal radiation.* All surfaces at finite temperature emit energy in the form of electromagnetic waves. Hence, even in the absence of an intervening medium, there is a net heat transfer because of the radiation between two surfaces at different temperatures.

## 1.2 Dimensions and Units

Any physical quantity can be characterized by *dimensions.* The magnitudes assigned to the dimensions are called *units.* Some basic dimensions such as mass $m$, length $L$, time $t$, and temperature $T$ are selected as *primary* dimensions, while others such as velocity $V$, energy $E$, and volume $\mathbb{V}$, expressed in terms of the primary dimensions are called *secondary* or *derived* dimensions. Table 1.1 gives the common system of units and their symbols for the primary dimensions.
In this book only SI system of units are used. However, other systems of units

mentioned in Table 1.1 are equally applicable to all the equations.

### 1.2.1 Dimensional Homogeneity

The law of dimensional homogeneity states that, "an analytically derived equation representing a physical phenomenon must be valid for all systems of units." Thus, the equation for kinetic energy $\frac{1}{2}mV^2$ is properly stated for any system of units. This explains why all natural phenomena proceed completely in accordance with man-made units, and hence the fundamental equations representing such events should have validity for any system of units. Thus, all equations must be dimensionally homogeneous, and consequently all relations derived from these equations must also be dimensionally homogeneous. For this to be valid for all systems of units, it is essential that each grouping in an equation has the same dimensional representation.

### Example 1.1

Examine the following dimensional representation of an equation:

$$L = T^2 + T$$

where $L$ denotes length and $T$ the time. Changing the units of length from meters to feet will change the value of the left-hand side, but would not affect the right-hand side, thus making the equation invalid in the new system of units. This kind of equations will not be considered in this book and only dimensionally homogeneous equations will be considered.

Some standard multipliers which are commonly used in SI units are listed in Table 1.2.

**Table 1.2** Multiplier factors for SI units

| Multiplier | Prefix | Abbreviation |
|---|---|---|
| $10^{12}$ | tera | T |
| $10^{9}$ | giga | G |
| $10^{6}$ | mega | M |
| $10^{3}$ | kilo | k |
| $10^{2}$ | hecto | h |
| $10^{1}$ | deca | da |
| $10^{-1}$ | deci | d |
| $10^{-2}$ | centi | c |
| $10^{-3}$ | milli | m |
| $10^{-6}$ | micro | $\mu$ |
| $10^{-9}$ | nano | n |
| $10^{-12}$ | pico | p |
| $10^{-15}$ | femto | f |
| $10^{-18}$ | atto | a |

**Table 1.3** Quantities used in heat transfer

| Quantity | Symbol (units) |
|---|---|
| Force | $F$ (newton) |
| Mass | $m$ (kilogram) |
| Time | $t$ (second) |
| Length | $L$ (meter) |
| Temperature | $T$ (°C or kelvin) |
| Energy | $E$ (joule) |
| Power | $P$ (watt) |
| Thermal conductivity | $k$ (watt/(meter °C)) |
| Convective heat transfer coefficient | $h$ (watt/(square meter °C)) |
| Specific heat | $c$ (joule/(kilogram °C)) |
| Heat flux | $\dot{q}$ (watt/square meter) |

A list of commonly encountered symbols in heat transfer are listed in Table 1.3.

## 1.3 Closed and Open Systems

A *system* is "an identified quantity of matter or an identified region in space chosen for a study." The region outside the system is called *surrounding.* The real or imaginary surface that separates the system from its surrounding is called the *boundary.* A system, its boundary and the surrounding are illustrated in Figure 1.2. The boundary of a system may be fixed or movable.

### 1.3.1 Closed System (Control Mass)

A closed system consists of a fixed amount of mass (say a gas of certain amount of mass), and no mass can cross its boundary. That is, no mass can enter or leave a closed system. A closed system is shown in Figure 1.3. Mass cannot cross the boundaries of a closed system, but energy can cross the boundaries. Because of this, a closed system is also referred to as *control mass.*

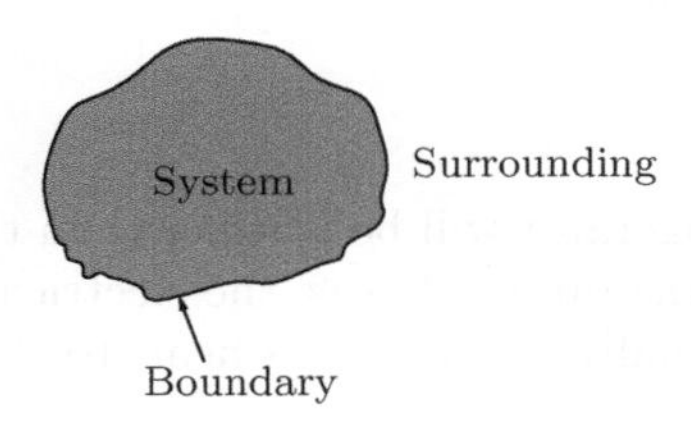


**Figure 1.2**
System, its boundary and the surrounding.

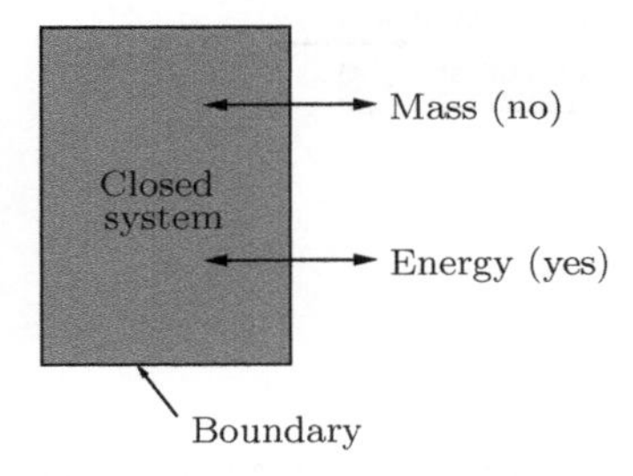


**Figure 1.3**
A closed system.

## Example 1.2

There are two identical rooms in a house. One has a refrigerator in it and the other has no refrigerator. When the doors and windows of the rooms are closed, will the room with the refrigerator be warmer or cooler than the other room? Explain your answer.

## Solution

The room with the refrigerator will be warmer than the other, because of the heat rejected by the refrigerator. That is, the electrical energy supplied to run the refrigerator is eventually dissipated as heat, to the room air.

### 1.3.2 Isolated System

An isolated system is a special case of closed system where even energy is not allowed to cross the boundary. Examine the piston-cylinder device shown in Figure 1.4. Here the gas in the cylinder is the system. There is no mass transfer across the boundaries, thus, it is a closed system. Energy may cross the boundary, and part of the boundary may move. Everything outside the gas, including the piston and the cylinder is the surroundings. If, as a special case, even energy is not allowed to cross the system boundaries, then it is referred to as an *isolated system*.

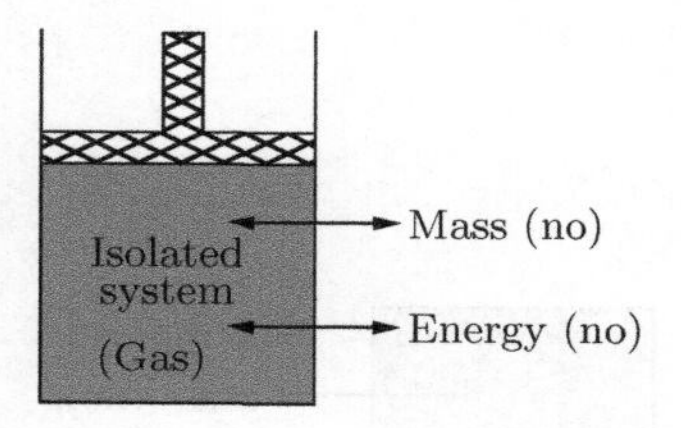


**Figure 1.4**
A piston-cylinder device.

### 1.3.3 Open System (Control Volume)

An open system is a properly selected region in space. It usually encloses a device which involves mass flow, such as a nozzle, diffuser, compressor or turbine. Flow through these devices are best studied by selecting the region within the device as the control volume. Both energy and mass can cross the boundary of a control volume, which is called a control surface, but the shape of the control volume will remain unchanged. An open system is illustrated in Figure 1.5. The volume enclosed by the dashed line in the figure is the control volume.

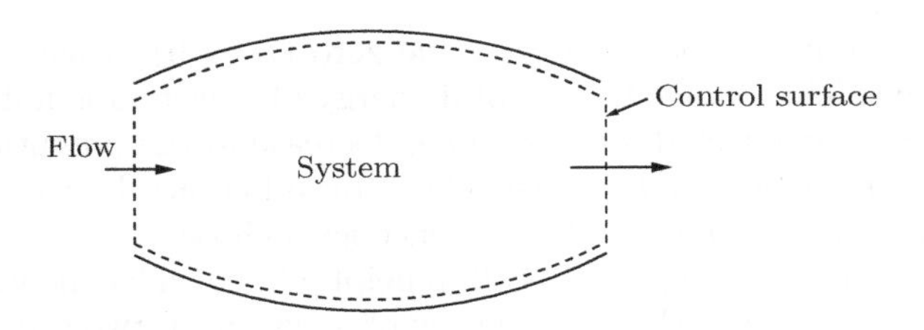


**Figure 1.5**
An open system.

From the definitions of closed and open systems, it is evident that:

A *control mass* or *closed system* is an identified quantity of matter, which may change its shape and position in space or time or both, but the quantity of matter will always be the same. A typical example for a control mass is a balloon filled with a light gas, such as hydrogen, released in atmosphere. The balloon will gain height and the volume of the balloon will increase with increase of altitude. Neglecting the binary diffusion of air and hydrogen through the balloon wall, the hydrogen mass in the balloon can be taken as the control mass system.

A *control volume* or *open system* is an identified shape, which may change its position in space or time or both but the shape will remain unchanged. The jet engine of an aircraft is a typical open system. The aircraft may run on the runway, take-off, climb, cruise at an altitude, descend, land and run on the runway and come to a halt. In these phases of travel from take-off to landing, the engine changes its position in space, the time rate of change of spacial position varies, and the mass flow rate of air, fuel, and the combustion products through the engine varies, but the shape of the engine remains unchanged.

## 1.4 Forms of Energy

Energy can exist in numerous forms—such as chemical, electrical, kinetic, potential, mechanical, thermal, and nuclear—and their sum constitutes the *total energy* $E$ of a system. The total energy of a system per unit mass is termed *specific energy* $e$. That is,

$$e = \frac{E}{m} \text{ (kJ/kg)}$$

where $m$ is the mass of the system.

Thermodynamics provides no information about the absolute value of the total energy of a system. It only deals with the change of the total energy of a system, which is of interest in engineering applications. Thus, the total energy of a system can be assigned value zero ($E = 0$) at some convenient reference point. The change in the total energy of a system is independent of the reference point selected. For example, decrease in the potential energy of a falling object depends only on the elevation difference between the initial and final positions and not on the reference level chosen.

In thermodynamic analysis, it is often helpful to consider the various forms of energy that constitute the total energy of a system in two groups, namely, *macroscopic* energy group and *microscopic* energy group. The macroscopic energies are those which a system possesses as a whole, with respect to some outside reference frame, such as the kinetic energy and potential energy. The microscopic forms of energy are those related to the molecular structure of a system and the degree of the molecular activity, and are independent of outside reference frames. The sum of all the microscopic forms of energy is called the *internal energy* $IE$ of a system, and is denoted by $U$. The macroscopic energy of a system is related to the motion and the influence of some external effects such as gravity, magnetism and so on.

The energy that a system possesses as a result of its motion relative to some reference frame is called the *kinetic energy* $KE$. When all parts of a system move with the same velocity $V$, the $KE$ is expressed as

$$KE = \frac{1}{2}mV^2 \text{ (kJ)}$$

where $m$ is the mass of the system. The kinetic energy per unit mass is

$$ke = \frac{V^2}{2} \text{ (kJ/kg)}$$

where $V$ is the velocity of the system relative to some fixed frame of reference.

The *potential energy* $PE$ is the energy that a system possesses as a result of its elevation in a gravitational field. It is expressed as

$$PE = mgz \text{ (kJ)}$$

where $g$ is the gravitational acceleration and $z$ is the elevation of the center of gravity of the system with respect to some arbitrarily chosen reference plane. The potential energy per unit mass is

$$pe = gz \text{ (kJ/kg)}$$

When the external effects such as magnetic, electric, surface tension, etc. are insignificant for the system under consideration, the total energy $E$ of the system consists of only $KE$, $PE$, and $IE$, and is expressed as

$$E = IE + KE + PE$$

That is,

$$E = U + \frac{1}{2}mV^2 + mgz \text{ (kJ)}$$

The total energy per unit mass $e = E/m$, also called the *specific energy* is given by

$$e = u + ke + pe$$

or

$$e = u + \frac{V^2}{2} + gz \text{ (kJ/kg)}$$

where $u = U/m$ is the specific internal energy. Most closed systems remain stationary during a process and thus experience no change in their kinetic and potential energies. Such closed systems with their velocity and elevation of the center of gravity remain unchanged during a process are referred to as *stationary systems.* For a stationary system the kinetic energy is zero and the potential energy is constant. Therefore, the change in the total energy $\Delta E$ of a stationary system is equal to the change in its internal energy $\Delta U$.

### 1.4.1 Internal Energy

Internal energy of a system is the sum of all the microscopic forms of energy of the system. It is the energy associated with the molecular structure and the molecular activity of the constituent particles of the system. It may be viewed as the sum of the kinetic and potential energies of the molecules.

In general, the molecules of a system move around with some velocity, vibrate with a frequency, and rotate about an axis during their random motion. Thus, the energies associated with a molecule are the translational, vibrational, and rotational energies. The sum of these energies constitutes the kinetic energy of a molecule.

The portion of the internal energy of a system, associated with the kinetic energy of the molecules is called the *sensible energy.* The translation, vibration and rotation of the molecules are proportional to the temperature of the system. Thus, at high temperatures the molecules would possess high kinetic energy, and as a result the system will have a high internal energy. The internal energy is also associated with the intermolecular forces between the molecules of the system. These are the forces that bind the molecules to each other, and they are the strongest in solids and the weakest in gases. When sufficient energy is added to the molecules of a solid or liquid, they will overcome these molecular forces and break away, turning the system to a gas. This is termed *phase-change process.* Because of the added energy, a system in a gas phase is at a higher internal energy level than when it exists in a solid or liquid phase. The internal energy associated with the phase of a system is called the *latent energy.*

The above changes of phase from solid to liquid or liquid to gas can occur without any change in the chemical composition of a system. Most thermodynamic problems fall into this category and we need not pay any attention to the forces binding the atoms in a molecule.

The internal energy associated with the bonds in a molecule is called the *chemical or bond energy.* During a chemical reaction, such as a combustion process, some chemical bonds are destroyed while others are formed. As a result the internal energy changes.

There is an enormous amount of internal energy associated with the bonds within the nucleus of the atom itself. This energy is called *nuclear energy* and it is released during nuclear reactions. Unless fusion or fission reaction is the process under consideration, the nuclear energy is of no consequence.

## Example 1.3

The body temperature of a person during a running exercise increases from 37°C to 38°C. If the person is 65 kg, determine the increase in the thermal energy content of his body, assuming the average specific heat of his body as 3600 J/(kg °C).

## Solution

Given, $m = 65$ kg, $\Delta T = 1°\text{C}$, $c_p = 3600$ J/(kg °C).

The change in the sensible internal energy of the person's body as a result of temperature rise, given by the first law of thermodynamics is

$$\begin{aligned}\Delta U &= mc_p \Delta T \\ &= 65 \times 3600 \times 1 \\ &= 234000\,\text{J} \\ &= \boxed{234\,\text{kJ}}\end{aligned}$$

## Example 1.4

A metallic ball of diameter 100 mm is to be heated from 20°C to 90°C. If the density and specific heat of the metal are 2.7 kg/m$^3$ and 900 J/(kg °C), determine the amount of heat that needs to be transferred to the metallic ball.

## Solution

Given, $D = 100$ mm, $\rho = 2.7$ kg/m$^3$, $c_p = 900$ J/(kg °C), $T_i = 20°\text{C}$ and $T_f = 90°\text{C}$.

The amount of heat to be added to the ball is equal to the change in its internal energy, given by

$$\Delta E = \Delta U = mc_p(T_f - T_i)$$

where $T_i$ and $T_f$ are the initial and final temperatures of the ball and $m$ is its mass.

The mass of the ball is

$$\begin{aligned} m &= \rho \mathbb{V} = \rho \left(\frac{\pi}{6} D^3\right) \\ &= 2.7 \times \left(\frac{\pi}{6} \times 0.1^3\right) \\ &= 1.414 \times 10^{-3}\,\text{kg} \end{aligned}$$

Therefore,

$$\begin{aligned} \Delta E &= (1.414 \times 10^{-3}) \times 900 \times (90 - 20) \\ &= \boxed{89.082\,\text{J}} \end{aligned}$$

## 1.5 Properties of a System

A *property* is any characteristic associated with a system. Some familiar properties we often encounter in the analysis of engineering systems are the pressure $p$, temperature $T$, volume $\mathbb{V}$, and mass $m$. In addition to these, viscosity, thermal conductivity, modulus of elasticity, thermal expansion coefficient, electric resistivity, and even velocity and elevation can also be treated as properties.

It is essential to note that, not all properties are independent. Some are defined in terms of other properties. For instance, density, defined as "mass per unit volume," is expressed in terms of mass and volume $\mathbb{V}$ as

$$\rho = \frac{m}{\mathbb{V}}\ (\text{kg/m}^3)$$

Sometimes, the density of a substance is given relative to that of a better known substance. Then it is called *specific gravity* or *relative density*, and is defined as the "the ratio of the density of a substance to the density of some standard substance at a specified temperature." Usually water at 4°C with $\rho_{\text{H}_2\text{O}} = 1000$ kg/m$^3$ is taken as the standard reference for expressing specific gravity. That is, the specific gravity $\rho_s$ of a substance of density $\rho$ is given by

$$\rho_s = \frac{\rho}{\rho_{\text{H}_2\text{O}}}$$

Note that, the specific gravity is *a dimensionless quantity.*

A more frequently used property in thermodynamics is the *specific volume* $v$. It is the reciprocal of density and is defined as the "volume per unit mass,"

$$v = \frac{\mathbb{V}}{m} = \frac{1}{\rho} \ (\mathrm{m^3/kg})$$

### Example 1.5

Determine the density and specific volume of 5 kg of a gas contained in a tank of volume 25 $\mathrm{m^3}$.

### Solution

Given, $m = 5$ kg, $\mathbb{V} = 25\ \mathrm{m^3}$. Therefore, the specific volume is

$$\begin{aligned} v &= \frac{\mathbb{V}}{m} = 25/5 \\ &= \boxed{5\ \mathrm{m^3/kg}} \end{aligned}$$

The density is

$$\begin{aligned} \rho &= \frac{1}{v} = \frac{1}{5} \\ &= \boxed{0.2\ \mathrm{kg/m^3}} \end{aligned}$$

At this stage it is essential to note that, in thermodynamics the atomic structure of a substance (thus the spaces between and within the molecules) is disregarded, and the substance is viewed to be continuous, homogeneous matter with no microscopic gaps or holes. In other words, the substance is treated as *continuum.* The continuum treatment is also referred to as macroscopic approach. This idealization is valid as long as we work with areas and lengths which are large compared to the intermolecular spacing. In situations such as highly rarefied (low-density) conditions, where the intermolecular spacing is much larger than any characteristic length under consideration, the continuum hypothesis is not valid, and the problem has to be analyzed by microscopic approach. The macroscopic and microscopic approaches are usually referred to as *integral* and *differential* approaches, respectively.

### 1.5.1 Intensive and Extensive Properties

Properties are usually classified as intensive and extensive. *Intensive* properties are those which are independent of the size of a system. For example, pressure, temperature and density are all intensive properties. *Extensive* properties are those whose values depend on the size or the extent of the system. Mass, volume and total energy are extensive properties. Extensive property per unit mass is termed *specific* property. Specific volume ($v = \mathbb{V}/m$), specific total energy ($e = E/m$) and specific internal energy ($u = U/m$) are specific properties.

## 1.6 State and Equilibrium

The properties of a system which do not undergo any change can be computed or measured throughout the entire system. They constitute a set of properties that completely describe the condition or the *state*. When all the properties of a system have fixed values, the system is said to be under an equilibrium state. Even if the value of one of the properties changes, the system will change to a different equilibrium state.

Thermodynamics deals with equilibrium states. The word equilibrium refers to a *state of balance*. In an equilibrium state, there are no unbalanced potentials or driving forces within the system. A system is said to be in *equilibrium* when its properties do not change with time. There are many types of equilibrium, such as mechanical equilibrium, thermal equilibrium, chemical equilibrium, and so on. A system is said to be in thermodynamic equilibrium only when the conditions of all the relevant types of equilibrium are satisfied. For example, a system is in *thermal equilibrium* if the temperature is the same throughout the system. That is, there is no temperature gradient, which is the driving potential for heat flow, within the system.

*Mechanical equilibrium* is related to pressure, and a system is said to be in mechanical equilibrium if there is no change in pressure with time at any point in the system. However, the pressure may vary within the system with elevation as a result of gravitational effects. But the higher pressure at a bottom layer is balanced by the extra weight it must carry, and, therefore, there is no imbalance of forces. Pressure variation as a result of the gravity is relatively small in most thermodynamic systems and is usually disregarded.

If a system involves two phases, it will be in *phase equilibrium* only when the mass of each phase reaches an equilibrium level and stays there. A system will be in *chemical equilibrium* if its chemical composition does not change with time, that is, no net chemical reaction occurs within the system. A system will be in *thermodynamic equilibrium* only when it satisfies the conditions for all modes of equilibrium.

## 1.7 Thermal and Calorical Properties

The equation $pv = RT$ or $p/\rho = RT$ is called *thermal equation of state*, where $p$, $T$ and $v (= 1/\rho)$ are *thermal properties* and $R$ is the gas constant. A gas which obeys the thermal equation of state is called *thermally perfect gas*. Any relation between the calorical properties, $u$, $h$ and $s$ and any two thermal properties is called *calorical equation of state*. In general, the thermodynamic properties (the properties which do not depend on process) can be grouped into thermal properties ($p$, $T$, $v$) and calorical properties ($u$, $h$, $s$). Two calorical state equations which are extensively used are the following [1].

$$u = u(T, v), \quad h = h(T, p)$$

where $u$ is the specific internal energy and $h$ is the specific enthalpy, which is a combination property, defined as the sum of internal energy and flow work ($pv$). That is, $h = u + pv$.

In terms of exact differentials, the above relations become

$$du = \left(\frac{\partial u}{\partial T}\right)_v dT + \left(\frac{\partial u}{\partial v}\right)_T dv \tag{1.1}$$

$$dh = \left(\frac{\partial h}{\partial T}\right)_p dT + \left(\frac{\partial h}{\partial p}\right)_T dp \tag{1.2}$$

For a constant volume process, Equation (1.1) reduces to

$$du = \left(\frac{\partial u}{\partial T}\right)_v dT$$

where $\left(\frac{\partial u}{\partial T}\right)_v$ is the specific heat at constant volume represented as $c_v$, therefore,

$$\boxed{du = c_v \, dT} \tag{1.3}$$

For an isobaric process, Equation (1.2) reduces to

$$dh = \left(\frac{\partial h}{\partial T}\right)_p dT$$

where $\left(\frac{\partial h}{\partial T}\right)_p$ is the specific heat at constant pressure represented by $c_p$, therefore,

$$\boxed{dh = c_p \, dT} \tag{1.4}$$

For a constant volume (isochoric) process, we have

$$\delta q = du = c_v \, dT \tag{1.5a}$$

where $\delta q$ is the heat addition. For a constant pressure (isobaric) process, $\delta q$ is given by

$$\delta q = dh = c_p \, dT \quad \text{or} \quad \delta q = dh = c_v \, dT + p \, dv \tag{1.5b}$$

For an adiabatic flow process ($\delta q = 0$), we have

$$dh = v \, dp \tag{1.5c}$$

The finite changes in the internal energy and enthalpy of an ideal gas during a process can be expressed approximately by using specific heat values at the average temperature as

$$\Delta u = c_{v,\text{ave}}\,\Delta T \quad \text{and} \quad \Delta h = c_{p,\text{ave}}\,\Delta T \tag{1.5d}$$

or

$$\Delta U = m\,c_{v,\text{ave}}\,\Delta T \quad \text{and} \quad \Delta H = m\,c_{p,\text{ave}}\,\Delta T \tag{1.5e}$$

where $m$ is the mass of the system.

From Equations (1.5) it can be inferred that,

1. If heat is added at constant volume, it only raises the internal energy.
2. If heat is added at constant pressure, it not only increases the internal energy but also does some external work, that is, heat addition at constant pressure increases the enthalpy.
3. If the change is adiabatic, the change in enthalpy is equal to external work $v\,dp$.

### 1.7.1 Specific Heat of an Incompressible Substance

A substance whose specific volume does not change with temperature or pressure is called an *incompressible substance.* The specific volumes of solids and liquids remains constant during a process, and thus they can be approximated as incompressible substances without affecting the accuracy significantly.

For incompressible substances, the constant volume and constant pressure specific heats are identical. Therefore, for incompressible substances such as solids and liquids, the subscripts for $c_v$ and $c_p$ can be dropped and both specific heats can be represented by a symbol $c$. The specific heats of incompressible substances depend on temperature only. Therefore, the change in the internal energy of solids and liquids can be expressed as

$$\Delta U = m\,c_{\text{ave}}\Delta T$$

where $c_{\text{ave}}$ is the average specific heat evaluated at the average temperature. This relation implies that, the internal energy change of a system that remain in a single phase (liquid, solid, or gas) during a process can be easily determined with average specific heat.

### 1.7.2 Thermally Perfect Gas

A gas is said to be thermally perfect when its internal energy and enthalpy are functions of temperature alone, that is, for a thermally perfect gas,

$$u = u(T), \qquad h = h(T) \tag{1.6a}$$

Therefore, from Equations (1.3) and (1.4), we get

$$c_v = c_v(T), \quad c_p = c_p(T) \tag{1.6b}$$

Further, from Equations (1.1), (1.2) and (1.6a), we obtain

$$\left(\frac{\partial u}{\partial v}\right)_T = 0, \quad \left(\frac{\partial h}{\partial p}\right)_T = 0 \tag{1.6c}$$

The important relations of this section are

$$\boxed{du = c_v\, dT, \quad dh = c_p\, dT}$$

These equations are universally valid so long as the gas is thermally perfect. Otherwise, in order to have equations of universal validity, we must add $\left(\frac{\partial u}{\partial v}\right)_T dv$ to the first equation and $\left(\frac{\partial h}{\partial p}\right)_T dp$ to the second equation.

The state equation for a thermally perfect gas is,

$$pv = RT$$

In the differential form, this equation becomes

$$p\, dv + v\, dp = R\, dT$$

Also,

$$\begin{aligned} h &= u + pv \\ dh &= du + p\, dv + v\, dp \end{aligned}$$

Therefore,

$$dh - du = p\, dv + v\, dp = R\, dT$$

that is,

$$R\, dT = c_p\, dT - c_v\, dT$$

Thus,

$$\boxed{R = c_p\,(T) - c_v(T)} \tag{1.7}$$

For thermally perfect gases, even though $c_p$ and $c_v$ are functions of temperature, their difference is a constant with reference to temperature [1].

## 1.8 The Perfect Gas

This is a still more specialization than thermally perfect gas. For a perfect gas, both $c_p$ and $c_v$ are constants and are independent of temperature [1], that is,

$$c_v = \text{constant} \neq c_v(T), \quad c_p = \text{constant} \neq c_p(T) \tag{1.8}$$

Such a gas with constant $c_p$ and $c_v$ is called a *calorically perfect gas*. Therefore, a perfect gas should be thermally as well as calorically perfect.

From the above discussions it is evident that,

1. A perfect gas must be both thermally and calorically perfect.
2. A perfect gas must satisfy both *thermal equation of state*; $p = \rho R T$ and *caloric equations of state*; $c_p = (\partial h/\partial T)_p$, $c_v = (\partial u/\partial T)_v$.
3. A calorically perfect gas must be thermally perfect, but a thermally perfect gas need not be calorically perfect. That is, thermal perfectness is a prerequisite for caloric perfectness.
4. For a thermally perfect gas, $c_p = c_p(T)$ and $c_v = c_v(T)$; that is, both $c_p$ and $c_v$ are functions of temperature. But even though the specific heats $c_p$ and $c_v$ vary with temperature, their ratio, $\gamma$ becomes a constant and independent of temperature, that is, $\gamma = \text{constant} \neq \gamma(T)$.
5. For a calorically perfect gas, $c_p$, $c_v$ as well as $\gamma$ are constants and independent of temperature.

## 1.9 Summary

*Heat transfer* is the science of energy in transit due to a temperature difference.
In a heat transfer process,

$$\text{Heat flow} = \frac{\text{Thermal potential difference}}{\text{Thermal resistance}}$$

• *Conduction* is an energy transfer process from more energetic particles of a substance to the adjacent, less energetic ones as a result of the interaction between the particles.
• *Convection* is the mode of heat transfer between a solid surface and the adjacent liquid or gas that is in motion.
• *Radiation* is the heat transfer mode in which the energy is emitted by matter in the form of electromagnetic waves (or photons) as a result of the changes in the electronic configurations of the atoms or molecules.

Any physical quantity can be characterized by *dimensions.* The magnitudes assigned to the dimensions are called *units.* The law of dimensional homogeneity states that, *an analytically derived equation representing a physical phenomenon must be valid for all systems of units.*

A *system* is an identified quantity of matter or an identified region in space chosen for a study. The region outside the system is called *surrounding.* The real or imaginary surface that separates the system from its surroundings is called the *boundary.* A closed system (control mass) consists of a fixed amount of mass (say a gas mass), and no mass can cross its boundary. Mass cannot cross the boundaries of a closed system, but energy can cross the boundaries. Isolated system is a special case of closed system where even energy is not allowed to cross the boundary. An open system (control volume) is a properly selected region in space, which may change its position and mass can enter and leave the region, but the size and shape of the system will remain invariant.

Energy can exist in numerous forms, such as chemical, electrical, kinetic, potential, mechanical, thermal, and nuclear, and their sum constitutes the *total energy* of a system. The total energy of a system per unit mass is termed *specific energy.* The *macroscopic* energies are those a system possesses as a whole, with respect to some outside reference frame, such as kinetic energy and potential energy. The *microscopic* forms of energy are those related to the molecular structure of a system and the degree of the molecular activity, and they are independent of outside reference frames. The sum of all the microscopic forms of energy is called the *internal* energy of a system. The energy that a system possesses as a result of its motion relative to some reference frame is called the *kinetic energy.* The *potential energy* is the energy that a system possesses as a result of its elevation in a gravitational field.

A closed system with its velocity and elevation of the center of gravity remain unchanged during a process is referred to as a *stationary system.* For a stationary system the kinetic energy is zero and the potential energy is constant. Therefore, the change in the total energy of a stationary system is equal to the change in its internal energy.

*Internal energy* is the sum of all the microscopic forms of energy of a system. The portion of the internal energy of a system, associated with the kinetic energy of the molecules is called the *sensible energy.* The translational, vibrational and rotational energies of the molecules are proportional to the temperature of the system. When sufficient energy is added to the molecules of a solid or liquid, they will overcome the molecular forces and break away, turning the system to a gas. This is termed *phase-change process.* The internal energy associated with the phase of a system is called the *latent energy.*

The internal energy associated with the bonds in a molecule is called the *chemical or bond energy.* There is an enormous amount of internal energy associated with the bonds within the nucleus of the atom itself. This energy is called *nuclear energy* and it is released during nuclear reactions.

A *property* is any characteristic associated with a system. Properties are usually classified as intensive and extensive. *Intensive properties* are those which are independent of the size of a system. *Extensive properties* are those whose values depend on the size or extent of the system.

Thermodynamics deals with equilibrium states. A system is said to be in thermodynamic equilibrium only when the conditions of all the relevant types of equilibrium are satisfied. *Mechanical equilibrium* is related to pressure, and a system is said to be in mechanical equilibrium if there is no change in pressure at any point in the system with time. If a system involves two phases, it will be in *phase equilibrium* only when the mass of each phase reaches an equilibrium level and stays there. A system will be in *chemical equilibrium* if its chemical composition does not change with time, that is, no net chemical reaction occurs.

## 1.10 Exercise Problems

1.1 A 1200-W air drier sucks air at 100 kPa and 22°C and delivers at 47°C. The exit area of the drier is 60 $cm^2$. Determine the exit velocity and the mass flow rate through the drier.

[**Ans.** 7.346 m/s, 0.048 kg/s]

1.2 Five liters of water in a water heater has to be heated from 22°C to 90°C. Assuming that the heater is perfectly insulated, determine the amount of energy that needs to be supplied to the water.

[**Ans.** 1399.9 kJ]

## Reference

1. Rathakrishnan, E., *Applied Gas Dynamics*, John Wiley, Hoboken, NJ, 2010.

# Chapter 2

# Conduction Heat Transfer

## 2.1 Introduction

Conduction is the mode of heat transfer due to *temperature gradient* in a stationary medium. The medium may be a solid or a fluid. When a temperature gradient exists in a body, there will be energy transfer from the high-temperature region to the low-temperature region. This kind of energy transfer is termed conduction and the heat transfer per unit area (normal to the direction of heat flow) is proportional to the temperature gradient. That is, the heat flux $\dot{q}$ (the amount of heat transferred per unit area per unit time) is given by

$$\dot{q}_x \propto \frac{\partial T}{\partial x}$$

or

$$\boxed{\dot{q}_x = -k\,\frac{\partial T}{\partial x}} \tag{2.1}$$

where $k$ is the proportionality constant, $\dot{q}_x$ is the heat flux and $\partial T/\partial x$ is the temperature gradient in the direction of heat flow. The positive constant $k$ is called the *thermal conductivity* of the material, and the minus sign indicates that the heat must flow from a higher-temperature region to a lower-temperature region. Equation (2.1) is called the *Fourier's law* of heat conduction. The thermal conductivity has the units of watts per meter kelvin [W/(m K)]. It can be expressed as [W/(m °C)]. This is because, $k$ appears in the relation involving temperature difference.

Consider a solid slab of thickness $b$, in the $y$-direction, and extending to infinity in the $x$- and $z$-directions, as shown in Figure 2.1. Let the initial temperature of the slab be uniform at $T_0$, throughout. Let at time $t = 0$ the temperature of the surface at $y = 0$ is raised to $T_s$ and held constant at that level, and the surface at $y = b$ is held at temperature $T_0$ throughout. Because of the temperature gradient established by the difference between $T_s$

and $T_0$, heat flows in the positive $y$-direction. Some heat is retained by the intermediate layers in the slab, thereby raising their temperature, while the remaining heat flows on.

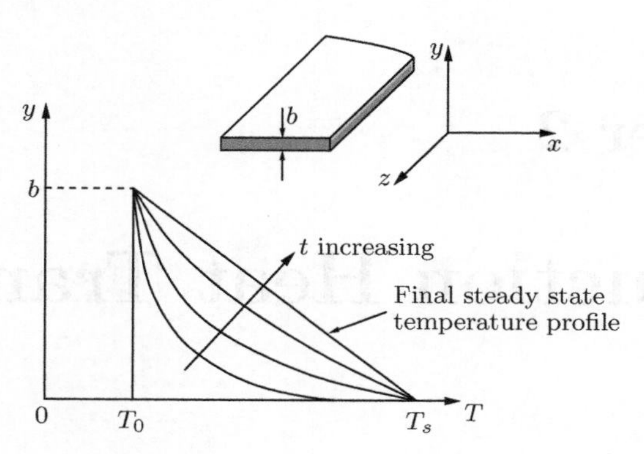


**Figure 2.1**
Heat conduction through a slab.

The temperature pattern across the slab changes with time, as shown in Figure 2.1, till a steady linear temperature distribution is established. At the steady state, the heat flux across all $xz$ surfaces is the same because no energy coming in is retained by any layer in the slab. That is, at all layers ($xz$-plane of the slab) the heat flux enters from the bottom face is conducted away through the top face, to the surrounding. This steady state heat flux is given by the Fourier's law of heat conduction [Equation (2.1)] as

$$\dot{q}_y = k\, \frac{T_s - T_0}{b}$$

or in the differential form

$$\dot{q}_y = -k\, \frac{\partial T}{\partial y}$$

where the subscript $y$ refers to the direction of heat flow.

For an isotropic medium (material properties are the same in all directions) the heat fluxes along $x$- and $z$-directions are given, respectively, by

$$\dot{q}_x = -k\, \frac{\partial T}{\partial x} \quad \text{and} \quad \dot{q}_z = -k\, \frac{\partial T}{\partial z}$$

For an isotropic medium the thermal conductivity is independent of direction, that is, $k_x = k_y = k_z$.

Recognizing that the heat flux is a vector quantity, we can write a more general statement of the heat conduction rate equation (Fourier's law) in the following form.

$$\dot{q} = -k \bigtriangledown T = -k \left( i\, \frac{\partial T}{\partial x} + j\, \frac{\partial T}{\partial y} + k\, \frac{\partial T}{\partial z} \right) \tag{2.2}$$

where the Laplacian (or *del*) $\bigtriangledown$ is the three-dimensional operator,

$$\bigtriangledown \equiv i\,\frac{\partial}{\partial x} + j\,\frac{\partial}{\partial y} + k\,\frac{\partial}{\partial z}$$

and $T(x, y, z)$ is the scalar temperature field. Equation (2.2) is the differential form of the Fourier's law of heat conduction. It implies that, the heat flux is in the direction perpendicular to the isothermal surfaces. An alternative form of Fourier's law is therefore,

$$\boxed{\dot{q}_n = -k\,\frac{\partial T}{\partial n}} \tag{2.3}$$

where $\dot{q}_n$ is the heat flux in a direction $n$, which is normal to an isotherm, as shown for the two-dimensional case in Figure 2.2.

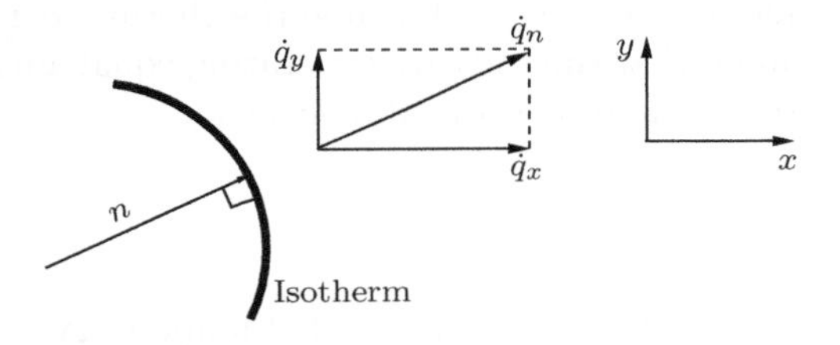


**Figure 2.2**
The heat flux vector normal to an isotherm in a two-dimensional coordinate system.

**Example 2.1**

If the opposite faces of a large aluminum plate of thickness 5 mm are maintained at 300°C and 100°C, compute the heat flux through the plate, assuming $k = 215$ W/(m °C) at the mean temperature of 200°C.

## Solution

The length and width of the plate are very large compared to its thickness. Therefore, the heat transfer in the thickness direction can be treated as one-dimensional. Therefore, by Fourier's law,

$$\dot{q} = -k\,\frac{dT}{dx}$$

This may be integrated to give the heat flux through the plate as

$$\dot{q} \;=\; -k\,\frac{\Delta T}{\Delta x}$$

$$= -(215) \times \frac{100 - 300}{5 \times 10^{-3}}$$

$$= \boxed{8.6\,\mathrm{MW/m^2}}$$

Note that, the thermal conductivity $k$ is given as W/(m °C). This implies that $k$ can be expressed as W/(m K) or W/(m °C).

### Example 2.2

In an experiment, two identical cylindrical rods of 4 cm diameter and 9 cm length are used in an electric heater. In each rod, two differential thermocouples are placed 4.2 cm apart. After an initial transient, the electric heater is observed to draw 0.4 ampere at 220 volts, and both the thermocouples read a temperature difference of 13°C. Determine the thermal conductivity of the rod material, assuming that the rods are generating equal amount of heat and the heat conduction in the rods is one-dimensional.

## Solution

Given, voltage $V = 220$ volts, current $I = 0.4$ ampere, $\Delta = 13$°C, $d = 0.04$ m, $L = 0.042$ m. The electrical power consumed by the cylindrical rods and converted to heat is

$$\dot{W}_{\mathrm{ele}} = VI = 220 \times 0.4$$

$$= 88\,\mathrm{W}$$

The heat flow rate through each rod is

$$\dot{Q} = \frac{1}{2}\dot{W}_{\mathrm{ele}} = \frac{1}{2} \times 88$$

$$= 44\,\mathrm{W}$$

since only half of the heat generated is flowing through each rod. The heat transfer area is the area normal to the direction of heat flow, which is the cross-sectional area $A$ of the cylindrical rods in this problem. Thus,

$$A = \frac{\pi d^2}{4}$$

$$= \frac{\pi (0.04)^2}{4}$$

$$= 1.26 \times 10^{-3}\,\mathrm{m^2}$$

Noting that the temperature drops by 13°C within 4.2 cm in the direction of heat flow, the thermal conductivity can be determined from the relation

$$\dot{Q} = kA\frac{\Delta T}{L}$$

Thus, the thermal conductivity of the rod material is

$$\begin{aligned} k &= \frac{\dot{Q}L}{A\Delta T} \\ &= \frac{44 \times 0.042}{(1.26 \times 10^{-3}) \times 13} \\ &= \boxed{112.8\,\mathrm{W/(m\ ^\circ C)}} \end{aligned}$$

## 2.2 Conduction Heat Transfer in a Stationary Medium

Consider heat conduction in a stationary medium or in a medium moving with an uniform velocity. A solid necessarily satisfies this criterion. In such a case, both the system and control volume approaches are similar, because no mass crosses a surface fixed in a coordinate frame moving with the same velocity as the control volume or system under consideration. Consider an infinitesimally small element in a stationary medium, as shown in Figure 2.3, as the control volume. Let $P(x, y, z)$ be a corner of the element with faces parallel to the coordinates planes, as shown. The volume chosen is infinitesimally small and hence the properties in the volume can be approximated by their values at the point $P(x, y, z)$.

By conservation of energy, the rate of change of thermal energy contained within the elemental volume will be equal to the rate of transfer of energy into the volume by conduction plus the rate of production of energy within the elemental volume.

The rate of change of thermal energy contained within the system is

$$\text{mass} \times \text{specific heat} \times \text{rate of change of temperature} = [\rho\ (\delta x \delta y \delta z)]\ c\ \frac{\partial T}{\partial t}$$

where $\rho$ is the density, $c$ is the specific heat, $T$ is the temperature, and $t$ is the time.

### 2.2.1 Energy Flux by Conduction

Consider the face PQRS, in Figure 2.3, with its normal in the negative $x$-direction. By Fourier's law of heat conduction, the heat flux through face

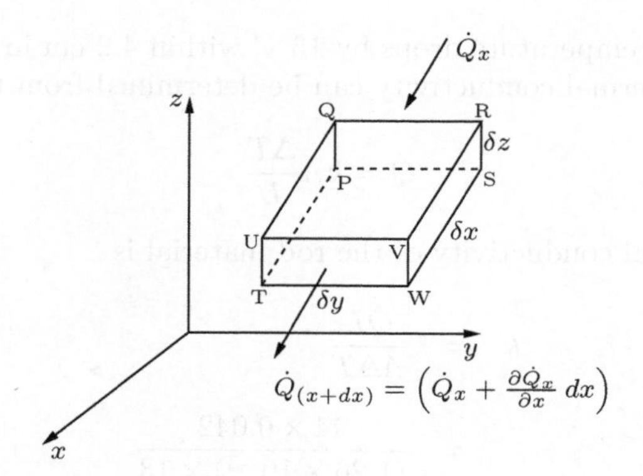


**Figure 2.3**
Differential control volume, $(\delta x \delta y \delta z)$, for conduction analysis in Cartesian coordinates.

PQRS (in the $x$-direction) is

$$\dot{q}_x = -k\,\frac{\partial T}{\partial x}$$

and the rate of heat transfer into the control volume is

$$\dot{Q}_x = \left(-k\,\frac{\partial T}{\partial x}\right)(\delta y \delta z)$$

where $k$ is the thermal conductivity and is independent of direction for isotropic materials, $(\delta y \delta z)$ is the area normal to the heat flow direction.

The heat flux across the face TUVW is

$$\dot{q}_{(x+dx)} = -\left[k\frac{\partial T}{\partial x} + \frac{\partial}{\partial x}\left(k\frac{\partial T}{\partial x}\right)\delta x\right]$$

Therefore, the rate of heat transfer from the control volume in the $x$-direction is

$$\dot{Q}_{(x+dx)} = -\left[k\frac{\partial T}{\partial x} + \frac{\partial}{\partial x}\left(k\frac{\partial T}{\partial x}\right)\delta x\right](\delta y \delta z)$$

Thus, the net energy added to the control volume in the $x$-direction is given by

$$-\left[(\text{Heat energy})_{\text{out}} - (\text{Heat energy})_{\text{in}}\right] = \frac{\partial}{\partial x}\left[k\frac{\partial T}{\partial x}\right](\delta x \delta y \delta z)$$

Therefore, the total heat conducted into the elemental volume from all the three directions is

$$\left[\frac{\partial}{\partial x}\left(k\frac{\partial T}{\partial x}\right) + \frac{\partial}{\partial y}\left(k\frac{\partial T}{\partial y}\right) + \frac{\partial}{\partial z}\left(k\frac{\partial T}{\partial z}\right)\right](\delta x \delta y \delta z)$$

that is,

$$\left(\nabla\cdot(k\nabla T)\right)(\delta x\delta y\delta z)$$

where the operator $\nabla$ is

$$\nabla \equiv i\frac{\partial}{\partial x} + j\frac{\partial}{\partial y} + k\frac{\partial}{\partial z}$$

and $i$, $j$, $k$ are the unit vectors along $x$-, $y$- and $z$-directions, respectively. When there is heat generation or heat absorption within the element volume, their contribution must also be taken into account in the energy balance. In most cases, such heat sources or sinks are specified as volumetric sources or sinks, hence, their strength is usually given as W/m$^3$.

If $\dot{g}_{\text{th}}$ represents the strength of a heat source, with units W/m$^3$, the total contribution of heat by the heat source from the volume element is $\dot{g}_{\text{th}}(\delta x\delta y\delta z)$. Usually $\dot{g}_{\text{th}}$ is taken as +ve for heat source and −ve for heat sink. Therefore, for conservation of energy,

$$\left[\rho\,(\delta x\delta y\delta z)\right]c\frac{\partial T}{\partial t} = \left[\nabla\cdot(k\nabla T)\right](\delta x\delta y\delta z) + \dot{g}_{\text{th}}(\delta x\delta y\delta z) \tag{2.4}$$

that is,

$$\boxed{\rho c\frac{\partial T}{\partial t} = \nabla\cdot(k\nabla T) + \dot{g}_{\text{th}}} \tag{2.5}$$

Equation (2.5) is the general form of the *heat diffusion equation.* This equation, usually known as the *heat equation*, provides the basic tool for heat conduction analysis. Solving the heat equation, we can obtain the temperature distribution $T(x, y, z)$ as a function of time.

The thermal conductivity $k$ being a material property depends on the temperature and other physical properties of the material. Even though $k$ varies significantly when the temperature interval is large, as an approximation it is treated as a constant in many cases. In our discussion in this book, $k$ will be taken as the constant evaluated at some mean temperature of the problem, unless otherwise specified. With $k$ as constant, Equation (2.5) becomes

$$\boxed{\frac{\partial T}{\partial t} = \frac{k}{\rho c}\nabla^2 T + \frac{\dot{g}_{\text{th}}}{\rho c}} \tag{2.6}$$

The group $(\frac{k}{\rho c})$ is called *thermal diffusivity* and is denoted by $\alpha$. The Laplacian operator

$$\nabla^2 \equiv \frac{\partial^2}{\partial x^2} + \frac{\partial^2}{\partial y^2} + \frac{\partial^2}{\partial z^2}$$

in Cartesian coordinates.

In cylindrical and spherical coordinates, shown in Figure 2.4, $\nabla^2$ can be expressed as follows.

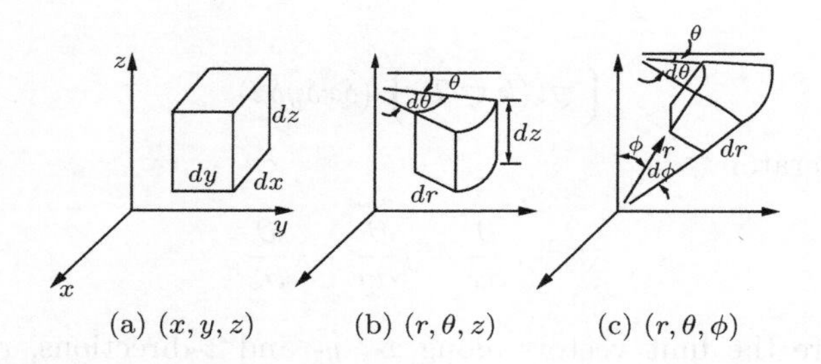


**Figure 2.4**
Elemental volume in (a) Cartesian coordinates, (b) cylindrical coordinates, (c) spherical coordinates.

In *cylindrical* coordinates

$$\nabla^2 = \frac{\partial^2}{\partial r^2} + \frac{1}{r}\frac{\partial}{\partial r} + \frac{1}{r^2}\frac{\partial^2}{\partial \theta^2} + \frac{\partial^2}{\partial z^2}$$

In *spherical* coordinates

$$\nabla^2 = \frac{1}{r^2}\left[\frac{\partial}{\partial r}\left(r^2\frac{\partial}{\partial r}\right) + \frac{1}{\sin\theta}\frac{\partial}{\partial \theta}\left(\sin\theta\frac{\partial}{\partial \theta}\right) + \frac{1}{\sin^2\theta}\left(\frac{\partial^2}{\partial \phi^2}\right)\right]$$

Thus, in the Cartesian, cylindrical, and spherical coordinates, respectively, the heat equation becomes

$$\boxed{\frac{\partial^2 T}{\partial x^2} + \frac{\partial^2 T}{\partial y^2} + \frac{\partial^2 T}{\partial z^2} + \frac{\dot{g}_{\text{th}}}{k} = \frac{1}{\alpha}\frac{\partial T}{\partial t}} \tag{2.6a}$$

$$\boxed{\frac{\partial^2 T}{\partial r^2} + \frac{1}{r}\frac{\partial T}{\partial r} + \frac{1}{r^2}\frac{\partial^2 T}{\partial \theta^2} + \frac{\dot{g}_{\text{th}}}{k} = \frac{1}{\alpha}\frac{\partial T}{\partial t}} \tag{2.6b}$$

$$\boxed{\frac{1}{r^2}\frac{\partial^2}{\partial r^2}\left(r^2\frac{\partial T}{\partial r}\right) + \frac{1}{r^2\sin\phi}\frac{\partial}{\partial \phi}\left(\sin\phi\frac{\partial T}{\partial \phi}\right) + \frac{1}{r^2\sin^2\phi}\frac{\partial^2 T}{\partial \theta^2} + \frac{\dot{g}_{\text{th}}}{k} = \frac{1}{\alpha}\frac{\partial T}{\partial t}} \tag{2.6c}$$

## Example 2.3

An aluminum cooking pan kept on a 1000-W heating plate is used for cooking. The pan diameter is 250 mm and the bottom is 25 mm thick. During steady operation, 90 percent of the heat generated in the heating plate is transferred to the pan bottom, and the temperature of the inner surface of the pan bottom is 100°C. Determine the heat flux at the bottom of the pan. What is the boundary condition at the pan bottom, during the steady state operation?

### Solution

Given, $D = 250$ mm, $L = 25$ mm.

Heat supplied to the cooking pan is

$$\dot{Q} = 0.9 \times 1000$$
$$= 900\,\text{W}$$

The heat flux at the pan bottom ($x = 0$) is

$$\dot{q}_0 = \frac{\text{Heat transfer rate}}{\text{Bottom surface area}}$$
$$= \frac{\dot{Q}}{A_{\text{bottom}}}$$
$$= \frac{900}{\pi \times 0.125^2}$$
$$= \boxed{18.33\,\text{kW/m}^2}$$

The boundary condition is

$$\text{at } x = 0,\ T_0 = 100°\text{C}$$

## 2.3 Thermal Conductivity

Thermal conductivity $k$ is *a measure of heat flow* in a given material. The numerical value of the thermal conductivity indicates how fast heat will flow in a given material. The thermal conductivity has units of watts per meter per degree celsius when the heat flow is expressed in watts. From kinetic theory, we know that energy transfer is taking place due to the molecular motion in the medium. Thus, the faster the molecular motion, the faster will be the transport of energy. Therefore, the thermal conductivity of a gas should depend on the temperature of the gas, because the molecular motion is dictated by the temperature. A simplified analytical treatment shows that, the thermal conductivity of a gas varies with the square root of the absolute temperature. This is analogous to the speed of sound in a perfect gas, which also varies as the square root of the absolute temperature. Variation of the thermal conductivity of some gases with temperature is shown in Figure 2.5.

It is seen that, the sensitivity of the thermal conductivity $k$ with temperature $T$ for helium He and hydrogen $H_2$ are large compared to air, oxygen $O_2$ and carbon dioxide $CO_2$. For most gases at moderate pressures, the thermal conductivity is a function of the temperature alone. However, when the pressure of the gas becomes of the order of its critical pressure or when non-ideal-gas behavior is encountered, both pressure and temperature will influence the thermal conductivity.

The physical mechanism of thermal-energy conduction in liquids is qualitatively the same as in gases. But the situation is more complex because the molecules in liquids are more closely spaced and molecular force fields exert a strong influence on the energy exchange in the collision process.

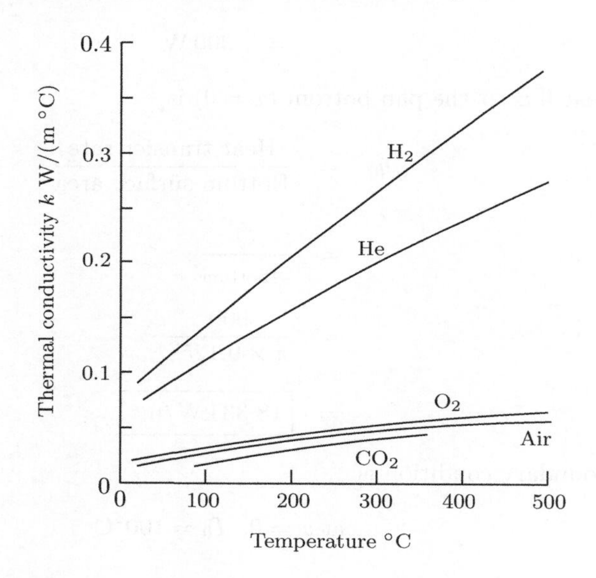


**Figure 2.5**
Thermal conductivity variation with temperature, for some gases.

The thermal conductivity of some liquids as a function of temperature is shown in Figure 2.6.

In solids, the thermal energy may be conducted through lattice vibration and by transport of free electrons. In good electrical conductors, a large number of free electrons move about in the lattice structure of the material. These electrons carry thermal energy from a high-temperature region to a low-temperature region. Energy may also be transmitted as vibrational energy in the lattice structure of the material. But usually this mode of energy transfer is not as large as the energy transfer associated with electron transport. Because of this reason, good electrical conductors are good heat conductors. For example, copper, silver and aluminum are good conductors of both electricity

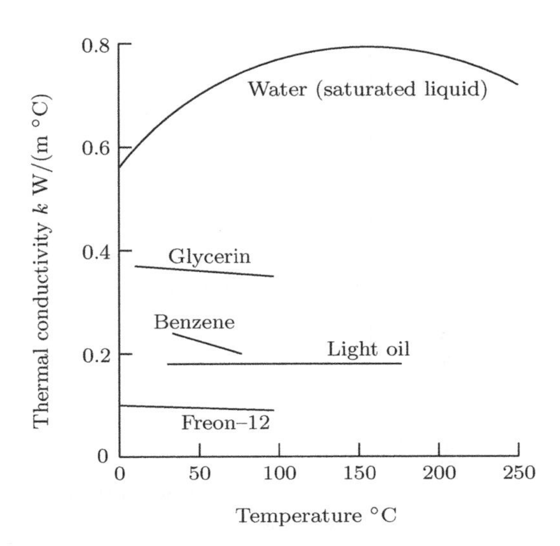


**Figure 2.6**
Thermal conductivity variation with temperature, for some liquids.

and heat. Similarly, all electrical insulators are usually good heat insulators. Thermal conductivity of some metals as a function of temperature are shown in Figure 2.7.

The thermal conductivity of insulating materials are very low. For example, the $k$ of glass wool and window glass are 0.038 W/(m °C) and 0.78 W/(m °C), respectively. At high temperatures, the energy transfer through an insulating material may involve many modes, such as conduction through the fibrous or porous solid material, conduction through the air trapped in the void spaces, and also as radiation at sufficiently high temperatures.

At this stage, it would be inspiring to know that an important challenge faced by researchers in the area of cryogenic engineering is the storing of cryogenic liquids, such as liquid hydrogen and liquid oxygen over long periods of time and transporting them to long distances. These cryogenic liquids need to be stored at very low temperatures of the order of around −250°C. Such applications led to the development of *super-insulations* for use at very low temperatures. For effective insulation, these super-insulations consist of multiple layers of highly reflective materials separated by insulating spacers. The entire system is evacuated to minimize the conduction through air. With such an arrangement, it is possible to minimize the thermal conductivities to as low as 0.3 mW/(m °C). To get an idea about the thermal conductivity, a gross summary of the thermal conductivities of some insulating materials at cryogenic temperatures are given in Figure 2.8.

To get an idea about the relative orders of magnitude, the conductivities

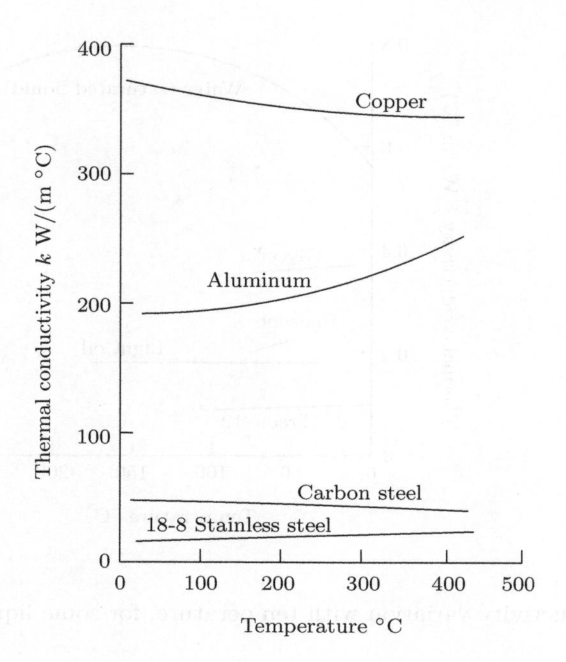


**Figure 2.7**
Thermal conductivity of some metals.

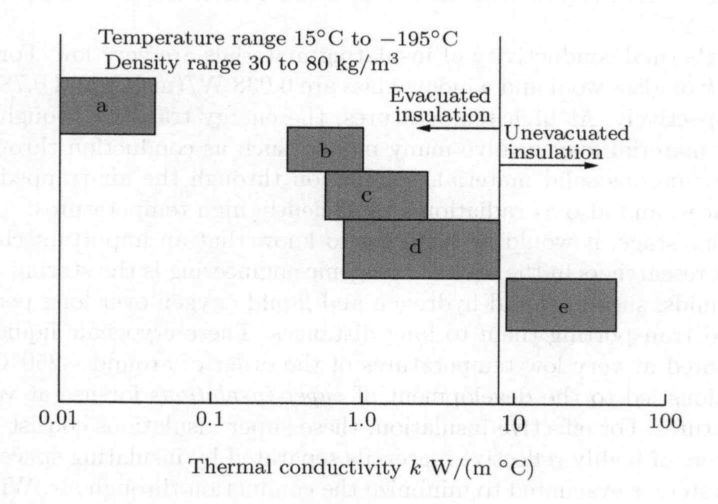


**Figure 2.8**
Thermal conductivities of some cryogenic insulation materials: (a) multilayer insulation, (b) opacified powders, (c) glass fibers, (d) powders, (e) foams, powders and fibers.

of some popular materials are given in Table 2.1.

## 2.4 Boundary and Initial Conditions

The heat diffusion equation [Equation (2.5)] is valid for all homogeneous media. But in heat conduction concerning composite media in which the thermophysical properties of one homogeneous region may differ significantly from those of others, the problem has to be solved by patching solutions of the heat diffusion equation for different regions of the homogeneous medium. To do this patching, the problem must satisfy the following conditions at the interfaces.

1. The temperatures in two adjoining regions should be identical.
2. The heat flux in one medium at an interface should be the same as that in the other side of the interface.

The second condition is obviously satisfied since the heat conducted to an interface from one region must be conducted away from the interface into the other region. Thus, at an interface

$$T^{I} = T^{II} \qquad \text{and} \qquad \dot{q}^{I} = \dot{q}^{II} \tag{2.7}$$

where superscripts $I$ and $II$ refer to regions 1 and 2, respectively.

With Fourier's law [Equation (2.1)], the second condition in Equation (2.7) becomes

$$-\left[k\,\frac{\partial T}{\partial n}\right]^{I} = -\left[k\,\frac{\partial T}{\partial n}\right]^{II} \tag{2.8}$$

where $n$ is the direction normal to the interface.

In addition to the conditions given by Equations (2.7) and (2.8), at the interface, we need the conditions at the boundaries of the region of interest, to solve the heat diffusion equation. One type of boundary condition specifies the temperature at the whole boundary. This occurs when there is boiling or condensation at the boundary. In the other type, the heat flux at the boundary is specified such that

$$\dot{q}_n = -k\,\frac{\partial T}{\partial n}\bigg|_{\text{at boundary}} \tag{2.9}$$

For insulated boundaries, $\dot{q}_n = 0$. We can also use $\partial T/\partial n = 0$ at the plane of symmetry in a problem.

At the solid-liquid interface, we can specify one more boundary condition. When a solid at a uniform temperature $T$ is immersed in a liquid bath at temperature $T_\infty$ and convection coefficient $h$, the boundary condition can be obtained by equating the heat flux given by Fourier's law to the heat flux

**Table 2.1** Thermal conductivity of some materials at 0°C

| Material | $k$ W/(m °C) |
|---|---|
| *Gases:* | |
| Air | 0.024 |
| Carbon dioxide | 0.0146 |
| Helium | 0.141 |
| Hydrogen | 0.175 |
| Water vapor (saturated) | 0.0206 |
| *Liquids:* | |
| Ammonia | 0.54 |
| Freon 12, $CCl_2F_2$ | 0.073 |
| Lubricating oil, SAE 50 | 0.147 |
| Mercury | 8.21 |
| Water | 0.556 |
| *Metals:* | |
| Aluminum (pure) | 202 |
| Copper (pure) | 385 |
| Carbon steel, 1% C | 43 |
| Chrome-nickel steel (18% Cr, 8% Ni) | 16.3 |
| Iron (pure) | 73 |
| Lead (pure) | 35 |
| Nickel (pure) | 93 |
| Silver (pure) | 410 |
| *Nonmetallic solids:* | |
| Glass, window | 0.78 |
| Glass wool | 0.038 |
| Magnesite | 4.15 |
| Maple or oak (wood) | 0.17 |
| Marble | 2.08 - 2.94 |
| Sandstone | 1.83 |
| Sawdust | 0.059 |
| Quartz, parallel to axis | 41.6 |

given by Newton's law of cooling at the boundary of the solid. That is, at the boundary, by Fourier's law of heat conduction,

$$\dot{q}_n = -k \frac{\partial T}{\partial n}$$

and by Newton's law of cooling,

$$\dot{q}_n = h (T_\infty - T)$$

Thus, we have

$$-k \frac{\partial T}{\partial n} = h (T_\infty - T)$$

or

$$k \frac{\partial T}{\partial n} = h (T - T_\infty) \tag{2.10}$$

Note that, when $T > T_\infty$, the heat from the surface of the solid will be convected away from the solid. This convection will cool the surface of the solid. This cooling will cause conduction of heat from the interior to the surface of the solid. Thus, the conduction within the solid and convection at the surface of the solid will be in the direction away from the center of the solid. On the other hand, when $T < T_\infty$, heat from the liquid will be convected to the solid surface. This convection will raise the surface temperature of the solid and establish a temperature gradient at the surface. This will induce heat conduction from the surface to the interior of the solid. Thus, for this case the conduction and convection will be in the direction towards the center of the solid.

The film heat transfer coefficient $h$ is to be determined from the theory of convection. In unsteady heat conduction problems, we also have to specify the initial temperature throughout the region of interest.

## Example 2.4

Steam flows through a pipe of diameter 60 mm, wall thickness 10 mm and length 30 m. The thermal conductivity of the pipe is 20 W/(m °C). If the inner and outer surfaces of the pipe are at 150°C and 60°C, respectively, obtain a general expression for the temperature distribution inside the pipe, under steady conditions, and find the rate of heat loss from the steam through the pipe to the surrounding.

## Solution

Given, $D_i = 60$ mm, $D_o = 80$ mm, $L = 30$ m, $k = 20$ W/(m °C), $T_i = 150$°C and $T_o = 60$°C.

The heat transfer through the pipe wall can be treated as one-dimensional, because there is thermal symmetry about the centerline and there is no variation of temperature in the axial direction. Thus, $T = T(r)$.

The governing equation for this case, given by Equation (2.6b) is

$$\frac{\partial^2 T}{\partial r^2} + \frac{1}{r}\frac{\partial T}{\partial r} = 0$$

that is,

$$\frac{d}{dr}\left(r\frac{dT}{dr}\right) = 0$$

The boundary conditions are

$$T(r_i) = T_i = 150°\text{C}$$

$$T(r_o) = T_o = 60°\text{C}$$

Integrating the governing equation with respect to $r$, we get

$$r\frac{dT}{dr} = c_1$$

where $c_1$ is an arbitrary constant. This can be arranged as

$$\frac{dT}{dr} = \frac{c_1}{r}$$

Integrating again with respect to $r$, we get

$$T = c_1 \ln r + c_2 \qquad (a)$$

where $c_1$ and $c_2$ are integration constants and they can be evaluated with the boundary conditions.
With the boundary conditions, we get

$$T_i = c_1 \ln r_i + c_2$$

$$T_o = c_1 \ln r_o + c_2$$

Solving for $c_1$ and $c_2$, we get

$$c_1 = \frac{T_o - T_i}{\ln(r_o/r_i)}$$

$$c_2 = T_i - \frac{T_o - T_i}{\ln(r_o/r_i)} \ln r_i$$

Substituting these $c_1$ and $c_2$ into Equation (a), we obtain, the general expression for the temperature distribution inside the pipe as

$$\boxed{T(r) = \frac{\ln(r/r_i)}{\ln(r_o/r_i)}(T_o - T_i) + T_i}$$

The rate of heat loss from the steam is the total rate of heat conducted through the pipe. This can be determined from Fourier's law as

$$\begin{aligned}
\dot{Q}_{\text{Cylinder}} &= -kA\frac{dT}{dr} \\
&= -k(2\pi rL)\frac{c_1}{r} = -2k\pi Lc_1 \\
&= 2k\pi L\frac{T_o - T_i}{\ln(r_o/r_i)} \\
&= 2\pi \times 20 \times 30 \times \frac{150 - 60}{\ln(0.04/0.03)} \\
&= \boxed{1179.4\,\text{kW}}
\end{aligned}$$

## 2.5 Summary

Conduction is the mode of heat transfer due to *temperature gradient* in a stationary medium. When a temperature gradient exists in a body, there will be energy transfer from the high-temperature region to the low-temperature region. The heat flux $\dot{q}$ (the amount of heat transferred per unit area) is

$$\dot{q}_x \propto \frac{\partial T}{\partial x}$$

or

$$\boxed{\dot{q}_x = -k\,\frac{\partial T}{\partial x}}$$

where $k$ is the *thermal conductivity* of the material.

Steady-state heat flux is given by Fourier's law of heat conduction as

$$\dot{q}_y = k\,\frac{T_s - T_0}{b}$$

or in the differential form

$$\dot{q}_y = -k\,\frac{\partial T}{\partial y}$$

where the subscript $y$ refers to the direction of heat flow.

For an isotropic medium, the heat fluxes along $x$- and $z$-directions are given, respectively, by

$$\dot{q}_x = -k\,\frac{\partial T}{\partial x} \quad \text{and} \quad \dot{q}_z = -k\,\frac{\partial T}{\partial z}$$

An alternative form of Fourier's law is

$$\boxed{\dot{q}_n = -k\,\frac{\partial T}{\partial n}}$$

where $\dot{q}_n$ is the heat flux in the direction $n$.

The rate of change of thermal energy contained within the system is

mass × specific heat × rate of change of temperature $= [\rho\,(\delta x \delta y \delta z)]\,c\,\frac{\partial T}{\partial t}$

The total heat conducted into the elemental volume is

$$\left(\frac{\partial}{\partial x}\left(k\frac{\partial T}{\partial x}\right)+\frac{\partial}{\partial y}\left(k\frac{\partial T}{\partial y}\right)+\frac{\partial}{\partial z}\left(k\frac{\partial T}{\partial z}\right)\right)(\delta x \delta y \delta z)$$

When there is heat generation or absorption within the element volume, their contribution must also be taken into account in the energy balance. In most cases such heat sources or sinks may be specified as volumetric sources or sinks, that is, their strength may be specified per unit volume.

If $\dot{g}_{th}$ represents a source strength, the total contribution from the volume element is $\dot{g}_{th}(\delta x \delta y \delta z)$. Therefore, for conservation of energy,

$$\boxed{[\rho\,(\delta x \delta y \delta z)]\,c\,\frac{\partial T}{\partial t} = [\nabla.\,(k \nabla T)]\,(\delta x \delta y \delta z) + \dot{g}_{th}(\delta x \delta y \delta z)}$$

that is,

$$\boxed{\rho\,c\,\frac{\partial T}{\partial t} = \nabla.\,(k \nabla T) + \dot{g}_{th}}$$

This is the general form of *heat diffusion equation.* This equation, usually known as the *heat equation*, provides the basic tool for heat conduction analysis.

With $k$ as constant, the heat diffusion equation becomes

$$\boxed{\frac{\partial T}{\partial t} = \frac{k}{\rho c}\,\nabla^2 T + \frac{\dot{g}_{th}}{\rho c}}$$

Thermal conductivity $k$ is *a measure of heat flow* in a given material. The numerical value of the thermal conductivity indicates how fast heat will flow in a given material. The thermal conductivity of a gas varies with the square root of the absolute temperature. This is analogous to the speed of sound in a perfect gas.

The physical mechanism of thermal-energy conduction in liquids is qualitatively the same as in gases. But the situation is more complex because the molecules in liquids are more closely spaced and molecular force fields exert a strong influence on the energy exchange in the collision process.

In solids, the thermal energy may be conducted through lattice vibration and by transport of free electrons.

The thermal conductivity of insulating materials are very low. At high temperatures, the energy transfer through an insulating material may involve many modes, such as conduction through the fibrous or porous solid material, conduction through the air trapped in the void spaces, and also as radiation at sufficiently high temperatures.

The heat diffusion equation is valid for all homogeneous media. But in heat conduction concerning composite media in which the thermophysical properties of one homogeneous region may differ significantly from those of others, the problem has to be solved by patching solutions of the heat diffusion equation for homogeneous medium in different regions. To do this patching, the problem must satisfy the following conditions at the interfaces.

1. The temperatures in two adjoining regions should be identical.
2. The heat flux in one medium at an interface should be the same as that in the other side of the interface.

The second condition is obviously satisfied since the heat conducted to an interface from one region must be conducted away from the interface into the other region. Thus, at an interface

$$T^I = T^{II} \qquad \text{and} \qquad \dot{q}^I = \dot{q}^{II}$$

At a solid-liquid interface the boundary condition is

$$k\,\frac{\partial T}{\partial n} = h\,(T - T_\infty)$$

## 2.6 Exercise Problems

2.1 A temperature difference of 420°C prevails steadily across an insulator layer of thickness 25 mm. If the thermal conductivity of the insulator material is 0.04 W/(m °C), determine the heat transfer rate per square meter area.

[**Ans.** 672 W/m$^2$]

2.2 A fiber glass layer of 10-cm thickness is laid over a wall surface of a hot chamber to limit the heat loss to 80 W/m$^2$ for a temperature difference of 160°C across the layer. Find the thermal conductivity of the fiber glass.

[**Ans.** 0.05 W/(m °C)]

2.3 An insulating material of thermal conductivity 0.04 W/(m °C) is to be used to insulate an ice box. At a temperature difference of 45°C across the walls of the box, if the heat loss should not exceed 40 W/m$^2$, what should be the thickness of the insulation?

[**Ans.** 45 mm]

2.4 A hot chamber made of 50 mm thick walls of thermal conductivity 1.2 W/(m °C), maintains temperatures 230°C and 20°C on either side of its walls. If the total surface area of the chamber walls is 6 $m^2$, determine the heat transfer rate across the walls.

[**Ans.** 30.24 kW]

2.5 The heat loss through a 5-cm thick insulating layer is 200 W/$m^2$. If the thermal conductivity of the layer material is 0.13 W/(m °C) and the hot surface temperature is 200°C, find the temperature of the cold surface?

[**Ans.** 123.08°C]

2.6 The inside and outside temperatures of a 15-mm thick and 90-cm high glass window of thermal conductivity 0.8 W/(m °C) are 22°C and −10°C, respectively. If the heat loss through the window is 2000 kJ per hour, determine its width.

[**Ans.** 36 cm]

2.7 The wall of an industrial furnace is constructed with 20-cm thick fire-clay bricks having thermal conductivity of 1.75 W/(m K). During steady-state operation, the temperatures of the inner and outer surfaces are 1200 and 900 K, respectively. If the wall is 2.5 m wide and 0.75 m high, determine the rate of heat loss through the wall.

[**Ans.** 4921.88 W]

2.8 An insulated steam pipe passes through a room in which both the air and walls are at 20°C. The outer diameter of the pipe is 100 mm, and its surface temperature and emissivity are 250°C and 0.8, respectively. If the coefficient associated with free convection heat transfer from the pipe surface to air is 15 W/($m^2$ K), compute the rate of heat loss from the surface per unit length of the pipe.

[**Ans.** 2046 W/m]

# Chapter 3

# One-Dimensional, Steady-State Conduction

## 3.1 Introduction

Application of Fourier's law of heat conduction for the calculation of heat flow in some simple, one-dimensional systems is discussed in this chapter. The term "one-dimensional" refers to the fact that only one spacial coordinate is needed to describe the spatial variation of the dependent variables. Several physical shapes may fall in the category of one-dimensional system. For instance,

- A cylindrical or a spherical body can be treated as one-dimensional system, when the temperature in the body is a function of only the radial distance, and is independent of the azimuth angle or axial distance.
- In some two-dimensional problems, the effect of one of the space coordinates may be so small compared to the other and hence can be neglected. Such two-dimensional heat-flow problems may be approximated as one-dimensional problems.

In these cases the governing equations reduce to simple form, leading to easy solutions.

## 3.2 Steady-State Conduction in a Plane Wall

Consider heat conduction through a plane wall in the direction ($x$-direction) normal to the wall, as shown in Figure 3.1. The left and right surfaces of the plate are at temperatures $T_1$ and $T_2$, respectively.

At steady state, the temperature $T$ distribution in the plane wall of finite thickness $L$ in the $x$-direction, and extending to infinity in the $y$- and $z$-directions, is a function of $x$ only. That is, the problem is essentially a

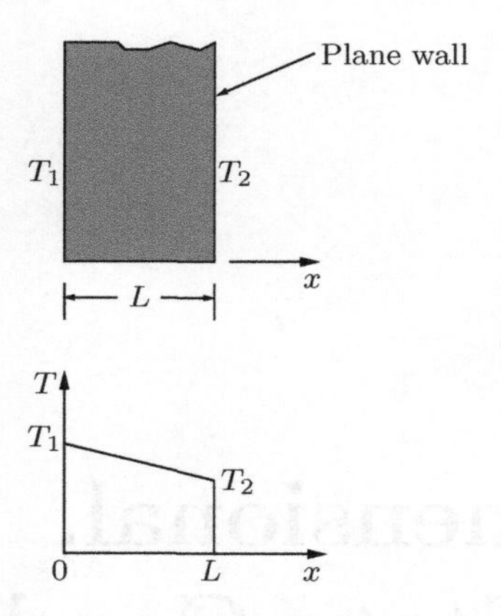


**Figure 3.1**
Heat transfer through a plane wall.

one-dimensional conduction in the $x$-direction. Therefore, for this problem of steady-state heat transfer through a plane wall in the limit $0 < x < L$ and without any heat source ($\dot{S}_{\text{th}}$), the heat equation [Equation (2.6)],

$$\frac{\partial T}{\partial t} = \frac{k}{\rho c} \nabla^2 T + \frac{\dot{g}_{\text{th}}}{\rho c}$$

reduces to

$$\boxed{\frac{d^2 T}{dx^2} = 0} \tag{3.1}$$

If the thermal conductivity $k$ of the wall material is assumed to be constant, Equation (3.1) can be integrated twice to obtain the following general equation for temperature distribution in the slab, as

$$T(x) = c_1 x + c_2 \tag{3.2}$$

where $c_1$ and $c_2$ are integration constants. Note that the governing equation (3.1) may also be obtained by the argument that, at steady state, whatever heat flux enters an element must leave it, hence

$$\frac{d}{dx}\left(\dot{q}_x\right) = \frac{d}{dx}\left(-k\,\frac{\partial T}{\partial x}\right) = 0$$

that is,

$$k\,\frac{\partial^2 T}{\partial x^2} = 0$$

or

$$\frac{d^2 T}{dx^2} = 0$$

The constants $c_1$ and $c_2$ in the linear temperature profile [Equation (3.2)] have to be determined by the thermal conditions at the boundary. The boundary conditions for the present problem of heat transfer through the plane wall are the following.

$$\text{at} \quad x = 0, \quad T = T_1$$

$$\text{at} \quad x = L, \quad T = T_2$$

Using these boundary conditions, from Equation (3.2), we get

$$\begin{aligned} T_1 &= 0 + c_2 \\ T_2 &= c_1 L + c_2 \end{aligned}$$

Therefore,

$$c_2 = T_1$$

$$c_1 = \frac{T_2 - T_1}{L}$$

Substituting these into the general solution [Equation (3.2)], we get the temperature distribution as

$$\boxed{T(x) = (T_2 - T_1)\,\frac{x}{L} + T_1} \tag{3.3}$$

From this result it is evident that, for one-dimensional, steady-state conduction in a plane wall with no heat generation ($\dot{g}_{\text{th}} = 0$) and constant thermal conductivity ($k$ = constant), the temperature varies linearly with $x$.

Now we may use Fourier's law of heat conduction [Equation (2.1)], and the temperature distribution given by Equation (3.3), to determine the conduction heat flux. That is,

$$\dot{q}_x = -k\,\frac{dT}{dx} = \frac{k}{L}\,(T_1 - T_2) \tag{3.4}$$

If the cross-sectional area ($yz$) of the slab wall is $A$, the heat current (heat transfer rate) becomes

$$\boxed{\dot{Q}_x = A\,\dot{q}_x = \frac{kA}{L}\,(T_1 - T_2)} \tag{3.5}$$

Note that, the heat flux $\dot{q}_x$ is a constant and independent of $x$.

## Example 3.1

An insulated steel rod of 30-mm diameter and 0.5-m length connects two hot metallic plates maintained at 100°C and 80°C, as shown in Figure E3.1. The rod is heated electrically by discharging electric energy at the rate of 20 W. Assuming the thermal conductivity of the rod to be 40 W/(m K), determine the heat flux at the ends of the rod, the maximum temperature in the rod, and the location of the temperature maximum, at steady-state conduction.

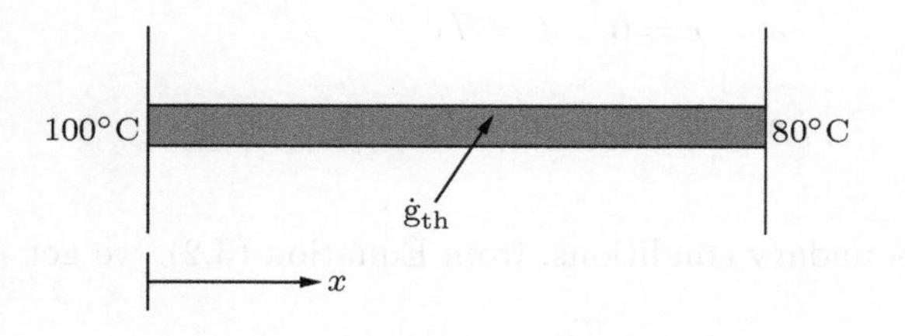


**Figure E3.1**
Conduction in an insulated rod.

## Solution

The governing equation for conduction, given by Equation (2.6), is

$$\frac{\partial T}{\partial t} = \frac{k}{\rho c} \nabla^2 T + \frac{\dot{g}_{th}}{\rho c}$$

For the given steady-state, one-dimensional conduction, this reduces to

$$k \frac{d^2T}{dx^2} + \dot{g}_{th} = 0 \qquad (i)$$

where $\dot{g}_{th}$ is the heat generated per unit volume, given by

$$\begin{aligned} \dot{g}_{th} &= \frac{20}{AL} = \frac{20}{(\pi D^2/4)\, L} \\ &= \frac{20 \times 4}{\pi \times 0.03^2 \times 0.5} \\ &= 56588.4\,\text{W/m}^3 \end{aligned}$$

The governing equation [Equation (*i*)] gives

$$\frac{d^2T}{dx^2} = -\frac{\dot{g}_{th}}{k}$$

Integrating this, we get

$$T = -\frac{\dot{g}_{th}}{k}\frac{x^2}{2} + c_1\, x + c_2$$

where $c_1$ and $c_2$ are integration constants.

At $x = 0$, $T = 100°\text{C}$, therefore,

$$c_2 = 100$$

At $x = 0.5$ m, $T = 80°\text{C}$, therefore,

$$80 = -\frac{56588.4}{40} \times \frac{0.5^2}{2} + 0.5\,c_1 + 100$$

This gives, $c_1 = 313.68$. Thus,

$$T = -\frac{\dot{g}_{\text{th}}}{k}\frac{x^2}{2} + 313.68\,x + 100$$

For maximum temperature, $dT/dx = 0$ and $d^2T/dx^2 < 0$. That is,

$$\frac{dT}{dx} = -\frac{\dot{g}_{\text{th}}}{k}\,x + 313.68 = 0$$

and

$$\frac{d^2T}{dx^2} = -\frac{\dot{g}_{\text{th}}}{k} < 0$$

Solving for $x$, we get

$$\begin{aligned} x &= \frac{313.68 \times 40}{56588.4} \\ &= \boxed{0.222\,\text{m}} \end{aligned}$$

Thus, the maximum temperature becomes

$$\begin{aligned} T_{\text{max}} &= -\left(\frac{56588.4}{40} \times \frac{0.222^2}{2}\right) + (313.68 \times 0.222) + 100 \\ &= \boxed{134.78°\text{C}} \end{aligned}$$

The heat flux at $x = 0$ is

$$\begin{aligned} \dot{q}\Big|_{x=0} &= -k\left(\frac{dT}{dx}\right)_{x=0} \\ &= -k\left(-\frac{\dot{g}_{\text{th}}}{k}\,x + 313.68\right)_{x=0} \\ &= -313.68\,k = -313.68 \times 40 \\ &= \boxed{-12.55\,\text{kW/m}^2} \end{aligned}$$

At steady state, the heat flux throughout the rod should be the same. Thus, the heat flow through the rod is $12.55\,\text{kW/m}^2$, the maximum temperature in the rod is 134.78°C and the location of temperature maximum is 0.222 m from the end maintained at 100°C.

### 3.2.1 Thermal Resistance

For the problem of heat conduction through a plane wall discussed in Section 3.2, instead of considering the left and right surfaces at constant temperatures $T_1$ and $T_2$, if the wall has constant temperature $T_1$ at $x = 0$ and at $x = L$ it is in contact with a fluid of temperature $T_{2f}$, the boundary condition at $x = L$ is given by Newton's law of cooling as

$$-k \left.\frac{dT}{dx}\right|_{x=L} = h_2 \left( \left. T \right|_{x=L} - T_{2f} \right) \tag{3.6}$$

where $h_2$ is the convection heat-transfer coefficient of the fluid. With the new boundary conditions, the constant $c_2$ in Equation (3.2) becomes

$$c_2 = T_1$$

Thus, Equation (3.2) becomes

$$T = c_1 x + T_1$$

Differentiating with respect to $x$, we get

$$\frac{dT}{dx} = c_1$$

Substituting the above expressions for $T$ and $dT/dx$ into Equation (3.6), we get

$$-k\, c_1 = h_2 \left( c_1 L + T_1 - T_{2f} \right)$$

Solving, we get

$$c_1 = -\frac{h_2 \left( T_1 - T_{2f} \right)}{h_2 L + k}$$

Substituting the above $c_1$ and $c_2$ into Equation (3.2), we get

$$T = T_1 - \frac{h_2 \left( T_1 - T_{2f} \right)}{h_2 L + k}\, x$$

The conduction heat current becomes

$$\dot{Q}_x = -kA \frac{dT}{dx} = \frac{kAh_2}{h_2 L + k} \left( T_1 - T_{2f} \right)$$

$$\boxed{\dot{Q}_x = \frac{T_1 - T_{2f}}{\dfrac{L}{kA} + \dfrac{1}{h_2 A}}} \tag{3.7}$$

We know that,

$$\text{Heat flow} = \frac{\text{Thermal potential difference}}{\text{Thermal resistance}}$$

That is,

$$\dot{Q}_{\text{th}} = \frac{\Delta T}{R_{\text{th}}}$$

For heat conduction through the wall shown in Figure 3.1, by Equation (3.7), we get

$$\dot{Q}_x = \frac{kA}{L}(T_1 - T_2)$$

Comparing the above two equations, we get the thermal resistance $R_{\text{th}}$ of the wall as

$$\boxed{R_{\text{th}} = \frac{L}{kA}} \tag{3.8}$$

This may be visualized as equivalent to the flow of current through a circuit where the driving potential is $(T_1 - T_2)$ and the resistance is $\frac{L}{kA}$.

We can draw a circuit diagram for a wall of area $A$, thickness $L$ and conductivity $k$, as shown in Figure 3.2.

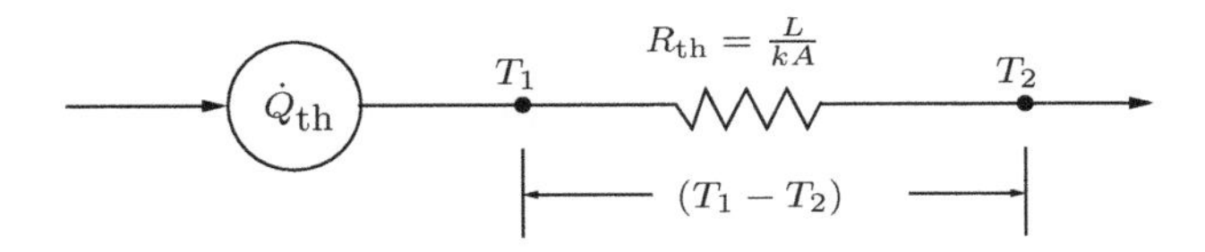


**Figure 3.2**
Equivalent electrical circuit for heat transfer through a wall.

For the case investigated above where a fluid film was assumed at the wall surface at $x = L$, the appropriate driving potential is $(T_1 - T_{2f})$ and the resistance $R_{\text{th}}$ is given by

$$\boxed{R_{\text{th}} = \frac{L}{kA} + \frac{1}{h_2 A}} \tag{3.9}$$

For this case, as seen from Equation (3.9), the resistance of the slab material $L/(kA)$ and the resistance of the fluid film $1/(h_2A)$ are two resistances in series, since the current through one should be the same as that through the other, and the overall driving potential $(T_1 - T_{2f})$ is the sum of the potential drops $(T_1 - T_2)$ and $(T_2 - T_{2f})$ in the two layers. Therefore, the equivalent electrical circuit for this case is as shown in Figure 3.3.
The resistance of the fluid film above is $R_f = 1/(h_2A)$. Thus, for a composite wall made up of a number of layers as shown in Figure 3.4, the resultant resistance is given by

$$R_{\text{th}} = \frac{1}{h_1 A} + \frac{L_{12}}{k_{12}A} + \frac{L_{23}}{k_{23}A} + \ldots.. + \frac{L_{(n-1)n}}{k_{(n-1)n}A} + \frac{1}{h_n A} \tag{3.10}$$

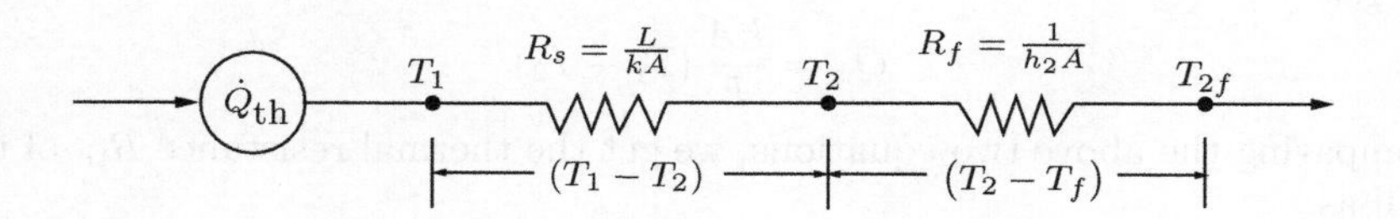


**Figure 3.3**
Equivalent electrical circuit for heat transfer through a plane wall with a fluid film at one of its surfaces.

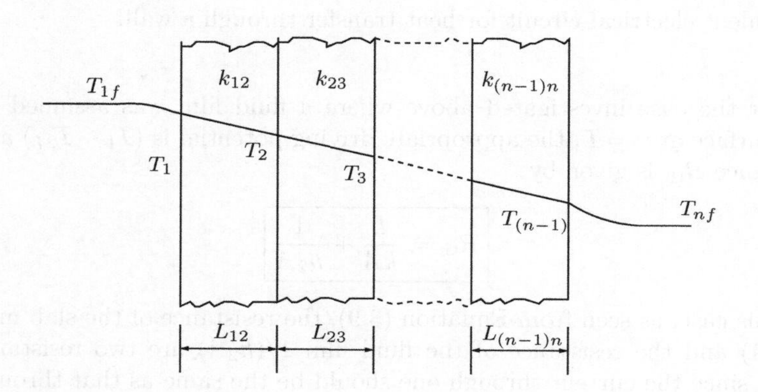


**Figure 3.4**
Temperature variation across a series composite wall.

and the heat current will be

$$\dot{Q}_{\text{th}} = \frac{T_{1f} - T_{nf}}{R_{\text{th}}} \tag{3.11}$$

The temperature at the interfaces can be determined by finding the temperature drops across each layer. Thus,

$$T_{1f} - T_1 = \dot{Q}_{\text{th}}\,\frac{1}{h_1 A}$$

$$(T_2 - T_1) = \dot{Q}_{\text{th}}\,\frac{L_{12}}{k_{12} A}$$

etc.

The overall heat transfer coefficient $U$, for a composite wall, may be defined as

$$\dot{Q}_{\text{th}} = UA\,(T_{1f} - T_{nf}) \tag{3.12}$$

where

$$U = \frac{1}{\dfrac{1}{h_1} + \dfrac{L_{12}}{k_{12}} + \dfrac{L_{23}}{k_{23}} + \ldots.. + \dfrac{L_{(n-1)n}}{k_{(n-1)n}} + \dfrac{1}{h_n}}$$

## Example 3.2

A 450-mm thick brick wall is plastered on one side with 30-mm thick concrete. The thermal conductivity of brick and concrete are 0.7 W/(m K) and 0.92 W/(m K), respectively. If the temperature of the exposed brick face is 35°C and that of concrete is 10°C, determine the heat loss per hour through a wall which is 5 m long and 3.6 m high. Also, determine the interface temperature.

## Solution

Given, $L_1 = 0.45$ m, $L_2 = 0.03$ m, $k_1 = 0.7$ W/(m K), $k_2 = 0.92$ W/(m K).

The heat flux across the wall is given by

$$\dot{Q} = A\,\dot{q} = A\,\frac{T_1 - T_2}{\dfrac{L_1}{k_1} + \dfrac{L_2}{k_2}}$$

where subscript 1 refers to brick and subscript 2 refers to concrete. Therefore,

$$\begin{aligned}\dot{Q} &= (5 \times 3.6)\,\frac{35 - 10}{\dfrac{0.45}{0.7} + \dfrac{0.03}{0.92}} \\ &= 666.2\,\text{J/s}\end{aligned}$$

Heat loss per hour is

$$\begin{aligned}\dot{Q} &= 666.2 \times 3600 \\ &= \boxed{2398.32\,\text{kJ}}\end{aligned}$$

*Interface Temperature $T_{\text{int}}$:*

For the brick wall, we have

$$\dot{Q} = \frac{k_1 A(T_1 - T_{\text{int}})}{L}$$

or

$$666.2 = \frac{0.7(5 \times 3.6)}{0.45}(35 - T_{\text{int}})$$

This gives

$$\begin{aligned}T_{\text{int}} &= 35 - 23.79 \\ &= \boxed{11.21\,°\text{C}}\end{aligned}$$

*Alternate Method:*

For the concrete we have

$$\dot{Q} = \frac{k_2 A(T_{\text{int}} - T_2)}{L}$$

or

$$666.2 = \frac{0.92(5 \times 3.6)}{0.03}(T_{\text{int}} - 10)$$

This gives

$$\begin{aligned}T_{\text{int}} &= 1.207 + 10 \\ &= \boxed{11.21\,°\text{C}}\end{aligned}$$

## Example 3.3

A composite window separating an oven cavity from the room air consists of two high-temperature plastics (A and B) of thickness $L_{\text{A}} = 3L_{\text{B}}$ and thermal conductivities $k_{\text{A}} = 0.1$ W/(m K) and $k_{\text{B}} = 0.06$ W/(m K). During a self-cleaning process, the oven wall and air temperatures, $T_w$ and $T_a$ are 300°C, while the room temperature $T_\infty$ is 20°C. The inside convection and radiation heat transfer coefficients, $h_i$ and $h_r$, as well as the outside convection coefficient, $h_o$, are each approximately 22 W/(m$^2$ K). Determine the minimum window thickness, $L = L_{\text{A}} + L_{\text{B}}$, required to ensure a temperature of 40°C or less at the outer surface of the window.

## Solution

The schematic diagram of the heat transfer system desired and the thermal circuit for the process are given in Figure ES3.3.

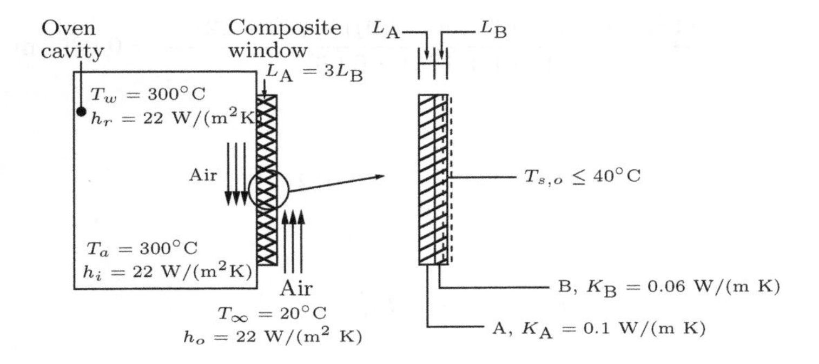


**Figure ES3.3**
Heat transfer system and the thermal circuit.

For energy balance,

$$\dot{E}_{\text{in}} = \dot{E}_{\text{out}}$$

From Equation (3.11), with $T_{1f} = T_w = T_a$ and $T_{2f} = T_{s,o}$, we have

$$\dot{E}_{\text{in}} = \dot{Q} = \frac{T_a - T_{s,o}}{R_{\text{th}}}$$

From Equation (3.12), we have

$$\dot{E}_{\text{out}} = \dot{Q} = h_o A \left(T_{s,o} - T_\infty\right)$$

The total thermal resistance $R_{\text{th}}$ between the oven and the outer surface of the window includes an equivalent resistance associated with convection and radiation, which acts in parallel, at the inner surface of the window, with conduction resistance of the window materials. Hence,

$$R_{\text{th}} = \left(\frac{1}{1/(h_i A)} + \frac{1}{1/(h_r A)}\right)^{-1} + \frac{L_{\text{A}}}{k_{\text{A}} A} + \frac{L_{\text{B}}}{k_{\text{B}} A}$$

or

$$R_{\text{th}} = \frac{1}{A}\left(\frac{1}{h_i + h_r} + \frac{L_{\text{A}}}{k_{\text{A}}} + \frac{L_{\text{A}}}{3k_{\text{B}}}\right)$$

Substituting into the energy balance equation, we get

$$\frac{T_a - T_{s,o}}{(h_i + h_r)^{-1} + (L_{\text{A}}/k_{\text{A}}) + (L_{\text{A}}/3k_{\text{B}})} = h_o(T_{s,o} - T_\infty)$$

This gives

$$L_{\mathrm{A}} = \frac{(1/h_o)(T_a - T_{s,o})/(T_{s,o} - T_\infty) - (h_i + h_r)^{-1}}{(1/k_{\mathrm{A}} + 1/3k_{\mathrm{B}})}$$

$$= \frac{(1/22)\,(300-40)/(40-20)) - (22+22)^{-1}}{1/0.1 + 1/(3 \times 0.06)} = 0.0365\,\mathrm{m}$$

and

$$L_{\mathrm{B}} = \frac{L_{\mathrm{A}}}{3} = 0.01217$$

Therefore,

$$L = L_{\mathrm{A}} + L_{\mathrm{B}}$$

$$= 0.0365 + 0.01217$$

$$= 0.04867\,\mathrm{m}$$

$$= \boxed{48.67\,\mathrm{mm}}$$

## Example 3.4

A composite plane wall made up of materials A and B has thickness $L_A = 50$ mm and $L_B = 25$ mm. The thermal conductivity of the materials are $k_A = 70$ W/(m °C) and $k_B = 100$ W/(m °C). The outer surface of wall A is perfectly insulated and inside the wall, heat is generated at a uniform rate of 2000 kW/m$^3$. The outer surface of wall B is cooled by a water stream at 20°C and $h = 1$ kW/(m$^2$ °C). Determine the temperatures at the insulated surface and the cooled surface, at steady-state condition. Assume the conduction to be one-dimensional and negligible contact resistance between the walls.

## Solution

The composite wall and the conditions given are schematically shown in Figure ES3.4.
The thermal resistance due to wall B and water is

$$R_{\mathrm{th}} = \left(\frac{L}{k}\right)_{\mathrm{B}} + \frac{1}{h_{\mathrm{w}}}$$

$$= \frac{0.025}{100} + \frac{1}{1000}$$

$$= 0.00125\,(\mathrm{m}^2\ {}^\circ\mathrm{C})/\mathrm{W}$$

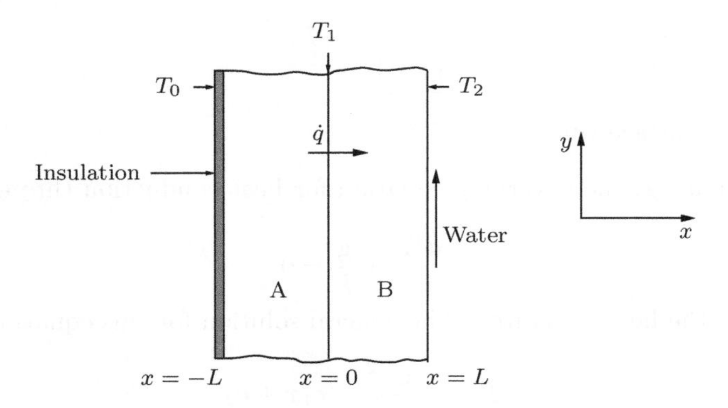


**Figure ES3.4**
Composite wall.

Heat generated inside the wall is $\dot{Q} = 2000$ kW/m$^3$. This heat is transferred through wall B of thickness 0.05 m. Thus, the heat flux through wall B becomes

$$\begin{aligned}\dot{q} &= \dot{Q} \times L \\ &= (2000 \times 10^3) \times 0.05 \\ &= 100000\,\mathrm{W/m^2}\end{aligned}$$

This heat flux is also equal to

$$\dot{q} = -\frac{\Delta T}{R_{\mathrm{B}}} = \frac{T_1 - T_2}{R_{\mathrm{B}}}$$

where $R_{\mathrm{B}}$ is the thermal resistance of wall B. Therefore,

$$\begin{aligned}\dot{q} &= 100 \times 10^3 \\ &= \frac{T_1 - T_2}{0.00125} \\ &= \frac{T_1 - 20}{0.00125} \\ T_1 &= (100 \times 10^3) \times 0.00125 + 20\end{aligned}$$

$$= \boxed{145°\text{C}}$$

The cooled surface will be at 145°C.

For constant $k_A$, the governing equation for heat conduction through wall $A$ becomes

$$\frac{d^2T}{dx^2} + \frac{\dot{g}}{k} = 0$$

where $\dot{g}$ is the heat generated. The general solution for this equation is

$$T = -\frac{\dot{g}\,x^2}{2k} + c_1 x + c_2$$

where $c_1$ and $c_2$ are integration constants. For the given conditions, we have $T(-L) = T_0$ and $T(L) = T_1$. With these conditions, the constants become

$$c_1 = \frac{T_1 - T_0}{2L}$$

$$c_2 = \frac{\dot{g}}{2k} L^2 + \frac{T_1 + T_0}{2}$$

Therefore, the general solution becomes

$$T(x) = \frac{\dot{g}\,L^2}{2k}\left(1 - \frac{x^2}{L^2}\right) + \frac{T_1 - T_0}{2}\frac{x}{L} + \frac{T_1 + T_0}{2}$$

The heat flux at any point in the wall can be determined using this relation and Fourier's law. However, with heat generation the heat flux is no longer independent of $x$.

This result gets simplified when both surfaces of wall $A$ are maintained at a common temperature $T_0 = T_1 \equiv T_{\text{mid}}$. The temperature distribution is then symmetrical about the mid-plane and is given by

$$T(x) = \frac{\dot{g}\,L^2}{2k}\left(1 - \frac{x^2}{L^2}\right) + T_1$$

The maximum temperature prevails at the mid-plane, that is, at $x = 0$. Thus,

$$T_{\text{mid}} = \frac{\dot{g}\,L^2}{2k} + T_1$$

$$= \frac{(2000 \times 10^3) \times (0.05)^2}{2 \times 70} + 145$$

$$= 35.7 + 145$$

$$= \boxed{180.7°\text{C}}$$

The insulated surface will be in thermal equilibrium at steady state, thus, $T_0 = 180.7°C$. The temperature in slab A will be uniform and in slab B the temperature will vary as shown in Figure ES3.4a.

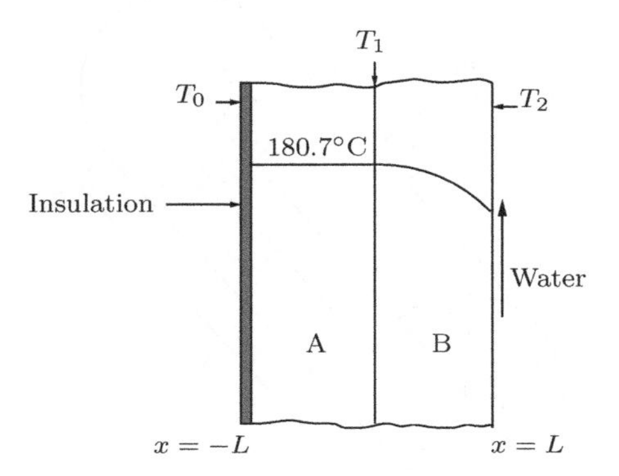


**Figure ES3.4a**
Temperature distribution in slabs A and B.

## 3.3 Heat Conduction across a Cylindrical Shell

Cylindrical and spherical systems often experience temperature gradients in the radial direction only and may therefore be treated as one-dimensional. Also, under steady-state conditions with no heat generation, such systems may be analyzed by using the standard method, which begins with the appropriate form of the heat conduction equation.

Consider a cylindrical shell in a cross-flow, shown in Figure 3.5.

For a sufficiently long cylinder, that is, the length is much larger than the radius, the end effects are negligible and the temperature is a function of the radius alone.

For steady-state radial conduction in the absence of heat sources, and with constant material properties, the heat equation

$$\frac{\partial T}{\partial t} = \frac{k}{\rho c} \nabla^2 T + \frac{\dot{g}_{\text{th}}}{\rho c}$$

reduces to

$$\nabla^2 T = 0 \tag{3.13}$$

In cylindrical coordinates

$$\nabla^2 = \frac{\partial^2}{\partial r^2} + \frac{1}{r}\frac{\partial}{\partial r} + \frac{1}{r^2}\frac{\partial^2}{\partial \theta^2} + \frac{\partial^2}{\partial z^2}$$

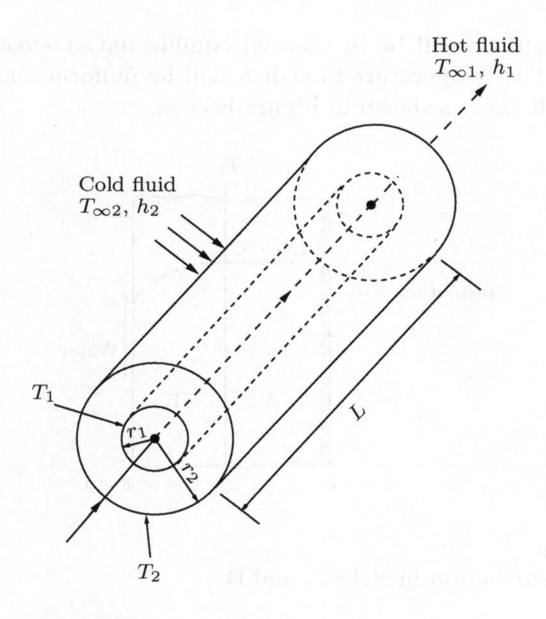


**Figure 3.5**
Hollow cylinder in a cross-flow.

Therefore, for the present problem, Equation (3.13) in cylindrical coordinates becomes

$$\frac{\partial^2 T}{\partial r^2} + \frac{1}{r}\frac{\partial T}{\partial r} = 0$$

since $\frac{\partial^2 T}{\partial \theta^2} = 0$ because of symmetry, and $\frac{\partial^2 T}{\partial z^2} = 0$ because $L$ is very large. Therefore, the temperature is function of only $r$, and the partial derivatives in the above equation become total derivatives, thus,

$$\frac{d^2 T}{dr^2} + \frac{1}{r}\frac{dT}{dr} = 0$$

that is,

$$\frac{d}{dr}\left(r\frac{dT}{dr}\right) = 0 \tag{3.14}$$

The boundary conditions are the following:

$$\text{at } r = r_1, \quad T = T_1$$

$$\text{at } r = r_2, \quad T = T_2$$

For the present problem, the heat flux $\dot{q}_r$ is not constant and decreases with increase of $r$ because of increase in area, but the heat current or heat transfer rate $\dot{Q}_r$ is constant in the radial direction.

Equation (3.14) may be integrated twice to obtain the general solution

$$T(r) = c_1 \ln r + c_2 \tag{3.15}$$

where $c_1$ and $c_2$ are integration constants to be evaluated with the above boundary conditions. With the boundary conditions, we get

$$T_1 = c_1 \ln r_1 + c_2$$
$$T_2 = c_1 \ln r_2 + c_2$$

Solving these two equations, we obtain

$$\begin{aligned} c_1 &= \frac{T_2 - T_1}{\ln\left(\dfrac{r_2}{r_1}\right)} \\ c_2 &= T_2 - \frac{T_2 - T_1}{\ln\left(\dfrac{r_2}{r_1}\right)} \ln r_2 \end{aligned}$$

Substituting $c_1$ and $c_2$ into Equation (3.15), we get

$$\boxed{T(r) = \frac{T_2 - T_1}{\ln\left(\dfrac{r_2}{r_1}\right)} \ln\left(\frac{r}{r_2}\right) + T_2} \tag{3.16}$$

The constants $c_1$ and $c_2$ can also be expressed as

$$\begin{aligned} c_1 &= \frac{T_2 - T_1}{\ln\left(\dfrac{r_2}{r_1}\right)} \\ c_2 &= T_1 - \frac{T_2 - T_1}{\ln\left(\dfrac{r_2}{r_1}\right)} \ln r_1 \end{aligned}$$

Substituting these $c_1$ and $c_2$, the temperature distribution can also be expressed as

$$\boxed{T(r) = \frac{T_2 - T_1}{\ln\left(\dfrac{r_2}{r_1}\right)} \ln\left(\frac{r}{r_1}\right) + T_1} \tag{3.16a}$$

Note that, the temperature distribution associated with radial conduction through a cylinder is logarithmic, not linear, as in the case of plane wall at identical conditions [Equation (3.3)].

The heat flux at any radial location $r$ is given by

$$\dot{q}_r = -k\frac{dT}{dr}$$

Using Equation (3.16) into the above, we get

$$\dot{q}_r = -\frac{k}{r}\,\frac{(T_2 - T_1)}{\ln\left(\dfrac{r_2}{r_1}\right)}$$

The heat current for a cylinder of length $L$ is

$$\dot{Q}_r = 2\pi r L\,\dot{q}_r = \frac{2\pi Lk\,(T_1 - T_2)}{\ln\left(\dfrac{r_2}{r_1}\right)} \tag{3.17}$$

The thermal resistance is given by

$$\boxed{R_{\text{th}} = \frac{\ln\left(\dfrac{r_2}{r_1}\right)}{2\pi Lk}} \tag{3.18}$$

Note that, the value of $\dot{Q}_r$ in Equation (3.17) is independent of $r$, hence the foregoing result could have been obtained by simply integrating the equation

$$\dot{Q}_r = A\dot{q}_r = -A\,k\frac{dT}{dr}$$

The area is $A = 2\pi rL$. Therefore,

$$\begin{aligned}\dot{Q}_r &= -2\pi rL\,k\frac{dT}{dr}\\ \dot{Q}_r\,\frac{dr}{r} &= -2\pi kL\,dT\end{aligned}$$

Now integration between the limits 1 to 2, we get

$$\begin{aligned}\dot{Q}_r\left[\ln r\right]_1^2 &= -2\pi kL\left[T\right]_1^2\\ \dot{Q}_r\ln\left(\frac{r_2}{r_1}\right) &= -2\pi kL\,(T_2 - T_1)\end{aligned}$$

or

$$\dot{Q}_r = \frac{2\pi kL\,(T_1 - T_2)}{\ln\left(\dfrac{r_2}{r_1}\right)}$$

Note that, this is same as Equation (3.17).

Consider now the composite cylindrical shell made up of two layers of material with conduction coefficients $k_{12}$ and $k_{23}$ over a cylindrical shell of radius $r_1$, shown in Figure 3.6.

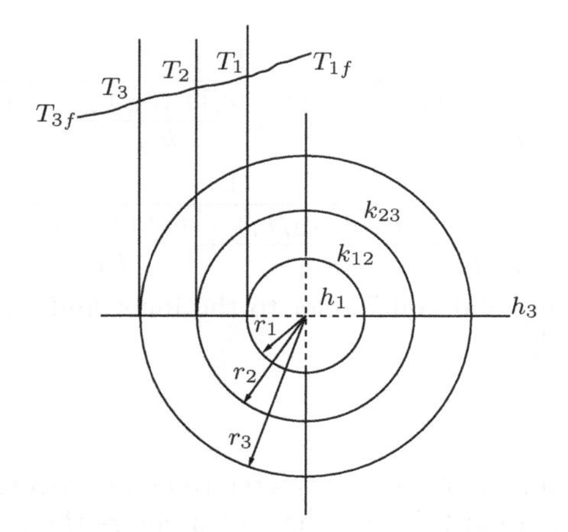


**Figure 3.6**
Temperature variation across a composite cylindrical shell.

For a convective film on a cylindrical surface of temperature $T_s$ and radius $r$, the heat current is given by

$$\dot{Q}_r = (2\pi r L)\, h\, (T_s - T_\infty)$$

where $T_\infty$ is the ambient fluid temperature, $h$ is the convection coefficient and $L$ is the length of the cylinder. Therefore, the film resistance is

$$R_{\text{th}} = \frac{1}{2\pi r L h}$$

Hence, the total resistance, due to the convection environment inside the cylindrical shell of radius $r_1$, the conduction through the layer of inside radius $r_1$ and outer radius $r_2$, the conduction through the layer of inside radius $r_2$ and outer radius $r_3$, and the convection environment over the entire outer surface of the composite cylindrical shell, is given by

$$R_{\text{th}} = \frac{1}{2\pi r_1 L h_1} + \frac{\ln\left(\dfrac{r_2}{r_1}\right)}{2\pi L k_{12}} + \frac{\ln\left(\dfrac{r_3}{r_2}\right)}{2\pi L k_{23}} + \frac{1}{2\pi r_3 L h_3} \tag{3.19}$$

We can also define an overall film transfer coefficient $U$, as

$$\dot{Q}_{\text{th}} = U\, A\, (T_{1f} - T_{3f})$$

But because of varying area in the $r$-direction, the area on which the $U$ is based has to be specified. Thus,

$$\dot{Q}_{\text{th}} = U_{\text{in}} A_{\text{in}}\, (T_{1f} - T_{3f}) = U_{\text{out}} A_{\text{out}}\, (T_{1f} - T_{3f}) \tag{3.20}$$

and

$$U_{\text{in}} = \frac{1}{\dfrac{1}{h_1} + \dfrac{r_1 \ln (r_2/r_1)}{k_{12}} + \dfrac{r_1 \ln (r_3/r_2)}{k_{23}} + \dfrac{r_1}{r_3}\dfrac{1}{h_3}} \tag{3.21}$$

$$U_{\text{out}} = \frac{1}{\dfrac{r_3}{r_1}\dfrac{1}{h_1} + \dfrac{r_3 \ln (r_2/r_1)}{k_{12}} + \dfrac{r_3 \ln (r_3/r_1)}{k_{23}} + \dfrac{1}{h_3}} \tag{3.22}$$

where subscripts "in" and "out" refer to the inner and outer surfaces of the cylinder, respectively.

## Example 3.5

A 150-m long steam pipe of outer diameter 100 mm conveys 1500 kg of steam per hour at a pressure of 2 MPa. The steam enters the pipe with a dryness fraction of 0.98 and has to exit with a minimum dryness fraction of 0.96. This is to be accomplished by suitably lagging the pipe, the coefficient of thermal conductivity of the lagging being 0.075 W/(m K). Assuming the temperature drop across the steam pipe to be negligible, determine the minimum thickness of lagging required to meet the specified condition. Take the temperature of the outside surface of the lagging as 30°C.

## Solution

At 2 MPa, the specific enthalpy of evaporation (from steam tables) is

$$h_{fg} = 1890.7\,\text{kJ/kg}$$

The heat loss per kg of steam passing through the pipe is

$$\begin{aligned} Q &= (0.98 - 0.96) \times 1890.7 \\ &= 37.814\,\text{kJ/kg} \end{aligned}$$

Therefore, the heat loss per second through the entire pipe length is

$$\begin{aligned} \dot{Q} &= \frac{37.814}{3600} \times 1500 \\ &= 15.756\,\text{kW} \end{aligned}$$

But $\dot{Q}$ is also given by Equation (3.17), as

$$\dot{Q} = \frac{2\pi Lk(T_1 - T_2)}{\ln (r_2/r_1)}$$

This gives

$$\ln (r_2/r_1) = \frac{2\pi Lk(T_1 - T_2)}{\dot{Q}}$$

Saturation temperature at 2 MPa (from steam tables) = 212.42°C. Therefore,

$$\begin{aligned}\ln\left(\frac{r_2}{r_1}\right) &= \frac{2\pi \times 150 \times 0.075(212.42 - 30)}{15.756 \times 10^3} \\ &= 0.8184\end{aligned}$$

or

$$\frac{r_2}{r_1} = 2.267$$

Hence,

$$\begin{aligned}r_2 &= 2.267\, r_1 \\ &= 2.267 \times \frac{100}{2} \\ &= 113.35\,\text{mm}\end{aligned}$$

Therefore, the minimum thickness of lagging required is

$$\begin{aligned}r_2 - r_1 &= 113.35 - 50 \\ &= \boxed{63.35\,\text{mm}}\end{aligned}$$

## Example 3.6

A 1.5 kW resistance heater wire of diameter 5 mm, length 0.7 m and thermal conductivity $k = 15$ W/(m°C) is used to boil water. If the outer surface temperature of the resistance wire is 110°C, determine the temperature at the center of the wire.

## Solution

For this problem we can assume the following: (i) Heat transfer is steady since there is no change with time; (ii) Heat transfer is one-dimensional. At steady state, the heat generated by the resistance wire $\dot{Q}_{\text{gen}}$ is dissipated to the water. Thus, the heat generated within the wire must be equal to heat conducted through the outer surface of the wire. That is,

$$\begin{aligned}-kA\frac{dT}{dr} &= \dot{g}\,(\pi r^2 L) \\ dT &= -\frac{\dot{g}}{2k}\,r dr\end{aligned}$$

where ġ is the heat generation per unit volume of the wire.

Integrating from wire center where $r = 0$ and $T(0) = T_0$ to the outer surface of the wire, where $r = r_o$ and $T(r_o) = T_s$, we get

$$T_0 = T_s + \frac{\dot{g}}{4k} r_o^2$$

The heat generated is

$$\begin{aligned} \dot{g} &= \frac{\dot{Q}_{\text{gen}}}{\mathbb{V}_{\text{wire}}} \\ &= \frac{1500}{\pi\,(0.0025)^2\,(0.7)} \\ &= 109.13 \times 10^6\ \text{W/m}^3 \end{aligned}$$

Thus, the temperature at the center of the wire becomes

$$\begin{aligned} T_0 &= 110 + \frac{109.13 \times 10^6}{4 \times 15} \times (0.0025)^2 \\ &= 110 + 11.4 \\ &= \boxed{121.4°\text{C}} \end{aligned}$$

### 3.3.1 Critical Thickness of Insulation

Let us consider a layer of insulation which might be installed around a circular pipe of radius $r_i$ and length $L$, as shown in Figure 3.7.

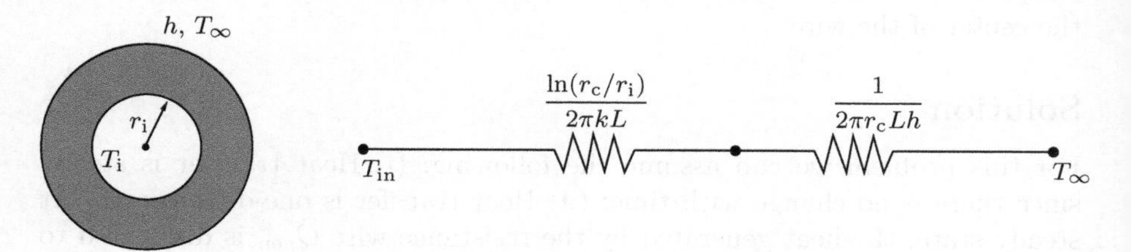


**Figure 3.7**
Insulation around a circular pipe and the thermal network.

The inner temperature of the insulation is constant at $T_i$ and the outer surface is exposed to a convection environment at $T_\infty$ and convection coefficient $h$. From the thermal network shown in Figure 3.7, the heat transferred across the insulation layer is given by

$$\dot{Q}_{\text{th}} = \frac{2\,\pi\,L\,(T_i - T_\infty)}{\frac{\ln\,(r_c/r_i)}{k} + \frac{1}{r_c\,h}} \tag{3.23}$$

Now, let $r_c$ be the outer radius of insulation for which the heat transfer is the maximum, that is, the resistance is minimum and $k$ is the thermal conductivity of the insulation material. The condition to be satisfied for maximum heat transfer rate $\dot{Q}_{\text{th}}$ is

$$\frac{d\dot{Q}_{\text{th}}}{dr_c} = 0$$

that is,

$$-\frac{2\,\pi\,L\,(T_i - T_\infty)\left(\frac{1}{kr_c} - \frac{1}{hr_c^2}\right)}{\left(\frac{\ln\,(r_c/r_i)}{k} + \frac{1}{r_c h}\right)^2} = 0$$

or

$$\frac{1}{kr_c} - \frac{1}{hr_c^2} = 0$$

which gives

$$\boxed{r_c = \frac{k}{h}} \tag{3.24}$$

This radius $r_c$ is called the *critical* radius of insulation. If the outer radius is less than the value given by the Equation (3.24), then the heat transfer will be increased by adding more insulation. For outer radii greater than the critical value, increase in insulation thickness will cause decrease of heat transfer. Thus, the heat transfer will be maximum when the thickness of the insulation is the critical thickness given by $k/h$.

The central concept is that for sufficiently small values of $h$ the convection heat loss may actually increase with the addition of insulation because of increased surface area.

Here it is essential to note that, the temperature at the inner radius $r_i$ of insulation is taken as the temperature at the inner surface of the pipe. This implies that, in this analysis, the inner and outer surfaces of the pipe are assumed to be at the same temperature $T_i$. This is possible only when the resistance to conduction through the pipe wall is zero. In other words, the resistance of the pipe wall is neglected here.

## 3.4 Steady Conduction in a Spherical Shell

Consider heat conduction in the hollow sphere shown in Figure 3.8. When the inner and outer surfaces of the spherical shell are maintained at two different constant temperatures, the symmetry dictates that the temperature has to be a function of the radius alone and that the heat flow has to be only in the radial direction.

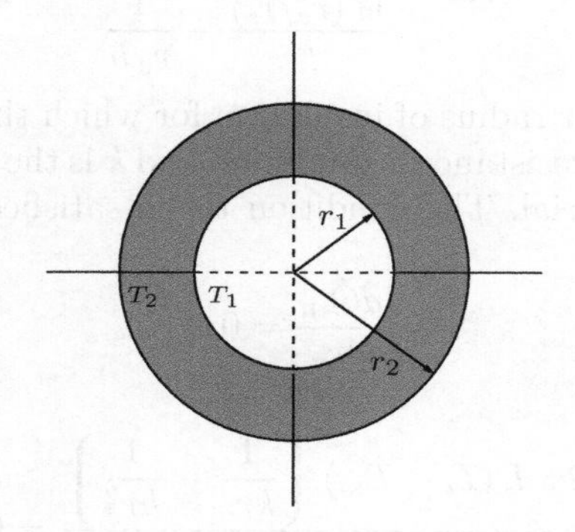


**Figure 3.8**
Conduction in a spherical shell.

For this problem, the steady-state constant-property heat conduction equation [Equation (2.6c)] in the absence of volumetric heat source becomes

$$\frac{d}{dr}\left(r^2\,\frac{dT}{dr}\right) = 0 \tag{3.25}$$

This gives, on integration, the temperature profile as

$$\boxed{T(r) = -\frac{c_1}{r} + c_2} \tag{3.26}$$

where $c_1$ and $c_2$ are integration constants to be evaluated with the boundary conditions. Equation (3.26) implies that, under steady state, the heat current at any radial location must be the same.
The boundary conditions are

$$\text{at} \quad r = r_1, \quad T = T_1$$

$$\text{at} \quad r = r_2, \quad T = T_2$$

Using these boundary conditions in Equation (3.26), we get

$$T_1 = -\frac{c_1}{r_1} + c_2$$

$$T_2 = -\frac{c_1}{r_2} + c_2$$

Solving these for $c_1$ and $c_2$, we get

$$T_2 - T_1 = c_1\left(\frac{1}{r_1} - \frac{1}{r_2}\right)$$

$$
\begin{aligned}
&= \frac{c_1}{r_1 r_2}(r_2 - r_1) \\
c_1 &= \frac{(T_2 - T_1)\, r_1 r_2}{r_2 - r_1}
\end{aligned}
$$

Now, substituting this $c_1$ into the $T_1$ expression above, we get

$$
\begin{aligned}
T_1 &= -\frac{(T_2 - T_1)\, r_2}{r_2 - r_1} + c_2 \\
c_2 &= T_1 + \frac{r_2}{r_2 - r_1}\,(T_2 - T_1)
\end{aligned}
$$

Substituting these expressions for $c_1$ and $c_2$ into Equation (3.26), we get

$$
\begin{aligned}
T(r) &= -\frac{r_1 r_2}{(r_2 - r_1)\, r}\,(T_2 - T_1) + T_1 + \frac{r_2}{r_2 - r_1}\,(T_2 - T_1) \\
&= T_1 + \frac{T_2 - T_1}{r_2 - r_1}\left(-\frac{r_1 r_2}{r} + r_2\right) \\
&= T_1 + \frac{(T_2 - T_1)\, r_1 r_2}{\left(\dfrac{1}{r_1} - \dfrac{1}{r_2}\right) r_1 r_2}\left(-\frac{1}{r} + \frac{1}{r_1}\right)
\end{aligned}
$$

Thus, the temperature profile becomes

$$
\boxed{T(r) = T_1 + \frac{(T_2 - T_1)}{\dfrac{1}{r_1} - \dfrac{1}{r_2}}\left(\frac{1}{r_1} - \frac{1}{r}\right)} \tag{3.27}
$$

The temperature profile can also be expressed as follows.

Using boundary conditions in Equation (3.26), we get

$$
\begin{aligned}
T_1 &= -\frac{c_1}{r_1} + c_2 \\
T_2 &= -\frac{c_1}{r_2} + c_2
\end{aligned}
$$

Subtracting $T_1$ from $T_2$, we get

$$
\begin{aligned}
T_2 - T_1 &= c_1\left(\frac{1}{r_1} - \frac{1}{r_2}\right) \\
c_1 &= \frac{(T_2 - T_1)}{\dfrac{1}{r_1} - \dfrac{1}{r_2}}
\end{aligned}
$$

From the relation for $T_2$, we have

$$\begin{aligned} c_2 &= T_2 + \frac{c_1}{r_2} \\ &= T_2 + \frac{(T_2 - T_1)}{r_2\left(\frac{1}{r_1} - \frac{1}{r_2}\right)} \end{aligned}$$

Now, substituting these $c_1$ and $c_2$ into Equation (3.26), we get

$$T(r) = -\frac{(T_2 - T_1)}{r\left(\frac{1}{r_1} - \frac{1}{r_2}\right)} + T_2 + \frac{(T_2 - T_1)}{r_2\left(\frac{1}{r_1} - \frac{1}{r_2}\right)}$$

or

$$\boxed{T(r) = T_2 + \frac{(T_2 - T_1)}{\frac{1}{r_1} - \frac{1}{r_2}}\left(\frac{1}{r_2} - \frac{1}{r}\right)} \qquad (3.27a)$$

Note that the temperature profiles in Equations (3.27) and (3.27$a$) are identical. This can be easily verified with the boundary conditions. That is, at $r = r_1$, the temperature should be $T_1$ and at $r = r_2$, the temperature should be $T_2$. It is seen that both the equations satisfy these boundary conditions.

The heat current across the spherical shell is given by

$$\begin{aligned} \dot{Q}_r &= A\,\dot{q}_r \\ &= A \times \left(-k\frac{dT}{dr}\right) \\ &= (4\pi\, r^2) \times \left(-k\frac{dT}{dr}\right) \\ &= -(4\pi\, r^2 k)\left(\frac{dT}{dr}\right) \end{aligned}$$

From Equation (3.27), we have

$$\frac{dT}{dr} = \frac{T_2 - T_1}{\frac{1}{r_1} - \frac{1}{r_2}}\left(\frac{1}{r^2}\right)$$

Substituting this into the above $\dot{Q}_r$ relation, we get

$$\dot{Q}_r = -(4\pi\, r^2 k) \times \left(\frac{1}{r^2}\right) \times \left(\frac{T_2 - T_1}{\frac{1}{r_1} - \frac{1}{r_2}}\right)$$

This simplifies to

$$\boxed{\dot{Q}_r = \frac{4\pi k\,(T_1 - T_2)}{\dfrac{1}{r_1} - \dfrac{1}{r_2}}} \tag{3.28}$$

This heat current expression can also be applied as follows.

The heat current across the spherical shell is given by

$$\begin{aligned} \dot{Q}_r &= A\,\dot{q}_r \\ &= A \times \left(-k\frac{dT}{dr}\right) \\ &= (4\pi\, r^2) \times \left(-k\frac{dT}{dr}\right) \end{aligned}$$

This can be arranged as

$$\dot{Q}_r \frac{dr}{r^2} = -4\pi k\, dT$$

Integrating between the inner and outer surfaces of the spherical shell, we get

$$\begin{aligned} \dot{Q}_r \int_1^2 \frac{dr}{r^2} &= -4\pi k \int_1^2 dT \\ \dot{Q}_r \left[-\frac{1}{r}\right]_1^2 &= -4\pi k \left[T\right]_1^2 \\ \dot{Q}_r \left[\frac{1}{r_1} - \frac{1}{r_2}\right] &= -4\pi k \left[T_2 - T_1\right] \\ \dot{Q}_r &= \frac{-4\pi k\ (T_2 - T_1)}{\dfrac{1}{r_1} - \dfrac{1}{r_2}} \end{aligned}$$

That is

$$\boxed{\dot{Q}_r = \frac{4\pi k\ (T_1 - T_2)}{\dfrac{1}{r_1} - \dfrac{1}{r_2}}}$$

This is the same as Equation (3.28). This can be expressed as

$$\dot{Q}_r = \frac{(T_1 - T_2)}{\dfrac{1}{4\pi k}\left(\dfrac{1}{r_1} - \dfrac{1}{r_2}\right)}$$

Thus, the thermal resistance of the shell becomes

$$\boxed{R_{\text{th}} = \frac{1}{4\pi\, k}\left(\frac{1}{r_1} - \frac{1}{r_2}\right)} \tag{3.29}$$

If there is a convective film at a spherical surface of radius $r$, the heat current should be

$$\dot{Q}_r = 4\pi\, r^2\, h\, (T_s - T_\infty)$$

where $T_s$ and $T_\infty$ are the temperatures of the solid surface and the ambient fluid, respectively, and $h$ is the convection coefficient. Thus, the film resistance is given by

$$\boxed{R_{\text{th}} = \frac{1}{4\pi r^2\, h}} \tag{3.30}$$

From the discussions in this chapter it is evident that,

- Heat transfer problems of composite shells can be tackled in a fashion analogous to the earlier problems, by considering the thermal resistances in series and drawing the appropriate thermal circuit.

- The formula for thermal conduction in spheres and spherical shells are very useful in simplifying the analysis of many practical problems.

- Three-dimensional flow of heat from small three-dimensional objects can quite often be reasonably approximated by a simple, one-dimensional flow.

- The radius of the idealized sphere is taken so as to equal the surface area of the object under study.

- The formula for spherical shell can also be used to estimate the lower bound on the convective film transfer coefficients. As has been indicated, small objects can be approximated as spheres, in such applications.

- The surrounding fluid will resemble a spherical shell of inner diameter $r$ and infinite outer diameter.

Neglecting any motion of fluid, the only mechanism of heat transfer will be conduction, and the resistance of this shell to heat flow will be just $1/(4\pi r k)$, which is obtained with $r = r_1$ and $r_2 = \infty$ in Equation (3.29). In actual case, the thermal resistance will be lower than this because of the reinforcing effect of fluid motion. Thus, $1/(4\pi r k)$ is the upper bound on the resistance of the surrounding fluid.

If $T_s$ and $T_\infty$ are the surface and ambient temperatures, respectively, the heat current is given by

$$\dot{Q}_r = 4\pi\, r\, k\, (T_s - T_\infty)$$

The heat current is also alternatively given by introducing the film transfer coefficient $h$, as

$$\dot{Q}_r = \left(4\pi\, r^2\right) h\, (T_s - T_\infty)$$

From the above two equations, we get

$$\boxed{h = \frac{k}{r}} \tag{3.31}$$

The heat transfer coefficient is usually expressed as a dimensionless number, $(hL)/k$ called the *Nusselt number* Nu, where $L$ is the characteristic length of the body. Therefore, for a sphere of diameter $D$, the Nusselt number becomes

$$\boxed{\text{Nu}_D = \frac{hD}{k}} \tag{3.32}$$

With $k/h = r$, from Equation (3.32), for sphere $\text{Nu}_D = 2$. This is the minimum value of the Nusselt number for the fluid film around a spherical shell.

Note that Nu $= 2$ is only a lower bound on the Nusselt number for heat transfer about three-dimensional objects. The actual value in most cases will be higher.

### 3.4.1 Critical Thickness of Insulation for a Spherical Shell

Heat transfer in general is defined as the ratio of thermal potential to thermal resistance. For the spherical shell of inner radius $r_1$, outer radius $r_c$, with a constant inside temperature of $T_1$, kept in an atmosphere at temperature $T_\infty$, shown in Figure 3.9, the thermal potential is $(T_1 - T_\infty)$ and the thermal conduction resistance of the spherical shell is

$$\frac{\frac{1}{r_1} - \frac{1}{r_c}}{4\pi k}$$

where $k$ is the conduction coefficient of the spherical shell.

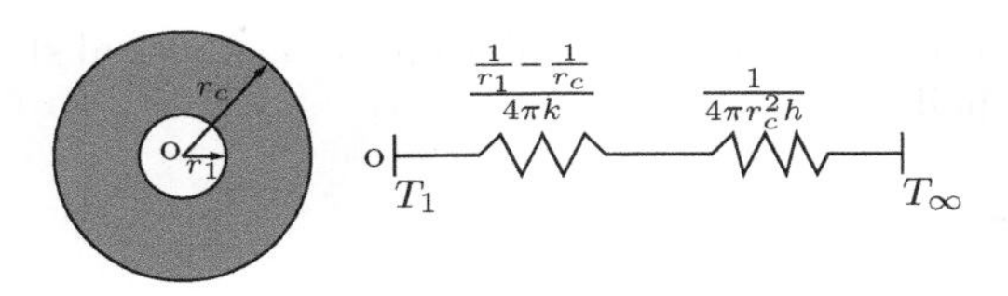


**Figure 3.9**
Conduction in a spherical shell.

The convective resistance due to the atmosphere of convection heat transfer coefficient $h$, surrounding the spherical shell is

$$\frac{1}{4\pi r_c^2 h}$$

These two resistances are in series as shown in Figure 3.9. Thus, the total resistance is

$$R_{\text{th}} = \frac{\dfrac{1}{r_1} - \dfrac{1}{r_c}}{4\pi k} + \frac{1}{4\pi r_c^2 h}$$

The heat transfer through the spherical shell is

$$Q_{\text{th}} = \frac{T_1 - T_\infty}{\left[\dfrac{\dfrac{1}{r_1} - \dfrac{1}{r_c}}{4\pi k} + \dfrac{1}{4\pi r_c^2 h}\right]}$$

For maximum heat transfer the total resistance $R_{\text{th}}$ should be minimum. Therefore,

$$\frac{dR_{\text{th}}}{dr_c} = 0$$

that is,

$$\frac{d}{dr_c}\left(\frac{\dfrac{1}{r_1} - \dfrac{1}{r_c}}{4\pi k} + \frac{1}{4\pi r_c^2 h}\right) = 0$$

$$\frac{1}{4\pi k r_c^2} - \frac{2}{4\pi h r_c^3} = 0$$

This gives the critical thickness of insulation for a spherical shell as

$$r_c = \boxed{\frac{2k}{h}}$$

Note that, the critical thickness of insulation for a spherical shell is twice that of a cylindrical shell.

## Example 3.7

A small hemispherical oven is built with an inner layer of insulating firebrick of 100 mm thickness, and an outer covering of 85 percent magnesia of 50 mm thickness. The inner surface of the oven is at 1000°C, the convection heat transfer coefficient at the outside surface is 12 W/(m$^2$ K) and the room temperature is 25°C. Calculate the rate of heat loss through the hemisphere if the inner radius is 1.5 m. Take the thermal conductivities of firebrick and 85 percent magnesia as 0.3 W/(m K) and 0.06 W/(m K), respectively.

## Solution

For the insulating brick:

From Equation (3.29), for the hemisphere the thermal resistance is

$$\begin{aligned} R_{\text{firebrick}} &= \frac{1}{2\pi k}\left(\frac{1}{r_1} - \frac{1}{r_2}\right) \\ &= \frac{1}{2\pi \times 0.3} \times \left(\frac{1}{1.5} - \frac{1}{1.60}\right) \\ &= 0.0321\,\text{K/W} \end{aligned}$$

For 85 percent magnesia:

$$\begin{aligned} R_{\text{magnesia}} &= \frac{1}{2\pi \times 0.06} \times \left(\frac{1}{1.6} - \frac{1}{1.65}\right) \\ &= 0.0502\,\text{K/W} \end{aligned}$$

For the outside air film:

$$\begin{aligned} R_{\text{film}} &= \frac{1}{2\pi r^2 h} \\ &= \frac{1}{2\pi \times 1.65^2 \times 12} \\ &= 0.00487\,\text{K/W} \end{aligned}$$

The total resistance is

$$\begin{aligned} R_{\text{th}} &= 0.0321 + 0.0502 + 0.00487 \\ &= 0.08717\,\text{K/W} \end{aligned}$$

Thus, the rate of heat loss through the hemisphere is

$$\begin{aligned} \dot{Q} &= \frac{T_{\text{in}} - T_{\text{out}}}{R_{\text{th}}} \\ &= \frac{1000 - 25}{0.08717} \\ &= \boxed{11.185\,\text{kW}} \end{aligned}$$

## 3.5 Heat Transfer from Extended Surface

The term *extended surface* is commonly used to refer to a solid that experiences energy transfer by conduction within its boundaries, as well as energy transfer by convection (and/or radiation) between its boundaries and the surroundings. Although there are many different situations that involve combined conduction-convection effects, the most frequent application is one in which an extended surface is specially used to *enhance* the heat transfer rate between a solid and the adjoining fluid. Such an extended surface is called a *fin*.

### 3.5.1 Fins

The heat which is conducted through a body must frequently be removed (or delivered) by some convection process. For example, the heat loss by conduction through a furnace wall must be dissipated to the surroundings through convection. In heat-exchanger applications, a finned-tube arrangement might be used to remove heat from a hot liquid. The heat transfer from the liquid to the finned tube is by convection. Thus, an analysis of combined *conduction-convection* heat transfer process is important from a practical point of view. Fin is one such practical system involving combined conduction-convection process of heat transfer. Since increased surface area will result in increased heat transfer, fins are designed to have a large surface area. As seen above, basically fins are extended surfaces from the main body from which heat has to be removed. Fins are designed to have different types of geometry depending upon the application. Some of the popular types of fins are shown in Figure 3.10.

Consider the one-dimensional fin exposed to a surrounding fluid at a temperature $T_\infty$, as shown in Figure 3.11. The temperature at the base of the fin is $T_b$. Consider an element of the fin of thickness $dx$.
For energy balance on the element considered,

Energy inflow through left-face = Energy outflow through right-face + Energy lost by convection through its surface area.

$$\begin{aligned}
\text{Energy flow into the left-face} &= -k\,A\,\frac{dT}{dx} \\
\text{Energy goes out at the right-face} &= -k\,A\,\frac{dT}{dx}\bigg|_{x+dx} \\
&= -k\,A\left(\frac{dT}{dx} + \frac{d^2T}{dx^2}\,dx\right)
\end{aligned}$$

$$\text{Energy lost by convection} = h\,P\,dx\,(T - T_\infty)$$

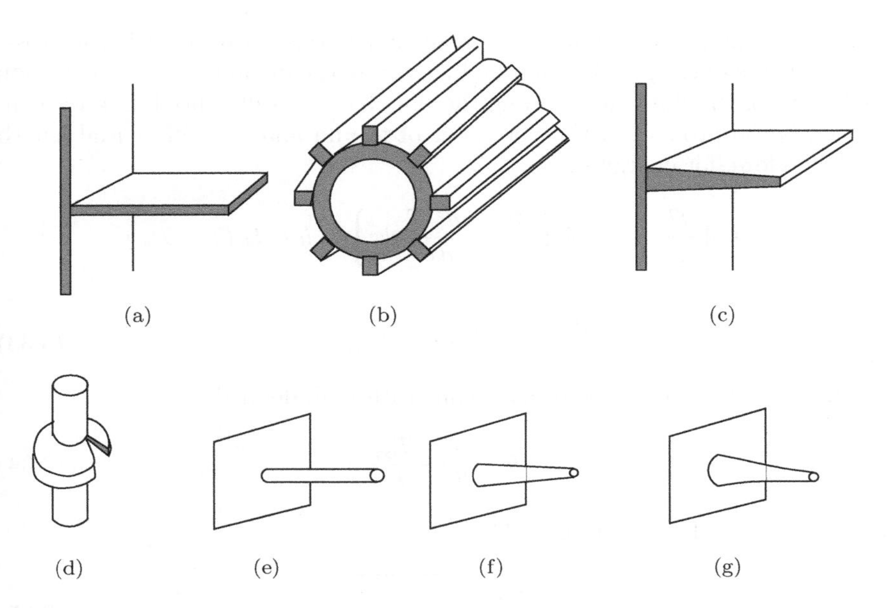


**Figure 3.10**
Different types of finned surfaces. (a) longitudinal fin of rectangular profile, (b) cylindrical tube equipped with fins of rectangular profile, (c) longitudinal fin of trapezoidal profile, (d) cylindrical tube equipped with radial fin of rectangular profile, (e) cylindrical spine (pin), (f) truncated conical spine fin, (g) parabolic spine fin.

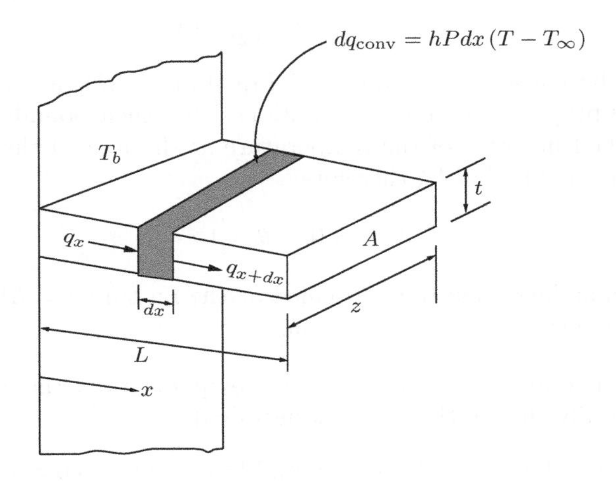


**Figure 3.11**
One-dimensional conduction-convection through a rectangular fin.

where $P$ is the perimeter of the element under consideration, $A$ is the cross-sectional area of the pin, $k$ is the conductivity of the fin and $h$ is the convection coefficient of the fluid surrounding the fin. Here the differential area for convection is the product of the perimeter of the fin and the differential length $dx$. Therefore, for energy balance,

$$-k\,A\,\frac{dT}{dx} = -k\,A\left(\frac{dT}{dx} + \frac{d^2\,T}{dx^2}\,dx\right) + h\,P\,dx\,(T - T_\infty)$$

that is,

$$\frac{d^2T}{dx^2} - \frac{h\,P}{k\,A}\,(T - T_\infty) = 0 \tag{3.33}$$

By introducing a non-dimensional temperature, $\theta$, defined as

$$\theta = \frac{T - T_\infty}{T_b - T_\infty} \tag{3.34}$$

we can express the Equation (3.33) as

$$\boxed{\frac{d^2\,\theta}{dx^2} - m^2\,\theta = 0} \tag{3.35}$$

where

$$m^2 = \frac{h\,P}{k\,A} \tag{3.36}$$

Equation (3.35) is a linear, homogeneous, second-order differential equation with constant coefficient. Its general solution is of the form

$$\theta(x) = c_1 e^{mx} + c_2 e^{-mx} \tag{3.37}$$

To evaluate the constants $c_1$ and $c_2$ of Equation (3.37), it is necessary to specify the appropriate boundary conditions. One such boundary condition may be specified in terms of the temperature at the *base* of the fin ($x = 0$). At the base of the fin $T = T_b$, therefore,

$$\text{at} \;\; x = 0, \quad \theta = 1$$

The second boundary condition, specified at the fin tip ($x = L$), may correspond to the following physical situation.

- Case 1: The fin is very long, and the temperature at the end of the fin is essentially that of the surroundings fluid.
- Case 2: The fin is of finite length and loses heat by convection from its end.
- Case 3: The end of the fin is insulated so that $\frac{dT}{dx} = 0$.

For case 1 above, the boundary conditions are

$$\text{at} \quad x = 0, \qquad \theta = 1$$

$$\text{at} \quad x = \infty, \qquad \theta = 0$$

Using the first boundary condition into Equation (3.37), we get

$$c_1 + c_2 = 1$$

For case 3 above, the boundary conditions are

$$\text{at} \quad x = 0, \qquad \theta = 1$$

$$\text{at} \quad x = L, \qquad \frac{d\theta}{dx} = 0$$

Using the condition; at $x = L$, $\frac{d\theta}{dx} = 0$, into Equation (3.37), we get

$$c_1 m e^{mL} - c_2 m e^{-mL} = 0$$

Solving the above two relations for $c_1$ and $c_2$, we get

$$c_1 = \frac{1}{1 + e^{2mL}}$$

and

$$c_2 = \frac{1}{1 + e^{-2mL}}$$

Therefore, Equation (3.37) for $\theta$ becomes

$$\theta = \frac{e^{mx}}{1 + e^{2mL}} + \frac{e^{-mx}}{1 + e^{-2mL}} = \frac{\cosh\left[m\left(L - x\right)\right]}{\cosh\left(mL\right)} \tag{3.38}$$

The total amount of heat dissipated by the fin can be calculated by integrating the convective heat loss over the fin, thus

$$\dot{Q}_{\text{th}} = \int_0^L P\,h\,\left(T_b - T_\infty\right)\,dx = \left(T_b - T_\infty\right)\int_0^L P\,h\,\theta\,dx \tag{3.39}$$

The total heat dissipated from the fin may also be obtained using the equation for the temperature distribution, since

$$\dot{Q}_{\text{th}} = -k\,A\left.\frac{dT}{dx}\right|_{x=0} = -k\,A\left(T_b - T_\infty\right)\left.\frac{d\theta}{dx}\right|_{x=0}$$

Substituting for $(d\theta/dx)_{x=0}$ from Equation (3.38), we get

$$\boxed{\dot{Q}_{\text{th}} = k\,A\,m\left(T_b - T_\infty\right)\tanh\left(mL\right)}$$

The thermal resistance of the fin, therefore, is

$$\boxed{R_{\text{th}} = \frac{1}{k\,A\,m\tanh{(mL)}}} \tag{3.40}$$

This resistance is the integrated resistance of the fin material and the associated fluid film around the fin.

It is important to note that in the present study it has been assumed that substantial temperature gradient occurs only in the $x$-direction. This assumption will be satisfied only when the fin is sufficiently thin. For most of the fins of practical interest, error introduced by this assumption is less than 1%.

The overall accuracy of practical fin calculations will usually be limited by uncertainties in the values of the convection coefficient $h$. It is worthwhile to note that the convection coefficient is seldom uniform over the entire surface of the fin, as has been assumed above. If severe nonuniform behavior is encountered, numerical techniques must be employed to solve the problem.

### 3.5.2 Infinitely Long Fin

For a sufficiently long fin of *uniform* cross-section, the temperature at the fin tip will approach the environmental temperature $T_\infty$, that is, $T_{\text{fin-tip}} = T_\infty$ and thus, $\theta$ will approach zero. Thus, the boundary condition for such a fin is

$$\theta(L) = T(L) - T_\infty = 0, \qquad \text{as } L \to \infty$$

This condition will be satisfied by the function $e^{-\,mx}$, but not by the positive function $e^{mx}$ since it tends to $\infty$ as $x$ becomes large. Therefore, the general solution in this case will consist of a constant multiple of $e^{-\,mx}$. The value of the constant multiple is determined from the requirement that, at the fin base where $x = 0$ the value of $\theta = \theta_b$. Noting that, $e^{-\,mx} = e^0 = 1$, the proper value of the constant is $\theta_b$, and the solution function we are looking for is $\theta(x) = \theta_b\, e^{-\,mx}$. This function satisfies the differential equation as well as the requirements that the solution reduces to $\theta_b$ at the fin base and approaches zero at the fin tip for large $x$. Noting that, $\theta = (T - T_\infty)/(T_b - T_\infty)$ and $m = \sqrt{hP/kA}$, the variation of temperature along the infinitely long fin becomes

$$\boxed{\theta = e^{-\,mx} = e^{-\,x\,\sqrt{hP/kA}}} \tag{3.41}$$

The temperature along the fin length in this case decreases *exponentially* from $T_b$ to $T_\infty$. The steady rate of *heat transfer* from the entire fin can be determined from Fourier's law of heat conduction as

$$\boxed{\dot{Q}_{\text{long-fin}} = -\,kA\left.\frac{dT}{dx}\right|_{x=0} = \sqrt{hPkA}\;(T_b - T_\infty)} \tag{3.42}$$

**Table 3.1**

| $mL$ | 0.1 | 0.2 | 0.5 | 1.0 | 1.5 | 2.0 | 3.0 | 4.0 | 5.0 |
|---|---|---|---|---|---|---|---|---|---|
| $\tanh(mL)$ | 0.100 | 0.197 | 0.462 | 0.762 | 0.905 | 0.964 | 0.995 | 0.999 | 1.00 |

To get an idea about treating a fin to be infinitely long, let us compare heat transfer from a fin of finite length to heat transfer from an infinitely long fin under identical conditions. The ratio of these two heat transfers is

$$\frac{\dot{Q}_{\text{finite-fin}}}{\dot{Q}_{\text{infinite-fin}}} = \frac{\sqrt{hPkA}\,(T_b - T_\infty)\tanh(mL)}{\sqrt{hPkA}\,(T_b - T_\infty)}$$

or

$$\frac{\dot{Q}_{\text{finite-fin}}}{\dot{Q}_{\text{infinite-fin}}} = \tanh(mL) \tag{3.43}$$

Now let us examine the values of $mL$ and $\tanh(mL)$ listed in Table 3.1.

It is seen that the heat transfer from a fin increases with $mL$ almost linearly at first, but it reaches a plateau later and reaches unity asymptotically for large values of $mL$, that is, for the infinitely long fin at about $mL = 5$. Therefore, a fin with $L = m/5$ can be considered to be an infinitely long fin.

### 3.5.3 Fin Efficiency or Fin Effectiveness

Owing to the thermal resistance offered by the fin material, the temperature in the fin decreases from $T_b$ at $x = 0$ with increasing $x$ (that is, along the length). Therefore, the heat flux rate at different locations of the fin are different. To indicate the effectiveness of a fin in transferring a given quantity of heat, a new parameter called fin efficiency, $\eta$ is defined as

$$\eta = \frac{\text{Actual heat transfer}}{\text{Heat which would be transferred if the entire fin area were at base temperature}}$$

that is,

$$\eta = \frac{\dot{Q}_{\text{th}}}{\dot{Q}_{\text{th,ideal}}} = \frac{k\,A\,m\,(T_b - T_\infty)\tanh(mL)}{h\,(T_b - T_\infty)\,L\,P} \tag{3.44}$$

or

$$\boxed{\eta = \frac{\tanh(mL)}{(mL)}} \tag{3.45}$$

since $m^2 = hP/kA$. It is seen from Equation (3.44) that, for small values of $mL$ the fin efficiency approaches unity.

At this stage, it is essential to note that the fin efficiency given by Equation (3.44) is for the fin with its end insulated (case 3 discussed above).

The fin resistance can now be expressed in terms of $\eta$ as follows. By Equation (3.40), the resistance is

$$\begin{aligned} R_{\text{th}} &= \frac{1}{kAm\tanh(mL)} \\ &= \frac{1}{kAm\,\eta\,mL} \\ &= \frac{1}{k\,A\,L\,\eta\,m^2} \\ &= \frac{1}{k\,A\,L\,\eta\left(\frac{h\,P}{k\,A}\right)} \end{aligned}$$

that is,

$$\boxed{R_{\text{th}} = \frac{1}{hPL\,\eta}} \tag{3.46}$$

## Example 3.8

A very long rod of 50 mm diameter has one of its ends maintained at 130°C. The surface of the rod is exposed to ambient air at 20°C with a convection heat transfer coefficient of 9 W/(m$^2$ K). Calculate the heat loss from the rod if its thermal conductivity is 390 W/(m K).

## Solution

Given, $d = 0.05$ m, $T_b = 130$°C, $T_\infty = 20$°C, $h = 9$ W/(m$^2$ K), $k = 390$ W/(m K).

When the rod is very long, the dimensionless temperature $\theta(L) = 0$. That is,

$$\theta = \frac{T - T_\infty}{T_b - T_\infty} = 0$$

because as $L \to \infty$, $T \to T_\infty$.
The heat dissipated from the fin is given by the equation

$$\dot{Q}_{\text{th}} = kAm\,(T_b - T_\infty)$$

Thus,

$$\begin{aligned}\dot{Q}_{\text{th}} &= kA\sqrt{\frac{hP}{kA}}\,(T_b - T_\infty) \\ &= \sqrt{hPkA}\,(T_b - T_\infty)\end{aligned}$$

Therefore,

$$\begin{aligned}\dot{Q}_{\text{th}} &= \sqrt{9 \times (\pi \times 0.05) \times 390 \times \left(\frac{\pi}{4} \times 0.05^2\right)} \times (130 - 20) \\ &= \boxed{114.45\,\text{W}}\end{aligned}$$

## Example 3.9

A very long copper rod of 5 mm diameter has one of its ends maintained at 100°C. The surface of the rod is exposed to ambient air at 25°C with $h = 100$ W/(m$^2$ K). Taking $k$ for the rod as 398 W/(m K), (a) determine the heat loss from the rod and (b) estimate the minimum length of the rod required to regard that as infinitely long.

## Solution

Given, $d = 0.005$ m, $T_B = 100$°C, $T_\infty = 25$°C.

(a) The heat dissipated from the fin is

$$\begin{aligned}\dot{Q}_{\text{th}} &= kA\sqrt{\frac{hP}{kA}}\,(T_b - T_\infty) \\ &= \sqrt{h\,P\,k\,A}\,(T_b - T_\infty) \\ &= \sqrt{100 \times (\pi \times 0.005) \times 398 \times \left(\frac{\pi \times 0.005^2}{4}\right)} \times (100 - 25) \\ &= \boxed{8.31\,\text{W}}\end{aligned}$$

(b) A fin can be considered infinitely long if $mL \geq 5$. Therefore, the length of an infinite fin $L \geq 5/m$. Given that,

$$\begin{aligned}m &= \sqrt{(h\,P)/(k\,A)} \\ &= \sqrt{100 \times (\pi \times 0.005)/[398 \times (\pi \times 0.005^2/4)]}\end{aligned}$$

$$= 14.18$$

Thus,

$$L \geq 5/14.18 \geq \boxed{0.35\,\text{m}}$$

A fin of diameter 5 mm can be regarded as infinitely long if its length is more than 350 mm.

*Note:* In this case the length limit of the infinite fin estimated is conservative. However, a fin can be declared infinitely long even when $mL \geq 3.0$ for which $\tan(mL) \geq 0.995$. In such a case $L \geq 3/14.18 = 0.21$ m. In fact, the length can be less than this also if some amount of heat transfer at the fin tip is acceptable.

## Example 3.10

In a heating system, steam flows through tubes of 3 cm outer diameter. The tube surface is at a steady temperature of 120°C. To improve the heat transfer from the tube, circular aluminum fins of outer diameter 6 cm and thickness 2 mm are attached to the tube at equal intervals with a gap of 6 mm between the fins. Heat is transferred to the surrounding air at 20°C, with a combined heat transfer coefficient of 60 W/(m$^2$ °C). If the conductivity of aluminum is 180 W/(m °C) and the efficiency of the fins is 0.95, determine the increase in heat transfer from the tube per meter of its length as a result of these fins.

## Solution

When there are no fins, the heat transfer from the tube per meter of its length is

$$\dot{Q}_{\text{without-fin}} = hA\,(T_b - T_\infty)$$

$$= 60 \times (\pi d_1 L) \times (120 - 20)$$

$$= 60 \times (\pi \times 0.03 \times 1) \times 100$$

$$= 565.5\,\text{W}$$

Area of each fin is

$$A_{\text{fin}} = 2\pi\left(r_2^2 - r_1^2\right) + 2\pi\,r_2 t$$

$$= 2\pi\left(0.03^2 - 0.015^2\right) + 2\pi \times 0.03 \times 0.002$$

$$= \quad 0.00462\,\text{m}^2$$

The fins are like concentric annular disks fitted to the tube and $r_2$ is the outer radius of the fins, $r_1$ is their inner radius, which is the same as the outer radius of the tube and $t$ is the fin thickness. Heat transfer from a fin is

$$\begin{aligned}\dot{Q}_{\text{fin}} &= \eta_{\text{fin}}\,\dot{Q}_{\text{fin,max}} \\ &= \eta_{\text{fin}}\,h\,A_{\text{fin}}\,(T_b - T_\infty) \\ &= 0.95 \times 60 \times 0.00462 \times (120 - 20) \\ &= 26.33\,\text{W}\end{aligned}$$

Heat transfer from the tube through the space of 6 mm between successive fins is

$$\begin{aligned}\dot{Q}_{\text{unfin}} &= h\,A_{\text{unfin}}\,(T_b - T_\infty) \\ &= 60 \times (\pi \times 0.03 \times 0.006) \times (120 - 20) \\ &= 3.39\,\text{W}\end{aligned}$$

For the given fin thickness and inter-fin spacing, 125 fins and 125 spacings can be taken as the appropriate numbers. Thus, there are 125 fins and 125 inter-fin spacings per meter length of the tube. Therefore, the total heat transfer per meter length of the finned tube is

$$\begin{aligned}\dot{Q}_{\text{total,fin}} &= 125\left(\dot{Q}_{\text{fin}} + \dot{Q}_{\text{unfin}}\right) \\ &= 125 \times (26.33 + 3.39) \\ &= 3715\,\text{W}\end{aligned}$$

Therefore, the increase of heat transfer per unit meter of tube length due to fins is

$$\Delta\dot{Q} = \dot{Q}_{\text{total,fin}} - \dot{Q}_{\text{total,without-fin}}$$

$$= 3715 - 565.5$$

$$= \boxed{3149.5\,\text{W}}$$

Note: The number of fins of 2 mm thickness with 6 mm gas in-between them, over the tube length 1 m is 125. This is arrived at, as follows.

Total length of the tube = 1 m.

Total length over which 125 fins can be positioned is $125 \times 2 = 250$ mm.

The spacing between 125 fins is $124 \times 6 = 744$ mm.

Thus, the total length for accommodating 125 fins and 124 spacings is only 0.994 m. Hence, 125 spacing used in the calculation is only an approximation.

## Example 3.11

Hot water flows through a cast iron pipe of diameter 30 mm, wall thickness 2.5 mm and length 15 m. The temperature of the water flowing at an average velocity of 1.5 m/s decreases from 70°C to 67°C, from inlet to the exit of the pipe exposed to an environment at 15°C. The conduction coefficient of the pipe is 52 W/(m °C) and the convection coefficient at the inner surface of the pipe is 400 W/(m$^2$ °C). Determine the combined convection and radiation heat transfer coefficient at the outer surface of the pipe. Take the specific heat for water as 4180 m$^2$/(s$^2$ °C).

## Solution

Given, $D_i = 30$ mm, $D_o = 35$ mm, $L = 15$ m, $T_\infty = 15$°C, $\Delta T = 3$°C, $k = 52$ W/(m °C), $h_i = 400$ W/(m$^2$ °C).

There are three resistances in series in this problem, as shown in Figure ES3.11.

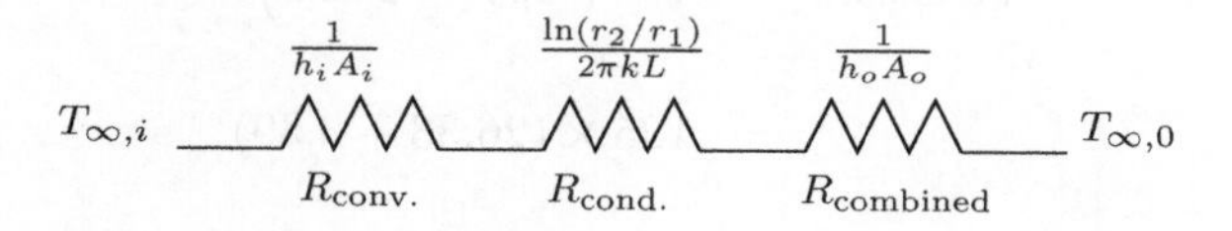


**Figure ES3.11**
Three resistances in series.

For water, $c_p = 4180$ J/(kg °C). The mass flow rate of water through the pipe is

$$\begin{aligned}\dot{m} &= \rho A V \\ &= 10^3 \times (\pi \times 0.015^2) \times 1.5 \\ &= 1.06\,\text{kg/s}\end{aligned}$$

The heat transfer rate is

$$\begin{aligned}\dot{Q} &= \dot{m}\, c_p \Delta T \\ &= 1.06 \times 4180 \times 3 \\ &= 13292.4\,\text{W}\end{aligned}$$

The convective resistance is

$$\begin{aligned}R_{\text{conv.}} &= \frac{1}{h_i A_i} \\ &= \frac{1}{400 \times (\pi \times 0.03 \times 15)} \\ &= 0.00177\,^{\circ}\text{C/W}\end{aligned}$$

The conduction resistance is

$$\begin{aligned}R_{\text{cond.}} &= \frac{\ln{(r_2/r_1)}}{2\pi k L} \\ &= \frac{\ln{(1.75/1.5)}}{2 \times \pi \times 52 \times 15} \\ &= 0.0000315\,^{\circ}\text{C/W}\end{aligned}$$

The heat transfer can be expressed as

$$\dot{Q} = \frac{T_{\infty,i} - T_{\infty,o}}{R_{\text{total}}}$$

where $T_{\infty,i}$ is the average of the inlet and exit temperatures of water, that is $(70 + 67)/2 = 68.5°\text{C}$ and $T_{\infty,o} = 15°\text{C}$, and

$$R_{\text{total}} = R_{\text{conv.}} + R_{\text{cond.}} + \frac{1}{h_{\text{combined}} A_o}$$

Therefore,

$$13292.4 = \frac{68.5 - 15}{0.00177 + 0.0000315 + \frac{1}{h_{\text{combined}} A_o}}$$

$$0.00177 + 0.0000315 + \frac{1}{h_{\text{combined}} A_o} = \frac{68.5 - 15}{13292.4}$$

$$\frac{1}{h_{\text{combined}} A_o} = 0.00402 - 0.00177 - 0.0000315$$

$$= 0.00222$$

$$h_{\text{combined}} = \frac{1}{A_o \times 0.00222}$$

$$= \frac{1}{(\pi \times 0.035 \times 15) \times 0.00222}$$

$$= \boxed{273.11 \text{ W/(m}^2 \text{ °C)}}$$

## 3.6 The Conduction Shape Factor

In a two-dimensional system where only two temperature limits are involved, a conduction shape factor $S$ may be defined such that, the heat transfer rate $\dot{Q}$ from the system to the surrounding can be expressed as

$$\dot{Q} = kS\,(\Delta T)_{\text{overall}} \tag{3.47}$$

where $k$ is the thermal conductivity of the medium surrounding the system and $(\Delta T)_{\text{overall}} = T_{\text{system}} - T_{\text{surrounding}}$. The value of the shape factor $S$ has been worked out for some geometries and listed in Table 3.2. An elaborate list of shape factors for a variety of one may refer to Hahne and Grigull [1].

For a three-dimensional wall, as in a furnace, separate shape factors are used to calculate the heat flow through the edges and corner sections. When all the interior dimensions are greater than one-fifth of the wall thickness,

$$S_{\text{wall}} = \frac{A}{L},\; S_{\text{edge}} = 0.54D,\; S_{\text{corner}} = 0.15L$$

where $A$ is the wall area, $L$ is the wall thickness, and $D$ is the length of the edge. These dimensions are illustrated in Figure 3.12.
Note: The inverse hyperbolic cosine in Table 3.2 can be expressed as

$$\cosh^{-1} x = \ln\left(x \pm \sqrt{x^2 - 1}\right)$$

**Table 3.2** Conduction shape factors of some specific geometries

| Physical system | Shape factor | Restrictions |
|---|---|---|
| Isothermal cylinder burried in semi-infinite medium | 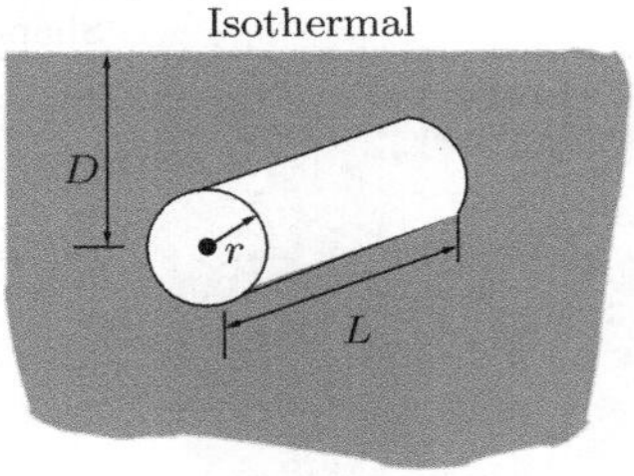 $\frac{2\pi L}{\cosh^{-1}(D/r)}$ | $L > r$ |
| | $\frac{2\pi L}{\ln(2D/r)}$ | $L >> r$, $D > 3r$ |
| | $\frac{2\pi L}{\ln\left(\frac{L}{r}\right)\left[1 - \frac{\ln(L/2D)}{\ln(L/r)}\right]}$ | $D >> r$, $L >> D$ |
| Isothermal sphere of radius $r$ burried in infinite medium |  $4\pi r$ | |
| Isothermal sphere of radius $r$ burried in semi-infinite medium | 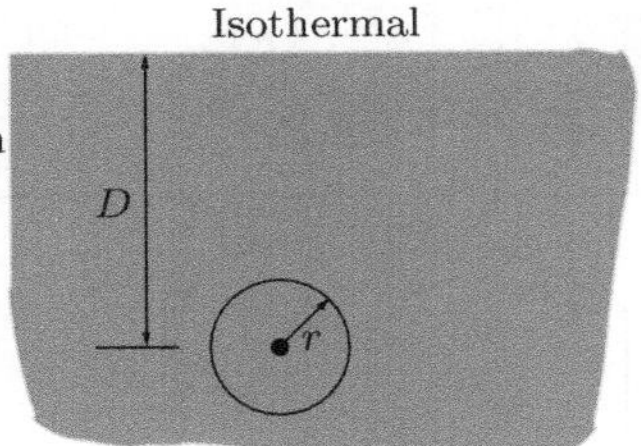 $\frac{4\pi r}{1-r/2D}$ | |
| Conduction between two isothermal cylinders burried in infinite medium | 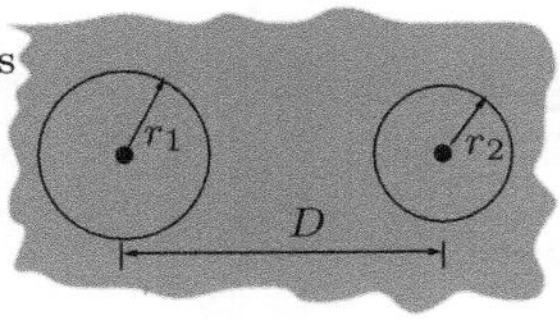 $\frac{2\pi L}{\cosh^{-1}\left(\frac{D^2-r_1^2-r_2^2}{2r_1r_2}\right)}$ | $L >> R$, $L >> D$ |

**Table 3.2** (Continued.)

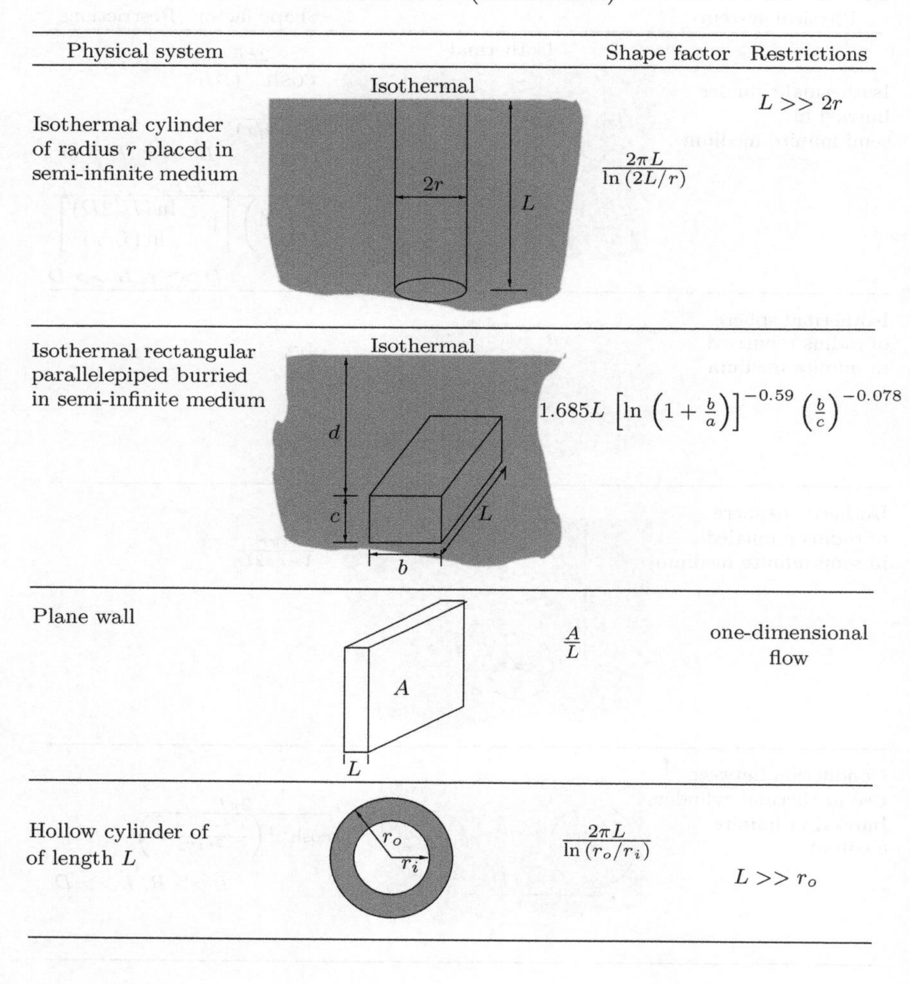

| Physical system | Shape factor | Restrictions |
|---|---|---|
| Isothermal cylinder of radius $r$ placed in semi-infinite medium | $\frac{2\pi L}{\ln(2L/r)}$ | $L >> 2r$ |
| Isothermal rectangular parallelepiped burried in semi-infinite medium | $1.685L\left[\ln\left(1+\frac{b}{a}\right)\right]^{-0.59}\left(\frac{b}{c}\right)^{-0.078}$ | |
| Plane wall | $\frac{A}{L}$ | one-dimensional flow |
| Hollow cylinder of of length $L$ | $\frac{2\pi L}{\ln(r_o/r_i)}$ | $L >> r_o$ |

**Table 3.2** (Continued.)

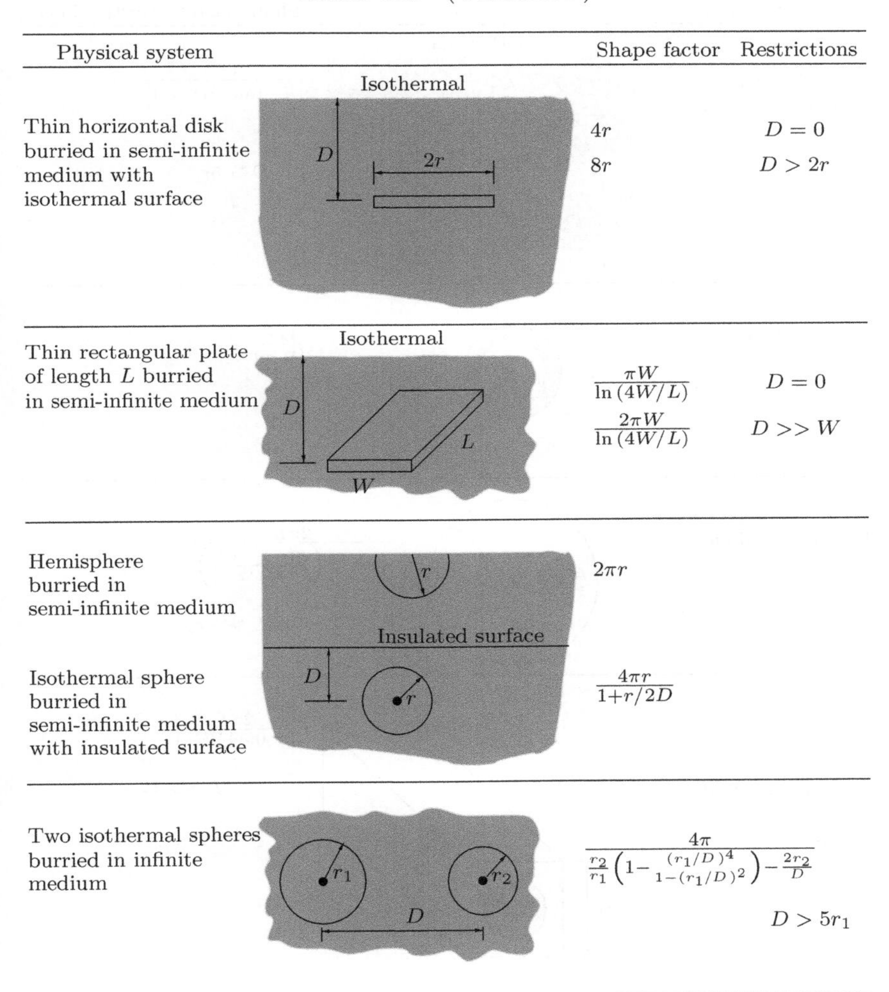


| Physical system | Shape factor | Restrictions |
|---|---|---|
| Thin horizontal disk burried in semi-infinite medium with isothermal surface | $4r$ | $D = 0$ |
| | $8r$ | $D > 2r$ |
| Thin rectangular plate of length $L$ burried in semi-infinite medium | $\frac{\pi W}{\ln(4W/L)}$ | $D = 0$ |
| | $\frac{2\pi W}{\ln(4W/L)}$ | $D >> W$ |
| Hemisphere burried in semi-infinite medium | $2\pi r$ | |
| Isothermal sphere burried in semi-infinite medium with insulated surface | $\frac{4\pi r}{1+r/2D}$ | |
| Two isothermal spheres burried in infinite medium | $\frac{4\pi}{\frac{r_2}{r_1}\left(1-\frac{(r_1/D)^4}{1-(r_1/D)^2}\right)-\frac{2r_2}{D}}$ | $D > 5r_1$ |

**Table 3.2** (Continued.)

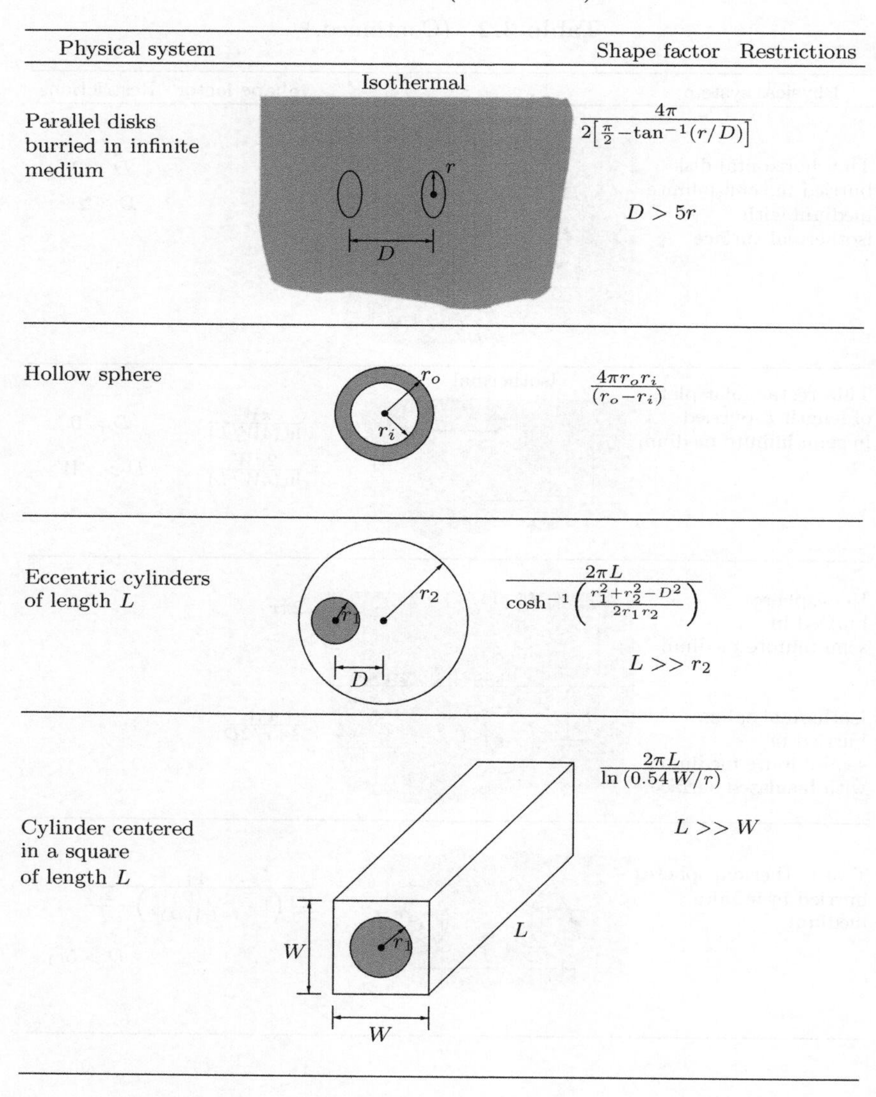

| Physical system | | Shape factor | Restrictions |
|---|---|---|---|
| Parallel disks burried in infinite medium | Isothermal; $r$; $D$ | $\frac{4\pi}{2\left[\frac{\pi}{2}-\tan^{-1}(r/D)\right]}$ | $D > 5r$ |
| Hollow sphere | $r_o$; $r_i$ | $\frac{4\pi r_o r_i}{(r_o-r_i)}$ | |
| Eccentric cylinders of length $L$ | $r_1$; $r_2$; $D$ | $\frac{2\pi L}{\cosh^{-1}\left(\frac{r_1^2+r_2^2-D^2}{2r_1r_2}\right)}$ | $L >> r_2$ |
| Cylinder centered in a square of length $L$ | $r_1$; $W$; $W$; $L$ | $\frac{2\pi L}{\ln(0.54\,W/r)}$ | $L >> W$ |

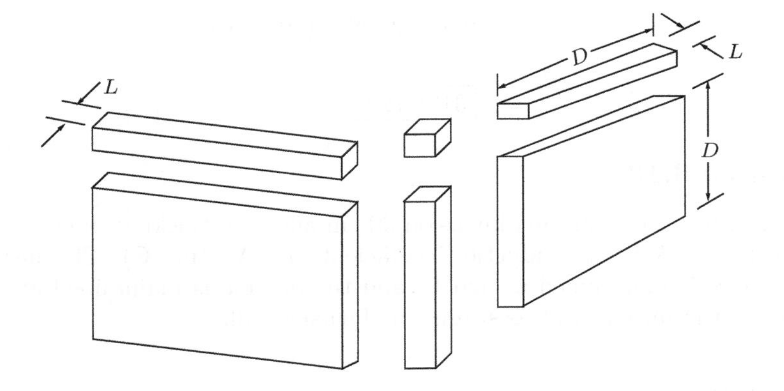


**Figure 3.12**
Illustration of parameters for calculating shape factors of three-dimensional shapes.

The use of shape factors for solving problems is illustrated in Examples 3.12 to 3.14, for their geometries.

## Example 3.12

A 100-cm diameter pipe of length 10 m carrying water at 90°C is laid at 0.95 m below the earth surface. If the earth surface is at 10°C and the thermal conductivity of earth is 0.8 W/(m °C), calculate the rate of heat loss from the pipe.

## Solution

Given, $L = 10$ m, $D = 0.95$ m, $r = 0.5$ m, that is, $D < 3r$, therefore, the shape factor can be calculated using the equation given in Table 3.2. Thus,

$$\begin{aligned} S &= \frac{2\pi L}{\cosh^{-1}(D/r)} \\ &= \frac{2\pi \times 10}{\cosh^{-1}(0.95/0.5)} \\ &= 49.98\,\text{m} \end{aligned}$$

The rate of heat loss from the pipe is

$$\dot{Q} = kS\Delta T$$

$$= 0.8 \times 49.98 \times (90 - 10)$$

$$= \boxed{3198.72\,\text{W}}$$

## Example 3.13

A cubical furnace of inside dimension 30 cm and wall thickness 6 cm is made of fireclay brick with conduction coefficient of 1 W/(m °C). The inside of the furnace is maintained at 520°C, and the outside is maintained at 50°C. Calculate the rate of heat loss from the furnace wall.

## Solution

The shape factor for this cubical shape can be determined by adding the shape factors of the walls, edges, and the corners.

Given, $A = 0.3 \times 0.3$ m$^2$, $L = 0.06$ m, $D = 0.3$ m. Therefore, the shape factors for the walls, edges and corners are

$$S_{\text{walls}} = \frac{A}{L} = \frac{0.3 \times 0.3}{0.06}$$

$$= 0.15\,\text{m}$$

$$S_{\text{edges}} = 0.54\,D = 0.54 \times 0.3$$

$$= 0.162\,\text{m}$$

$$S_{\text{corners}} = 0.15\,L = 0.15 \times 0.06$$

$$= 0.009\,\text{m}$$

For the given furnace, there are 6 walls, 12 edges, and 8 corners. Thus the total shape factor is

$$S_{\text{total}} = 6 \times 0.15 + 12 \times 0.162 + 8 \times 0.009$$

$$= 0.9 + 1.944 + 0.072$$

$$= 2.916\,\text{m}$$

Thus, the total heat loss from the furnace to the surroundings is

$$\begin{aligned}\dot{Q} &= kS\Delta T \\ &= 1 \times 2.916 \times (520 - 50) \\ &= \boxed{1370.52\,\text{W}}\end{aligned}$$

## Example 3.14

A spherical tank of diameter 1.5 m containing radioactive material is buried in the earth. The distance from the earth surface to the center of the tank is 2.5 m. The surface of the tank is maintained at 120°C as a result of radioactive decay while the earth surface is at a uniform temperature of 12°C. Taking the conduction coefficient of earth as 0.8 W/(m °C), calculate the rate of heat generated in the tank by radioactive decay.

## Solution

Given, $D = 2.5$ m, $r = 0.75$ m, $T_s = 120$°C, $T_\infty = 12$°C, $k = 0.8$ W/(m °C).

The shape factor for the spherical tank, from Table 3.2, is

$$\begin{aligned}S &= \frac{4\pi r}{1 - r/2D} \\ &= \frac{4\pi \times 0.75}{1 - 0.75/(2 \times 2.5)} \\ &= 11.09\,\text{m}\end{aligned}$$

The heat loss from the tank is

$$\begin{aligned}\dot{Q} &= Sk\,(T_s - T_\infty) = 11.09 \times 0.8 \times (120 - 12) \\ &= \boxed{958.18\,\text{W}}\end{aligned}$$

For maintaining the surface temperature of the spherical tank steady at 120°C, the heat generated in the tank by radioactive decay should be 958.18 W.

# 3.7 Summary

Application of Fourier's law for heat flow in some simple one-dimensional system is discussed in this chapter. The term "one-dimensional" refers to the fact that only one coordinate is needed to describe the spatial variation of the dependent variables.

In the steady state, the temperature $T$ in a plane wall is a function of $x$ only. Therefore, the heat equation

$$\frac{\partial T}{\partial t} = \frac{k}{\rho c} \bigtriangledown^2 T + \frac{\dot{g}_{\text{th}}}{\rho c}$$

reduces to

$$\frac{d^2T}{dx^2} = 0$$

For heat transfer through the plane wall, the temperature distribution is

$$\boxed{T(x) = (T_2 - T_1)\frac{x}{L} + T_1}$$

For one-dimensional, steady-state conduction in a plane wall with no heat generation and constant thermal conductivity, the temperature varies linearly with $x$, and the heat current (heat transfer rate) is given by

$$\boxed{\dot{Q}_x = A\,\dot{q}_x = \frac{kA}{L}(T_1 - T_2)}$$

For heat conduction through a plane wall with temperatures $T_1$ at $x = 0$ and $T_{2f}$ at $x = L$, the conduction heat current becomes

$$\boxed{\dot{Q}_x = \frac{T_1 - T_{2f}}{\dfrac{L}{kA} + \dfrac{1}{h_2 A}}}$$

The resistance $R_{\text{th}}$ is given by

$$R_{\text{th}} = \frac{L}{kA} + \frac{1}{h_2 A}$$

For this case, the resistance of the slab material and the resistance of the fluid film are two resistances in series, since the current through one should be the same as that through the other.

For a composite wall made up of a number of layers, the resultant resistance is given by

$$R_{\text{th}} = \frac{1}{h_1 A} + \frac{L_{12}}{k_{12} A} + \frac{L_{23}}{k_{23} A} + \ldots.. + \frac{L_{(n-1)n}}{k_{(n-1)n} A} + \frac{1}{h_n A}$$

and the heat current through the wall is

$$\dot{Q}_{\text{th}} = \frac{T_{1f} - T_{nf}}{R_{\text{th}}}$$

The overall heat transfer coefficient for a composite wall, $U$, may be defined as

$$\dot{Q}_{\text{th}} = UA\,(T_{1f} - T_{nf})$$

where

$$U = \frac{1}{\dfrac{1}{h_1} + \dfrac{L_{12}}{k_{12}} + \dfrac{L_{23}}{k_{23}} + \ldots\ldots + \dfrac{L_{(n-1)n}}{k_{(n-1)n}} + \dfrac{1}{h_n}}$$

For a sufficiently long cylinder, that is, length is much larger than the radius, the end effects are negligible and the temperature is a function of the radius alone.

For steady-state radial conduction in the absence of heat sources, and with constant material properties, the heat equation

$$\frac{\partial T}{\partial t} = \frac{k}{\rho c}\,\nabla^2 T + \frac{\dot{g}_{\text{th}}}{\rho c}$$

reduces to

$$\nabla^2 T = 0$$

In cylindrical coordinates

$$\nabla^2 = \frac{\partial^2}{\partial r^2} + \frac{1}{r}\frac{\partial}{\partial r} + \frac{1}{r^2}\frac{\partial^2}{\partial \theta^2} + \frac{\partial^2}{\partial z^2}$$

The boundary conditions are the following.

$$\text{At} \qquad r = r_1, \quad T = T_1$$

$$\text{At} \qquad r = r_2, \quad T = T_2$$

The temperature distribution associated with radial conduction through a cylinder is

$$\boxed{T(r) = \frac{T_2 - T_1}{\ln\left(\dfrac{r_2}{r_1}\right)}\,\ln\left(\frac{r}{r_2}\right) + T_2}$$

The heat current for a cylinder of length $L$ is

$$\dot{Q}_r = 2\pi r L \dot{q}_r = \frac{2\pi L k\,(T_1 - T_2)}{\ln\left(\dfrac{r_2}{r_1}\right)}$$

The thermal resistance is given by

$$\boxed{R_{\text{th}} = \frac{\ln\left(\dfrac{r_2}{r_1}\right)}{2\pi L k}}$$

For a convective film on a cylindrical surface of temperature $T_s$ and radius $r$, the heat current is given by

$$\dot{Q}_r = (2\pi r L)\, h\, (T_s - T_\infty)$$

The film resistance is

$$R_{\text{th}} = \frac{1}{2\pi r L h}$$

The overall film transfer coefficient $U$ is defined as

$$\dot{Q}_{\text{th}} = U\, A\, (T_{1f} - T_{3f})$$

But because of varying area in the $r$–direction, the area on which the $U$ is based has to be specified. Thus,

$$\dot{Q}_{\text{th}} = U_{\text{in}}\, A_{\text{in}}\, (T_{1f} - T_{3f}) = U_{\text{out}}\, A_{\text{out}}\, (T_{1f} - T_{3f})$$

and

$$U_{\text{in}} = \frac{1}{\dfrac{1}{h_1} + \dfrac{r_1 \ln (r_2/r_1)}{k_{12}} + \dfrac{r_1 \ln (r_3/r_2)}{k_{23}} + \dfrac{r_1}{r_3}\dfrac{1}{h_3}}$$

$$U_{\text{out}} = \frac{1}{\dfrac{r_3}{r_1}\dfrac{1}{h_1} + \dfrac{r_3 \ln (r_2/r_1)}{k_{12}} + \dfrac{r_3 \ln (r_3/r_1)}{k_{23}} + \dfrac{1}{h_3}}$$

For an insulated circular pipe, the heat transfer is

$$\dot{Q}_{\text{th}} = \frac{2\,\pi\, L\,(T_i - T_\infty)}{\dfrac{\ln (r_c/r_i)}{k} + \dfrac{1}{r_c\, h}}$$

Now, let $r_c$ be the outer radius of insulation for which the heat transfer is the maximum, that is, the resistance is minimum. The condition to be satisfied for maximum heat transfer rate $\dot{Q}_{\text{th}}$ is

$$\frac{d\dot{Q}_{\text{th}}}{dr_c} = 0$$

this gives

$$\boxed{r_c = \frac{k}{h}}$$

The radius $r_c$ is called the *critical radius of insulation.*

For conduction in hollow sphere, the governing equation is

$$\frac{d}{dr}\left(r^2\,\frac{dT}{dr}\right) = 0$$

The temperature profile becomes

$$T = T_1 + \frac{T_2 - T_1}{\dfrac{1}{r_1} - \dfrac{1}{r_2}} \left( \frac{1}{r_1} - \frac{1}{r} \right)$$

and the heat current across the spherical shell is given by

$$\dot{Q}_r = -\left(4\pi\, r^2\right) k \,\frac{dT}{dr} = \frac{4\pi k \,(T_1 - T_2)}{\dfrac{1}{r_1} - \dfrac{1}{r_2}}$$

The thermal resistance of the shell becomes

$$\boxed{R_{\text{th}} = \frac{1}{4\pi\, k} \left( \frac{1}{r_1} - \frac{1}{r_2} \right)}$$

If there is a convective film at a spherical surface of radius $r$, the heat current should be

$$\dot{Q}_r = 4\pi\, r^2\, h\,(T_s - T_\infty)$$

where $T_s$ and $T_\infty$ are the temperatures of the solid surface and the ambient fluid, respectively.
Thus, the film resistance is given by

$$\boxed{R_{\text{th}} = \frac{1}{4\pi r^2\, h}}$$

The critical radius of insulation for a spherical shell is

$$\boxed{r_c = \frac{2k}{h}}$$

The heat transfer coefficient is usually expressed as a dimensionless number, $(hL)/k$ called the *Nusselt number*, Nu, where $L$ is the characteristic length of the body. Therefore, for sphere

$$\boxed{\text{Nu}_D = \frac{hD}{k}}$$

Fins are extended surfaces from the main body from which heat has to be removed. For a one-dimensional fin with temperature $T_b$ at its base, exposed to a surrounding fluid at a temperature $T_\infty$, the energy balance results in

$$\boxed{\frac{d^2\,\theta}{dx^2} - m^2\,\theta = 0}$$

where

$$\theta = \frac{T - T_\infty}{T_b - T_\infty}$$

and

$$m^2 = \frac{hP}{kA}$$

This is a linear, homogeneous, second-order differential equation with constant coefficient. Its general solution is of the form

$$\theta(x) = c_1 e^{mx} + c_2 e^{-mx}$$

The thermal resistance of fin is

$$\boxed{R_{\text{th}} = \frac{1}{kAm \tanh mL}}$$

This resistance is the integrated resistance of the fin material and the associated fluid film around the fin.

The variation of temperature along the infinitely long fin is

$$\boxed{\theta = e^{-mx} = e^{-x\sqrt{hP/kA}}}$$

The steady rate of *heat transfer* from the entire fin can be determined from Fourier's law of heat conduction as

$$\boxed{\dot{Q}_{\text{long fin}} = -kA \left.\frac{dT}{dx}\right|_{x=0} = \sqrt{hPkA}\,(T_b - T_\infty)}$$

Temperature in the fin decreases from $T_b$ at $x = 0$ with increasing $x$ (that is, along the length). Therefore, the flux rate at different locations of the fin are different. Fin efficiency, $\eta$ is

$$\boxed{\eta = \frac{\tanh mL}{mL}}$$

The fin resistance in terms of $\eta$ is

$$\boxed{R_{\text{th}} = \frac{1}{hPL\,\eta}}$$

In a two-dimensional system where only two temperature limits are involved, a conduction shape factor $S$ may be defined such that the heat transfer rate $\dot{Q}$ from the system to the surrounding can be expressed as

$$\dot{Q} = kS\,(\Delta T)_{\text{overall}}$$

where $k$ is the thermal conductivity of the medium surrounding the system and $(\Delta T)_{\text{overall}} = T_{\text{system}} - T_{\text{surrounding}}$.

For a three-dimensional wall, as in a furnace, separate shape factors are used to calculate the heat flow through the edges and corner sections. When all the interior dimensions are greater than one-fifth of the wall thickness,

$$S_{\text{wall}} = \frac{A}{L}, \; S_{\text{edge}} = 0.54D, \; S_{\text{corner}} = 0.15L$$

where $A$ is the wall area, $L$ is the wall thickness, and $D$ is the length of the edge.

## 3.8 Exercise Problems

3.1 A 100-mm diameter steam pipe is covered with two layers of insulation. The inner layer is 50 mm thick and has $k_1 = 0.08$ W/(m K). The outer layer is 30 mm thick and has $k_2 = 1.1$ W/(m K). The pipe conveys steam at a pressure of 2 MPa with 30°C superheat. The outside temperature of the insulation is 30°C. If the steam pipe is 25 m long, determine (a) the heat lost per hour, and (b) the interface temperature of the insulation, neglecting the temperature drop across the steam pipe.

[**Ans.** (a) 13.49 MJ, (b) 35.69°C]

3.2 A wall of a room measuring 3 m by 6 m is made of concrete material. The thickness of the wall is 300 mm. The inside temperature of the wall is 20°C and the outside temperature is 3°C. Determine the heat loss through the wall, taking the thermal conductivity $k$ for the concrete as 0.8 W/(m °C).

[**Ans.** 816 W]

3.3 A 250-mm thick brick wall has temperatures of 40°C and 20°C on either side of it. If the thermal conductivity of the brick is 0.52 W/(m K), calculate the rate of heat transfer per unit area of the wall surface.

[**Ans.** 41.6 W/m$^2$]

3.4 A steel pipe of 100 mm inner diameter and 114 mm outer diameter carries steam at 260°C. The pipe has two layers of insulation. The inner layer is 40 mm thick having $k = 0.09$ W/(m K), and the outer layer is 60 mm thick having $k = 0.07$ W/(m K). The heat transfer coefficients for the inside of the pipe and the outer layer of insulation are 550 and 15 W/(m$^2$ K), respectively. The thermal conductivity of steel is 50 W/(m K). If the atmospheric temperature is 15°C, calculate (a) the rate of heat loss by the steam per unit length of the pipe, and (b) the temperature at the outside surface of the insulation.

[**Ans.** (a) 116.18 W, (b) $T_{\text{out}} = 22.85$°C]

3.5 An aluminum fin 2 mm thick, 100 mm long and 70 mm wide is attached to a body to be cooled. The base temperature of the fin $T_b = 270$°C. If the thermal conductivity of aluminum is 190 W/(m K), find the cooling capacity of the fin if the conductivity of the atmospheric air at 20°C is 10 W/(m$^2$ K).

[**Ans.** 49 W]

3.6 The rate of heat transfer through a 3 m × 4 m brick wall of thickness 40 cm is 800 W. If the inner and outer surfaces of the wall are maintained at

35°C and 8°C, respectively, determine the thermal conductivity of the brick wall.

[**Ans.** 0.988 W/(m °C)]

3.7 The inner and outer surfaces of a 2 m × 2 m window of 6 mm thick glass are at 10°C and 2°C, respectively, in winter. If the thermal conductivity of the glass is 0.78 W/(m °C), determine the amount of heat loss, in kJ, through the window over a period of 8 hours. How much would be the reduction in heat loss if the window thickness is doubled?

[**Ans.** 119.808 MJ, 59.904 MJ]

3.8 In a nuclear reactor, cylindrical uranium fuel rods of diameter 30 mm generate heat at a rate of $5 \times 10^7$ W/m$^3$. If the thermal conductivity of uranium is 27.6 W/(m K), determine the temperature difference between the center and surface of the fuel rod.

[**Ans.** 101.9°C]

3.9 The two sides of a large plane wall of thickness 0.3 m, surface area 10 m$^2$, and thermal conductivity 1.2 W/(m °C) are maintained at constant temperatures of 100°C and 60°C. Determine the temperature at the middle of the wall and the rate of heat conduction through the wall under steady conditions.

[**Ans.** 80°C, 1600 W]

3.10 A heat flux meter attached to the inner surface of a refrigerator door indicates a heat flux of 30 W/m$^2$ through the door. The average thermal conductivity of the refrigerator door is 0.09 W/(m °C). If the temperatures of the inner and outer surfaces of the door are 5°C and 20°C, respectively, find the door thickness.

[**Ans.** 4.5 cm]

3.11 Consider a large plane wall of thickness 0.3 m and thermal conductivity 2.5 W/(m °C). The left side of the wall is subjected to a net heat flux of 925 W/m$^2$, while the temperature at that surface is 92°C. Assuming constant thermal conductivity and no heat generation in the wall, determine the temperature on the right surface of the wall.

[**Ans.** − 19°C]

3.12 In a nuclear reactor, heat is generated uniformly at a rate of $6.5 \times 10^7$ W/m$^3$, by cylindrical uranium rods of 6 cm diameter and 1.2 m length. Determine the rate of heat generation in each rod.

[**Ans.** 220.5 kW]

3.13 Heat at a uniform rate of $5 \times 10^6$ W/m$^2$ is generated in a large stainless steel plate of thickness 0.3 m. Assuming the plate is losing heat from both sides, determine the heat flux at the surface of the plate during a steady heat transfer process.

[**Ans.** 75000 W/m$^2$]

3.14 A soldering iron heating rod is 10 mm in diameter and 50 mm long. If the electric power at the rate of 100 W is supplied to the rod, determine the rate of heat generation in the rod per cubic centimeter.

[**Ans.** 25.5 W/cm$^3$]

3.15 A glass window of a room maintained at 25°C of thickness 12 mm has −12°C on its outer surface. Determine the heat loss through this window.

[**Ans.** 2402.6 W/m$^2$]

3.16 In a large copper plate of thickness 2 cm, heat is generated uniformly at a rate of 3 MW/m$^3$. Assuming the plate is losing heat on both sides, determine the heat flux from the surface of the plate during steady-state heat transfer.

[**Ans.** 30,000 W/m$^2$]

3.17 The base plate of a domestic iron box is supplied with 800 W. The base plate area and thickness are 160 cm$^2$ and 6 mm. The inner surface of the plate is subjected to uniform heat flux by the resistance heaters inside of it. When steady-state operating conditions are reached, the outer surface temperature of the plate remains at 85°C. Determine the inner surface temperature of the plate, at this state. Assume $k$ to be 20 W/(m °C).

[**Ans.** 100°C]

3.18 A 100-W electric bulb has 0.5 mm filament of length 50 mm. The diameter of the bulb dome is 80 mm and wall thickness is 1 mm. Determine (a) the heat flux at the filament surface and (b) if it is glowing in a room at 20°C, what will be the bulb surface temperature at steady-state operation?

[**Ans.** (a) $1.27 \times 10^6$ W/m$^2$, (b) 26.38°C]

3.19 A large plane wall of thickness 40 cm, thermal conductivity 2.3 W/(m °C), and surface area 25 m$^2$ is maintained at 80°C at its left surface. On the right surface heat is lost by convection to the surrounding air at 15°C. If the heat transfer coefficient of the surrounding air is 24 W/(m$^2$ °C), determine the rate of heat transfer through the wall at steady state.

[**Ans.** 7523.25 W]

3.20 A 2-m long electric wire of 6 mm diameter is insulated with a material with $k = 0.15$ W/(m °C). The thickness of insulation is 2 mm. When a current of 8 ampere passes through the wire, there is a drop of 8 volts along the wire. If the insulated wire is exposed to a room air at 22°C with $h = 10$ W/(m$^2$ °C), determine the temperature at the interface of the wire and the insulation, in steady operation. Check whether doubling the insulation thickness will increase or decrease the temperature at the interface.

[**Ans.** 141.1°C, doubling the insulation will result in decrease of interface temperature]

3.21 A steel rod of diameter 20 mm and length 400 mm is exposed to air at 20°C. One end of the rod is maintained at 135°C. Determine the heat loss from the rod, if the conduction coefficient of steel is 50 W/(m °C) and convection coefficient of air is 60 W/($m^2$ °C).

[**Ans.** 27.98 W]

3.22 Circular aluminum fins of thickness 1 mm and length 10 mm are attached on a tube of outer diameter 250 mm. The tube wall is maintained at 150°C and the heat from the tube is dissipated to the ambient atmosphere at 22°C, with convection heat transfer coefficient of 180 W/($m^2$ °C). If 100 fins are placed per meter length of the tube and the conduction heat transfer coefficient of fin material is 150 W/(m °C), assuming the radiation effect are negligible, (a) calculate the heat loss to the ambient air per meter length of the tube? (b) What will be the reduction in the heat transfer from the tube if fins were not there?

[**Ans.** (a) 54286 W, (b) 36190.4 W]

3.23 A steam pipe of diameter 40 mm is insulated with asbestos and fiberglass. Thicknesses of asbestos and fiberglass layers are 10 mm and 30 mm, respectively. The thermal conductivity of asbestos is 0.15 W/(m °C) and that of fiberglass is 0.05 W/(m °C). If the surface temperature of the steam pipe is constant at 330°C and the outer surface of the fiberglass layer is at 30°C, determine (a) the heat transfer rate per unit length of the pipe, and (b) the temperature at the interface of asbestos and fiberglass layers.

[**Ans.** (a) 114 W, (b) 280.8°C]

3.24 A 50-cm thick furnace wall has to be insulated to keep the heat loss from the furnace, less than 750 W/$m^2$, when the inside temperature is 1350°C and outside surface temperature of the insulation is 50°C. If the conduction coefficients of the wall and insulation material are 1 W/(m °C) and 0.08 W/(m °C), respectively, what should be the insulation layer thickness?

[**Ans.** 98.64 mm]

3.25 A double-pane glass window of height 0.8 m, width 1.5 m and thickness 4 mm, and a air gap of 10 mm in between the panes fixed to a room, maintains 20°C and −10°C on its inner and outer surfaces, respectively. The conduction coefficients of glass and air are 0.78 W/(m °C) and 0.026 W/(m °C), respectively. Assuming the convection coefficients on the inner and outer surfaces of the window as 10 W/($m^2$ °C) and 40 W/($m^2$ °C), respectively, determine the overall heat transfer coefficient and the heat loss through the window.

[**Ans.** 2.309 W/°C, 69.27 W]

3.26 Heat is conducted through a thin sheet of a circuit board made of copper and epoxy. The circuit board and the direction of heat conduction are shown in Figure P3.26. Treating the heat transfer as one-dimensional, in the direction marked in the figure, determine the percentage of heat transfer (a) along the copper layer, (b) along the epoxy layer, and (c) the effective conductivity of the board. Take the conductivity of copper and epoxy as 386 W/(m °C) and

0.26 W/(m °C), respectively.

[**Ans.** (a) 99.2 percent, (b) 0.8 percent, (c) 29.93 W/(m °C)]

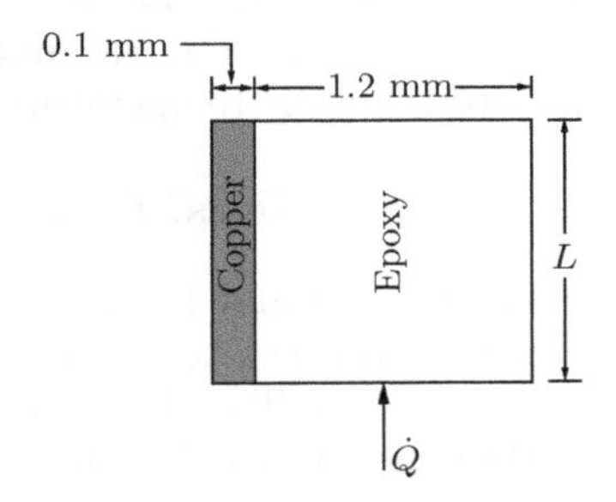


**Figure P3.26**
Thin circuit board.

3.27 The overall heat transfer coefficient of a 2-m high and 5-m long wall is 2.25 W/(m$^2$ °C). If a 0.5-m thick layer of insulation is added to the outer surface of the wall, determine the overall heat transfer coefficient of the insulated wall and the reduction of heat transfer caused by the insulation, when the temperature at the inner and outer surfaces of the wall are 25°C and −10°C, respectively, taking the conduction coefficient of the insulator as 0.02 W/(m °C).

[**Ans.** 0.0393 W/(m$^2$ °C), 98.25 percent]

3.28 Determine the steady-state heat flux through a 300-mm thick brick wall of thermal conductivity 0.7 W/(m °C), when the temperatures of its surfaces are 30°C and −12°C.

[**Ans.** 98 W/m$^2$]

3.29 A cubical box made of an insulating material is of sides 0.5 m and wall thickness 25 mm. The thermal conductivity of the material is 0.04 W/(m °C). If the temperature difference between the inside and outside of the box is 25°C, neglecting the thermal resistance for the convective heat transfer at the inner and outer surfaces, determine the heat loss from the box, based on the outer surface area.

[**Ans.** 60 W]

3.30 Derive an expression for the temperature distribution $T(x)$, during a steady-state, one-dimensional heat conduction in a slab of thickness $L$ and constant thermal conductivity $k$, in which heat is generated at a constant rate of $\dot{g}_{th}$ W/m$^3$, slab surface at $x = 0$ is insulated and the surface at $x = L$ is maintained at 0°C. Find the temperature at the insulated surface if $k = 40$ W/(m°C), $\dot{g}_{th} = 10^6$ W/m$^3$ and $L = 0.1$ m.

[**Ans.** $T(x) = -\dfrac{\dot{g}_{th}}{k}\dfrac{x^2}{2} + 125$, 125°C]

3.31 A large glass window of a room is 10 mm thick and its thermal conductivity is 0.8 W/(m °C). The room air is maintained at 22°C and the heat transfer coefficient of the room air is 13 W/(m$^2$ °C). The outside air is at −10°C and its heat transfer coefficient is 44 W/(m$^2$ °C). Assuming steady heat transfer through the window glass, determine the temperature at the inner and outer surfaces of the glass.

[**Ans.** $T_{s,i} = 0.06$°C, $T_{s,o} = -3.52$°C]

3.32 A pipe of 100 mm inner diameter and 120 mm outer diameter and conduction coefficient 44 W/(m °C) carrying a hot gas is insulated with a material of thermal conductivity 0.5 W/(m °C). The thickness of insulation is 10 mm. Determine the percentage decrease of heat loss per unit length of the pipe because of insulation when the outer surfaces of the pipe and insulator are at 400°C and 25°C, respectively.

[**Ans.** 98.6%]

3.33 A 100-W power transistor is to be cooled by attaching a heat sink in the form of a flat fin normal to the surface of the transistor. (a) If the transistor temperature should not exceed 95°C when the ambient temperature is 22°C, what should be the limiting thermal resistance of the fin? (b) If the fin length is 150 mm, determine its width, assuming that the temperature of the fin is uniform at 95°C and the heat from it is convected to the surrounding air of convection coefficient 15 W/(m$^2$ °C). Assume that, the contact resistance between transistor and fin is negligible.

[**Ans.** (a) $\leq$ 1.37 W/(m °C), (b) 0.304 m]

3.34 A spherical tank of radius 1 m containing radioactive material is buried in earth. The distance between the earth surface and tank center is 2 m and the thermal conductivity of the earth is 1 W/(m °C). If the heat release due to radioactive decay is steady at 950 W, calculate the temperature of the tank surface, if the earth surface is at a constant temperature of 22°C.

[**Ans.** 78.72°C]

3.35 A hot water pipe of diameter 100 mm and length 8 m is laid 600 mm below the earth surface at a constant temperature of 10°C. When the water temperature is constant at 80°C, the rate heat loss from the pipe is 1 kW. Determine the thermal conductivity of the pipe material.

[**Ans.** 0.903 W/(m °C)]

3.36 A cylindrical storage tank of radius 500 mm and length 2.5 m is buried horizontally in the earth, with the tank axis at 2 m below the earth surface. If the tank surface is at a constant temperature of 70°C and the earth surface is at 20°C, determine the heat loss from the tank, neglecting the loss at the end-faces. Take the thermal conductivity of the earth as 1.2 W/(m °C).

[**Ans.** 453.24 W]

3.37 A 20-mm radius pipe carrying hot water at 70°C is embedded eccentrically inside a long, cylindrical concrete block of radius 80 mm, as shown in Figure P3.37. The outside temperature of the concrete cylinder is 20°C and the thermal conductivity of the concrete is 0.76 W/(m °C). Determine the rate of heat loss per unit length of the pipe.

[**Ans.** 223.82 W]

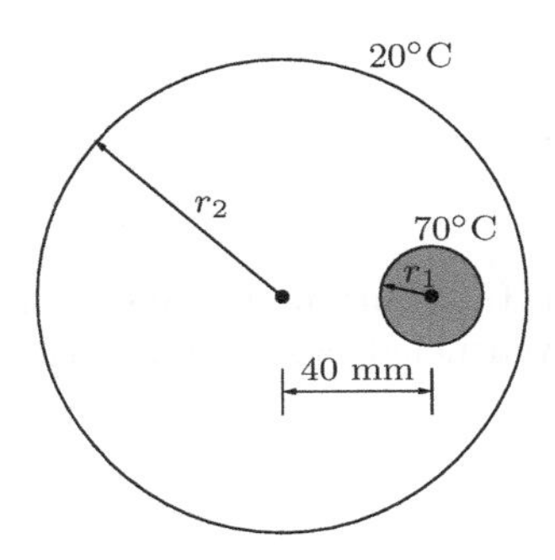


**Figure P3.37**
Hot water pipe embedded in a concrete block.

3.38 Liquid nitrogen at −198°C is stored in a thin-walled spherical tank of diameter 500 mm, insulated to a thickness of 25 mm. When the tank is exposed to ambient air at 30°C, the rate of heat transferred to the tank is 13 W. If the convection heat transfer coefficient of air is 20 W/(m$^2$ K), determine the thermal conductivity of the insulation.

[**Ans.** 0.00165 W/(m K)]

3.39 A long metallic rod of diameter 10 mm is attached to a wall at a uniform temperature of 200°C. If 20 W of heat is dissipated from the rod by convection to the surrounding air at 22°C, with a convection coefficient of 15 W/(m$^2$ °C), determine the conduction coefficient of the rod material.

[**Ans.** 341 W/(m K)]

3.40 A metallic rod of diameter 10 mm, length 300 mm and thermal conductivity 65 W/(m °C), attached horizontally to a large tank at a constant temperature of 200°C dissipates heat to the ambient air at 20°C, and with a convection coefficient of 15 W/(m$^2$ °C). Determine the temperature of the rod at 100 mm and 200 mm from the tank wall.

[**Ans.** 88.94°C, 46.28°C]

3.41 The inner and outer surfaces of a thick-walled cylindrical tube of radius 50 mm and wall thickness 50 mm are maintained at 300°C and 100°C, respectively. In the temperature range from 100°C to 300°C, the thermal conductivity of the tube is governed by

$$k(T) = 0.5\left(1 + \frac{T}{1000}\right)$$

where $T$ is in degree celcius. Determine the rate of heat transfer per unit length of the tube.

[**Ans.** 1087.5 W/m]

# Reference

1. Hahne, E., and Grigull, U., "Formfaktor und formwiederstand der stationaren mehrdimensionalen warmeleitung," *Int. J. Heat Mass Transfer*, 1974 Vol. 18, p. 751, 1975.

# Chapter 4

# Unsteady Heat Conduction

## 4.1 Introduction

We know that the temperature of a solid body subjected to a sudden change of environment takes some time to attain an equilibrium. For example, if the surface temperature of a system is altered, the temperature at each point in the system will also begin to change. The changes will continue till a *steady-state* temperature distribution is reached. Consider a hot metal billet that is removed from a furnace and exposed to a cool air stream. Energy is transferred by convection and radiation from the billet surface to its surroundings. Energy transfer by conduction also occurs from the interior of the metal to the surface, and the temperature of each point in the billet decreases until a steady-state condition is reached. Such time-dependent effects occur in many industrial heating and cooling processes.

In the transient heating and cooling process, which takes place in the interim period before equilibrium is established, the analysis must be modified to take into account the change in the *initial condition* of this body with time, and the boundary conditions must be adjusted to match the physical situation which is apparent in the unsteady-state heat transfer problem.

In the absence of any heat source, the governing equation for the unsteady heat conduction becomes

$$\boxed{\frac{\partial T}{\partial t} = \alpha \nabla^2 T} \tag{4.1}$$

where $\alpha = \dfrac{k}{\rho c}$ is called the *thermal diffusivity*. This equation needs both the boundary and initial conditions for its solution.

## 4.2 Transient Conduction in an Infinite Wall

For our study here, let us consider problems involving only one space dimension, that is, one-dimensional problems only. Examine the plane wall of thickness $2L$, which extends infinitely in the other two dimensions ($y$- and $z$-directions) as shown in Figure 4.1.

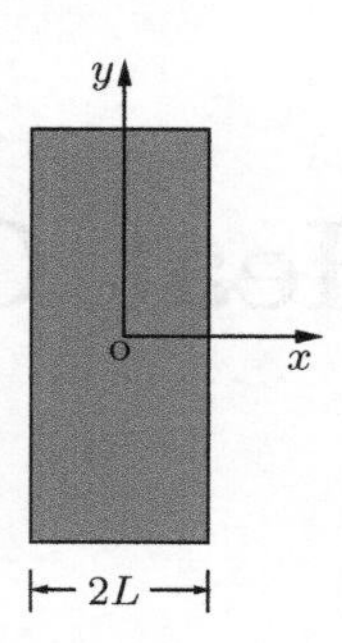


**Figure 4.1**
An infinite wall immersed in a fluid.

Let the slab be initially at temperature $T_0$, and at time $t = 0$ it is immersed in a fluid bath at temperature $T_\infty$. The governing equation for the unsteady heat conduction, Equation (4.1), for this case reduces to

$$\frac{\partial T}{\partial t} = \alpha \frac{\partial^2 T}{\partial x^2} \tag{4.2}$$

The initial condition is

$$T(x, t) = T_0, \text{ for } t < 0 \tag{4.3}$$

The boundary conditions at $x = -L$ and $x = L$ are given by the Newton's law of cooling as

$$-k\frac{\partial T}{\partial x} = h\left(T_\infty - T\right), \quad \text{at } x = -L \tag{4.4a}$$

$$-k\frac{\partial T}{\partial x} = h\left(T - T_\infty\right), \quad \text{at } x = L \tag{4.4b}$$

with the positive (+) flux at $x = -L$ and $x = +L$ as $h\left(T_\infty - T\right)$ and $h\left(T - T_\infty\right)$, respectively.

Equation (4.2) can be solved to get the functional relation of the temperature $T$ with all other influencing parameters. Also, we can express the temperature $T$ as a function of all the parameters influencing $T$ with dimensional analysis, as follows.

The temperature $T$ can be expressed in the functional form as

$$T = T\left(x, t, L, \rho, c, k, h, T_0, T_\infty\right) \tag{4.5}$$

where $x$ is the coordinate along the wall thickness (Figure 4.1), $t$ is the time, $2L$ is the wall thickness, $\rho$, $c$, $k$ are the density, specific heat and thermal conductivity of the wall material, respectively, and $h$ is the convection heat transfer coefficient of the fluid in which the wall is immersed, $T_0$ is the initial temperature of the wall and $T_\infty$ is the temperature of the fluid.

From Equation (4.5), the nondimensional form of the temperature, defined as

$$\theta = \frac{T - T_\infty}{T_0 - T_\infty}$$

can be expressed after carrying out the dimensional analysis as

$$\theta = \theta\left(\frac{x}{L}, \frac{kt}{\rho c L^2}, \frac{hL}{k}\right) \tag{4.6}$$

where $\frac{x}{L}$, $\frac{kt}{\rho c L^2}$, and $\frac{hL}{k}$ are independent nondimensional groups.

### 4.2.1 The Dimensionless Groups

The nondimensional group

$$\frac{hL}{k} = \text{Bi} \quad \text{is called the } \textit{Biot number}$$

and

$$\frac{kt}{\rho c L^2} = \frac{\alpha t}{L^2} = \text{Fo} \quad \text{is called the } \textit{Fourier number.}$$

The Biot number can also be expressed as

$$\begin{aligned} \text{Bi} &= \frac{hL}{k} = \frac{h}{k/L} \\ &= \frac{\text{Heat transfer at the surface of solid}}{\text{Internal conductance of solid across length } L} \end{aligned}$$

That is, Bi is the ratio of heat transfer coefficient to the unit conductance of solid over the characteristic length $L$.

The Fourier number is essentially a dimensionless time, and it can be expressed as

$$\begin{aligned} \text{Fo} &= \frac{\alpha t}{L^2} = \frac{k\,(1/L)\,L^2}{\rho c\, L^3/t} \\ &= \frac{\text{Rate of heat conduction across } L \text{ in volume } L^3}{\text{Rate of heat storage in volume } L^3} \end{aligned}$$

That is, Fourier number, Fo, is a measure of the rate of heat conduction in comparison with the rate of heat storage in a given volume element. Therefore, the larger the Fo the deeper would be the penetration of heat into a solid over a given time period.

The dimensionless temperature $\theta$ in Equation (4.6) depends on the slab geometry $x/L$, Biot number Bi $(hL/k)$ and Fourier number Fo $(\alpha t/L^2)$. This functional relation of the nondimensional temperature $\theta$ [Equation (4.6)] is extremely useful for analyzing experimental data. These dimensionless groups may also be obtained from the governing equation [Equation (4.2)] and the boundary conditions [Equations (4.4a) and (4.4b)], as follows.

Let the half-thickness $L$ of the wall be the characteristic length, and $(T_0 - T_\infty)$ be the characteristic temperature difference. Therefore, we can write expressions for the dimensionless length $x^*$, and dimensionless temperature $\theta$ as

$$x^* = \frac{x}{L}$$

$$\theta = \frac{T - T_\infty}{T_0 - T_\infty}$$

In terms of $x^*$ and $\theta$, the first and second derivatives of temperature $T$ with respect to $x$ are the following.

From the definition of $\theta$, we have the temperature $T$ as

$$T = (T_0 - T_\infty)\,\theta + T_\infty$$

Differentiating $T$ with respect to $x$, we get

$$\begin{aligned}\frac{\partial T}{\partial x} &= (T_0 - T_\infty)\,\frac{\partial \theta}{\partial x} \\ &= (T_0 - T_\infty)\,\frac{\partial \theta}{\partial \left(\frac{x}{L}\right) L} \\ &= \frac{(T_0 - T_\infty)}{L}\,\frac{\partial \theta}{\partial x^*}\end{aligned}$$

Differentiating $\dfrac{\partial T}{\partial x}$ with respect to $x$, we get

$$\begin{aligned}\frac{\partial^2 T}{\partial x^2} &= (T_0 - T_\infty)\,\frac{\partial^2 \theta}{\partial x^2} \\ &= (T_0 - T_\infty)\,\frac{\partial^2 \theta}{\partial \left(\frac{x}{L}\right)^2 L^2}\end{aligned}$$

$$= \frac{(T_0 - T_\infty)}{L^2} \frac{\partial^2 \theta}{\partial x^{*2}}$$

There is no apparent characteristic time for this problem. Let $\tau$ be an arbitrary characteristic time. Then the dimensionless time becomes

$$t^* = \frac{t}{\tau}$$

Now, differentiating

$$T = (T_0 - T_\infty)\,\theta + T_\infty$$

with respect to time $t$, we get

$$\begin{aligned} \frac{\partial T}{\partial t} &= (T_0 - T_\infty)\,\frac{\partial \theta}{\partial t} \\ &= (T_0 - T_\infty)\,\frac{\partial \theta}{\partial \left(\frac{t}{\tau}\right) \tau} \\ &= \frac{(T_0 - T_\infty)}{\tau} \frac{\partial \theta}{\partial t^*} \end{aligned}$$

Substituting the above expressions for $\partial T/\partial t$ and $\partial^2 T/\partial x^2$ into the governing Equation (4.2), we get the equation in terms of the nondimensional length $x^*$, nondimensional temperature $\theta$, and nondimensional time $t^*$, as

$$\frac{(T_0 - T_\infty)}{\tau} \frac{\partial \theta}{\partial t^*} = \alpha \frac{(T_0 - T_\infty)}{L^2} \frac{\partial^2 \theta}{\partial x^{*2}}$$

that is,

$$\boxed{\frac{\partial \theta}{\partial t^*} = \frac{\alpha \tau}{L^2} \frac{\partial^2 \theta}{\partial x^{*2}}} \tag{4.7}$$

In Equation (4.7), $\theta$, $t^*$, and $x^*$ are dimensionless quantities, hence the group $\alpha\tau/L^2$ must also be dimensionless. Furthermore, $\tau$ is an arbitrary characteristic time; therefore, we can assume it to be equal to $L^2/\alpha$. With $\tau = L^2/\alpha$, Equation (4.7) simplifies to

$$\boxed{\frac{\partial \theta}{\partial t^*} = \frac{\partial^2 \theta}{\partial x^{*2}}} \tag{4.8}$$

where $t^* = \alpha t/L^2$ is the dimensionless time.

The boundary conditions given by Equations (4.4a) and (4.4b), in terms of $\theta$ and $x^*$ become

$$\frac{-k}{L} \frac{\partial \theta}{\partial x^*} = -h\,\theta, \quad \text{at } x^* = -1$$

that is,

$$\boxed{\frac{\partial \theta}{\partial x^*} = \frac{hL}{k}\theta, \quad \text{at } x^* = -1} \tag{4.9}$$

Similarly,

$$\boxed{\frac{\partial \theta}{\partial x^*} = -\frac{hL}{k}\theta, \quad \text{at } \quad x^* = 1} \tag{4.10}$$

The solution of Equation (4.8) with boundary conditions given by Equations (4.9) and (4.10) describes the problem completely.

From Equations (4.8) to (4.10) it is seen that two problems will result in identical solutions if the dimensionless parameter $\frac{hL}{k}$, namely the Biot number Bi, is the same for both the problems. Therefore, Bi is the *similarity parameter* for this problem. Thus,

$$\boxed{\theta = \theta\,(x^*, t^*, \text{Bi})} \tag{4.11}$$

This is the same as Equation (4.6). This result can be expressed in a useful graphical form. One of the popular forms is the Heisler chart, given in Figures 4.2 and 4.3, presented by M. P. Heisler in 1947 [1]. The chart in Figure 4.2 is meant for determining temperature $T_0$ at the center of the slab, at a given time $t$. The chart in Figure 4.3 is to determine the temperature in other locations of the slab, in terms of $T_0$. Figure 4.4 gives the variation of dimensionless heat transfer as a function of dimensionless time, for various values of Biot number for a slab of thickness $2L$. This was devised by Gröber and hence termed Gröber's chart. Here, $Q$ is the amount of heat energy lost by the slab over a time period $t$ during the transient heat transfer, and $Q_0$ is the initial internal energy of the slab relative to the ambient temperature, given by

$$Q_0 = \rho \mathbb{V} c\,(T_i - T_\infty)$$

where $\rho$, $\mathbb{V}$ and $c$ are the density, volume and specific heat of the slab, respectively, $T_i$ is the initial temperature of the slab, and $T_\infty$ is the temperature of the environment to which the slab is losing heat.

Figure 4.4 is to determine the total amount of heat transferred in time $t$. Figures 4.2 - 4.4 are valid for Fourier number (or dimensionless time $t^*$) greater than 2.

These charts can also be used for a slab in which one surface is insulated, in which case $L$ equals the thickness of the slab, and $x^* = 0$ corresponds to the insulated face.

Figure 4.3 gives the ratio of the temperature at any $x^*$ to the temperature at the center line $x^* = 0$, both at the same Fourier number and same time.

It is important to note that, the case of $1/\text{Bi} = k/(hL) = 0$ corresponds to $h \to \infty$, which corresponds to the case of specified surface temperature $T_\infty$. In other words, $1/\text{Bi} = 0$ corresponds to the situation when the surface of the body is suddenly exposed to an environment at temperature $T_\infty$ at $t = 0$, and

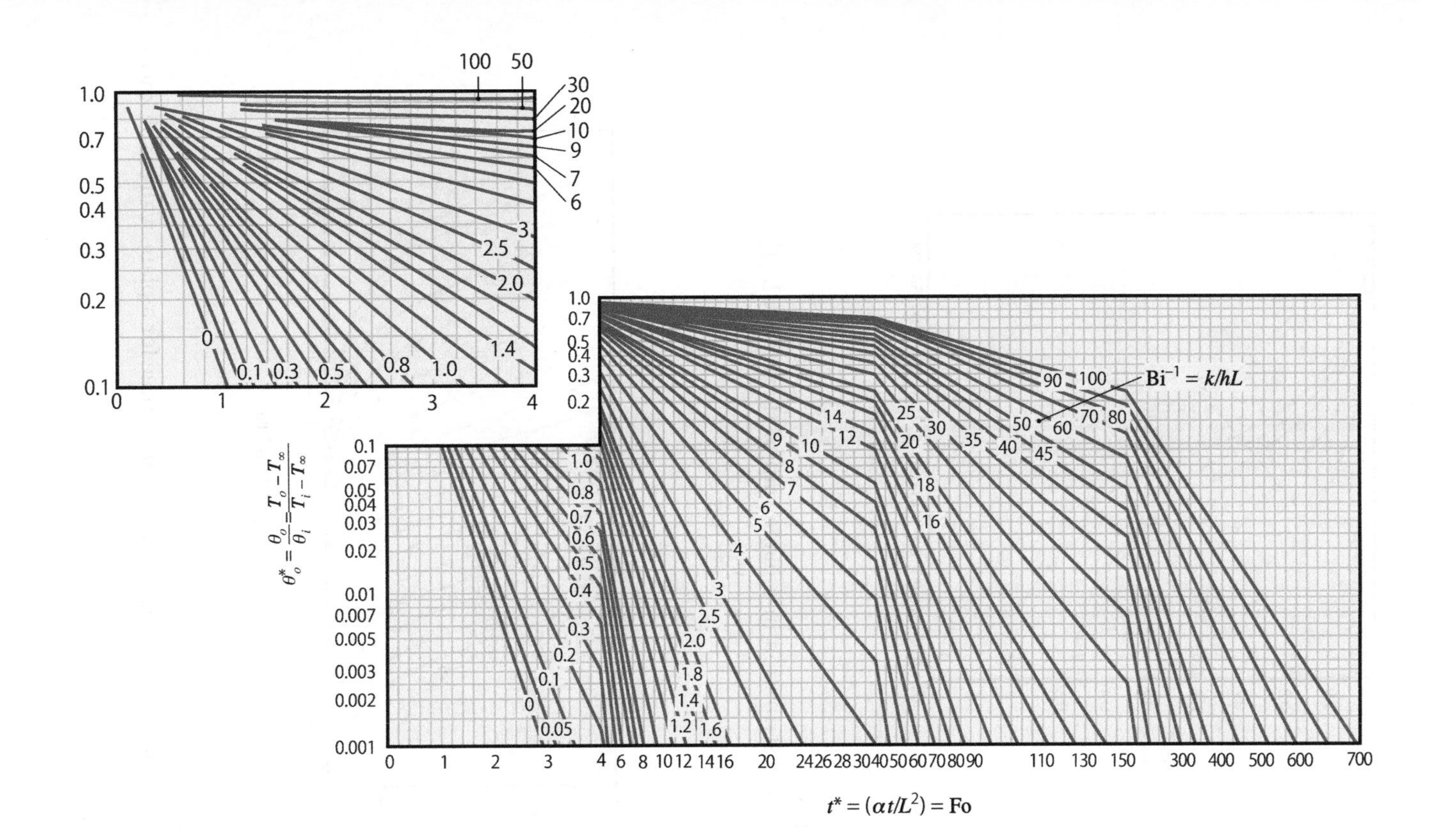


**Figure 4.2**
Mid-plane temperature as a function of time for a plane wall of thickness $2L$ [1]

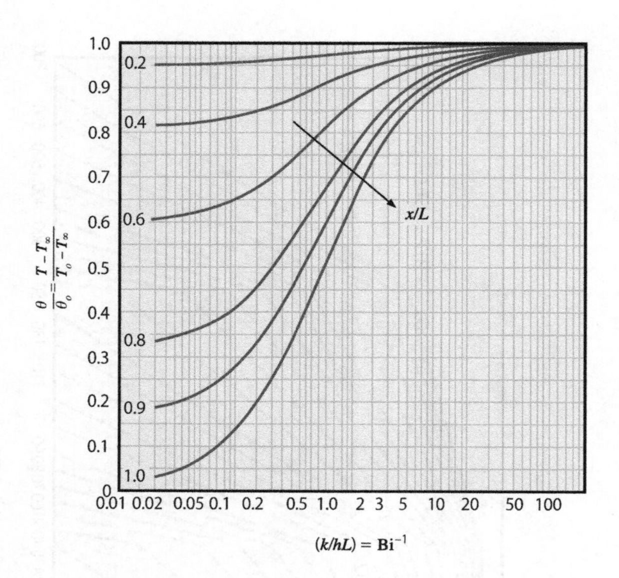


**Figure 4.3**
Temperature distribution in a plane wall of thickness $2L$ [1].

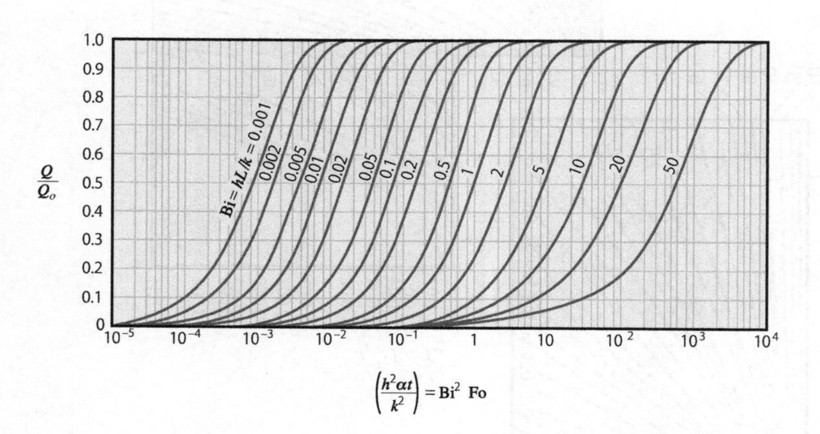


**Figure 4.4**
Dimensionless heat loss $Q/Q_0$ as a function of time for a plane wall of thickness $2L$ [2].

kept at $T_\infty$ at all times. This kind of heat transfer problems can be solved by setting $h$ to infinity.

The temperature of the body changes from its initial temperature $T_i$ to the temperature $T_\infty$ of the surroundings at the end of the transient heat transfer process. Thus, the maximum amount of heat that a body can gain (or lose if $T_i > T_\infty$) is identically equal to the change in the energy content of the body, that is,

$$\boxed{Q_{\text{max}} = mc\,(T_\infty - T_i)}$$

where $m$ is the mass of the body and $c$ is its specific heat.

For one-dimensional transient conduction through infinitely long cylinders and spheres immersed in a bath, the relevant nondimensional groups are the same as those for the slab problem, with radius $r$ of the cylinder or sphere as the characteristic length.

Heisler's charts for cylinders of radius $r_o$ are given in Figures 4.5 and 4.6.

Internal energy change as a function of time for an infinite cylinder of radius $r_o$ or the variation of dimensionless heat transfer as a function of dimensionless time, for various values of Biot number, is shown in Figure 4.7.

Dimensionless temperature distribution in a sphere of radius $r_o$, as a function of nondimensional time, and the dimensionless temperature variation in a sphere with Biot number, for various values of $r/r_o$, are shown in Figures 4.8 and 4.9, respectively.

The internal energy change as a function of time for a sphere of radius $r_o$ is shown in Figure 4.10.

The Heisler's and Gröber charts are valid for bodies which are initially at a uniform temperature and without any heat source inside, and the temperature and convection heat transfer coefficients of the surrounding medium in which the body is immersed are constant and uniform.

## Example 4.1

An iron sphere of diameter 80 mm, initially at a uniform temperature of 200°C is suddenly exposed to ambient air at 30°C with convection coefficient 510 W/(m$^2$ °C). Taking the density, specific heat, conduction coefficient and thermal diffusivity for the sphere as 8000 kg/m$^3$, 460 J/(kg °C), 60 W/(m °C) and $1.6\times 10^{-5}$ m$^2$/s, respectively, calculate the temperature at the sphere center, and at a depth of 5 mm from the surface at 1 minute after exposure to the convection environment. Also, calculate the heat removed from the sphere in this 1-minute duration.

## Solution

Given, $r_0 = 40$ mm, $T_i = 200$°C, $T_\infty = 30$°C, $h = 510$ W/(m$^2$ °C), $\rho = 8000$ kg/m$^3$, $c = 460$ J/(kg °C), $k = 60$ W/(m °C), $\alpha = 1.6 \times 10^{-5}$ m$^2$/s, $x = 5$

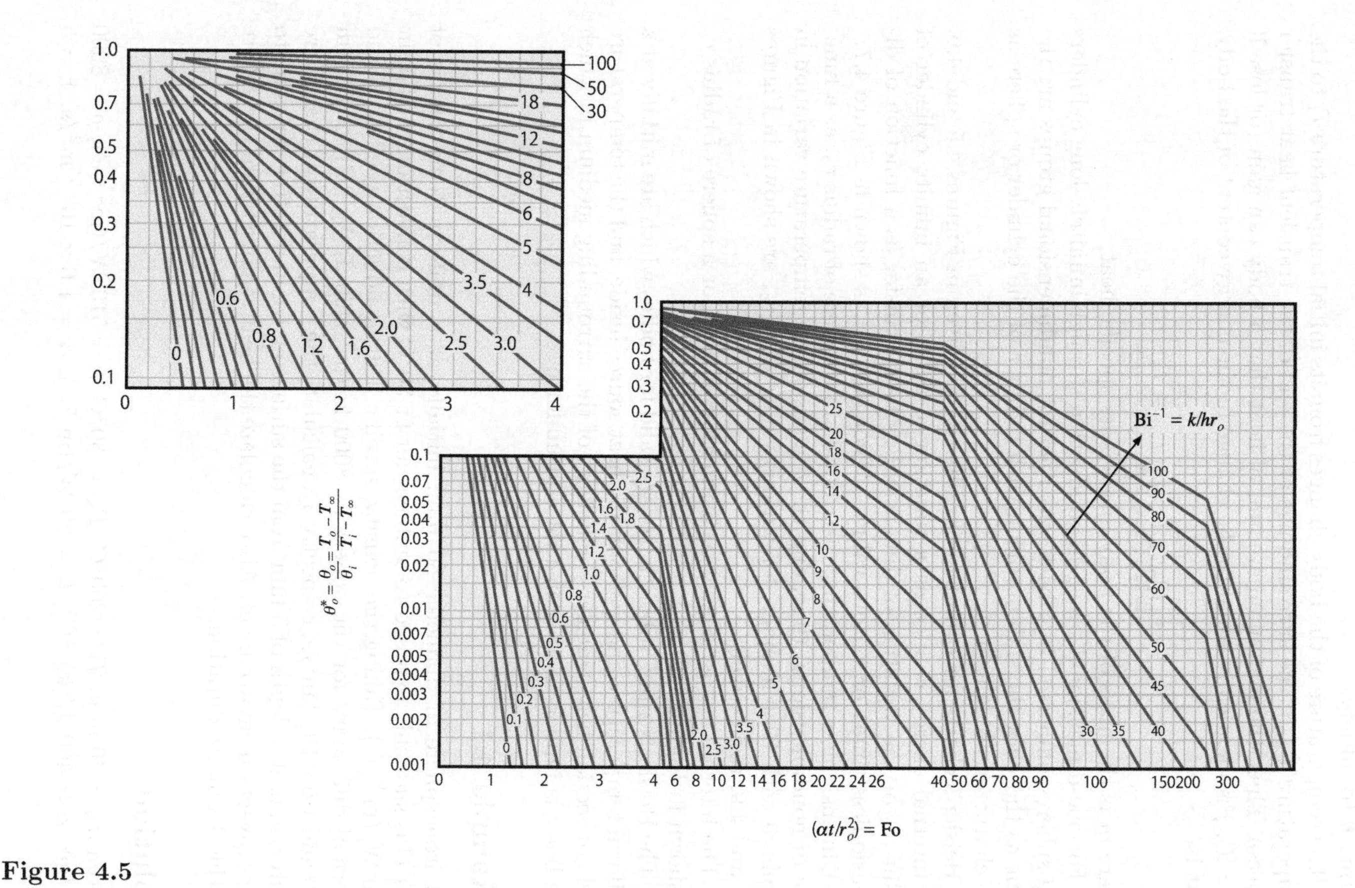


**Figure 4.5**
Centerline temperature as a function of time for an infinite cylinder of radius $r_o$ [1].

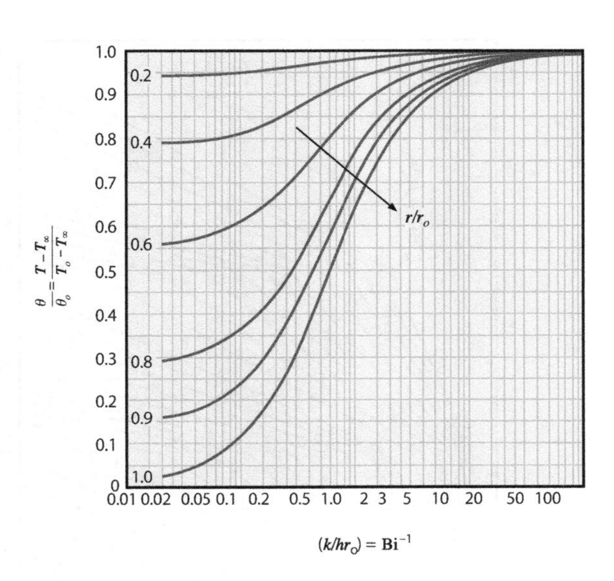


**Figure 4.6**
Temperature distribution in an infinite cylinder of radius $r_o$ [1].

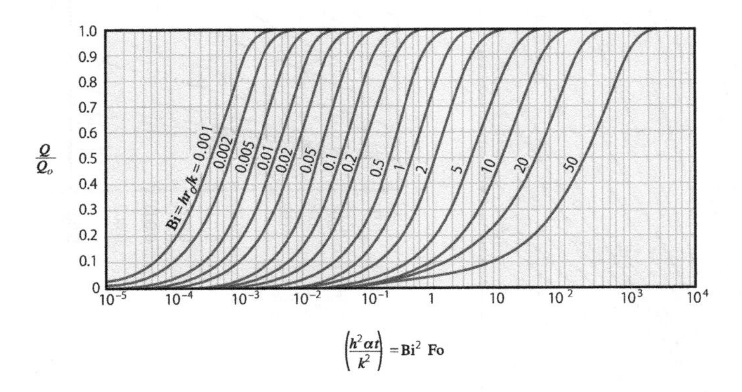


**Figure 4.7**
Dimensionless heat loss $Q/Q_0$ in an infinite cylinder of radius $r_o$ with time [2].

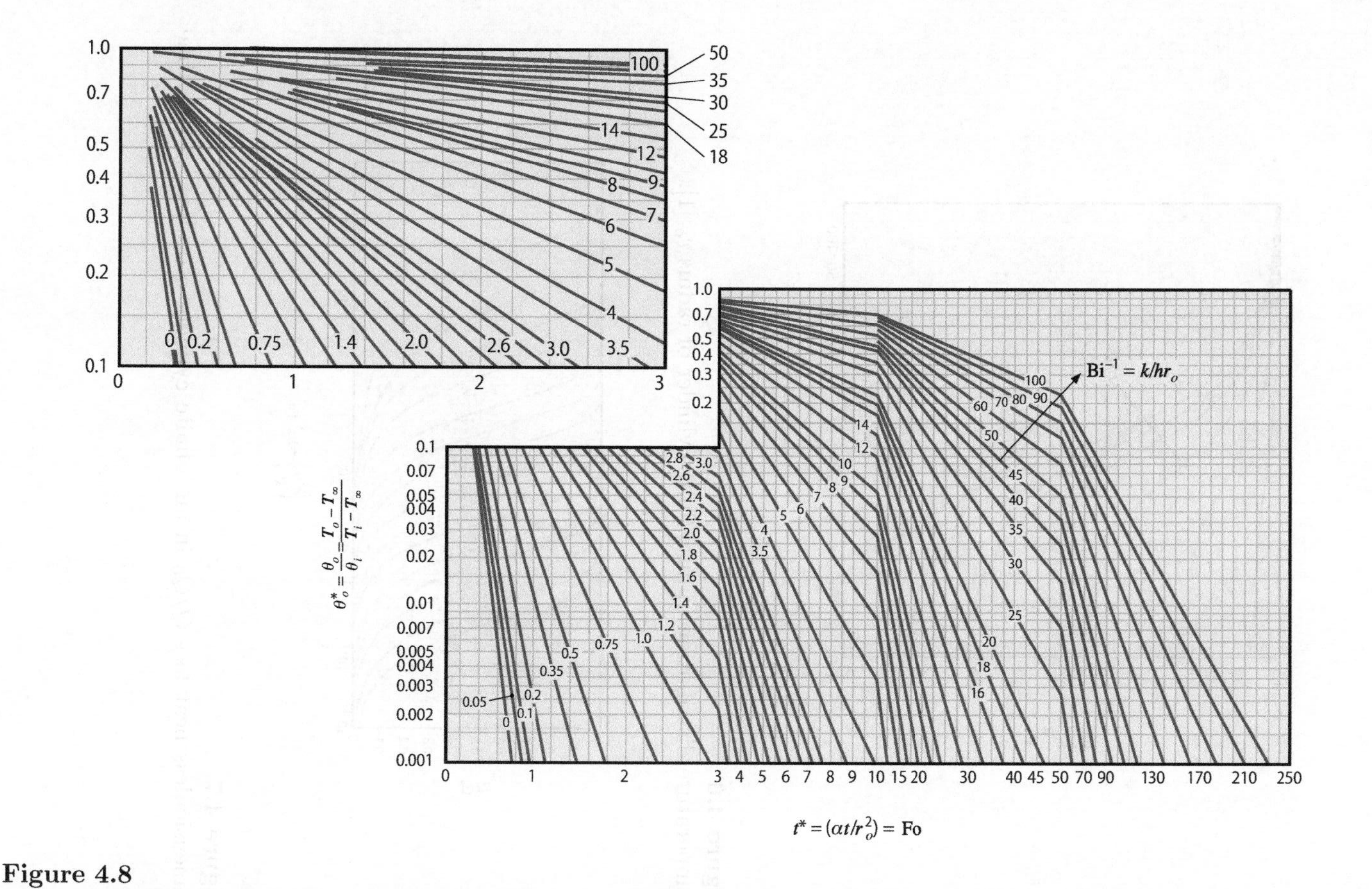


**Figure 4.8**
Center temperature distribution as a function of time in a sphere of radius $r_o$ [1].

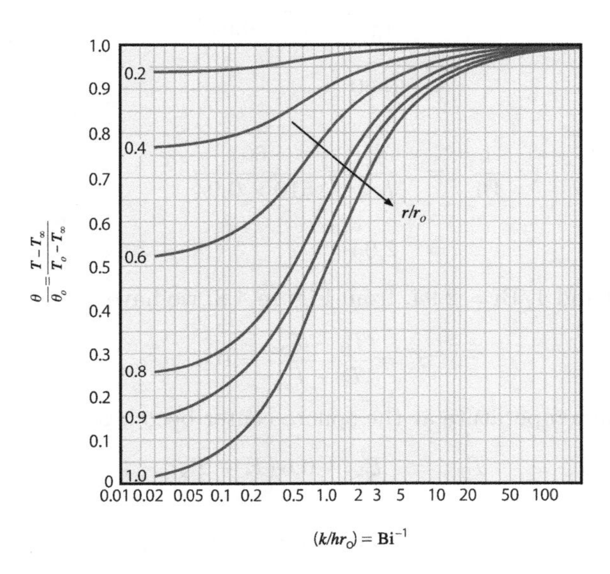


**Figure 4.9**
Temperature distribution in a sphere of radius $r_o$ [1].

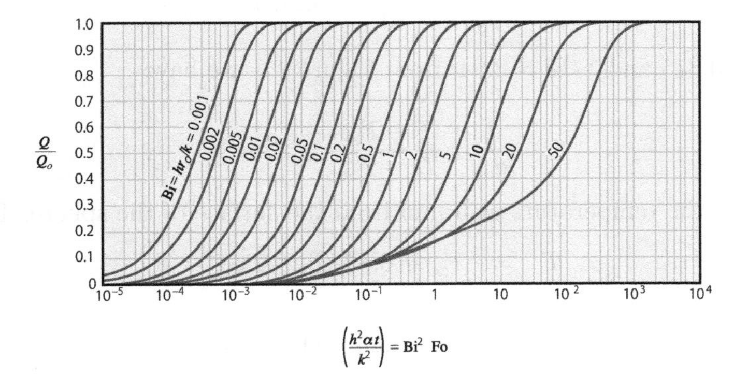


**Figure 4.10**
Dimensionless heat loss $Q/Q_0$ as a function of temperature in a sphere of radius $r_o$ [2].

mm, $t = 60$ s. Therefore,

$$t^* = \frac{\alpha t}{r_0^2} = \frac{(1.6 \times 10^{-5}) \times 60}{0.04^2} = 0.60$$

$$\frac{1}{\text{Bi}} = \frac{k}{hr_0} = \frac{60}{510 \times 0.04} = 2.94$$

For $t^* = 0.60$ and $1/\text{Bi} = 2.94$, from Figure 4.8, we have,

$$\theta^* = \frac{T_0 - T_\infty}{T_i - T_\infty} = 0.6$$

Therefore, the temperature at the sphere center becomes

$$T_0 = T_\infty + 0.6\,(T_i - T_\infty) = 30 + 0.6 \times (200 - 30) = \boxed{132°\text{C}}$$

Given,

$$\frac{r}{r_0} = \frac{40 - 5}{40} = 0.875$$

For $\dfrac{r}{r_0} = 0.875$ and $1/\text{Bi} = 2.94$, from Figure 4.9, we have

$$\frac{\theta}{\theta_0} = \frac{T - T_\infty}{T_0 - T_\infty} = 0.85$$

where $T$ is the temperature at 5 mm from the surface of the sphere. Thus,

$$T = T_\infty + 0.85\,(T_0 - T_\infty) = 30 + 0.85 \times (132 - 30) = \boxed{116.7°\text{C}}$$

For the given problem, $\text{Bi} = 1/2.94 = 0.34$ and

$$\frac{h^2 \alpha t}{k^2} = \frac{510^2 \times (1.6 \times 10^{-5}) \times 60}{60^2} = 0.0694$$

For Bi = 0.34 and $\dfrac{h^2\alpha t}{k^2} = 0.0694$, from Figure 4.10, we have

$$\frac{Q}{Q_0} = 0.45$$

where

$$\begin{aligned} Q_0 &= mc\Delta T \\ &= \rho\left(\frac{4}{3}\pi r_0^3\right) c\left(T_i - T_\infty\right) \\ &= \left[8000 \times \left(\frac{4}{3}\pi \times 0.04^3\right) \times 460\right] \times (200 - 30) \\ &= 167712.46\,\text{J} \end{aligned}$$

Thus, the heat loss from the sphere is

$$\begin{aligned} Q &= 0.45 Q_0 \\ &= 0.45 \times 167712.46 \\ &= \boxed{75470.6\,\text{J}} \end{aligned}$$

### 4.2.2 Transient Heating of Bodies with Negligible Internal Resistance

From Figure 4.3, it is seen that as the Biot number Bi decreases, the temperature at all points in the body tends to assume the temperature at the center line. Thus, for very low Bi the internal gradients of temperature are very small, compared to the overall temperature difference $(T_0 - T_\infty)$. That is, when Bi is small the entire body may be assumed to be at a uniform temperature $T$, which is a function of time. In other words, for small values of Bi, the entire body can be assumed to be at a uniform temperature at any given instant of time.

Now, for a body of volume $\mathbb{V}$ immersed in a fluid, energy balance requires that the heat conducted into the body across the fluid film around the body be equal to the rate of change of the thermal content of the body, that is,

$$hA\left(T_\infty - T\right) = \rho \mathbb{V} c \frac{dT}{dt} \tag{4.12}$$

where $A$ is the surface area of the body and $h$ is the convection coefficient of the fluid film surrounding the body.

Now, let us assume that at time $t < 0$, the body was at temperature $T_0$. Here $t < 0$ implies the period before immersing the body into the fluid at temperature $T_\infty$, at time $t = 0$.

The dimensionless temperature $\theta$ for this problem is

$$\theta = \frac{T - T_\infty}{T_0 - T_\infty}$$

where $T$ is the temperature at any instant of time $t$. Rearranging this, we have

$$T = (T_0 - T_\infty)\,\theta + T_\infty$$

Differentiating $T$ with respect to $t$, we have

$$\frac{dT}{dt} = (T_0 - T_\infty)\,\frac{d\theta}{dt}$$

Substituting this $\frac{dT}{dt}$ into Equation (4.12), we get

$$hA\,(T_\infty - T) = \rho \mathbb{V}\,c\,(T_0 - T_\infty)\,\frac{d\theta}{dt}$$

$$\frac{d\theta}{dt} = -\left(\frac{T - T_\infty}{T_0 - T_\infty}\right)\frac{hA}{\rho \mathbb{V}\,c}$$

That is,

$$\frac{d\theta}{dt} = -\frac{hA}{\rho \mathbb{V} c}\theta \tag{4.13}$$

This can be expressed as

$$\frac{d\theta}{\theta} = -\frac{hA}{\rho \mathbb{V} c}dt$$

Integrating, we get

$$\ln\,\theta = -\frac{hA}{\rho \mathbb{V} c}t + \text{constant}$$

The constant of integration can be obtained with the initial condition. At $t = 0$,

$$\theta = \frac{T_0 - T_\infty}{T_0 - T_\infty} = 1$$

Therefore, we have

$$\ln\,(1) = 0 + \text{constant}$$

or

$$\text{constant} = 0$$

Thus,

$$\ln \theta = -\frac{hA}{\rho \mathbb{V} c} t$$

or

$$\boxed{\theta = e^{-\frac{hA}{\rho \mathbb{V} c} t}} \tag{4.14}$$

Thus, the difference between the temperatures of the body and fluid decreases exponentially with the characteristic temperature $\tau$, given by

$$\boxed{\tau = \frac{\rho \mathbb{V} c}{hA}} \tag{4.15}$$

# 4.3 Lumped System Analysis

Lumped system is that in which the temperature of the system varies with time but remains uniform throughout the system at any specified instant of time. In heat transfer analysis, some bodies behave like a "lump" whose interior temperature remains uniform at all times during a heat transfer process. The temperature of such bodies can be taken to be a function of time only, $T(t)$. Heat transfer analysis that utilizes this idealization is known as *lumped system analysis.*

By Equation (4.14), the temperature of a lumped system at time $t$, $T(t)$ becomes

$$\frac{T(t) - T_\infty}{T_i - T_\infty} = e^{-bt} \tag{4.14a}$$

where $b = hA/(\rho \mathbb{V} c)$ and $T_i$ is the initial temperature of the system, before exposure to the surrounding environment at temperature $T_\infty$. The temperature $T(t)$ of a body at time $t$, or alternatively, the time $t$ required for the temperature to reach a specified value $T(t)$ can be determined using Equation (4.14a). Equation (4.14a) shows that, the temperature of a body approaches the ambient temperature $T_\infty$ exponentially. The temperature of the body changes rapidly at the beginning, but rather slowly later on (something similar to that shown in Figure 4.14). A large value of $b$, in Equation (4.14b), indicates that the body will approach the environment temperature $T_\infty$ in a short time. That is, the larger the value of the exponent $b$, the faster would be the temperature decay of the lumped system.

Once the temperature $T(t)$ at time $t$ is available from Equation (4.14), the rate of convection heat transfer $\dot{Q}(t)$ between the body and the environment at that time can be determined from Newton's law of cooling,

$$\boxed{\dot{Q}(t) = hA\left(T(t) - T_\infty\right) \text{ W}}$$

The maximum energy transfer possible when a body at temperature $T_i$ is exposed at $t = 0$ to an environment at temperature $T_\infty$ is the amount of heat transfer between the body and the environment from $t = 0$ to the instant $t = \infty$ at which the body comes to thermal equilibrium with the environment. In other words, the *total amount* of heat transfer $Q$ between the body, initially at a uniform temperature of $T_i$, before exposure to the surrounding, and the surrounding medium at $T_\infty$ is simply the change in the energy content of the body, over the time interval from the beginning of exposure at $t = 0$ to $t = \infty$, at which thermal equilibrium between the body and the surrounding is established. That is,

$$\boxed{Q = mc\,(T_i - T_\infty)\ \text{kJ}}$$

This expression can be obtained by substituting the $T(t)$ relation from Equation (4.14a) into the $\dot{Q}(t)$ expression above and integrating it from $t = 0$ to $t \to \infty$. That is,

$$\frac{dQ}{dt} = hA\left(T(t) - T_\infty\right)$$

From Equation (4.14a),

$$T(t) = (T_i - T_\infty)\,e^{-bt} + T_\infty$$

Substituting this, we get

$$\begin{aligned}\frac{dQ}{dt} &= hA\left((T_i - T_\infty)\,e^{-bt} + T_\infty - T_\infty\right)\\ &= hA\,(T_i - T_\infty)\,e^{-bt}\\ dQ &= hA\,(T_i - T_\infty)\,e^{-bt}\,dt\end{aligned}$$

Integrating between $t = 0$ to $t = \infty$, we get

$$\begin{aligned}\int dQ &= \int_0^\infty hA\,(T_i - T_\infty)\,e^{-bt}\,dt\\ Q &= hA\,(T_i - T_\infty)\int_0^\infty e^{-bt}\,dt\\ &= hA\,(T_i - T_\infty)\left[\frac{1}{b}\right]\\ &= hA\,(T_i - T_\infty)\left[\frac{\rho \mathbb{V} c}{hA}\right]\\ &= \rho \mathbb{V} c\,(T_i - T_\infty)\end{aligned}$$

But $\rho\mathbb{V} = m$, thus,

$$\boxed{Q = mc\,(T_i - T_\infty)}$$

### 4.3.1 Convection Boundary Condition

In many practical environments the transient heat-conduction problem is connected with a boundary condition at the surface of the solid. For such problems the boundary conditions for the differential equation governing the heat transfer should be modified to account for the convection heat transfer at the surface. For the semi-infinite solid (Figure 4.1) problem, this would be given by

$$hA\left(T_\infty - T\right)_{x=0} = -kA\left.\frac{\partial T}{\partial x}\right|_{x=0} \tag{4.16}$$

Schneider [3] solved this involved equation and expressed the temperature distribution as

$$\frac{T - T_i}{T_\infty - T_i} = 1 - \operatorname{erf}(\chi) - \left[\exp\left(\frac{hx}{k} + \frac{h^2\alpha t}{k^2}\right)\right]\left[1 - \operatorname{erf}\left(\chi + \frac{h\sqrt{\alpha t}}{k}\right)\right] \tag{4.17}$$

where $\chi = x/(2\sqrt{\alpha t})$, $T_i$ is the initial temperature of the solid, $T_\infty$ is the environmental temperature, and $t$ is the time. This solution is presented in graphical form in Figure 4.11.

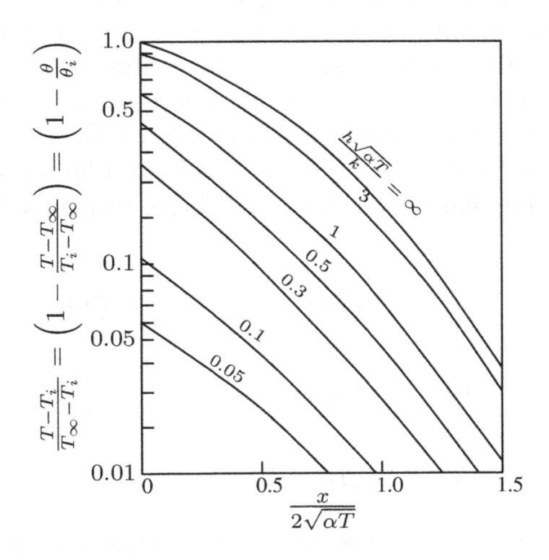


**Figure 4.11**
Temperature distribution in a semi-infinite solid with convection boundary condition.

The temperature distribution relation given by Equation (4.17) has been solved for many geometries of practical importance. Some of the shapes of interest from an application point of view are plates with its thickness significantly smaller than the width and length, cylinder with its diameter smaller compared to its length, and sphere. The thin plate, slender cylinder and sphere, described above, are schematically shown in Figure 4.12.

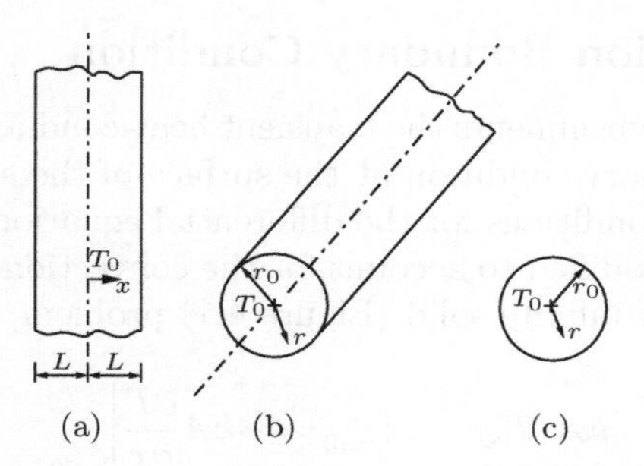


**Figure 4.12**
One-dimensional solids initially at uniform Temperature $T_i$, suddenly subjected to convection environment at $T_\infty$, (a) infinite plate of thickness $2L$, (b) infinite cylinder of radius $r_0$, (c) sphere of radius $r_0$.

The temperature distribution in the geometries shown in Figure 4.12 have been presented by Heisler [1]. In all these cases, the temperature of the convection environment is designated as $T_\infty$, and the temperature at the center plane (at $x = 0$) of infinite slab, at the center line (at $r = 0$) of slender cylinder and at the center (at $r = 0$) of sphere is $T_0$. At time $t = 0$, each solid is assumed to be at a uniform temperature of $T_i$. Temperature distribution in these solids are given in Figures 4.2 and 4.3, Figures 4.5 and 4.6, and Figures 4.8 and 4.9, for plate, cylinder and sphere, respectively. In these plots,

$$\theta = (T(x,t) - T_\infty) \text{ or } (T(r,t) - T_\infty)$$

$$\theta_i = (T_i - T_\infty)$$

$$\theta_0 = (T_0 - T_\infty)$$

If a centerline temperature is to be determined, only one chart would be sufficient to obtain a value for $\theta_0$ and then $T_0$, using the $\theta_0$ obtained. To determine an off-center temperature, two charts are required to calculate the product as

$$\frac{\theta}{\theta_i} = \left(\frac{\theta_0}{\theta_i}\right) \times \left(\frac{\theta}{\theta_0}\right)$$

The heat loss for the infinite plate, infinite cylinder, and sphere are given in Figures 4.4, 4.7 and 4.10, respectively. In these plots, $Q_0$ is the initial internal energy content of the body with reference to the environment temperature, given as

$$Q_0 = \rho c \mathbb{V} (T_i - T_\infty) = \rho c \mathbb{V} \theta_i \tag{4.18}$$

and $Q$ is the actual heat lost by the body in time $t$, given by

$$Q = \rho c \mathbb{V} \left(T_t - T_\infty\right)$$

where $T_t$ temperature at time $t$.

## Example 4.2

A semi-infinite aluminum cylinder of 70 mm diameter initially at 180°C is suddenly exposed to a convection environment at 50°C with convection heat transfer coefficient of 500 W/(m$^2$ °C). Taking the conduction coefficient and the thermal diffusivity of aluminum as 210 W/(m °C) and $8.4 \times 10^{-5}$ m$^2$/s, respectively, determine the temperatures at the axis and surface of the cylinder at 120 mm from the end, at 80 seconds after exposing to the convection environment.

## Solution

This problem may be solved by combining the solutions of an infinite cylinder $[C(\theta)]$ and semi-infinite slab $[S(X)]$.

Given, $r_0 = 35$ mm, $k = 210$ W/(m °C), $h = 500$ W/(m$^2$ °C), $\alpha = 8.4 \times 10^{-5}$ m$^2$/s, $x = 120$ mm, $t = 80$ s, $T_i = 180$°C and $T_\infty = 50$°C.

Thus,

$$\frac{h\sqrt{\alpha t}}{k} = \frac{500 \times \sqrt{(8.4 \times 10^{-5})(80)}}{210} = 0.195$$

$$\frac{x}{2\sqrt{\alpha t}} = \frac{0.12}{2 \times \sqrt{(8.4 \times 10^{-5})(80)}} = 0.732$$

From Figure 4.11, for $\frac{h\sqrt{\alpha t}}{k} = 0.195$ and $\frac{x}{2\sqrt{\alpha t}} = 0.732$, we have

$$1 - \frac{T - T_\infty}{T_i - T_\infty} = 0.05$$

Therefore,

$$\frac{T - T_\infty}{T_i - T_\infty} = 1 - 0.05 = 0.95$$

But

$$\frac{T - T_\infty}{T_i - T_\infty} = \left(\frac{T - T_\infty}{T_0 - T_\infty}\right) \times \left(\frac{T_0 - T_\infty}{T_i - T_\infty}\right)$$

$$= (\theta) \times \left(\frac{1}{\theta_i}\right)$$

Thus,

$$\left(\frac{\theta}{\theta_i}\right)_{\text{semi-infinite slab}} = 0.95$$

$$= S(X)$$

For the infinite cylinder the temperature ratios at the axis and surface are to be determined. We have

$$\frac{k}{hr_0} = \frac{210}{500 \times 0.035}$$

$$= 12$$

$$\frac{\alpha t}{r_0^2} = \frac{8.4 \times 10^{-5} \times 80}{0.035^2}$$

$$= 5.49$$

For $\dfrac{k}{hr_0} = 12$ and $\dfrac{\alpha t}{r_0^2} = 5.49$, from Figure 4.5, the temperature ratio at the axis ($r = 0$) is

$$\frac{\theta_0}{\theta_i} = 0.55$$

Now, for $\dfrac{k}{hr_0} = 12$ and $\dfrac{r}{r_0} = 1$, from Figure 4.6, the surface temperature ratio is

$$\frac{\theta}{\theta_0} = 0.95$$

Thus, at $r = r_0$,

$$\frac{\theta}{\theta_i} = \left(\frac{\theta}{\theta_0}\right) \times \left(\frac{\theta_0}{\theta_i}\right)$$

$$= 0.95 \times 0.55$$

$$= 0.522$$

$$= C(\theta)$$

Now, combining the solutions for the semi-infinite cylinder and infinite slab, at the surface with $r = r_0$, we have

$$\begin{aligned}\left(\frac{\theta}{\theta_i}\right)_{\text{surface of cyl.}} &= C(\theta) \times S(X) \\ &= 0.522 \times 0.95 \\ &= 0.5\end{aligned}$$

At the center with $r = 0$,

$$\begin{aligned}\left(\frac{\theta}{\theta_i}\right)_{\text{center of cyl.}} &= 0.55 \times 0.95 \\ &= 0.5225\end{aligned}$$

Thus, the temperature at the axis of the cylinder, at 80 seconds after exposure to convection environment is

$$\begin{aligned}T_{\text{axis}} &= T_\infty + 0.5225(T_i - T_\infty) \\ &= 50 + 0.5225 \times (180 - 50) \\ &= \boxed{117.93^\circ\text{C}}\end{aligned}$$

The temperature at the surface of the cylinder at 80 seconds after exposure to convection environment is

$$\begin{aligned}T_{\text{surface}} &= T_\infty + 0.5(T_i - T_\infty) \\ &= 50 + 0.50 \times (180 - 50) \\ &= \boxed{115^\circ\text{C}}\end{aligned}$$

### 4.3.2 Criteria for Lumped System Analysis

From the discussions in Section 4.2.1, it is evident that the first step in establishing a criterion for the applicability of the lumped system analysis is to define a *characteristic length* as

$$L = \frac{\text{Volume of the system}}{\text{Exposed area}} = \frac{\mathbb{V}}{A}$$

and Biot number Bi as

$$\text{Bi} = \frac{hL}{k}$$

The Biot number can be expressed as

$$\text{Bi} = \frac{h}{k/L}\frac{\Delta T}{\Delta T} = \frac{\text{Convection at the body surface}}{\text{Conduction within the body}}$$

The Biot number can also be expressed as

$$\text{Bi} = \frac{L/k}{1/h} = \frac{\text{Conduction resistance within the body}}{\text{Convection resistance at the body surface}}$$

Thus, the Biot number is *the ratio of the resistance to heat conduction within a body to the resistance to heat convection surrounding the body.* The conduction coefficient $k$ and the convection coefficient $h$ in the Biot number expression are the conduction coefficient of the material of the solid body and the convective heat transfer coefficient of the fluid surrounding the body, respectively.

When a solid body is heated by a hotter fluid surrounding it, heat is first *convected* from the fluid to the surface of the body, and subsequently *conducted* from the body surface to the interior portions within the body. The Biot number is the *ratio* of the internal resistance of a body to *heat conduction* to the resistance to *heat convection* from the fluid to the body. Therefore, a small Biot number represents a small resistance to heat conduction, and thus a small temperature gradient within the body. In other words, it can be stated that, *when the Biot number is small, the body can be assumed to be at a uniform temperature throughout.*

Lumped system analysis which assumes a *uniform* temperature throughout the body, is valid only when the thermal resistance of a body to heat conduction is zero. Thus, lumped system analysis is *exact* when $\text{Bi} = 0$, and for $\text{Bi} > 0$ it is only *approximate.* Thus, the smaller the Bi the more accurate is the lumped system analysis. Therefore, it is essential to identify the limiting value of Bi below which the analysis can be taken as reasonably accurate. It is generally accepted that, lumped system analysis is *applicable* if the "temperature within the body relative to the surroundings, that is, $(T - T_\infty)$ remains within 5 percent of each other, even for well-rounded geometries such as a spherical ball." This marginal difference of temperature is found to correspond to $\text{Bi} \approx 0.1$. Thus, when $\text{Bi} < 0.1$, the variation of temperature with location within the body will be marginal and can reasonably be approximated to be uniform.

## Example 4.3

A 3-cm diameter steel ball is initially at a uniform temperature of 300°C. It is suddenly placed in a controlled environment in which the temperature is maintained at 80°C. If the convection heat transfer coefficient of the environment is 10 W/(m$^2$ K), calculate the time required for the ball to attain a temperature of 100°C. For steel, $c = 0.46$ kJ/(kg K) and $k = 35$ W/(m K). Take the density of steel to be 7800 kg/m$^3$.

## Solution

Given, $T_0 = 300°\text{C}$, $T_\infty = 80°\text{C}$, $d = 0.03$ m, $h = 10$ W/(m$^2$ K). Therefore,

$$\begin{aligned} \theta &= \frac{T - T_\infty}{T_0 - T_\infty} \\ &= \frac{100 - 80}{300 - 80} \\ &= 0.09 \end{aligned}$$

$$\begin{aligned} \frac{hA}{\rho \mathbb{V} c} &= \frac{h\,(4\pi r^2)}{\rho\left(\frac{4}{3}\pi r^3\right) c} \\ &= \frac{10(4\pi \times 0.015^2)}{7800 \times \frac{4}{3}\pi(0.015^3) \times 460} \\ &= 5.574 \times 10^{-4}\,\text{s}^{-1} \end{aligned}$$

By Equation (4.14),

$$\theta = e^{-\left(\frac{hA}{\rho \mathbb{V} c} t\right)}$$

Taking log on both sides, we have

$$\ln \theta = -\frac{hA}{\rho \mathbb{V} c} t$$

$$\begin{aligned} \ln\,(0.09) &= -\left(5.574 \times 10^{-4}\right) t \\ -2.408 &= -\left(5.574 \times 10^{-4}\right) t \\ t &= \frac{2.408}{0.0005574} \\ &= 4320\,\text{s} \\ &= \boxed{1.2\,\text{h}} \end{aligned}$$

## Example 4.4

Determine the response time of mercury in a glass thermometer of 3 mm bulb diameter in an environment with heat transfer coefficient of 10 W/(m$^2$ K). Take $k = 10$ W/(m K), and $\alpha = 5 \times 10^{-5}$ m$^2$/s, for mercury.

## Solution

Given, $d = 0.003$ m, $h = 10$ W/(m$^2$ K), $k = 10$ W/(m K), $\alpha = \dfrac{k}{\rho c} = 5 \times 10^{-5}$ m$^2$/s.

Therefore,

$$
\begin{aligned}
\mathrm{Bi} &= \frac{hL}{k} = \frac{10 \times 1.5 \times 10^{-3}}{10} \\
&= 1.5 \times 10^{-3}
\end{aligned}
$$

where $L$ is the radius of the bulb. The Bi is very small; therefore, the characteristic time, by Equation (4.15), is

$$
\begin{aligned}
\tau &= \frac{\rho \mathbb{V} c}{hA} = \frac{\mathbb{V} k}{Ah\left(\dfrac{k}{\rho c}\right)} \\
&= \frac{\mathbb{V} k}{Ah\alpha} \\
&= \frac{\frac{4}{3}\pi\left(1.5 \times 10^{-3}\right)^3}{4\pi\left(1.5 \times 10^{-3}\right)^2} \times \frac{10}{10 \times \left(5 \times 10^{-5}\right)} \\
&= \boxed{10\,\mathrm{s}}
\end{aligned}
$$

The time required for the thermometer to come to an equilibrium with the system, in which the bulb is placed, is 10 seconds.

### 4.3.3 Penetration Depth-Significance of Thermal Diffusivity

Examine the heat conduction through a semi-infinite solid, shown in Figure 4.13.

Since such a solid extends to infinity in all but one direction, it is characterized by a single identifiable surface at the left face of the solid, as shown in Figure 4.13. If a sudden change of temperature is imposed at this surface (at $x = 0$), transient, one-dimensional conduction will take place within the

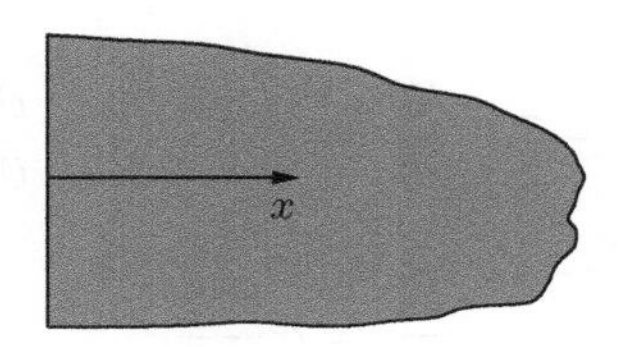


**Figure 4.13**
Heat conduction through a semi-infinite solid.

solid. Let the initial temperature of the body be $T_0$. At time $t = 0$, the face at $x = 0$ is raised to temperature $T_s$ and maintained at the same level. For this one-dimensional, unsteady conduction in a semi-infinite solid, the governing Equation (4.1) becomes

$$\frac{\partial T}{\partial t} = \alpha \frac{\partial^2 T}{\partial x^2} \tag{4.18}$$

The initial conditions are

$$T(x,t) = T_0, \quad \text{for} \quad t < 0$$

$$T(0,t) = T_s, \quad \text{for} \quad t \geq 0$$

The boundary condition is

$$T(x,t) = T_0, \quad \text{for} \quad x \to \infty$$

The dimensionless temperature $\theta$ is

$$\theta = \frac{(T - T_0)}{(T_s - T_0)}$$

Therefore,

$$T = (T_s - T_0)\,\theta + T_0$$

Differentiating $T$ with respect to time $t$, we get

$$\frac{\partial T}{\partial t} = (T_s - T_0)\,\frac{\partial \theta}{\partial t}$$

Now, differentiating $T$ with respect to $x$, we get

$$\frac{\partial T}{\partial x} = (T_s - T_0)\,\frac{\partial \theta}{\partial x}$$

$$\frac{\partial^2 T}{\partial x^2} = (T_s - T_0)\,\frac{\partial^2 \theta}{\partial x^2}$$

Substituting these $\frac{\partial T}{\partial t}$ and $\frac{\partial^2 T}{\partial x^2}$ into Equation (4.18), we get

$$(T_s - T_0)\frac{\partial \theta}{\partial t} = \alpha (T_s - T_0)\frac{\partial^2 \theta}{\partial x^2}$$

That is,

$$\frac{\partial \theta}{\partial t} = \alpha \frac{\partial^2 \theta}{\partial x^2} \tag{4.19}$$

The new initial and boundary conditions are

$$\theta(x,t) = 0, \quad \text{for} \quad t < 0$$

$$\theta(0,t) = 1, \quad \text{for} \quad t \geq 0$$

and

$$\theta(x,t) = 0, \quad \text{as} \quad x \to \infty$$

From Equation (4.19), we see that $\theta$ is a function of $x$, $t$ and $\alpha$. But $\theta$ is nondimensional; therefore, $x$, $t$ and $\alpha$ also should occur only as a dimensionless group. By inspection, we can see that $x/\sqrt{\alpha t}$ is the only such possible group. Therefore, let $\eta = x/\sqrt{4\,\alpha t}$, where 4 is introduced in the denominator for convenience. In terms of $\eta$, Equation (4.19) becomes

$$\frac{d^2\theta}{d\eta^2} + 2\eta\frac{d\theta}{d\eta} = 0 \tag{4.20}$$

For the given conditions, the solution of Equation (4.20) is

$$\boxed{\theta = 1 - \frac{2}{\sqrt{\pi}}\int_0^{\eta} e^{-\eta^2}\,d\eta} \tag{4.21}$$

But, by definition,

$$\frac{2}{\sqrt{\pi}}\int_0^{\eta} e^{-\eta^2}\,d\eta$$

is the error function erf $(\eta)$. Thus,

$$\theta = 1 - \text{erf}\,(\eta)$$

The right-hand side of Equation (4.21) is the *complimentary* error function, erfc $(\eta)$, defined as

$$\text{erfc}\,(\eta) \equiv 1 - \text{erf}\,(\eta)$$

where erf $(\eta)$ is the error function. From Equation (4.21) it is seen that, at $\eta = 0$, $\theta = 1$ but $\theta$ tends to zero as $\eta$ becomes large.

For $\eta = 1.8$, $\theta = 0.01$. Thus, at a depth of $x/\sqrt{4\alpha t} = 1.8$ or at $x = 1.8\sqrt{4\alpha t}$, the body temperature is almost at the initial temperature value. That is, the temperature $T_s$ imposed at the surface at $x = 0$ has penetrated only up to a distance of $x = 1.8\sqrt{4\alpha t}$, from the surface, in time $t$. This depth up to which the heat conduction due to the temperature gradient $(T_s - T_i)$ imposed at $x = 0$ has reached is called the *penetration depth*. The penetration depth increases as the square root of time $t$. The transient temperature distribution in a semi-infinite solid (for constant surface temperature) is given in Figure 4.14.

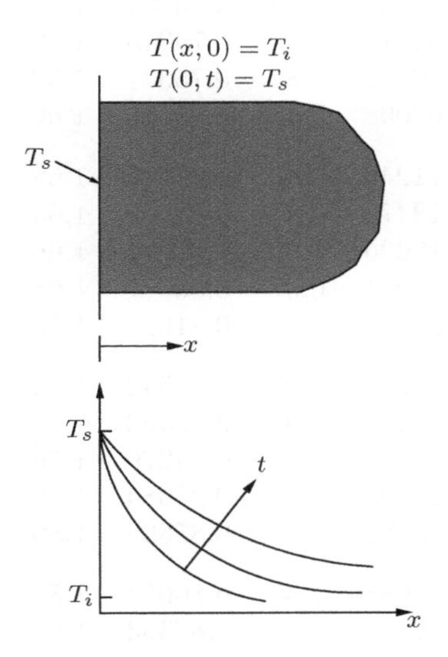


**Figure 4.14**
Transient temperature distribution in a semi-infinite solid for constant surface temperature distribution.

The error function erf $(\eta)$, defined as

$$\text{erf}\,(\eta) = \frac{2}{\sqrt{\pi}} \int_0^{\eta} e^{-\eta^2}\, d\eta$$

is evaluated for different values of $\eta$ as listed in Table 4.1.

### 4.3.4 Surface Heat Flux from a Semi-Infinite Solid

Equation (4.21) can be expressed as

$$\begin{aligned} \text{erf}\,(\eta) &= 1 - \theta \\ &= 1 - \frac{T - T_0}{T_s - T_0} \\ &= \frac{T_s - T}{T_s - T_0} \end{aligned}$$

**Table 4.1** The error function

| $\eta$ | erf $\eta$ | $\eta$ | erf $\eta$ | $\eta$ | erf $\eta$ |
|---|---|---|---|---|---|
| 0.00 | 0.00000 | 0.76 | 0.71754 | 1.52 | 0.96841 |
| 0.02 | 0.02256 | 0.78 | 0.73001 | 1.54 | 0.97059 |
| 0.04 | 0.04511 | 0.80 | 0.74210 | 1.56 | 0.97263 |
| 0.06 | 0.06762 | 0.82 | 0.75381 | 1.58 | 0.97455 |
| 0.08 | 0.09008 | 0.84 | 0.76514 | 1.60 | 0.97636 |
| 0.10 | 0.11246 | 0.86 | 0.77610 | 1.62 | 0.97804 |
| 0.12 | 0.13476 | 0.88 | 0.78669 | 1.64 | 0.97962 |
| 0.14 | 0.15695 | 0.90 | 0.79691 | 1.66 | 0.98110 |
| 0.16 | 0.17901 | 0.92 | 0.80677 | 1.68 | 0.98249 |
| 0.18 | 0.20094 | 0.94 | 0.81627 | 1.70 | 0.98379 |
| 0.20 | 0.22270 | 0.96 | 0.82542 | 1.72 | 0.98500 |
| 0.22 | 0.24430 | 0.98 | 0.83423 | 1.74 | 0.98613 |
| 0.24 | 0.26570 | 1.00 | 0.84270 | 1.76 | 0.98719 |
| 0.26 | 0.28690 | 1.02 | 0.85084 | 1.78 | 0.98817 |
| 0.28 | 0.30788 | 1.04 | 0.85865 | 1.80 | 0.98909 |
| 0.30 | 0.32683 | 1.06 | 0.86614 | 1.82 | 0.98994 |
| 0.32 | 0.34913 | 1.08 | 0.87333 | 1.84 | 0.99074 |
| 0.34 | 0.36936 | 1.10 | 0.88020 | 1.86 | 0.99147 |
| 0.36 | 0.38933 | 1.12 | 0.88079 | 1.88 | 0.99216 |
| 0.38 | 0.40901 | 1.14 | 0.89308 | 1.90 | 0.99279 |
| 0.40 | 0.42839 | 1.16 | 0.89910 | 1.92 | 0.99338 |
| 0.42 | 0.44749 | 1.18 | 0.90484 | 1.94 | 0.99392 |
| 0.44 | 0.46622 | 1.20 | 0.91031 | 1.96 | 0.99443 |
| 0.46 | 0.48466 | 1.22 | 0.91553 | 1.98 | 0.99489 |
| 0.48 | 0.50275 | 1.24 | 0.92050 | 2.00 | 0.99532 |
| 0.50 | 0.52050 | 1.26 | 0.92524 | 2.10 | 0.99702 |
| 0.52 | 0.53790 | 1.28 | 0.92973 | 2.20 | 0.99813 |
| 0.54 | 0.55494 | 1.30 | 0.93401 | 2.30 | 0.99885 |
| 0.56 | 0.57162 | 1.32 | 0.93806 | 2.40 | 0.99931 |
| 0.58 | 0.58792 | 1.34 | 0.94191 | 2.50 | 0.99959 |
| 0.60 | 0.60386 | 1.36 | 0.94556 | 2.60 | 0.99976 |
| 0.62 | 0.61941 | 1.38 | 0.94902 | 2.70 | 0.99986 |
| 0.64 | 0.63459 | 1.40 | 0.95228 | 2.80 | 0.99992 |
| 0.66 | 0.64938 | 1.42 | 0.95538 | 2.90 | 0.99995 |
| 0.68 | 0.66278 | 1.44 | 0.95830 | 3.00 | 0.99997 |
| 0.70 | 0.67780 | 1.46 | 0.96105 | 3.20 | 0.99999 |
| 0.72 | 0.69143 | 1.48 | 0.96365 | 3.40 | 0.99999 |
| 0.74 | 0.70468 | 1.50 | 0.96610 | 3.60 | 1.00000 |

or

$$\text{erf}\,(\eta) = \frac{T - T_s}{T_0 - T_s}$$

where $T$ is the temperature at a depth $x$ at any given time $t$, that is, $T = T(x,t)$, $T_0$ is the initial temperature and $T_s$ is the new temperature to which the surface is suddenly exposed. Thus,

$$\frac{T(x,t) - T_s}{T_0 - T_s} = \frac{2}{\sqrt{\pi}} \int_0^{x/\sqrt{4\alpha t}} e^{-\eta^2}\, d\eta \tag{4.21$a$}$$

The heat flow at any $x$ position may be obtained from

$$\dot{Q}_x = -kA\frac{\partial T}{\partial x}$$

Performing the partial differentiation of Equation (4.21a), we get

$$\frac{\partial T}{\partial x} = (T_0 - T_s)\,\frac{2}{\sqrt{\pi}}\,e^{-x^2/(4\alpha t)}\,\frac{\partial}{\partial x}\left(\frac{x}{\sqrt{4\alpha t}}\right)$$

that is,

$$\frac{\partial T}{\partial x} = \frac{(T_0 - T_s)}{\sqrt{\pi \alpha t}}\,e^{-x^2/(4\alpha t)} \tag{4.22}$$

Substituting this into the heat flux relation above, we get

$$\dot{Q}_x = -\frac{kA\,(T_0 - T_s)}{\sqrt{\pi \alpha t}}e^{-x^2/(4\alpha t)}$$

Therefore, the heat flow at the surface where $x = 0$ becomes

$$\dot{Q}_0 = \frac{kA\,(T_s - T_0)}{\sqrt{\pi \alpha t}} \tag{4.23}$$

The surface heat flux $\dot{q}_0 = \dot{Q}_0/A$ can be determined by evaluating the temperature gradient at $x = 0$ from Equation (4.22). A plot of the temperature distribution for a semi-infinite solid will be as shown in Figure 4.14.

## Example 4.5

A large aluminum slab at a uniform temperature of 180°C is exposed to a cool environment, which suddenly lowers the surface temperature of the slab to 50°C. Determine the total heat removed from the slab per unit surface area, when the temperature at a depth 20 mm from the surface has dropped to 100°C. Take the conduction coefficient and thermal diffusivity of the slab as 210 W/(m °C) and $8.4 \times 10^{-5}$ m$^2$/s, respectively.

## Solution

Given, $k = 210$ W/(m °C), $\alpha = 8.4 \times 10^{-5}$ m$^2$/s, $T_0 = 180$°C, $T_s = 50$°C, $x = 20$ mm and $T(x,t) = 100$°C. Therefore,

$$\begin{aligned} \theta &= \frac{T(x,t) - T_s}{T_0 - T_s} \\ &= \frac{100 - 50}{180 - 50} \\ &= 0.3846 \end{aligned}$$

But

$$\theta = 1 - \text{erf}\left(\frac{x}{\sqrt{4\alpha t}}\right)$$

Therefore,

$$\begin{aligned} \text{erf}\left(\frac{x}{\sqrt{4\alpha t}}\right) &= 1 - 0.3846 \\ &= 0.6154 \end{aligned}$$

From Table 4.1, for this value of error function, we have

$$\eta = \frac{x}{\sqrt{4\alpha t}} = 0.61226$$

Thus,

$$\begin{aligned} t &= \frac{x^2}{4\alpha \times 0.61226^2} \\ &= \frac{0.02^2}{0.61226^2} \times \frac{1}{4 \times (8.4 \times 10^{-5})} \\ &= 3.176\,\text{s} \end{aligned}$$

That is, in 3.176 seconds the temperature of the slab at a depth of 20 mm from the surface decreases to 100°C.

The amount of heat removed from the slab, per unit surface area in 3.176 seconds, is obtained by integrating Equation (4.23), as

$$\begin{aligned} \frac{Q_0}{A} &= \int_0^t \frac{k\,(T_s - T_0)}{\sqrt{\pi \alpha t}}\, dt \\ &= 2k\,(T_s - T_0)\sqrt{\frac{t}{\pi\alpha}} \end{aligned}$$

$$= \quad 2 \times 210 \times (50 - 180) \times \sqrt{\frac{3.176}{\pi \times (8.4 \times 10^{-5})}}$$

$$= \quad \boxed{-599\,\mathrm{kJ/m^2}}$$

## 4.3.5 The Significance of Thermal Diffusivity $\alpha$

For solids with high thermal diffusivity $\alpha$, the conductivity is large but the heat capacity per unit volume is small. Therefore, when a temperature gradient is imposed at the surface of a body for which $\alpha$ is high, only a small amount of the heat flowing in is retained by the layers within the solid and the rest of the heat is conducted inwards. Thus, the process of conduction in a medium with low thermal resistance results in larger penetration depth compared to a solid for which $\alpha$ is small.

By Equation (4.21), the time $t_p$ required for the temperature $T_s$ imposed at the surface of a semi-infinite slab to reach a depth, say $l$, from the surface is given by $\eta = 1.8$. That is,

$$\frac{l}{\sqrt{4\alpha t_p}} = 1.8$$

Thus, the penetration time becomes

$$\boxed{t_p = \frac{l^2}{13\alpha}} \tag{4.24}$$

That is, the penetration time $t_p$ varies as $l^2/\alpha$.

## Example 4.6

The surface of a large block of steel with $k = 45$ W/(m K) and $\alpha = 1.4 \times 10^{-5}$ m$^2$/s, initially at a uniform temperature of 30°C, is subjected to a heat flux by suddenly raising the surface temperature to 200°C. Calculate the temperature at a depth of 1 cm from the surface, after the elapse of 30 seconds.

## Solution

Given, $T_0 = 30$°C, $T_s = 200$°C, $t = 30$ s and $x = 0.01$ m.

Therefore,

$$\eta \quad = \quad \frac{x}{\sqrt{4\alpha t}}$$

$$= \quad \frac{0.01}{\sqrt{4 \times 1.4 \times 10^{-5} \times 30}}$$

$$= \quad 0.24$$

The value of erf (0.24) from Table 4.1 is 0.2657.

By Equation (4.21),

$$\theta = 1 - \text{erf}\,(\eta)$$

Thus,

$$\begin{aligned}
\theta &= \frac{T - T_0}{T_s - T_0} \\
&= 1 - \text{erf}\,(0.24) \\
&= 1 - 0.2657 \\
&= 0.7343
\end{aligned}$$

Therefore,

$$\begin{aligned}
\frac{T - 30}{200 - 30} &= 0.7343 \\
T &= 0.7343 \times (200 - 30) + 30 \\
&= \boxed{154.83°\text{C}}
\end{aligned}$$

## Example 4.7

In a place, the ground is covered with snow at a temperature of $-10°$C, continuously for two months in winter. The average soil properties at that place are $k = 0.4$ W/(m °C) and $\alpha = 0.15 \times 10^{-6}$ m$^2$/s. Assuming an initial uniform ground temperature of 15°C at the beginning of winter when the snow covers the ground, determine the minimum depth from the ground surface at which a water pipe is to be laid to prevent freezing of water.

## Solution

Given, $T_\infty = -10°$C, $T_0 = 15°$C, $t = 2$ months.

When the environmental temperature remains below 0°C for a prolonged period of time, the water pipe lines become a major concern, since water expands on freezing, leading to bursting of water pipes. But for underground water pipes, the soil remains relatively warm during this situation, and it takes weeks for the freezing temperature to reach the water pipe laid underground.

Thus, the soil essentially serves as an insulation and protects the water from freezing.

The soil temperature is affected only by the thermal conditions at the surface. Thus, the soil can be considered to be a semi-infinite medium with a steady surface temperature of $-10°C$. The water in the pipe will not freeze as long as the temperature is above $0°C$. Thus, the problem is essentially finding the time required for the water pipe surface temperature to decrease from $15°C$ to $0°C$. The nondimensional temperature is

$$\begin{aligned}\theta &= \frac{T - T_\infty}{T_0 - T_\infty} \\ &= \frac{0 - (-10)}{15 - (-10)} \\ &= 0.4\end{aligned}$$

By Equation (4.21),

$$\theta = \text{erfc}\,(\eta) = 1 - \text{erf}\,(\eta)$$

Therefore,

$$\begin{aligned}0.4 &= \text{erfc}\,(\eta) = 1 - \text{erf}\,(\eta) \\ \text{erf}\,(\eta) &= 1 - 0.4 \\ &= 0.6\end{aligned}$$

From error function table (Table 4.1), for $\text{erf}\,(\eta) = 0.6$, we have $\eta = 0.595$. But

$$\eta = \frac{x}{\sqrt{4\alpha t}}$$

Therefore,

$$\begin{aligned}x &= \eta\sqrt{4\alpha t} \\ &= 0.595 \times \sqrt{4 \times (0.15 \times 10^{-6}) \times (60 \times 24 \times 3600)} \\ &= \boxed{1.05\,\text{m}}\end{aligned}$$

Note that, the time $t$ is expressed as seconds in the above calculation.

The water pipe should be laid at a depth of at least 105 cm below the surface to avoid freezing of water during winter with a constant surface temperature of $-10°C$ for a two months' duration.

## Example 4.8

A 40-mm thick large brass plate initially at 20°C is heated by passing it through an oven maintained at 500°C. The plate remains in the oven for 10 minutes. The combined convection and radiation heat transfer coefficient is 120 W/(m$^2$ °C). Taking the properties of brass at room temperature to be $k = 110$ W/(m °C), $\rho = 8500$ kg/m$^3$, $c = 380$ J/(kg °C), and $\alpha = 34 \times 10^{-6}$ m$^2$/s, determine the surface temperature of the plate when it comes out of the oven.

## Solution

The temperature at a specified location at a given time can be determined from Heisler's chart. Given that, the half-thickness of the plate $L = 0.02$ m. For the given brass plate, we have

$$\begin{aligned} \frac{1}{\text{Bi}} &= \frac{k}{hL} \\ &= \frac{110}{120 \times 0.02} \\ &= 45.83 \end{aligned}$$

$$\begin{aligned} \text{Fo} &= t^* = \frac{\alpha t}{L^2} \\ &= \frac{34 \times 10^{-6} \times (10 \times 60)}{0.02^2} \\ &= 51 \end{aligned}$$

For these values of Bi$^{-1}$ and Fo, from Figure 4.2, we have

$$\theta^* = \frac{T_0 - T_\infty}{T_i - T_\infty} = 0.34$$

Given that, $T_\infty = 500$°C and $T_i = 20$°C. Thus,

$$\begin{aligned} T_0 &= 0.34\,(T_i - T_\infty) + T_\infty \\ &= 0.34 \times (20 - 500) + 500 \\ &= -163.2 + 500 \\ &= \boxed{336.8°\text{C}} \end{aligned}$$

The surface temperature of the plate will be 336.8°C when it comes out of the oven.

## 4.4 Multidimensional Systems

In Section 4.2, transient conduction in an infinite wall of finite thickness, in a long cylinder and sphere were discussed, treating the heat transfer process as one-dimensional. Therefore, the Heisler charts in Section 4.2 are meant for obtaining the temperature distribution in infinite plates, long cylinders and spheres. But for handling heat transfer problems associated with a wall with height and length comparable to its thickness or a cylinder with length not very large compared to its diameter, additional space coordinates need be considered while solving the governing equation, because for such geometries the one-dimensional approximation is no longer valid. Hence, the Heisler charts in Section 4.2 no longer apply. Therefore, to solve the heat transfer problems associated with these multidimensional systems, it is necessary to seek some appropriate method. This situation can be comfortably handled by combining the solutions for the one-dimensional systems, studied in Section 4.2.

Let us examine the infinite rectangular bar of finite thickness and finite width, but length extending to infinity, shown in Figure 4.15.

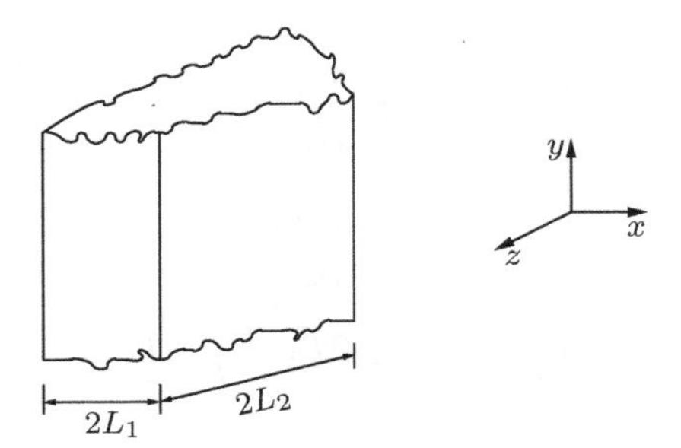


**Figure 4.15**
Infinite rectangular bar.

This rectangular bar can be visualized as a body made up of two infinite plates of thickness $2L_1$ and $2L_2$. The governing equation for transient conduction [Equation (4.1)] for this case becomes

$$\frac{\partial^2 T}{\partial x^2} + \frac{\partial^2 T}{\partial z^2} = \frac{1}{\alpha}\frac{\partial T}{\partial t} \tag{4.25}$$

To use the separation of variable method to solve this equation, let us assume a product solution of the form

$$T(x, z, t) = X(x) \times Z(z) \times \theta(t)$$

It can be shown that, the dimensionless temperature distribution may be expressed as a product of the solutions for two plate problems of thickness $2L_1$ and $2L_2$, that is,

$$\left(\frac{T - T_\infty}{T_i - T_\infty}\right) = \left(\frac{T - T_\infty}{T_i - T_\infty}\right)_{2L_1 \text{ plate}} \times \left(\frac{T - T_\infty}{T_i - T_\infty}\right)_{2L_2 \text{ plate}} \tag{4.26}$$

where $T_i$ is the initial temperature of the bar and $T_\infty$ is the temperature of the environment to which the bar is exposed.

For the two infinite plates considered, the governing equations would be

$$\frac{\partial^2 T_1}{\partial x^2} = \frac{1}{\alpha}\frac{\partial T_1}{\partial t}, \quad \frac{\partial^2 T_2}{\partial z^2} = \frac{1}{\alpha}\frac{\partial T_2}{\partial t} \tag{4.27}$$

Let us assume the solutions for these equations as

$$T_1 = T_1(x,t), \quad T_2 = T_2(z,t) \tag{4.28}$$

Now, let us show that the product of solution to Equation (4.25) can be formed by the product of the functions of $T_1$ and $T_2$, in Equation (4.28). That is,

$$T(x,z,t) = T_1(x,t) \times T_2(z,t) \tag{4.29}$$

Thus, we have

$$\begin{aligned}
\frac{\partial^2 T}{\partial x^2} &= T_2 \frac{\partial^2 T_1}{\partial x^2} \\
\frac{\partial^2 T}{\partial z^2} &= T_1 \frac{\partial^2 T_2}{\partial z^2} \\
\frac{\partial T}{\partial t} &= T_1 \frac{\partial T_2}{\partial t} + T_2 \frac{\partial T_1}{\partial t}
\end{aligned}$$

Using Equation (4.27), we have the expression for $\dfrac{\partial T}{\partial t}$ as

$$\frac{\partial T}{\partial t} = \alpha\, T_1 \frac{\partial^2 T_2}{\partial z^2} + \alpha\, T_2 \frac{\partial^2 T_1}{\partial x^2}$$

Substituting these into Equation (4.25), we get

$$T_2 \frac{\partial^2 T_1}{\partial x^2} + T_1 \frac{\partial^2 T_2}{\partial z^2} = \frac{1}{\alpha}\left(\alpha\, T_1 \frac{\partial^2 T_2}{\partial z^2} + \alpha\, T_2 \frac{\partial^2 T_1}{\partial x^2}\right)$$

Canceling $\alpha$ on the right-hand side, we get

$$\text{LHS} \equiv \text{RHS}$$

This shows that, the product solution assumed [Equation (4.29)] satisfies the governing equation for multidimensional systems, given by Equation (4.25).

This implies that, the dimensionless temperature distribution for the infinite rectangular bar in Figure 4.15 may be expressed as the product of the solutions for two plate problems of thickness $2L_1$ and $2L_2$.

In an analogous manner as above, the solution for the three-dimensional blocks of semi-infinite plate and infinite rectangular bar, shown in Figure 4.16, may be expressed as a product of the three infinite-plate solutions for plates having the thicknesses as the three sides of the block.

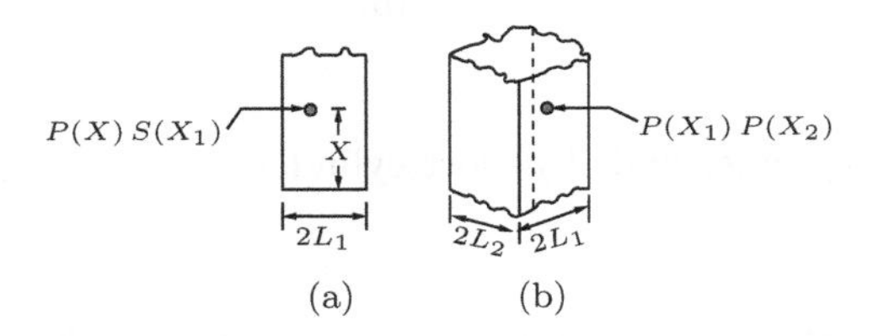


**Figure 4.16**
Semi-infinite plate and (b) infinite rectangular bar.

Similarly, a solution for a cylinder of finite length, such as the semi-infinite rectangular bar and rectangular parallelepiped, shown in Figure 4.17, can be expressed as a product of the solutions of an infinite cylinder and an infinite plate having thickness equal to the length of the cylinder. In the same way, the solution for the semi-infinite cylinder and short cylinder shown in Figure 4.18 can be obtained combining the solutions of an infinite-cylinder and an infinite-plate.

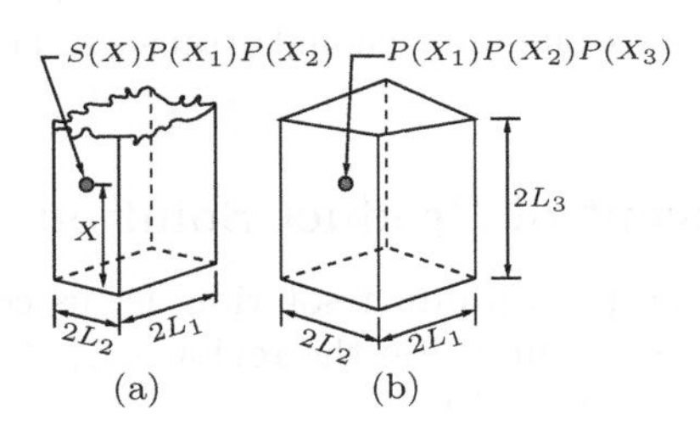


**Figure 4.17**
(a) Semi-infinite rectangular bar and (b) rectangular parallelepiped.

In Figures 4.16 - 4.18,

$$C(\theta) \;=\; \text{solution for infinite cylinder}$$

$$P(X) \;=\; \text{solution for infinite plate}$$

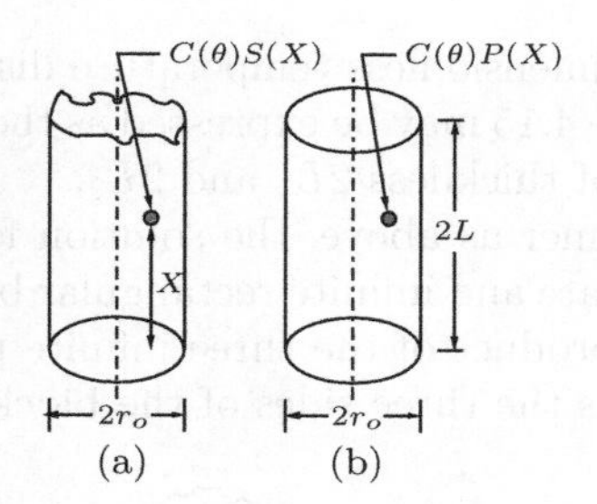


**Figure 4.18**
(a) Semi-infinite cylinder and (b) short cylinder.

$$S(X) \quad = \quad \text{solution for semi-infinite solid}$$

## 4.5 Product Solution

When the temperature gradients in more than one direction within a solid are considerable, the problem essentially becomes a two-dimensional heat conduction. For such problems, when there is no heat generation within the solid, the problems can be solved by combining the solutions of one-dimensional transient-temperature charts, by treating the problem as a suitable combination of two one-dimensional problems. Such an approach, called *method of product solution*, is applicable if the solution of a two-dimensional, time-dependent heat conduction problem can be shown to be equivalent to the product solutions of two one-dimensional, transient heat conduction problems.

### 4.5.1 The Concept of Product Solution

To understand the concept of product solution, let us consider a rectangular bar of sides $2L_1$ and $2L_2$, confined in the region $-L_1 \leq x \leq +L_1$ and $-L_2 \leq y \leq +L_2$, as shown in Figure 4.19.

Initially the slab is at a uniform temperature of $T_i$. At time $t = 0$, all the surfaces of the bar are suddenly exposed to a convection environment at a constant temperature $T_\infty$. In terms of the dimensionless temperature

$$\theta(x, y, t) = \frac{T(x, y, t) - T_\infty}{T_i - T_\infty}$$

the governing equation for the heat conduction in the rectangular bar becomes

$$\frac{\partial^2 \theta}{\partial x^2} + \frac{\partial^2 \theta}{\partial y^2} = \frac{1}{\alpha} \frac{\partial \theta}{\partial t} \tag{4.30}$$

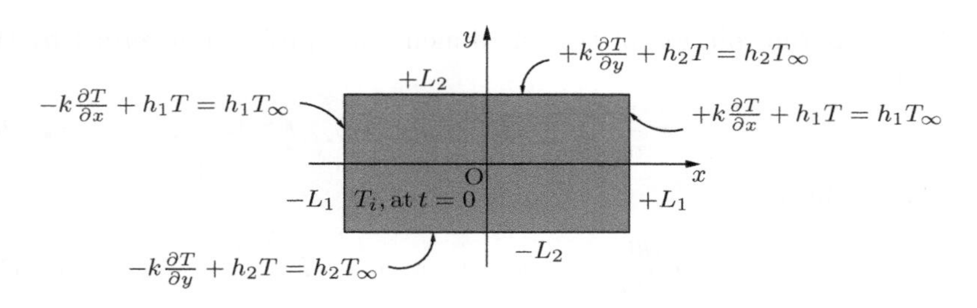


**Figure 4.19**
Transient heat conduction in a rectangular bar.

This is valid for $-L_1 < x < +L_1$, $-L_2 < y < +L_2$ and $t > 0$. Thus, at the boundary surfaces at $-L_1$, $+L_1$, $-L_2$ and $+L_2$, we have

$$-k\frac{\partial \theta}{\partial x} + h_1\theta = 0, \text{ at } x = -L_1 \tag{4.30a}$$

$$+k\frac{\partial \theta}{\partial x} + h_1\theta = 0, \text{ at } x = +L_1 \tag{4.30b}$$

$$-k\frac{\partial \theta}{\partial y} + h_2\,\theta = 0, \text{ at } x = -L_2 \tag{4.30c}$$

$$+k\frac{\partial \theta}{\partial y} + h_2\,\theta = 0, \text{ at } x = +L_2 \tag{4.30d}$$

Also,

$$\theta = 1, \text{ for } t = 0 \tag{4.30e}$$

The solution for this two-dimensional problem can be expressed as a product of the solutions $\theta_1(x,t)$ and $\theta_2(y,t)$ of two one-dimensional problems, in the form

$$\theta\,(x,y,t) = \theta_1(x,t) \times \theta_2(y,t)$$

where $\theta_1(x,t)$ is the solution of the one-dimensional problem governed by the equation

$$\frac{\partial^2\theta_1}{\partial x^2} = \frac{1}{\alpha}\frac{\partial \theta_1}{\partial t}, \quad -L_1 < x < +L_1,\ t > 0 \tag{4.31a}$$

with the conditions

$$-k\frac{\partial \theta_1}{\partial x} + h_1\theta_1 = 0, \text{ at } x = -L_1 \tag{4.31b}$$

$$+k\frac{\partial \theta_1}{\partial x} + h_1\theta_1 = 0, \text{ at } x = +L_1 \tag{4.31c}$$

$$\theta_1 = 0, \text{ for } t = 0 \tag{4.31d}$$

and $\theta_2(y,t)$ is the solution of the one-dimensional problem governed by the equation

$$\frac{\partial^2\theta_2}{\partial y^2} = \frac{1}{\alpha}\frac{\partial\theta_2}{\partial t}, \; -L_2 < y < +L_2, \; t > 0 \tag{4.32a}$$

with the conditions

$$-k\frac{\partial\theta_2}{\partial y} + h_2\,\theta_2 = 0, \;\text{at}\, y = -L_2 \tag{4.32b}$$

$$+k\frac{\partial\theta_2}{\partial y} + h_2\,\theta_2 = 0, \;\text{at}\, y = +L_2 \tag{4.32c}$$

$$\theta_2 = 0, \;\text{for}\; t = 0 \tag{4.32d}$$

The above one-dimensional problems for the functions $\theta_1(x,t)$ and $\theta_2(y,t)$ are exactly the same as that whose solution is given by the transient temperature chart in Figure 4.2. The solution of the two-dimensional heat conduction problems defined by Equation (4.30), for the rectangular regions $-L_1 < x < +L_1$ and $-L_2 < y < +L_2$, can be constructed as the product of the solutions for a rectangular bar made up of the product of the solutions of two one-dimensional, time-dependent problems of slabs whose solutions are illustrated in Figure 4.20.

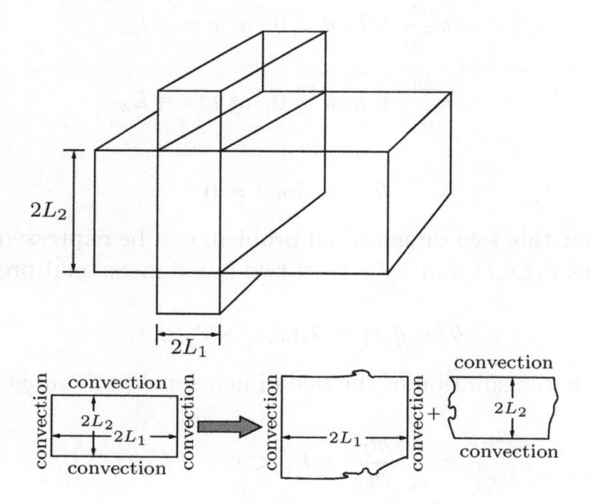


**Figure 4.20**
Decomposition for product solution of two-dimensional heat conduction in a rectangular bar.

This basic idea developed with reference to a rectangular bar can be extended to other configurations. For example, the two-dimensional transient-temperature distribution $T(r,z,t)$ in a solid cylinder of radius $r$ and finite length $2L$, initially at a uniform temperature $T_i$ and suddenly subjected to convection at the boundaries, can be expressed in the dimensionless form as

$$\theta(r, z, t) = \frac{T(r, z, t) - T_\infty}{T_i - T_\infty}$$

where $T_\infty$ is the temperature of the convection environment at the surface of the cylinder. The solution for this problem can be expressed as the product of the solutions of two one-dimensional problems, consisting of a slab (Figure 4.2) and a long cylinder (Figure 4.5). The decomposition of the solutions for a finite solid cylinder is illustrated in Figure 4.21.

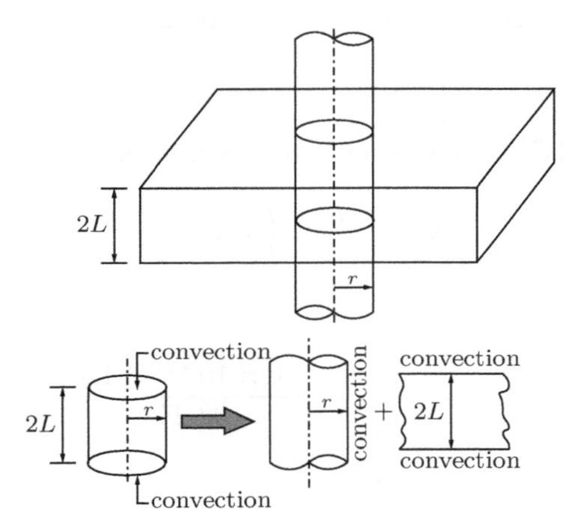


**Figure 4.21**
Decomposition for product solution of two-dimensional heat conduction in a finite cylinder.

## Example 4.9

A rectangular metallic bar of cross-section 50 mm by 40 mm with thermal conductivity 60 W/(m °C) and thermal diffusivity $1.6 \times 10^{-5}$ m$^2$/s, initially at a uniform temperature of 225°C is suddenly exposed to a convecting environment of convection coefficient 500 W/(m$^2$ °C) and temperature 25°C. Determine the temperature at the center of the bar at 2 minutes from the beginning of exposure to convecting environment.

## Solution

This problem can be solved, treating the bar as two slabs of thickness $2L_1 = 50$ mm and $2L_2 = 40$ mm.

Given, $k = 60$ W/(m °C), $\alpha = 1.6 \times 10^{-5}$ m$^2$/s, $h = 500$ W/(m$^2$ °C), $T_\infty = 25$°C, $T_i = 225$°C, $t = 120$ s.

For the slab with $2L_1 = 50$ mm,

$$t^* = \frac{\alpha t}{L_1^2} = \frac{(1.6 \times 10^{-5}) \times 120}{0.025^2} = 3.07$$

$$\frac{1}{\text{Bi}} = \frac{k}{hL_1} = \frac{60}{500 \times 0.025} = 4.8$$

For $t^* = 3.07$ and $1/\text{Bi} = 4.8$, from Figure 4.2, we have

$$\theta_{1,0}^* = \frac{T_0 - T_\infty}{T_i - T_\infty} \approx 0.6$$

For the slab with $2L_2 = 40$ mm,

$$t^* = \frac{\alpha t}{L_2^2} = \frac{(1.6 \times 10^{-5}) \times 120}{0.02^2} = 4.8$$

$$\frac{1}{\text{Bi}} = \frac{k}{hL_2} = \frac{60}{500 \times 0.02} = 6$$

For $t^* = 4.8$ and $1/\text{Bi} = 6$, from Figure 4.2, we have

$$\theta_{2,0}^* = \frac{T_0 - T_\infty}{T_i - T_\infty} \approx 0.5$$

Therefore,

$$\theta_0^* = \theta_{1,0}^* \times \theta_{2,0}^* = 0.6 \times 0.5 = 0.3$$

Thus,

$$\frac{T_0 - T_\infty}{T_i - T_\infty} = 0.3$$

$$\frac{T_0 - 25}{225 - 25} = 0.3$$

$$T_0 = 0.3 \times (225 - 25) + 25$$

$$= \boxed{85°\text{C}}$$

## Example 4.10

A short stainless steel cylinder of diameter 80 mm and length 60 mm, initially at a uniform temperature of 600 K is suddenly immersed in an oil bath maintained at 300 K with convective heat transfer coefficient 500 W/(m$^2$ °C). Determine the temperature (a) at the center of the cylinder, marked as point A in Figure E4.10, (b) at point B at the center of its top surface and (c) at point C at the middle of the side surface, as shown in Figure E4.11, at the end of 3 minutes from the beginning of the cooling process. Take $k = 17.4$ W/(m °C), $\alpha = 4.19 \times 10^{-6}$ m$^2$/s and $c_p = 526$ J/(kg °C).

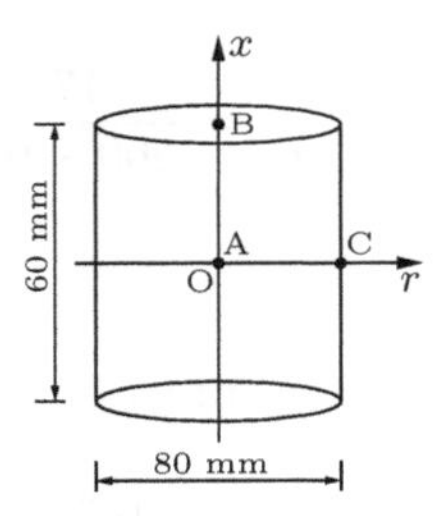


**Figure E4.10**
A short cylinder.

## Solution

Given, $T_i = 600$ K, $T_\infty = 300$ K, $h = 500$ W/(m$^2$ °C), $t = 180$ s.

(a) This problem can be solved by product solution by treating the finite cylinder as an infinite cylinder of radius $r_0 = 40$ mm and an infinite slab of thickness $2L = 60$ mm.

Treating the object as a slab, we have point A as the mid-point of the slab with $2L = 60$ mm. Therefore, the Fourier number and 1/Bi for the slab are

$$\frac{\alpha t}{L^2} = \frac{(4.19 \times 10^{-6}) \times 180}{0.03^2}$$

$$= 0.838$$

$$\frac{1}{\text{Bi}} = \frac{k}{hL} = \frac{17.4}{500 \times 0.03}$$

$$= 1.16$$

For $\frac{\alpha t}{L^2} = 0.838$, $1/\text{Bi} = 1.16$, from Figure 4.2,

$$\theta^*_{\text{slab}} = 0.6$$

Treating the object as a cylinder of radius $r_0 = 0.04$ m, we have point A at the axis of the cylinder. Therefore,

$$\frac{\alpha t}{r_0^2} = \frac{(4.19 \times 10^{-6}) \times 180}{0.04^2}$$

$$= 0.47$$

$$\frac{1}{\text{Bi}} = \frac{k}{hr_0} = \frac{17.4}{500 \times 0.04}$$

$$= 0.87$$

For $\frac{\alpha t}{L^2} = 0.47$, $1/\text{Bi} = 0.87$, from Figure 4.5,

$$\theta^*_{\text{cylinder}} = 0.55$$

Thus,

$$\theta^*_{\text{total}} = \theta_{\text{slab}} * \times \theta^*_{\text{cylinder}}$$

$$= 0.6 \times 0.55$$

$$= 0.33$$

Thus,

$$\theta^*_{\text{total}} = \frac{T_{\text{A}} - T_\infty}{T_i - T_\infty} = 0.33$$

$$T_{\text{A}} = 0.33 \times (600 - 300) + 300$$

$$= \boxed{399\,\text{K}}$$

(b) Point B is along the axis of the cylinder with $r = 0.04$ m, and at the surface of the slab with $x = L$. Therefore,

$$\frac{x}{L} = 1$$

$$\frac{1}{\text{Bi}} = \frac{k}{hL} = \frac{17.4}{500 \times 0.03} = 1.16$$

For $x/L = 1$ and $1/\text{Bi} = 1.16$, from Figure 4.3,

$$\frac{T - T_\infty}{T_0 - T_\infty} = 0.55$$

At time $t = 180$ s, $T_0 = 399$ K. Therefore,

$$\frac{T_\text{B} - 300}{399 - 300} = 0.55$$

$$T_\text{B} = 0.55 \times (399 - 300) + 300 = \boxed{354.5\,\text{K}}$$

(c) For point C, at the surface of the cylinder, at $r = r_0$ m and $L = 0.0$ m. Therefore,

$$r/r_0 = 1$$

$$\frac{1}{\text{Bi}} = \frac{k}{hr_0} = \frac{17.4}{500 \times 0.04} = 0.87$$

For $r/r_0 = 1$ and $1/\text{Bi} = 0.87$, from Figure 4.6,

$$\frac{T - T_\infty}{T_0 - T_\infty} = 0.46$$

Thus,

$$\frac{T_\text{C} - 300}{399 - 300} = 0.65$$

$$T_\text{C} = 0.46 \times (399 - 300) + 300 = \boxed{345.5\,\text{K}}$$

## 4.6 Summary

The temperature of a solid body subjected to a sudden change of environment takes some time to attain an equilibrium. The changes will continue till a *steady-state* temperature distribution is reached. In the transient heating and cooling process which takes place in the interim period before equilibrium is established, the analysis must be modified to take into account the change in the *initial condition* of this body with time, and the boundary conditions must be adjusted to match the physical situation which is apparent in the unsteady-state heat transfer problem.

The governing equation for the unsteady heat conduction, in the absence of heat generation within the system becomes

$$\boxed{\frac{\partial T}{\partial t} = \alpha \nabla^2 T}$$

For a plane wall of thickness $2L$, which extends infinitely in the other two dimensions ($y-$ and $z-$directions), the governing equation for the unsteady heat conduction, reduces to

$$\frac{\partial T}{\partial t} = \alpha \frac{\partial^2 T}{\partial x^2}$$

The initial condition is $T(x,t) = T_0$ , for $t < 0$, and the boundary condition at both $x = -L$ and $x = L$ are given by the Newton's law of cooling as

$$-k\frac{\partial T}{\partial x} = h\left(T_\infty - T\right), \quad \text{at } x = -L$$

$$-k\frac{\partial T}{\partial x} = h\left(T - T_\infty\right), \quad \text{at } x = L$$

The above governing equation can be solved to get the functional relation of $T$ with all other influencing parameters.

The temperature $T$ can be expressed in the functional form as

$$T = T\left(x, t, L, \rho, c, k, h, T_0, T_\infty\right)$$

The nondimensional form of temperature

$$\theta = \frac{T - T_\infty}{T_0 - T_\infty}$$

can be expressed after carrying out the dimensional analysis as

$$\theta = \theta\left(\frac{x}{L}, \frac{kt}{\rho c L^2}, \frac{hL}{k}\right)$$

The nondimensional group

$$\frac{hL}{k} = \text{Bi} \quad \text{is called the } \textit{Biot number}$$

and

$$\frac{kt}{\rho c L^2} = \frac{\alpha t}{L^2} = \text{Fo} \quad \text{is called the } \textit{Fourier number.}$$

conditions, as follows.

The governing equation for unsteady conduction through a slab, in terms of the nondimensional length, temperature and time becomes

$$\boxed{\frac{\partial \theta}{\partial t^*} = \frac{\alpha \tau}{L^2} \frac{\partial^2 \theta}{\partial x^{*2}}}$$

where $\theta$, $t^*$, and $x^*$ are dimensionless quantities, hence the group $\alpha\tau/L^2$ must also be dimensionless. Furthermore, $\tau$ is an arbitrary characteristic time, therefore, we can assume it to be equal to $L^2/\alpha$ and simplify the above equation to

$$\boxed{\frac{\partial \theta}{\partial t^*} = \frac{\partial^2 \theta}{\partial x^{*2}}}$$

where $t^* = \alpha t/L^2$.

The new boundary conditions are

$$\frac{-k}{L}\frac{\partial \theta}{\partial x^*} = -h\theta, \quad \text{at} \quad x^* = -1$$

that is,

$$\boxed{\frac{\partial \theta}{\partial x^*} = \frac{hL}{k}\theta, \quad \text{at} \quad x^* = -1}$$

Similarly,

$$\boxed{\frac{\partial \theta}{\partial x^*} = -\frac{hL}{k}\theta, \quad \text{at} \quad x^* = 1}$$

Two problems will result in identical solutions if the dimensionless parameter $hL/k$, namely the Bi is the same for both problems. Therefore, Bi is the *similarity parameter* for this problem. Thus,

$$\boxed{\theta = \theta\left(x^*, t^*, \text{Bi}\right)}$$

As Bi decreases, the temperature at all points in the body tends to the temperature at the center line. Thus, for very low Bi the internal gradients of temperature are very small compared to the total difference $(T_0 - T_\infty)$. That is, when Bi is small the body may be assumed to have a uniform temperature $T$, which is a function of time.

For a body of volume $\mathbb{V}$, the heat conducted into the body across the film equals the rate of change of thermal content of the body, that is,

$$hA\left(T_\infty - T\right) = \rho \mathbb{V} c \frac{dT}{dt}$$

where $A$ is the surface area of the body.

Now, let at time $t < 0$, the body was at temperature $T_0$. Therefore, the above equations in terms of $\theta = (T - T_\infty)\,/\,(T_0 - T_\infty)$ can be written as

$$\frac{d\theta}{dt} = -\frac{hA}{\rho \mathbb{V} c}\theta$$

$$\boxed{\theta = e^{-\frac{hA}{\rho \mathbb{V} c}t}}$$

The constant of integration has been set equal to ln (1) to accommodate the initial condition.

Thus, the difference in temperature of the body and fluid decreases exponentially with the characteristic temperature $\tau$, given by

$$\boxed{\tau = \frac{\rho \mathbb{V} c}{hA}}$$

Lumped system is that in which the temperature of the system varies with time but remains uniform throughout the system at any time. Heat transfer analysis that utilizes this idealization is known as *lumped system analysis.*

The temperature of a lumped system at time $t$, $T(t)$ is expressed as

$$\frac{T(t) - T_\infty}{T_i - T_\infty} = e^{-bt}$$

Once the temperature $T(t)$ at time $t$ is available, the rate of convection heat transfer between the body and the environment at that time can be determined from Newton's law of cooling as

$$\boxed{\dot{Q}(t) = hA\left(T(t) - T_\infty\right)\ \mathrm{W}}$$

The *total amount* of heat transfer between the body and the surrounding medium over the time interval from $t = 0$ to $t$ is simply the change in the energy content of the body,

$$\boxed{Q = mc\left(T(t) - T_i\right)\ \mathrm{kJ}}$$

The first step in establishing a criterion for the applicability of the lumped system analysis is to define a *characteristic length* as

$$L = \frac{\text{Volume of the system}}{\text{Exposed area}} = \frac{\mathbb{V}}{A}$$

and a Biot number Bi as

$$\mathrm{Bi} = \frac{hL}{k}$$

The Biot number can also be expressed as

$$\text{Bi} = \frac{h}{k/L}\frac{\Delta T}{\Delta T} = \frac{\text{Convection at the body surface}}{\text{Conduction within the body}}$$

or

$$\text{Bi} = \frac{L/k}{1/h} = \frac{\text{Conduction resistance within the body}}{\text{Convection resistance at the body surface}}$$

Lumped system analysis assumes a *uniform* temperature distribution throughout the body, which is valid only when the thermal resistance of a body to heat conduction is zero. Thus, lumped system analysis is *exact* when Bi $= 0$ and for Bi $> 0$ it is only *approximate.* Thus, the smaller the Bi, the more accurate is the lumped system analysis. Therefore, it is essential to identify the limiting value of Bi below which the analysis can be taken as reasonably accurate. It is generally accepted that lumped system analysis is *applicable* if the "temperature within the body relative to the surroundings, that is, $(T - T_\infty)$ remains within 5 percent of each other even for well-rounded geometries such as a spherical ball." Thus, when Bi $< 0.1$, the variation of temperature with location within the body will be marginal and can reasonably be approximated to be uniform.

In solids with high $\alpha$, such that conductivity is large but heat capacity per unit volume is small, less heat is retained for raising the temperature of the body and more heat is conducted inwards. This results in increased penetration depth.

Heisler charts are meant for obtaining the temperature distribution in infinite plates, long cylinders and spheres. But for handling heat transfer problems associated with a wall with height and length comparable to its thickness or a cylinder with length not very large compared to its diameter, additional space coordinates need be considered while solving the governing equation, because for such geometries the one-dimensional approximation is no longer valid. To solve the heat transfer problems associated with these multidimensional systems, it is necessary to seek some appropriate method. This situation can be comfortably handled by combining the solutions for the one-dimensional systems.

A finite rectangular bar can be visualized as a body made up of two infinite plates of thickness $2L_1$ and $2L_2$. For the two infinite plates considered, the governing equations would be

$$\frac{\partial^2 T_1}{\partial x^2} = \frac{1}{\alpha}\frac{\partial T_1}{\partial t}, \quad \frac{\partial^2 T_2}{\partial z^2} = \frac{1}{\alpha}\frac{\partial T_2}{\partial t}$$

The dimensionless temperature distribution for the infinite rectangular bar may be expressed as the product of the solutions for two plate problems of thickness $2L_1$ and $2L_2$.

In an analogous manner, the solution for the three-dimensional blocks of semi-infinite plate and infinite rectangular bar may be expressed as a product of the three infinite-plate solutions for plates having the thicknesses as the three sides of the block.

Similarly, a solution for a cylinder of finite length, such as the Semi-infinite rectangular bar and rectangular parallelepiped can be expressed as a product of the solutions of an infinite cylinder and an infinite plate having thickness equal to the length of the cylinder. In the same way, the solution for the semi-infinite cylinder and short cylinder can be obtained combining the solutions of an infinite-cylinder and an infinite-plate.

When the temperature gradients in more than one direction within a solid are considerable, the problem essentially becomes a two-dimensional heat conduction. For such problems, when there is no heat generation within the solid, the problems can be solved by combining the solutions of one-dimensional transient-temperature charts, by treating the problem as a suitable combination of two one-dimensional problems. Such an approach, called *method of product solution*, is applicable if the solution of a two-dimensional, time-dependent heat conduction problem can be shown to be equivalent to the product solutions of two one-dimensional, transient heat conduction problems.

## 4.7 Exercise Problems

4.1 A thermocouple of bulb diameter 1.2 mm measures the temperature of a gas stream with $k$ = 35 W/(m °C), $c_p$ = 320 J/(kg °C), $\rho$ = 8500 kg/m$^3$. The heat transfer coefficient between the gas and the thermocouple junction is 65 W/(m$^2$ °C). Determine the time required for the thermocouple to read 99 percent of the initial temperature difference.

[**Ans.** 38.54 s]

4.2 A cold storage room is used to store 100 fruit boxes of 60 kg each. The room gains heat through the walls and door at a rate of 0.6 kW. The fruit boxes enter the room at 30°C and are cooled to 10°C in 9 hours. Assuming the average specific heat of the fruit boxes as 3 kJ/(kg °C), determine the refrigeration load of the cold storage room.

[**Ans.** 11.7 kW]

4.3 A 40-mm thick iron plate at a uniform temperature of 200°C is suddenly exposed to air at 20°C. For steel, the conduction coefficient, specific heat, density and thermal diffusivity are 60 W/(m °C), 460 J/(kg °C), 8000 kg/m$^3$ and $1.65 \times 10^{-5}$ m$^2$/s, respectively. For air the convection coefficient is 500 W/(m$^2$ °C). Determine (a) the temperature at a depth of 10 mm from the surface of the plate at 60 seconds after it is exposed to air, and (b) the heat removed from the plate per unit area in this time period.

[**Ans.** (a) 138.8°C, (b) 5.83 MJ]

4.4 A large aluminum plate of thickness 100 mm at a uniform temperature of 220°C is suddenly exposed to an environment at 50°C with heat transfer coefficient 500 W/(m$^2$ °C). Calculate the temperature at a depth of 10 mm from one of the faces of the plate at 30 seconds after the plate has been exposed to the environment. For aluminum, $k = 215$ W/(m °C) and thermal diffusivity is $8.4 \times 10^{-5}$ m$^2$/s.

[**Ans.** 179.2°C]

4.5 A long aluminum cylinder of diameter 50 mm at a uniform temperature of 200°C is suddenly exposed to an environment with convection coefficient of 525 W/(m$^2$ °C). If the temperature at a depth of 12.5 mm from the cylinder surface is at 120°C after 60 seconds of exposure, determine the temperature of the environment. For aluminum, take $k = 215$ W/(m °C), thermal diffusivity as $8.4 \times 10^{-5}$ m$^2$/s.

[**Ans.** 70.13°C]

4.6 A rectangular bar of cross-section 60 mm by 30 mm, initially at a uniform temperature of 175°C is suddenly exposed to a convective environment at 25°C with a convection coefficient of 250 W/(m$^2$ °C). Determine the centerline temperature of the bar at 60 seconds after it is exposed to the environment. Take $k = 200$ W/(m °C), $\rho = 2700$ kg/m$^3$, $\alpha = 8.4 \times 10^{-5}$ m$^2$/s and $c_p = 890$ K/(kg °C).

[**Ans.** 127°C]

4.7 An iron rod of 60 mm diameter, initially at a uniform temperature of 800°C is immersed in an oil tank maintained at 50°C. If the convection coefficient at the rod surface in the oil is 400 W/(m$^2$ °C), determine the time from immersion at which the centerline temperature of the rod will cool to 60°C. Take $k = 60$ W/(m °C) and $\alpha = 2 \times 10^{-5}$ m$^2$/s for the iron rod.

[**Ans.** 9.375 minutes]

4.8 An iron sphere of diameter 100 mm, initially at a uniform temperature of 300°C is immersed in a coolant bath maintained at 50°C. The heat transfer coefficient between the sphere surface and the coolant is 1000 W/(m$^2$ °C). If the conduction coefficient and thermal diffusivity of the sphere are 60 W/(m °C) and $1.6 \times 10^{-5}$ m$^2$/s, respectively, determine the temperature at the center of the sphere after 3 minutes of immersion.

[**Ans.** 87.5°C]

4.9 An aluminum sphere of diameter 100 mm, initially at a uniform temperature is immersed in a liquid bath maintained at 80°C. The heat transfer coefficient at the sphere surface in the liquid is 1000 W/(m$^2$ °C). If the sphere center temperature after 80 seconds of immersion is 100°C, determine the temperature of the sphere before immersion. Take $k = 200$ W/(m °C) and $\alpha = 8.4 \times 10^{-5}$ m$^2$/s for the aluminum sphere.

[**Ans.** 205°C]

4.10 A thick steel slab with $k$ = 60 W/(m °C) and $\alpha = 1.6 \times 10^{-5}$ m$^2$/s is initially at a uniform temperature of 150°C. Its surface temperature is suddenly lowered to 20°C. By treating this as a one-dimensional transient heat conduction problem in a semi-infinite medium, determine the temperature at a depth of 20 mm from the surface at 60 seconds after the surface temperature is lowered.

[**Ans.** 65.5°C]

4.11 The surface of a large steel block initially at a uniform temperature of 22°C is suddenly exposed to a heat flux, raising its surface temperature to 200°C. Determine the temperature at a depth of 10 mm from the surface after 60 seconds from the exposure. Take the conduction coefficient and thermal diffusivity for steel as 45 W/(m °C) and $1.4 \times 10^{-5}$ m$^2$/s, respectively.

[**Ans.** 56.3°C]

4.12 A long copper wire of diameter 6 mm is initially at a uniform temperature of 10°C. It is suddenly exposed to an air stream at 37°C, with convection coefficient of 50 W/(m$^2$ °C). Determine the average temperature of the wire after 30 seconds of exposure to the air stream. Take $k$ = 390 W/(m °C), $\rho$ = 8900 kg/ m$^3$ and $c_p$ = 386 m$^2$/(s$^2$ °C) for the wire.

[**Ans.** 289.8 K]

4.13 A concrete cylinder of length 100 mm and diameter 100 mm is initially at a uniform temperature of 20°C. It is suddenly exposed to an environment of water vapor at 100°C and convection coefficient 8500 W/(m$^2$ °C). Determine the time required for the center of the cylinder to reach 40°C, treating this as an infinitely long cylinder. Assume $k$ = 1.2 W/(m °C) and $\alpha = 5.95 \times 10^{-7}$ m$^2$/s for the cylinder.

[**Ans.** 14 minutes]

4.14 A flat wall of thickness 500 mm, made of fireclay is initially at a uniform temperature of −73°C. One of its faces is suddenly exposed to a hot gas at 1200 K and the opposite face is perfectly insulated. The convection coefficient at the hot face is 7.5 W/(m$^2$ °C). Determine the time required for the temperature at the center of the wall to reach 600 K, treating the heat conduction through the wall as one-dimensional. What will be the temperature at the face of the insulated wall at this instant? For the fireclay wall, take $k$ = 1.1 W/(m °C), $c_p$ = 919 J/(kg °C), $\rho$ = 2310 kg/m$^3$ and $\alpha = 5.30 \times 10^{-7}$ m$^2$/s.

[**Ans.** 39.3 hours, 267°C]

4.15 A short steel cylinder of length 600 mm and diameter 300 mm, initially at a uniform temperature of 650 K, is immersed in an oil bath maintained at 300 K. The heat transfer coefficient at the surface of the cylinder in oil is 34 W/(m$^2$ °C). Determine the temperature at the center of the cylinder after one hour of immersion. Assume, $k$ = 50 W/(m °C), $c_p$ = 470 J/(kg °C), and $\alpha = 1.2 \times 10^{-5}$ m$^2$/s for the cylinder.

[**Ans.** 508 K]

4.16 A short aluminum cylinder of length 30 mm and diameter 60 mm initially at a uniform temperature of 175°C is suddenly exposed to a convection environment with heat transfer coefficient of 250 W/(m$^2$ °C) and temperature 25°C. Calculate the centerline temperature of the cylinder after 60 seconds from the beginning of exposure to the convection environment. Take $k = 200$ W/(m °C) and $\alpha = 8.4 \times 10^{-5}$ m$^2$/s for the cylinder.

[**Ans.** 98.5°C]

4.17 A semi-infinite corner of a concrete block, in the limits $0 \leq x < \infty$, $0 \leq y < \infty$, as shown in Figure P4.17, with $k = 1.3$ W/(m °C) and $\alpha = 8 \times 10^{-7}$ m$^2$/s is initially at a uniform temperature of 400 K. The surfaces of the concrete block are suddenly exposed to a convection environment with heat transfer coefficient of 100 W/(m$^2$ °C) and temperature of 300 K. Determine the temperature at point P shown, at 2 hours from the beginning of exposure to the convection environment.

[**Ans.** 300.09 K]

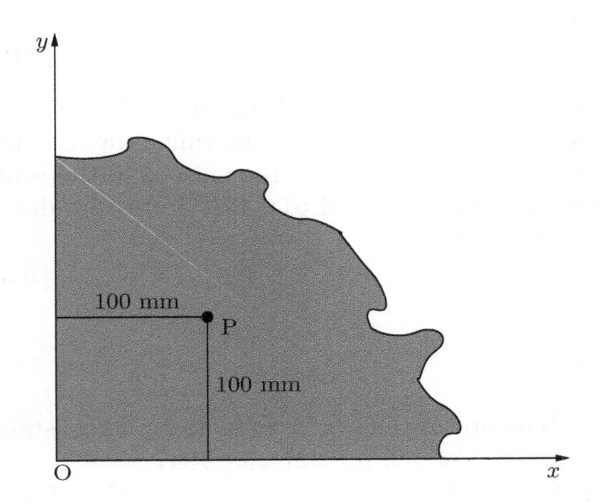


**Figure P4.17**
Semi-infinite corner of a concrete block.

4.18 A semi-infinite strip of fireclay brick slab, shown in Figure P4.18 is initially at a uniform temperature of 600 K. All the surfaces of the slab are suddenly exposed to a convection environment with heat transfer coefficient 100 W/(m$^2$ °C) and temperature 40°C. Taking $k = 1$ W/(m °C) and $\alpha = 5.4 \times 10^{-7}$ m$^2$/s, determine the temperature at point P shown, at 2 hours from the beginning of exposure to the convection environment.

[**Ans.** 315.87 K]

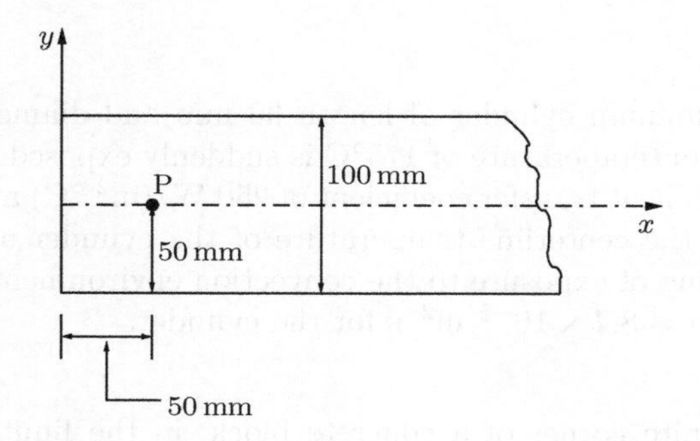


**Figure P4.18**
Semi-infinite fireclay brick slab.

4.19 An isothermal sphere of radius 100 mm, at a temperature of 100°C, is buried in a large body of earth at a constant temperature of 10°C. Calculate the rate of heat loss from the sphere, taking the conduction coefficient for earth as 1.2 W/(m °C).

[**Ans.** 135.756 W]

4.20 A 0.5-m thick brick wall of thermal diffusivity $4.72 \times 10^{-7}$ $m^2/s$ is initially at a uniform temperature of 300 K. Determine the time required for the temperature at a depth of 0.25 m from the surface of the wall to reach 425 K if the surface temperature is raised to 600 K and maintained, treating the wall as a semi-infinite solid.

[**Ans.** 27.9 hours]

# References

1. Heisler, M. P., "Temperature charts for induction and constant temperature heating," Trans. ASME, Vol. 69, pp. 227-236, 1947.
2. Gröber, H. S. Erk and Grigull, U., *Fundamentals of Heat Transfer*, McGraw-Hill Book Co., New York, 1961.
3. Schneider, P. J., *Conduction Heat Transfer*, Addison-Wesley Publishing Company, Inc., Reading, MA, 1955.

# Chapter 5

# Convective Heat Transfer

## 5.1 Introduction

So far we focused our attention on heat transfer by conduction and considered convection only to the extent that it provided a possible boundary condition for conduction problems. In Chapter 1, the term "convection" was used to describe energy transfer between a surface and a fluid moving over the surface. In convection heat transfer, although the mechanism of diffusion (random motion of fluid molecules) contributes to energy transfer, the dominant contribution is generally due to the bulk or gross motion of the fluid elements. Our interest in this chapter is to get ourselves introduced with the physical mechanism that causes convection heat transfer and the means to perform convective heat transfer calculations.

To get a feel for the physical mechanism of convection heat transfer and its relation to the conduction process, let us consider flow over a hot plate, shown in Figure 5.1.

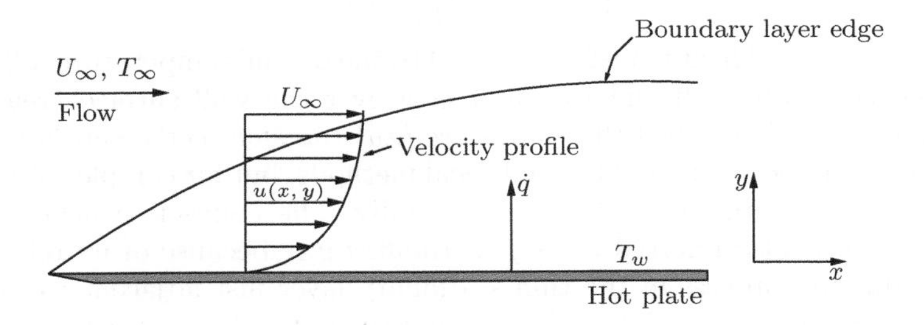


**Figure 5.1**
Convective heat transfer from a hot plate.

The plate is maintained at a constant temperature of $T_w$ and the temperature of the flowing fluid in the freestream (that is, in the undisturbed flow above the boundary layer) is $T_\infty$. The flow with velocity $U_\infty$ in the freestream is brought to rest (zero velocity) at the plate surface, due to no-slip condition caused by the viscosity of the fluid. The flow velocity assumes a profile as shown in Figure 5.1. When $T_w > T_\infty$, the heat transfer from the surface of the hot plate to the fluid flow has to be only through conduction, because the flow velocity at the wall is zero. Thus, the heat transferred from the plate to the fluid can be computed using Fourier's law of heat conduction [Equation (2.1)],

$$\dot{q} = -k\frac{\partial T}{\partial x}$$

where $k$ is the thermal conductivity of the fluid and $\partial T/\partial x$ is the temperature gradient at the plate surface.

At this stage it is natural to have the doubt that, when it is possible to calculate the heat transfer associated with the fluid flow over the heated plate, with heat conduction relation, what is the need for *heat convection* analysis? This question can be answered as follows. The temperature gradient at the plate surface is strongly influenced by the rate at which the heat from the surface is carried away by the flowing fluid. It can be comfortably visualized that the higher the flow velocity the higher would be the temperature gradient at the wall. That is, the temperature gradient at the wall depends on the flow velocity. Thus, even though the physical mechanism of heat transfer at the plate surface is conduction, the overall heat transfer associated with the flow over the plate is governed by the heat convected away by the fluid flow. Hence, for the analysis of convective heat transfer, it is essential to develop an expression relating the flow velocity and the temperature gradient.

The overall effect of the convection heat transfer process can be represented by Newton's law of cooling;

$$\boxed{\dot{Q} = h\,A\,(T_w - T_\infty)} \tag{5.1}$$

That is, the rate of heat transfer is related to the overall temperature difference between the plate wall and the fluid flow over the wall surface area $A$. In Equation (5.1), $h$ is called the *convective heat transfer coefficient.* For simple systems, $h$ can be calculated by analytical methods, but for complex situations it has to be determined only experimentally. The convective heat transfer coefficient $h$ is also referred to as *film conductance* because of its relation to the conduction process in the thin stationary layer just adjacent to the wall surface. From Equation (5.1), it is seen that the units of $h$ are watts per square meter per degree celsius [W/(m$^2$ °C)], when the heat flow is in watts. The convective heat transfer coefficient $h$ can also be expressed with units W/(m$^2$ K), since it is associated with the relation involving only the temperature difference, and not the temperature itself.

From the above discussion, it is evident that in addition to the thermal properties of the fluid, such as the thermal conductivity $k$, specific heat $c_p$, density $\rho$ and viscosity $\mu$ of the fluid, the flow velocity $U_\infty$ also has a strong influence on the convective heat transfer. If a heated plate is exposed to stagnant ambient air in a room (without an external flow), a movement of the air would be experienced as a result of the density gradient near the plate. This kind of convection heat transfer, where there is no *forced velocity* is referred to as *free* or *natural* convection, and it is caused by the *body force* due to the presence of *density gradient* in the flow field. The net effect is the *buoyancy force*, which induces free convection currents. In most cases, the density gradient is due to the temperature gradient, and the body force is due to the gravitational field. Convection heat transfer in fluid flows originates from an *external forcing* condition. For example, fluid motion may be induced by a fan or a pump, or it may result from the motion of a solid through the fluid. When a temperature gradient is present during such motion, *forced* convection heat transfer will occur. Boiling and condensation phenomena are also generally grouped under convective heat transfer. The approximate ranges of convection heat transfer coefficients for some specific cases are listed in Table 5.1.

## 5.2 The Convection Boundary Layers

Basically, the velocity boundary layer, the thermal boundary layer, and the concentration boundary layer are all convection boundary layers. The velocity boundary layer is of extent $\delta(x)$ and is characterized by the presence of *velocity gradient* or shear stress. The thermal boundary layer is of extent $\delta_t\,(x)$ and is characterized by *temperature gradient* and heat transfer. The concentration boundary layer is of extent $\delta_c\,(x)$ and is characterized by *concentration gradient* and species transfer. For engineering applications, the principal manifestations of these three boundary layers are the *skin friction*, *convection heat transfer* and *convection mass transfer*, respectively. Therefore, the key parameters associated with the velocity, thermal and concentration boundary layers, respectively, are the skin friction coefficient $c_f$, the convection heat transfer coefficient $h$, and the mass transfer convection coefficient $h_m$.

For fluid flow past any body, because of viscosity, there will always exist a velocity boundary layer, and hence surface friction. However, a thermal boundary layer and the associated convection heat transfer will prevail only when the body surface and the freestream are at different temperatures. Similarly, a concentration boundary layer and convection mass will transfer exist only when the concentration of a species at the body surface differs from its freestream concentration. Situations can arise in which all the three boundary layers are present. In such cases, all the three boundary layers rarely grow at the same rate. Usually the values of $\delta$, $\delta_t$ and $\delta_c$, at a given axial location $x$ are not the same.

**Table 5.1** Approximate values of convection heat transfer coefficient $h$

| Mode | $h$ W/(m$^2$ °C) |
|---|---|
| *Free convection* ($\Delta T = 30$°C): | |
| Vertical plate 300 mm high in air | 4.5 |
| 50 mm diameter cylinder (horizontal) in air | 6.5 |
| *Forced convection*: | |
| 2 m/s air flow over 200 mm square plate | 12 |
| 35 m/s air flow over 750 mm square plate | 75 |
| Air at 2 atm flowing over 25 mm diameter tube with 10 m/s | 65 |
| Air flow at 50 m/s across 50 mm diameter cylinder | 180 |
| *Boiling water*: | |
| In a pool | 2500 – 35,000 |
| Flowing through a tube | 5000 – 100,000 |
| *Condensation of water vapor*, at 1 atm: | |
| Vertical surfaces | 4000 – 11,300 |
| Outside of horizontal tube | 9500 – 25,000 |

An essential first step in the analysis of a convection heat transfer problem is to determine whether the boundary layer is *laminar* or *turbulent*. Surface friction and the convection heat transfer rates strongly depend on the nature of the boundary layer. In calculating the boundary layer behavior, it is often reasonable to assume that the transition begins at some location $x_c$ from the leading edge of the body. This location is determined by a dimensionless grouping of variables called *Reynolds number*, defined as

$$\mathrm{Re}_x \equiv \frac{\rho_\infty u_\infty x}{\mu}$$

where $\rho_\infty$ is the freestream flow density, $u_\infty$ is the freestream velocity of the flow, $\mu$ is the viscosity coefficient and $x$ is the characteristic length, which is usually the distance from the leading edge. The *critical* Reynolds number is the value of $\text{Re}_x$ for which the transition (change of flow from laminar to turbulent nature) begins, and for external flows it is known to vary from $10^5$ to $3 \times 10^6$, depending on the surface roughness, the turbulence level of the freestream and the nature of the pressure variation along the surface. A representative value of transition Reynolds number for a flat plate boundary layer, based on the distance $x$ from the leading edge is

$$\text{Re}_{x,c} = \frac{\rho_\infty u_\infty x}{\mu} = 5 \times 10^5$$

For a circular cylinder, the critical Reynolds number, based on diameter $d$ , is

$$\text{Re}_d = \frac{\rho_\infty u_\infty d}{\mu} = 1.66 \times 10^5$$

For pipe flow, the critical Reynolds number, based on mean velocity and inside diameter $d$, is

$$\text{Re}_d = \frac{\rho\, u_{\text{mean}} d}{\mu} = 2300$$

For a channel flow, the critical Reynolds number, based on mean velocity and height $h$, is

$$\text{Re}_h = \frac{\rho\, u_{\text{mean}} h}{\mu} = 2300$$

These values are generally taken as the critical Reynolds numbers for boundary layer calculations. It is essential to note that the transition occurs over a range of Reynolds number. Therefore, there is a lower limit for the Reynolds number at which the transition begins and ends at an upper limiting value of Reynolds number. Thus, there are two critical Reynolds numbers, namely the lower critical Reynolds number and upper critical Reynolds number. The *lower critical Reynolds number* is that below which the entire flow field is laminar and *the upper critical Reynolds number* is that above which the entire flow field is turbulent.

## 5.3 The Convection Heat Transfer Equations

Examine the simultaneous development of velocity, thermal, and concentration boundary layers over the solid surface illustrated in Figure 5.2. To simplify our analysis, let us assume the flow to be two-dimensional, steady, and with constant thermal conductivity, and $x$ is in the direction along the surface and $y$ is normal to the surface.

### 5.3.1 The Velocity Boundary Layer

The conservation law that is pertinent to the velocity boundary layer is that *matter can neither be created nor destroyed.* Considering a differential control

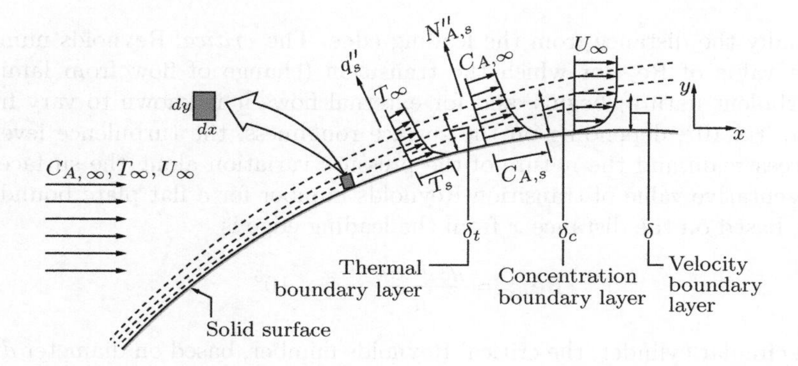


**Figure 5.2**
Development of velocity, thermal, and concentration boundary layers over an arbitrary surface.

volume of length $dx$, height $dy$ and depth unity, shown in Figure 5.3, the above conservation law, namely the mass conservation or continuity equation, can be expressed in differential form.

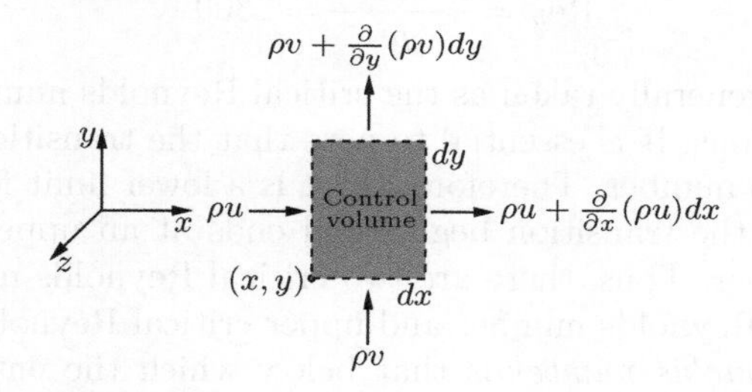


**Figure 5.3**
Differential control volume ($dx.dy.1$) for mass conservation in a two-dimensional velocity boundary layer.

Assuming that, there is no mass generation or absorption inside the control volume, for mass conservation,

**Mass flow into the control volume = Mass flow out of the control volume**

that is,

$$\rho u dy + \rho v dx = \left[\rho u + \frac{\partial}{\partial x}(\rho u)dx\right] dy + \left[\rho v + \frac{\partial}{\partial y}(\rho v)dy\right] dx$$

This reduces to

$$\left[\frac{\partial(\rho u)}{\partial x} + \frac{\partial(\rho v)}{\partial y}\right](dxdy) = 0$$

This simplifies to

$$\boxed{\frac{\partial(\rho u)}{\partial x} + \frac{\partial(\rho v)}{\partial y} = 0} \tag{5.2}$$

Equation (5.2) is a general expression of the *overall* mass conservation requirement, and it must be satisfied at every point in the velocity boundary layer. This equation applies for a single species fluid, as well as for mixtures in which species diffusion and chemical reactions may be occurring.

The second fundamental law that is pertinent to the velocity boundary layer is the Newton's second law of motion or momentum conservation law. For a differential control volume in the velocity boundary layer, Newton's second law states that, *the sum of all forces acting on the control volume must be equal to the net rate at which momentum leaves the control volume (outflow − inflow).*

Two kinds of forces may act on the fluid in the boundary layer. They are the *body forces*, which are proportional to the volume, and the *surface forces*, which are proportional to the surface area. Gravitational, centrifugal, magnetic, and/or electric fields may contribute to the total body force. The surface forces are due to the fluid static pressure and viscous stress. At any point in the boundary layer, the viscous stress may be resolved into two perpendicular components, which include a normal stress $\sigma_{ii}$ and a shear stress $\tau_{ij}$, as shown in Figure 5.4.

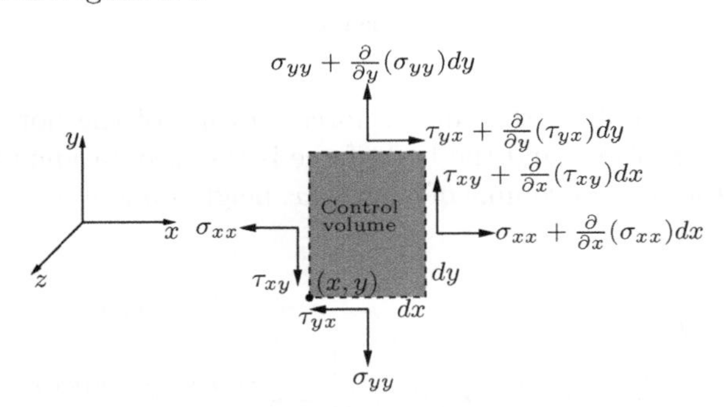


**Figure 5.4**
Normal ($\sigma_{xx}$, $\sigma_{yy}$) and shear ($\tau_{xy}$, $\tau_{yx}$) stresses acting on a differential control volume ($dx.dy.1$) in a two-dimensional velocity boundary layer.

A double subscript notation is used to specify the stress components. The first subscript indicates the orientation of the surface, by providing the direction of its outward normal, and the second subscript indicates the direction of the force component. Thus, $\tau_{xy}$ is the shear stress acting on the surface

$(dy \times 1)$, whose outward normal is in the $x$-direction. All the normal stress components shown in Figure 5.4 are positive in the sense that *both* the surface normal and the stress component are in the same direction. That is, they are either in the positive coordinate direction or in the negative coordinate direction. By this convention the normal (viscous) stresses are *tensile stresses.* In contrast, the static pressure $p$, which originates from an external force acting on the control volume, is a *compressive* stress.

Each of these stresses may change continuously along $x$- and $y$-directions. Using Taylor series expansion for the stresses, the *net surface force*, acting on the control volume, along the $x$- and $y$-directions may be expressed as

$$F_{s,x} = \left(\frac{\partial \sigma_{xx}}{\partial x} - \frac{\partial p}{\partial x} + \frac{\partial \tau_{yx}}{\partial y}\right) dxdy \tag{5.3}$$

$$F_{s,y} = \left(\frac{\partial \tau_{xy}}{\partial x} + \frac{\partial \sigma_{yy}}{\partial y} - \frac{\partial p}{\partial y}\right) dxdy \tag{5.4}$$

where $p$ is the static pressure.

For a steady-flow, by Newton's second law, we have,

**Change in momentum across the control volume = The net force acting on the control volume**

Also,

**The net force acting on the control volume = Surfaces forces + body forces**

For the present problem, the surface forces consist of the normal $(\sigma_{ii})$ and tangential $(\tau_{ij})$ stresses, and the body force is the gravitational force $(g)$.

Consider the control volume of length $dx$, height $dy$ and unit width, shown in Figure 5.5.

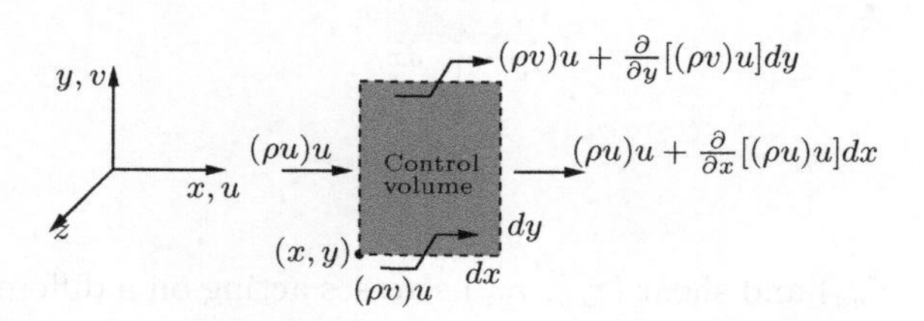


**Figure 5.5**
Momentum fluxes for control volume $(dx.dy.1)$ in a two-dimensional velocity boundary layer.

The sum of forces acting on the control volume, in the $x$-direction is

$$\left[\sigma_{xx} + \frac{\partial \sigma_{xx} dx}{\partial x}\right] dy - \sigma_{xx} dy + \left[\tau_{yx} + \frac{\partial \tau_{yx} dy}{\partial y}\right] dx - \tau_{yx} dx + X.(dxdy)$$

where $\sigma_{xx}$ is the normal stress, $\tau_{yx}$ is the tangential or shear stress and $X$ is the body force. This simplifies to

$$\left[X + \frac{\partial \sigma_{xx}}{\partial x} + \frac{\partial \tau_{yx}}{\partial y}\right] dxdy$$

The *net* rate at which $x$-momentum leaves the control volume is

$$\frac{\partial \left[(\rho u) u\right]}{\partial x} dxdy + \frac{\partial \left[(\rho v) u\right]}{\partial y} dydx$$

Equating the rate of change of $x$-momentum of the fluid to the sum of the forces acting in the $x$-direction, we get

$$\frac{\partial \left(\rho u^2\right)}{\partial x} + \frac{\partial \left(\rho uv\right)}{\partial y} = \frac{\partial \sigma_{xx}}{\partial x} + \frac{\partial \tau_{yx}}{\partial y} + X \tag{5.5}$$

where $X$ is the body force in the $x$-direction.

Equation (5.5) may be expressed in a more convenient form by expanding the derivatives on the left-hand side and substituting from the continuity equation [Equation (5.2)], as follows.

Expanding the left-hand side of Equation (5.5), we get

$$\left[u\frac{\partial(\rho u)}{\partial x} + u\frac{\partial(\rho u)}{\partial x} + v\frac{\partial(\rho u)}{\partial y} + u\frac{\partial(\rho v)}{\partial y}\right] = \frac{\partial \sigma_{xx}}{\partial x} + \frac{\partial \tau_{yx}}{\partial y} + X$$

Grouping the terms, we get

$$\left[u\frac{\partial(\rho u)}{\partial x} + v\frac{\partial(\rho u)}{\partial y}\right] + u\left[\frac{\partial(\rho u)}{\partial x} + \frac{\partial(\rho v)}{\partial y}\right] = \frac{\partial \sigma_{xx}}{\partial x} + \frac{\partial \tau_{yx}}{\partial y} + X$$

But by continuity equation,

$$\frac{\partial(\rho u)}{\partial x} + \frac{\partial(\rho v)}{\partial y} = 0$$

Therefore, for incompressible flows, Equation (5.5) simplifies to

$$\boxed{\rho\left(u\frac{\partial u}{\partial x} + v\frac{\partial u}{\partial y}\right) = \frac{\partial}{\partial x}\left(\sigma_{xx}\right) + \frac{\partial \tau_{yx}}{\partial y} + X} \tag{5.6}$$

Similarly, for the $y$-direction we can obtain

$$\boxed{\rho\left(u\frac{\partial v}{\partial x} + v\frac{\partial v}{\partial y}\right) = \frac{\partial \tau_{xy}}{\partial x} + \frac{\partial}{\partial y}\left(\sigma_{yy}\right) + Y} \tag{5.7}$$

where $Y$ is the body force in the $y$-direction.

Now it is essential to relate the viscous stresses in Equations (5.6) and (5.7) to other flow variables, in order to obtain a solution to these equations. These stresses are associated with the deformation of the fluid and are a function of the fluid viscosity and velocity gradients. Let us examine the deformation of a fluid element subjected to normal and shear stresses, illustrated in Figure 5.6. It is seen that, a normal stress produces a *linear deformation* of the fluid element, whereas a shear stress produces an *angular deformation.*

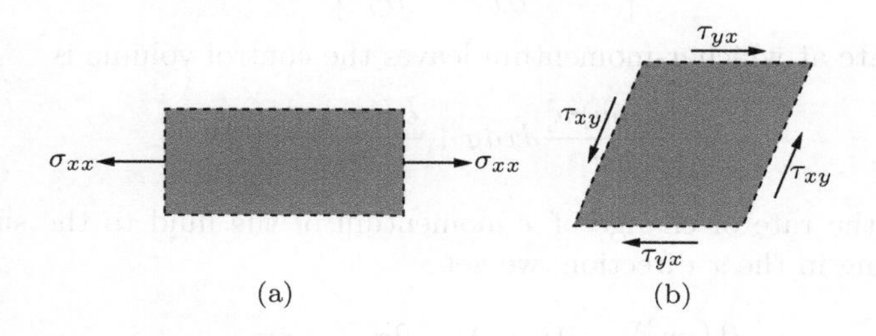


**Figure 5.6**
Deformation of a fluid element due to viscous stress, (a) linear deformation due to normal stress, (b) angular deformation due to shear stresses.

Moreover, the magnitude of a stress is proportional to the *rate* at which the deformation occurs. The deformation rate, in turn, is related to the viscosity of the fluid and the velocity gradient in the flow. For a *Newtonian fluid* (a fluid for which the shear stress is linearly proportional to the rate of angular deformation), the stresses are proportional to the velocity gradients, and the proportionality constant is the viscosity coefficient $\mu$ of the fluid. It can be shown that [1]

$$\sigma_{xx} = -p + 2\,\mu\frac{\partial u}{\partial x} - \frac{2}{3}\,\mu\left(\frac{\partial u}{\partial x} + \frac{\partial v}{\partial y}\right) \tag{5.8}$$

$$\sigma_{yy} = -p + 2\,\mu\frac{\partial v}{\partial y} - \frac{2}{3}\,\mu\left(\frac{\partial u}{\partial x} + \frac{\partial v}{\partial y}\right) \tag{5.9}$$

$$\tau_{xy} = \tau_{yx} = \mu\left(\frac{\partial u}{\partial x} + \frac{\partial v}{\partial y}\right) \tag{5.10}$$

Equation (5.2) and Equations (5.6) to (5.10) provide a complete representation of conditions in a two-dimensional velocity boundary layer, and the velocity field in the boundary layer may be determined by solving these equations.

Now, substituting the expressions for $\sigma_{xx}$, $\sigma_{yy}$ and $\tau_{xy}$ [Equations (5.8), (5.9), (5.10)] into Equations (5.6) and (5.7), we get the $x$- and $y$-momentum equations as

$$\rho\left(u\frac{\partial u}{\partial x}+v\frac{\partial u}{\partial y}\right)=\frac{\partial}{\partial x}\left[-p+2\,\mu\frac{\partial u}{\partial x}-\frac{2}{3}\,\mu\left(\frac{\partial u}{\partial x}+\frac{\partial v}{\partial y}\right)\right]+\mu\left[\frac{\partial}{\partial y}\left(\frac{\partial u}{\partial x}+\frac{\partial v}{\partial y}\right)\right]+X \tag{5.11a}$$

$$\rho\left(u\frac{\partial v}{\partial x}+v\frac{\partial v}{\partial y}\right)=\frac{\partial}{\partial y}\left[-p+2\,\mu\frac{\partial v}{\partial x}-\frac{2}{3}\,\mu\left(\frac{\partial u}{\partial x}+\frac{\partial v}{\partial y}\right)\right]+\mu\left[\frac{\partial}{\partial x}\left(\frac{\partial u}{\partial x}+\frac{\partial v}{\partial y}\right)\right]+Y \tag{5.11b}$$

But for incompressible flows, by continuity equation,

$$\frac{\partial u}{\partial x}+\frac{\partial v}{\partial y}=0$$

Therefore, for incompressible flows, the above $x$- and $y$-momentum equations become

$$\rho\left(u\frac{\partial u}{\partial x}+v\frac{\partial u}{\partial y}\right)=X-\frac{\partial p}{\partial x}+\mu\frac{\partial^2 u}{\partial x^2} \tag{5.12}$$

$$\rho\left(u\frac{\partial v}{\partial x}+v\frac{\partial v}{\partial y}\right)=Y-\frac{\partial p}{\partial y}+\mu\frac{\partial^2 v}{\partial y^2} \tag{5.13}$$

These two components of momentum equations can be combinedly written in vectorial form as

$$\rho V.\nabla V=F_B-\nabla p+\mu\nabla^2 V \tag{5.14}$$

where $F_B$ is the body force.

### 5.3.2 Thermal Boundary Layer

For applying the energy conservation principle to a differential control volume in the thermal boundary layer, as shown in Figure 5.7, it is essential to have a thorough understanding of the relevant physical processes.

Neglecting the potential energy effects, the energy per unit mass of the fluid includes the internal energy $e$ and the kinetic energy $V^2/2$, where $V^2=u^2+v^2$. Thus, the internal energy and kinetic energy are *advected* with the *bulk fluid* motion across the control surfaces. The net *rate* at which the kinetic energy *enters* the control volume in the $x$-direction is

$$\dot{E}_{\text{adv},x}-\dot{E}_{\text{adv},x+dx}\equiv\rho u\left(e+\frac{V^2}{2}\right)dy$$
$$-\left\{\rho u\left(e+\frac{V^2}{2}\right)+\frac{\partial}{\partial x}\left[\rho u\left(e+\frac{V^2}{2}\right)\right]dx\right\}dy$$

This simplifies to

$$\dot{E}_{\text{adv},x}-\dot{E}_{\text{adv},x+dx}=-\frac{\partial}{\partial x}\left\{\rho u\left(e+\frac{V^2}{2}\right)\right\}dxdy \tag{5.15}$$

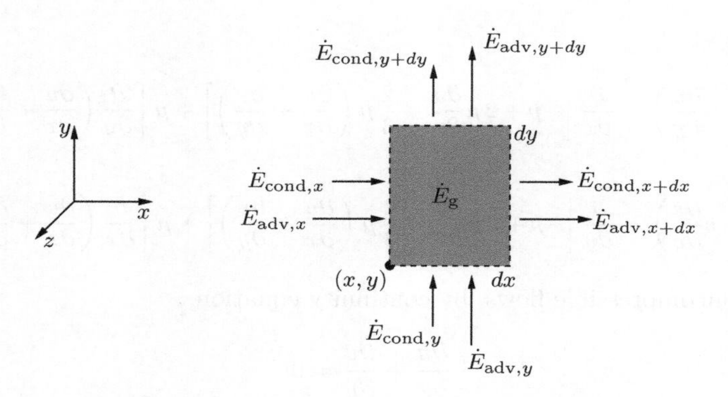


**Figure 5.7**
Differential control volume ($dx.dy.1$) for energy conservation in a two-dimensional thermal boundary layer.

Similarly, the rate at which the kinetic energy enters the control volume in the $y$-direction is

$$\dot{E}_{\text{adv},y} - \dot{E}_{\text{adv},y+dy} = -\frac{\partial}{\partial y}\left[\rho v\left(e + \frac{V^2}{2}\right)\right] dxdy \tag{5.15a}$$

Energy is also transported across the control surface by *molecular diffusion process*. Thus, the following are two modes of energy transport across the control surface.

• The first mode consists of energy transfer due to *conduction* and energy transfer due to the *diffusion* of species, say $A$ and $B$. But the energy transport through species diffusion is considerable only in boundary layers which are chemically reacting. Therefore, in the present analysis where the chemical reactions are not considered, the diffusion effect is neglected. The *net* transfer of energy by conduction, in the $x$-direction, into the control volume is

$$\dot{E}_{\text{cond},x} - \dot{E}_{\text{cond},x+dx} = -\left(k\frac{\partial T}{\partial x}\right) dy - \left[-k\frac{\partial T}{\partial x} - \frac{\partial}{\partial x}\left(k\frac{\partial T}{\partial x}\right) dx\right] dy$$

This simplifies to

$$\dot{E}_{\text{cond},x} - \dot{E}_{\text{cond},x+dx} = \frac{\partial}{\partial x}\left(k\frac{\partial T}{\partial x}\right) dxdy \tag{5.16}$$

The rate at which energy by conduction enters into the control volume in the $y$-direction can be shown as

$$\dot{E}_{\text{cond},y} - \dot{E}_{\text{cond},y+dy} = \frac{\partial}{\partial y}\left(k\frac{\partial T}{\partial y}\right) dxdy \tag{5.16a}$$

• The second mode is energy transfer by work interaction. In this mode, energy is transferred to and from the fluid in the control volume by *work* interaction involving the *body* and the *surface forces.* The rate at which work is done on the fluid in the control volume, by the forces acting in the $x$-direction may be expressed as

$$\dot{W}_{\text{net},x} = (Xu)\, dxdy + \frac{\partial}{\partial x}\left[(\sigma_{xx} - p)\, u\right] dxdy + \frac{\partial}{\partial y}(\tau_{yx} u)\, dxdy \tag{5.17}$$

where $X$ is the body force, $p$ is pressure, and $\sigma_{xx}$ and $\tau_{yx}$ are the normal and shear stresses, respectively. Equation (5.17) represents the work done by the body force, the pressure force and viscous forces.

In a similar manner, the rate at which work is done on the fluid in the control volume, by the forces acting in the $y$-direction becomes

$$\dot{W}_{\text{net},y} = (Yv)\, dydx + \frac{\partial}{\partial y}\left[(\sigma_{yy} - p)\, v\right] dydx + \frac{\partial}{\partial x}(\tau_{xy} v)\, dydx \tag{5.17a}$$

where $Y$ is the body force, $p$ is pressure, and $\sigma_{yy}$ and $\tau_{xy}$ are the normal and shear stresses, respectively.

Combining the energy transport into the control volume by kinetic energy, conduction, and due to work interaction involving body and surface forces, given by Equations (5.15) to (5.17), and the energy generated within the control volume, the energy conservation equation can be expressed as

$$-\frac{\partial}{\partial x}\left[\rho u\left(e + \frac{V^2}{2}\right)\right] - \frac{\partial}{\partial y}\left[\rho v\left(e + \frac{V^2}{2}\right)\right] + \frac{\partial}{\partial x}\left(k\frac{\partial T}{\partial x}\right) + \frac{\partial}{\partial y}\left(k\frac{\partial T}{\partial y}\right)$$
$$+ (Xu + Yv) - \frac{\partial}{\partial x}(pu) - \frac{\partial}{\partial y}(pv) + \frac{\partial}{\partial x}(\sigma_{xx} u + \tau_{xy} v)$$
$$+ \frac{\partial}{\partial x}(\tau_{yx} u + \sigma_{yy} v) + \dot{g} = 0 \tag{5.18}$$

where $\dot{g}$ is the rate at which energy is generated per unit volume. Equation (5.18) is the general form of *energy conservation equation.* Equation (5.18) implies that, in addition to energy transport by kinetic energy, conduction and work interaction, there may be energy generation within the control volume. If any such energy source is present, that contribution also should be accounted for in the energy balance.

The energy conservation equation (5.18) accounts for the conservation of both *mechanical* and *thermal* energies, in a thermal boundary layer. Therefore, it is rarely used for solving heat transfer problems. A more convenient form, termed the *thermal energy equation,* which is of direct application to heat transfer analysis can be obtained by multiplying Equations (5.6) and (5.7) with $u$ and $v$, respectively, and subtracting the results from Equation (5.18), as

$$\rho u\frac{\partial h}{\partial x}+\rho v\frac{\partial h}{\partial y}=\frac{\partial}{\partial x}\left(k\frac{\partial T}{\partial x}\right)+\frac{\partial}{\partial y}\left(k\frac{\partial T}{\partial y}\right)+\left(u\frac{\partial p}{\partial x}+v\frac{\partial p}{\partial y}\right)+\mu\Phi+\dot{\text{g}} \quad (5.19)$$

where $h$ is the enthalpy per unit mass of mixture, defined as, $h = e + p/\rho$ and $\mu\Phi$ is the *viscous dissipation*, defined as

$$\mu\Phi \equiv \mu\left[\left(\frac{\partial u}{\partial y}+\frac{\partial v}{\partial x}\right)^2+2\left[\left(\frac{\partial u}{\partial x}\right)^2+\left(\frac{\partial v}{\partial y}\right)^2\right]-\frac{2}{3}\left(\frac{\partial u}{\partial x}+\frac{\partial v}{\partial y}\right)^2\right] \quad (5.20)$$

The first term on the right-hand side of Equation (5.20) is due to the viscous shear stresses, and the remaining terms are due to the viscous normal stresses. Equation (5.20) gives the rate at which *mechanical energy is converted to thermal energy due to viscous effects in the fluid.*

The thermal energy equation [Equation (5.19)] can also be expressed in the following form, which is very much useful for convective heat transfer studies. In the present form, Equation (5.19) is

$$\rho u\frac{\partial h}{\partial x}+\rho v\frac{\partial h}{\partial y}=\left(u\frac{\partial p}{\partial x}+v\frac{\partial p}{\partial y}\right)+k\nabla^2 T+\mu\Phi+\dot{\text{g}} \quad (5.21)$$

But we know that, $h = h(T, p)$, where $p$ and $T$ are state variables. Therefore,

$$dh=\left(\frac{\partial h}{\partial T}\right)_p dT+\left(\frac{\partial h}{\partial p}\right)_T dp \quad (5.22)$$

Also, by definition,

$$\left(\frac{\partial h}{\partial T}\right)_p \equiv c_p \quad \text{and} \quad \left(\frac{\partial h}{\partial p}\right)_T=\frac{(1-\beta T)}{\rho}$$

where $\beta$ is the coefficient of thermal expansion defined as

$$\beta \equiv -\frac{1}{\rho}\left(\frac{\partial \rho}{\partial T}\right)_p$$

Thus,

$$dh=c_p\,dT+\frac{1-\beta T}{\rho}dp \quad (5.23)$$

Therefore,

$$\frac{\partial h}{\partial x}=c_p\frac{\partial T}{\partial x}+\frac{1-\beta T}{\rho}\frac{\partial p}{\partial x} \quad (5.24)$$

$$\frac{\partial h}{\partial y}=c_p\frac{\partial T}{\partial y}+\frac{1-\beta T}{\rho}\frac{\partial p}{\partial y} \quad (5.25)$$

Substituting these two expressions into Equation (5.21), we get

$$\boxed{\rho c_p \left( u\frac{\partial T}{\partial x} + v\frac{\partial T}{\partial y} \right) = \beta T \left( u\frac{\partial p}{\partial x} + v\frac{\partial p}{\partial y} \right) + k \nabla^2 T + \mu\Phi + \dot{g}} \tag{5.26}$$

This form of energy equation is more convenient for convective heat transfer analysis.

## 5.4 Approximation and Special Conditions

In the previous section, we have derived the equations which govern the physical processes that influence the conditions in the velocity and thermal boundary layers. However, only rarely all the terms in the governing equation need to be considered. Therefore, it is a usual practice to work with simplified forms of the equations in the studies of convective heat transfer. One such situation commonly encountered is that in which the two-dimensional boundary layer can be assumed to be *steady* (time-independent), *incompressible* ($\rho$ is constant), having constant properties ($k, \mu$, etc.) and negligible body forces ($X \approx 0, Y \approx 0$), and without energy generation ($\dot{g} = 0$).

In addition to the above assumptions, further simplifications can be made by using *boundary layer approximations.* We know that, the boundary layer is a thin layer adjacent to a solid surface, in which the flow velocity increases from zero to freestream level, as illustrated in Figure 5.1. Because the boundary layers are very thin, the following approximations can be applied.

• For velocity boundary layer,

$$u \gg v$$

$$\frac{\partial u}{\partial y} \gg \frac{\partial u}{\partial x}, \; \frac{\partial v}{\partial y}, \; \frac{\partial v}{\partial x}$$

• For thermal boundary layer,

$$\frac{\partial T}{\partial y} \gg \frac{\partial T}{\partial x}$$

The above approximations clearly demonstrate that, the velocity component in the direction along the surface is much larger than the component of velocity normal to the surface, and the velocity and temperature gradients normal to the surface are much larger than the gradients along the surface. In other words, the diffusion of momentum and heat in the direction normal to the surface is significantly larger than these transports along the surface.

• The normal stresses ($\sigma_{xx}$ and $\sigma_{yy}$) given by Equations (5.8) and (5.9) are negligible, and the single relevant shear stress component of Equation (5.10) reduces to

$$\tau_{xy} = \tau_{yx} = \mu \left( \frac{\partial u}{\partial y} \right) \tag{5.27}$$

• Furthermore, the conduction and the species diffusion rates in the $y$-direction are much larger than those in the $x$-direction.

With the above approximations, the continuity equation [Equation (5.2)] and the $x$-momentum equation [Equation (5.6)] reduce to

$$\boxed{\frac{\partial u}{\partial x} + \frac{\partial v}{\partial y} = 0} \tag{5.28}$$

$$\boxed{u\frac{\partial u}{\partial x} + v\frac{\partial u}{\partial y} = -\frac{1}{\rho}\frac{\partial p}{\partial x} + \nu\frac{\partial^2 u}{\partial y^2}} \tag{5.29}$$

where $\nu = \mu/\rho$ is the *kinematic viscosity*, it is also known as *momentum diffusivity*.

Also, from an order-of-magnitude analysis, using the velocity boundary layer approximations, it can be shown that the $y$-momentum equation [Equation (5.7)] reduces to [2]

$$\boxed{\frac{\partial p}{\partial y} = 0} \tag{5.30}$$

That is, the *pressure does not vary in the direction normal to the surface.* In other words, the static pressure in the freestream flow above the boundary layer is impressed through the boundary layer, to the surface of the body over which the boundary layer prevails. Hence, the static pressure $p$ in the boundary layer depends only on $x$, and is equal to the pressure in the freestream flow outside the boundary layer. The form of $p(x)$, which depends on the surface geometry, may then be obtained from a separate consideration of flow conditions in the freestream. Hence, as far as Equation (5.29) is concerned, $(\partial p/\partial x) = (dp/dx)$, and the pressure gradient may be treated as a known quantity.

With the above assumptions and simplifications, the energy equation [Equation (5.26)] reduces to

$$\boxed{u\frac{\partial T}{\partial x} + v\frac{\partial T}{\partial y} = \alpha\frac{\partial^2 T}{\partial y^2} + \frac{\nu}{c_p}\left(\frac{\partial u}{\partial y}\right)^2} \tag{5.31}$$

where $\alpha = k/(\rho c_p)$ is termed *thermal diffusivity.*

It is important to note that, in most situations the last term on the right-hand side of Equation (5.31) may be neglected relative to the terms which account for the advection (the terms on the left-hand side of the equation) and conduction (the first term of the right-hand side). But in flows such as sonic flows and high-speed motion of lubricating oils, the viscous dissipation could be considerable and hence must be accounted for.

Equations (5.28), (5.29) and (5.31) may be solved to determine the spatial variations of $u$, $v$, and $T$ in boundary layers. For an incompressible, constant property flow, Equations (5.28) and (5.29) are *uncoupled* from Equation (5.31). That is, Equations (5.28) and (5.29) may be solved for the velocity field, $u(x, y)$ and $v(x, y)$, without considering the energy equation [Equation (5.31)]. Using this $u(x, y)$, the velocity gradient $(\partial u/\partial y)_{y=0}$ and the shear stress can be calculated. But the temperature $T$ in the energy Equation (5.31) is coupled to the velocity field. Therefore, the velocity components $u(x, y)$ and $v(x, y)$ must be known for solving Equation (5.31) to obtain $T(x, y)$. Once the temperature distribution $T(x, y)$ is known, the convective heat transfer coefficient can be determined.

The following analysis will make use of the above equations to identify the key *boundary layer similarity parameters*, as well as important *analogies* between the momentum transfer and heat transfer.

## 5.5 Solving Convection Problems

For solving convection heat transfer problems, we have to consider continuity, momentum, and energy equations simultaneously. Except for very simple situations, it is extremely difficult to obtain analytical solutions with these equations. However, under simplified conditions, it is possible to deal with these equations comfortably. Another way to simplify these equations is to make use of order-of-magnitude analysis and neglect few terms in comparison with others.

### 5.5.1 Similarity Parameters in Heat Transfer

Consider the steady incompressible flow at temperature $T_\infty$ about a body with surface temperature $T_s$, as shown in Figure 5.8.

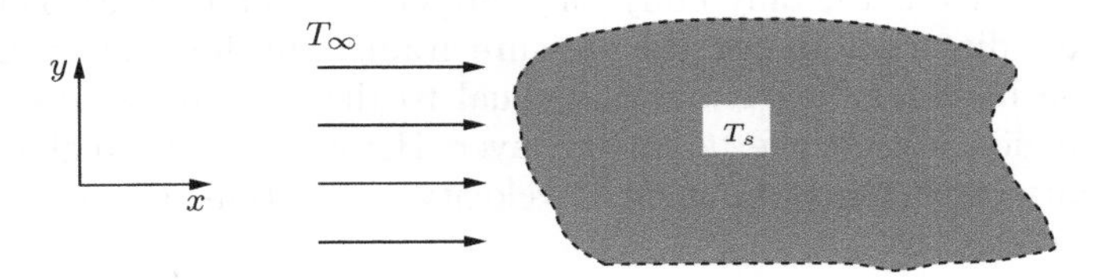


**Figure 5.8**
Incompressible flow past a body.

Let gravity be the only body force acting. For this problem, the governing equations are the following.

The continuity equation [from Equation (5.2)], for this flow with constant density, simplifies to

$$\frac{\partial u}{\partial x} + \frac{\partial v}{\partial y} = 0 \tag{5.32}$$

The $x$- and $y$-momentum equations for a steady, two-dimensional, incompressible flow of Newtonian fluid, in the Cartesian coordinate system, given by Equation (5.14), are

$$u\frac{\partial u}{\partial x} + v\frac{\partial u}{\partial y} = -\frac{1}{\rho}\frac{\partial p}{\partial x} + \frac{\mu}{\rho}\left(\frac{\partial^2 u}{\partial x^2} + \frac{\partial^2 u}{\partial y^2}\right) \tag{5.33}$$

$$u\frac{\partial v}{\partial x} + v\frac{\partial v}{\partial y} = g\beta\left(T - T_\infty\right) + \frac{\mu}{\rho}\left(\frac{\partial^2 v}{\partial x^2} + \frac{\partial^2 v}{\partial y^2}\right) \tag{5.34}$$

The terms on the left-hand side of Equation (5.33) represent the *inertia* forces, the first term on the right-hand side is the *pressure* force, and the second term on the right-hand side is the *viscous* force.

The energy equation [from Equation (5.26)] is

$$\rho\, c_p\left(u\frac{\partial v}{\partial x} + v\frac{\partial v}{\partial y}\right) = \beta\, T\left(u\frac{\partial p}{\partial x} + v\frac{\partial p}{\partial y}\right) + k\left(\frac{\partial^2 T}{\partial x^2} + \frac{\partial T}{\partial y^2}\right) + \mu\Phi \tag{5.35}$$

Note that, in the $y$-momentum equation [Equation (5.34)], $-\frac{1}{\rho}\frac{\partial p}{\partial y}$ is expressed as $g\beta\left(T - T_\infty\right)$, we can obtain the same as follows. Further, it is essential to note that, the $y$-momentum equation and the energy equation cannot be separated due to the heat (or temperature) term present in the momentum equation. Therefore, they must be solved simultaneously.

The $y$-momentum equation for the present problem is

$$u\frac{\partial v}{\partial x} + v\frac{\partial v}{\partial y} = -\frac{1}{\rho}\frac{\partial p}{\partial y} - g + \nu\frac{\partial^2 v}{\partial x^2} \tag{5.36}$$

This equation may be expressed in a more convenient form by noting that the gravitational force is the only body force present; therefore, there is no body force in the $x$-direction. Hence, the pressure gradient in the $y$-direction at any point in the boundary layer must be equal to the pressure gradient in the quiescent region *outside* the boundary layer. However, in the region outside the boundary layer, the component of velocity $v = 0$, thus, the $y$-momentum equation reduces to

$$\frac{\partial p}{\partial y} = -\rho_\infty\, g \tag{5.37}$$

Note that, the density is represented with subscript $\infty$, to indicate that condition outside the boundary layer is applied. Now, substituting this into the $y$-momentum Equation (5.36), we get

$$u\frac{\partial v}{\partial x} + v\frac{\partial v}{\partial y} = \frac{g}{\rho}\left(\rho_\infty - \rho\right) + \nu\,\frac{\partial^2 v}{\partial x^2} \tag{5.38}$$

The first term on the right-hand side of this equation is the buoyancy force, and this can cause a buoyant flow, since $\rho$ is a variable. The origin for the variation of density may be made more explicit by introducing a *volumetric thermal expansion coefficient*, $\beta$,

$$\beta = -\frac{1}{\rho}\left(\frac{\partial \rho}{\partial T}\right)_p$$

This *thermodynamic property* of the fluid provides a measure of the amount by which the density changes in response to a change in temperature at constant pressure. If it is expressed in the following approximate form,

$$\beta \approx -\frac{1}{\rho}\left(\frac{\rho_\infty - \rho}{T_\infty - T}\right)$$

it follows that,

$$(\rho_\infty - \rho) \approx \rho\beta\,(T - T_\infty)$$

This is the term used in Equation (5.34).

## 5.6 Boundary Layer Similarity Parameters

The continuity and momentum equations [Equations (5.32) - (5.34)] for the two-dimensional flow past the body shown in Figure 5.8 can be simplified by defining dimensionless lengths $x^*$, $y^*$, dimensionless velocity components $u^*$, $v^*$, dimensionless temperature $\theta$, and dimensionless pressure $p^*$.

The dimensionless independent variables are defined as

$$x^* \equiv \frac{x}{L} \quad \text{and} \quad y^* \equiv \frac{y}{L} \tag{5.39}$$

where $L$ is some *characteristic length* for the surface of interest (e.g., length of a flat plate). The dependent dimensionless variables ($u^*$, $v^*$) may be defined as

$$u^* \equiv \frac{u}{U} \quad \text{and} \quad v^* \equiv \frac{v}{U} \tag{5.40}$$

where $U$ is the freestream flow velocity upstream of the surface.

The dimensionless forms of temperature and pressure can be expressed as

$$\theta = \frac{T - T_s}{T_\infty - T_s} \tag{5.41}$$

$$p^* = \frac{p}{\frac{1}{2}\rho U^2} \tag{5.42}$$

where subscripts '$s$' and '$\infty$' refer to the plate surface and freestream flow, respectively.

Substituting Equations (5.39) to (5.42), the governing Equations (5.32) to (5.35) can be expressed in the dimensionless form as

$$\frac{\partial u^*}{\partial x^*} + \frac{\partial v^*}{\partial y^*} = 0 \tag{5.43}$$

$$u^* \frac{\partial u^*}{\partial x^*} + v^* \frac{\partial u^*}{\partial y^*} = -\frac{1}{\rho} \frac{\partial p^*}{\partial x^*} + \frac{\nu}{UL} \left( \frac{\partial^2 u^*}{\partial x^{*2}} + \frac{\partial^2 u^*}{\partial y^{*2}} \right) \tag{5.44}$$

$$u^* \frac{\partial v^*}{\partial x^*} + v^* \frac{\partial v^*}{\partial y^*} = -\frac{g\beta L}{v^2} (T_s - T_\infty) + \frac{\nu}{UL} \left( \frac{\partial^2 v^*}{\partial x^{*2}} + \frac{\partial^2 v^*}{\partial y^{*2}} \right) \tag{5.45}$$

$$u^* \frac{\partial \theta}{\partial x^*} + v^* \frac{\partial \theta}{\partial y^*} = -\frac{U^2}{2c_p (T_s - T_\infty)} \left[ \beta T_\infty + \beta (T_s - T_\infty) \theta \right] \left( u^* \frac{\partial p^*}{\partial x*} + v^* \frac{\partial p^*}{\partial y^*} \right)$$

$$+ \frac{k}{\rho_\infty c_p UL} \left[ \frac{\partial^2 \theta}{\partial x^{*2}} + \frac{\partial^2 \theta}{\partial y^{*2}} \right] + \frac{\mu U}{\rho_\infty c_p UL (T_s - T_\infty)} \Phi^* \tag{5.46}$$

where

$$\Phi^* = 2 \left( \left( \frac{\partial u^*}{\partial x^*} \right)^2 + \left( \frac{\partial v^*}{\partial y^*} \right)^2 \right) + \left[ \frac{\partial u^*}{\partial x^*} + \frac{\partial v^*}{\partial y^*} \right]^2$$

It is interesting to note that the following group of parameters,
$\nu/(UL)$, $g\beta L (T_s - T_\infty)/U^2$, $k/(\rho_\infty c_p UL)$, $\mu U/[\rho_\infty c_p L (T_s - T_\infty)]$ and $(U^2 \beta T_\infty)/[2c_p (T_s - T_\infty)]$
appearing in Equations (5.43) to (5.46), are all dimensionless groups.

These equations suggest that although conditions in the velocity and thermal boundary layers depend on the fluid and material properties, and the length scale $L$, this dependence may be simplified by grouping these variables in a functional form as follows.

$$\theta = \theta \left( x^*, y^*, \frac{\nu}{UL}, \frac{g\beta L \Delta T}{U^2}, \frac{k}{\rho_\infty c_p UL}, \frac{\mu U}{\rho_\infty c_p L \Delta T}, \frac{U^2 \beta T_\infty}{2c_p \Delta T} \right) \tag{5.47}$$

The dimensionless groups in Equation (5.47) are the similarity parameters associated with the present heat transfer process. We can express the above similarity parameters as follows.

$$\begin{aligned} \frac{g\beta L \Delta T}{U^2} &= \frac{g\beta L^3 \Delta T}{\frac{U^2 L^2}{\nu^2} \nu^2} \\ &= \frac{g\beta L^3 \Delta T}{\nu^2} \frac{1}{\mathrm{Re}^2} \\ &= \frac{\mathrm{Gr}}{\mathrm{Re}^2} \end{aligned}$$

where the group $\left(\frac{g\beta L^3 \Delta T}{\nu^2}\right)$ is called the *Grashof number* Gr, and $\left(\frac{UL}{\nu}\right)$ is the *Reynolds number* Re.

Also, the *Prandtl number* Pr and *Eckert number* Ec are defined as

$$\mathrm{Pr} = \frac{\mu c_p}{k} = \frac{\nu}{\alpha}$$

and

$$\mathrm{Ec} = \frac{U^2}{c_p \Delta T}$$

where $c_p$ is the specific heat of the fluid.

Using the Pr, Re, Ec and Gr, the governing Equations (5.44) to (5.46) can be written as

$$u^* \frac{\partial u^*}{\partial x^*} + v^* \frac{\partial u^*}{\partial y^*} = -\frac{1}{\rho}\frac{\partial p^*}{\partial x^*} + \frac{1}{\mathrm{Re}}\left(\frac{\partial^2 u^*}{\partial x^{*2}} + \frac{\partial^2 u^*}{\partial y^{*2}}\right) \tag{5.48}$$

$$u^* \frac{\partial v^*}{\partial x^*} + v^* \frac{\partial v^*}{\partial y^*} = \frac{Gr}{\mathrm{Re}^2}\theta + \frac{1}{\mathrm{Re}}\left(\frac{\partial^2 v^*}{\partial x^{*2}} + \frac{\partial^2 v^*}{\partial y^{*2}}\right) \tag{5.49}$$

$$u^* \frac{\partial \theta}{\partial x^*} + v^* \frac{\partial \theta}{\partial y^*} = \left(u^* \frac{\partial p^*}{\partial x^*} + v^* \frac{\partial p^*}{\partial y^*}\right) \frac{\mathrm{Ec}}{2}\Big[\beta T_\infty + \beta\left(T_s - T_\infty\right)\theta\Big]$$
$$+\frac{1}{\mathrm{PrRe}}\left(\frac{\partial^2 \theta}{\partial x^{*2}} + \frac{\partial^2 \theta}{\partial y^{*2}}\right) + \frac{\mathrm{Ec}}{\mathrm{Re}}\Phi^* \tag{5.50}$$

In functional form, the dimensionless temperature $\theta$ can be expressed as

$$\theta = \theta\Big(x^*,\ y^*,\ \mathrm{Re},\ \mathrm{Pr},\ \mathrm{Gr},\ \mathrm{Ec},\ \beta T_\infty,\ \beta\left(T_s - T_\infty\right)\Big) \tag{5.51}$$

The heat flux at the wall is given by

$$\dot{q}_{\mathrm{th}}\Big|_{\mathrm{surface}} = -k\left(\frac{\partial T}{\partial n}\right)_{\mathrm{surface}}$$

But this flux is convected away by the fluid; therefore, the heat flux at the wall is also given by

$$\dot{q}_{\mathrm{th}} = h\left(T_s - T_\infty\right)$$

Equating these two fluxes, we get the convection heat transfer coefficient as

$$h = -\frac{k\left(\frac{\partial T}{\partial n}\right)_{\mathrm{surface}}}{T_s - T_\infty} \tag{5.52}$$

In the nondimensional form, the film transfer coefficient becomes

$$\frac{hL}{k} = -\left(\frac{\partial \theta}{\partial n^*}\right)_{\mathrm{surface}} \tag{5.53}$$

where $L$ is the characteristic dimension of the body over which the flow is taking place and $n^* = n/L$. The group $\left(\dfrac{hL}{k}\right)$ is known as *Nusselt number* Nu, and the functional form of $\theta$ given by Equation (5.51) can be written as

$$\mathrm{Nu} = \mathrm{Nu}\left(x^*, y^*, \mathrm{Re}, \mathrm{Pr}, \mathrm{Gr}, \mathrm{Ec}, \beta T_\infty, \beta\left(T_s - T_\infty\right)\right) \tag{5.54}$$

The analysis of convection heat transfer aims at determining this functional relationship.

## 5.7 Physical Significance of Dimensionless Parameters

All the foregoing dimensionless parameters have physical interpretations that relate to conditions in the flow field.

- *Reynolds number*, $\mathrm{Re} = \dfrac{\rho U L}{\mu}$, may be interpreted as the ratio of inertia force to viscous force. For a differential control volume in a boundary layer, inertia forces are associated with the momentum flux of the fluid moving through the control volume. From Equation (5.5), it is evident that the inertia forces are of the form $\partial\left[(\rho u)\, u\right]/\partial x$, in which case an order of magnitude approximation gives the inertia force as $F_I \approx \rho U^2/L$. Similarly, the net shear force is of the form $\partial \tau_{yx}/\partial y = \partial\left[\mu\left(\partial u/\partial y\right)\right]/\partial y$, and may be approximated as $F_s \approx \mu U/L^2$. The ratio of the inertia force to viscous force is

$$\frac{F_I}{F_s} \approx \frac{\rho U^2/L}{\mu U/L^2} = \frac{\rho U L}{\mu} = \mathrm{Re}_L$$

Therefore, inertia forces dominate at large values of Re and viscous forces dominate when the Reynolds number is small.

- *Prandtl number*, $\mathrm{Pr} = \dfrac{\nu}{\alpha}$, may be interpreted as the ratio of momentum diffusivity $\nu$ to the thermal diffusivity $\alpha$. The Prandtl number may be viewed as a measure of the *relative effectiveness* of momentum transport and energy transport by diffusion in the velocity (momentum) and thermal boundary layers, respectively. For example, for gases, $\mathrm{Pr} \approx 1$; therefore, the energy and momentum transfers by diffusion are comparable. For liquid metals, $\mathrm{Pr} \ll 1$, hence the energy diffusion rate greatly exceeds the momentum diffusion rate. The opposite is true for oils, for which $\mathrm{Pr} \gg 1$. Thus, the value of Prandtl number Pr strongly influences the relative growth of the velocity and thermal boundary layers. In fact, for laminar boundary layers (in which transport by diffusion is not overshadowed by turbulent mixing), it is reasonable to expect that,

$$\frac{\delta}{\delta_t} \approx \mathrm{Pr}^n$$

where $\delta$ is the thickness of the velocity (momentum) boundary layer, $\delta_t$ is the thickness of the thermal boundary layer, and $n$ is a positive exponent. Hence, for a gas, $\delta_t \approx \delta$, for a liquid metal, $\delta_t > \delta$ and for an oil, $\delta_t < \delta$.

The above-mentioned thicknesses of momentum and thermal boundary layers are schematically shown in Figure 5.9, for (a) air with $\mathrm{Pr} \approx 1$, (b) for liquid metal with $\mathrm{Pr} > 1$ and (c) for oil flow with $\mathrm{Pr} < 1$.

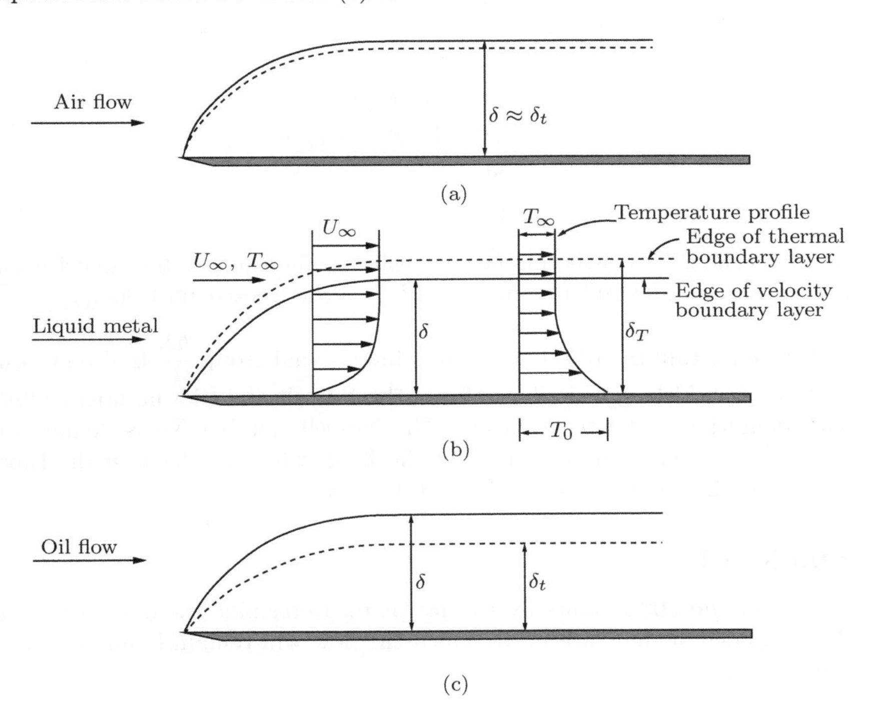


**Figure 5.9**
Comparison of momentum and thermal boundary layers for (a) air flow, (b) liquid metal flow, and (c) oil flow past a flat plate.

- *Grashof number*, $\mathrm{Gr} = \dfrac{g\beta L^3 \Delta T}{\nu^2}$, provides a measure of the ratio of buoyancy force to viscous force in the boundary layer. Its role in free convection is much the same as that of the Reynolds number in forced convection.

- *Eckert number*, $\mathrm{Ec} = \dfrac{U^2}{c_p \Delta T}$, provides a measure of kinetic energy of the flow relative to the enthalpy difference across the thermal boundary layer. It plays an important role in high-speed flows for which viscous dissipation is substantial.

- *Stanton number*, $\text{St} = \dfrac{h_x}{\rho c_p U_\infty}$, is a measure of the heat flux to the heat capacity of the fluid flow. The Stanton number can be expressed as the ratio of Nusselt number to the product of Reynolds number and Prandtl number, as follows.

$$\begin{aligned} \text{St} &= \frac{h_x}{\rho c_p U_\infty} \\ &= \frac{(h_x L)/k}{[(\rho U_\infty L)/\mu]\ [(\mu c_p)/k]} \\ &= \frac{\text{Nu}_x}{\text{Re}_x \text{Pr}} \end{aligned}$$

where, $h_x$ is the local convection heat transfer coefficient, $\rho$ is flow density, $c_p$ is specific heat at constant pressure, and $U_\infty$ is the freestream velocity.

Note: It is important to note that, the nondimensional group $\dfrac{hL}{k}$ is also known as *Biot number*. Although similar in form, the Nusselt and Biot numbers differ in both definition and interpretation. The Nusselt number Nu is defined in terms of the thermal conductivity $k$ of the fluid, whereas, the $k$ in the Biot number Bi is the thermal conductivity of the solid.

## Example 5.1

Air at 1 atm and 300 K flows over a flat plate. Determine the distance from the leading edge of the plate up to which the flow will remain laminar, when the flow speed is 30 m/s.

## Solution

Given, $p = 1$ atm, $T = 300$ K, $U = 30$ m/s. For air, the gas constant is $R = 287\ \text{m}^2/(\text{s}^2\ \text{K})$ and 1 atm $= 101325$ Pa.

By thermal equation of state, we have $p = \rho R T$. Therefore, the density of the flow becomes

$$\begin{aligned} \rho &= \frac{p}{RT} = \frac{101325}{287 \times 300} \\ &= 1.177\,\text{kg/m}^3 \end{aligned}$$

By Sutherland relation, we have the viscosity as [2]

$$\mu = 1.4 \times 10^{-6} \, \frac{T^{3/2}}{T + 111} \, \text{kg/(m s)}$$

Therefore, the viscosity at 300 K is

$$\begin{aligned} \mu &= 1.4 \times 10^{-6} \, \frac{300^{3/2}}{300 + 111} \\ &= 1.85 \times 10^{-5} \, \text{kg/(m s)} \end{aligned}$$

The Reynolds number of the flow, based on the length of the plate is

$$\text{Re}_x = \frac{\rho U x}{\mu} = \frac{1.177 \times 30 \times x}{1.85 \times 10^{-5}}$$

The critical Reynolds number for flat plate is $5 \times 10^5$. Therefore, if the Reynolds number, based on the plate length is less than this value, the flow will remain laminar. Thus, the distance $x_{\text{cri}}$ from the leading edge up to which the flow is laminar becomes

$$\begin{aligned} x_{\text{cri}} &= \frac{(5 \times 10^5) \times (1.85 \times 10^{-5})}{1.177 \times 30} \\ &= \boxed{0.262 \, \text{m}} \end{aligned}$$

## 5.8 Boundary Layer Concepts

We know that at very high Reynolds numbers the viscous effects are far smaller than the inertia effects, and may therefore be neglected. For a boundary layer over a flat plate, as we know from order of magnitude analysis, the momentum boundary layer thickness $\delta$ is given by [2]

$$\boxed{\frac{\delta}{L} \sim \frac{1}{\sqrt{\text{Re}_L}}} \tag{5.55}$$

That is, the boundary layer region where the viscous stresses are important is very thin.

For a flat plate boundary layer with zero pressure gradient, Equation (5.48) becomes,

$$u^* \frac{\partial u^*}{\partial x^*} + v^* \frac{\partial u^*}{\partial y^*} = \frac{1}{\text{Re}_L} \frac{\partial^2 u^*}{\partial y^{*2}} \tag{5.56}$$

Let us assume that the plate surface is maintained at temperature $T_s$, and $T_\infty$ is the static temperature of the fluid flow. The governing equation for temperature distribution along the plate, given by Equation (5.50), is

$$u^* \frac{\partial \theta}{\partial x^*} + v^* \frac{\partial \theta}{\partial y^*} = \left( u^* \frac{\partial p^*}{\partial x^*} + v^* \frac{\partial p^*}{\partial y^*} \right) \frac{\text{Ec}}{2} \left[ \beta T_\infty + \beta \left( T_s - T_\infty \right) \theta \right]$$

$$+ \frac{1}{\text{Pr}\,\text{Re}_L} \left( \frac{\partial^2 \theta}{\partial x^{*2}} + \frac{\partial^2 \theta}{\partial y^{*2}} \right) + \frac{\text{Ec}}{\text{Re}_L} \Phi^* \tag{5.57}$$

The viscous term in this equation is negligible because of the small magnitude of $\text{Ec}/\text{Re}_L$. The Prandtl number Pr for most gases is around unity. But on the other hand $\text{Re}_L$ is generally very large. Therefore, the conduction effects in gases are generally negligible.

Now, nondimensionalizing $y$ with the thermal boundary layer thickness $\delta_t$ and proceeding in the similar manner as in the momentum boundary layer analysis, we can show that, the thickness of the thermal boundary layer can be expressed as

$$\boxed{\frac{\delta_t}{L} \sim \frac{1}{(\text{Re}_L \text{Pr})^{\frac{1}{2}}}} \tag{5.58}$$

and the thermal boundary layer equation [Equation (5.57)] can be expressed as

$$u^* \frac{\partial \theta}{\partial x^*} + v^* \frac{\partial \theta}{\partial y^*} = \frac{1}{\text{Pr}\,\text{Re}_L} \frac{\partial^2 \theta}{\partial y^{*2}} \tag{5.59}$$

The right-hand side of Equation (5.59) has the term for conduction in the $y$-direction.

Now, dividing Equation (5.58) by Equation (5.55), we get the ratio of velocity boundary layer thickness $\delta$ to thermal boundary layer thickness $\delta_t$, as

$$\frac{\delta_t}{\delta} = \frac{1}{\sqrt{\text{Pr}}}$$

The term $1/\text{Pr}$ can be expressed as

$$\begin{aligned} \frac{1}{\text{Pr}} &= \frac{k}{\mu\, c_p} \\ &= \frac{k}{(\mu/\rho)\, c_p\, \rho} \\ &= \frac{k/(\rho\, c_p)}{(\mu/\rho)} \\ &= \frac{\alpha}{\nu} \end{aligned}$$

Thus,

$$\boxed{\frac{\delta_t}{\delta} = \frac{1}{(\text{Pr})^{\frac{1}{2}}} = \sqrt{\frac{\alpha}{\nu}}} \tag{5.60}$$

From Equation (5.60) it is seen that, if the thermal diffusivity $\alpha$ is more than the kinematic viscosity $\nu$, $\delta_t > \delta$, and vice versa. For Pr of the order unity, both velocity and thermal boundary layer thicknesses are of the same order, as shown in Figure 5.9(a).

For $\Pr = 1$, the momentum and energy equations for flow over flat plate are, respectively,

$$u^* \frac{\partial u^*}{\partial x^*} + v^* \frac{\partial u^*}{\partial y^*} = \frac{1}{\mathrm{Re}_L} \frac{\partial^2 u^*}{\partial y^{*2}} \tag{5.61}$$

$$u^* \frac{\partial \theta}{\partial x^*} + v^* \frac{\partial \theta}{\partial y^*} = \frac{1}{\mathrm{Re}_L} \frac{\partial^2 \theta}{\partial y^{*2}} \tag{5.62}$$

The boundary conditions are

$$u^* = v^* = 0, \quad \text{at} \quad y^* = 0$$

$$u^* \to 1, \quad \text{as} \quad y^* \to \infty$$

and

$$\theta = 0, \quad \text{at} \quad y^* = 0$$

$$\theta \to 1, \quad \text{as} \quad y^* \to \infty$$

Defining $\theta^* = (1 - \theta)$, Equation (5.62) and the above boundary conditions can be expressed as

$$u^* \frac{\partial \theta^*}{\partial x^*} + v^* \frac{\partial \theta^*}{\partial y^*} = \frac{1}{\mathrm{Re}_L} \frac{\partial^2 \theta^*}{\partial y^{*2}} \tag{5.63}$$

$$\theta^* = 1, \quad \text{at} \quad y^* = 0$$

$$\theta^* \to 0, \quad \text{as} \quad y^* \to \infty$$

Note that, the momentum Equation (5.61) and energy Equation (5.63), in terms of $\theta^*$, and their boundary conditions are identical.

Now, an interesting relation between the Nusselt number and skin friction coefficient $c_f$ can be obtained as follows.

By definition, the Nusselt number is

$$\mathrm{Nu}_x = \frac{hx}{k}$$

Dividing and multiplying the right-hand side, by $L$, we get

$$\mathrm{Nu}_x = \frac{hL}{k} \frac{x}{L}$$

But by Equation (5.53),

$$\frac{hL}{k} = -\left(\frac{\partial \theta}{\partial y^*}\right)_w$$

Therefore,

$$\mathrm{Nu}_x = -\frac{x}{L}\left(\frac{\partial \theta}{\partial y^*}\right)_w$$

By definition, $\theta^* = (1-\theta)$, therefore,

$$\theta = 1 - \theta^*$$

$$\frac{\partial \theta}{\partial y^*} = -\frac{\partial \theta^*}{\partial y^*}$$

Substituting this, the Nusselt number becomes

$$\mathrm{Nu}_x = \frac{hx}{k} = -\frac{x}{L}\left(\frac{\partial \theta}{\partial y^*}\right)_w = \frac{x}{L}\left(\frac{\partial \theta^*}{\partial y^*}\right)_w \tag{5.64}$$

The term $\left(\frac{\partial u^*}{\partial y^*}\right)_w$ can be arranged as

$$\begin{aligned}
\left(\frac{\partial u^*}{\partial y^*}\right)_w &= \left(\frac{\partial (u/U_\infty)}{\partial (y/L)}\right)_w \\
&= \frac{L}{U_\infty}\left(\frac{\partial u}{\partial y}\right)_w \\
&= \frac{L\left(\frac{1}{2}\rho {U_\infty}^2\right)}{U_\infty \mu}\frac{\mu\left(\frac{\partial u}{\partial y}\right)_w}{\frac{1}{2}\rho {U_\infty}^2} \\
&= \frac{1}{2}\left(\frac{\rho U_\infty L}{\mu}\right)\left(\frac{\mu\left(\frac{\partial u}{\partial y}\right)_w}{\frac{1}{2}\rho {U_\infty}^2}\right)
\end{aligned}$$

By Newton's law of viscosity,

$$\tau_w = \mu\left(\frac{\partial u}{\partial y}\right)_w$$

Therefore,

$$\left(\frac{\partial u^*}{\partial y^*}\right)_w = \frac{1}{2}\mathrm{Re}_L \frac{\tau_w}{\frac{1}{2}\rho {U_\infty}^2}$$

But,

$$\frac{\tau_w}{\frac{1}{2}\rho {U_\infty}^2} = C_f$$

Therefore,

$$\left(\frac{\partial u^*}{\partial y^*}\right)_w = \frac{1}{2}\mathrm{Re}_L\, c_f \tag{5.65}$$

where $c_f$ is the local skin friction coefficient. Since the dimensionless velocity $u^*$ and dimensionless temperature $\theta^*$ are identical for this case, it follows from Equations (5.64) and (5.65) that,

$$\begin{aligned}
\mathrm{Nu}_x &= \frac{x}{L}\left(\frac{\partial \theta^*}{\partial y^*}\right)_w \\
&= \frac{x}{L}\left(\frac{\partial u^*}{\partial y^*}\right)_w \\
&= \frac{x}{L}\frac{1}{2}\mathrm{Re}_L C_f \\
&= \frac{x}{L}\frac{1}{2}\frac{\rho U L}{\mu} C_f \\
&= \frac{1}{2}\frac{\rho U x}{\mu} C_f
\end{aligned}$$

That is,

$$\boxed{\mathrm{Nu}_x = \frac{1}{2}\mathrm{Re}_x c_f} \tag{5.66}$$

This result is usually expressed in the following form with a new nondimensional parameter, namely the *Stanton number* St, as follows.

By definition,

$$\begin{aligned}
\mathrm{St} &= \frac{h}{\rho c_p U} \\
&= \frac{h}{\frac{\rho U x}{\mu}}\frac{x}{\mu c_p} \\
&= \frac{hx}{k}\frac{k}{\mu c_p}\frac{1}{\frac{\rho U x}{\mu}} \\
&= \frac{\mathrm{Nu}_x}{\mathrm{Re}_x \mathrm{Pr}}
\end{aligned}$$

Substituting this into Equation (5.66) and assuming the Prandtl number Pr to be unity, we get

$$\boxed{\mathrm{St} = \frac{1}{2} c_f} \tag{5.67}$$

The Stanton number is also known as the *modified Nusselt number*. Equation (5.67) is known as the *Reynolds analogy*. We have seen that,

$$\mathrm{St} = \frac{\mathrm{Nu}}{\mathrm{Re}\ \mathrm{Pr}}$$

For Pr = 1, this becomes

$$\mathrm{St} = \frac{\mathrm{Nu}}{\mathrm{Re}}$$

Also, for Pr = 1, by Equation (5.67),

$$\mathrm{St} = \frac{1}{2}\,c_f$$

From these two equation, we get the Nusselt number as

$$\boxed{\mathrm{Nu} = \mathrm{St}\,\mathrm{Re} = \frac{1}{2}\mathrm{Re}\,c_f} \tag{5.68}$$

For laminar flow over a flat plate, the skin friction coefficient is given by [2]

$$c_f = \frac{0.664}{\sqrt{\mathrm{Re}_x}} \tag{5.68$a$}$$

This relation is valid for laminar flow with Reynolds number less than $5 \times 10^5$. The average drag coefficient over the plate length from $x = 0$ to $x = L$ is defined as

$$C_{fm} = \frac{1}{L}\int_0^L C_{fx}dx$$

It is important to note that, the *skin friction coefficient* is referred to as *drag coefficient.* This is because, for a streamlined body, such as a flat plat with flow over its surface, the skin friction drag is the major portion of the drag and the wake drag is negligible. The opposite to this is a bluff body, for which the wake drag is substantial and the skin friction drag is negligibly small. Thus, for laminar flow over a flat plate, the average drag coefficient becomes

$$\begin{aligned}
C_{fm} &= \frac{1}{L}\int_0^L \frac{0.664}{\sqrt{\mathrm{Re}_x}}dx \\
&= \frac{1}{L}\int_0^L \frac{0.664}{\sqrt{(\rho U x)/\mu}}dx \\
&= \frac{0.664}{L\sqrt{(\rho U)/\mu}}\int_0^L x^{-\frac{1}{2}}dx \\
&= \frac{0.664}{L\sqrt{(\rho U)/\mu}}\left[\frac{x^{\frac{1}{2}}}{\frac{1}{2}}\right]_0^L \\
&= \frac{0.664}{L\sqrt{(\rho U)/\mu}}\,2L^{\frac{1}{2}} \\
&= \frac{2 \times 0.664}{\sqrt{(\rho U L)/\mu}}
\end{aligned}$$

That is,

$$\boxed{C_{fm} = \frac{1.328}{\sqrt{\text{Re}_L}}} \tag{5.68b}$$

Substituting Equation (5.68a) into Equation (5.67), we get the local Stanton number as

$$\text{St}_x = \frac{0.332}{\sqrt{\text{Re}_x}}$$

Substituting Equation (5.68a) into Equation (5.68), we get and the local Nusselt number as

$$\boxed{\text{Nu}_x = 0.332\sqrt{\text{Re}_x}} \tag{5.69}$$

The Reynolds analogy is valid only for laminar flow over a flat plate with zero pressure gradient and $\text{Pr} = 1$.
Note: It is important to that, even though Reynolds analogy [Equation (5.67)] gives simple and useful relation for skin friction coefficient in terms of Stanton number, it is valid only when $\text{Pr} = 1$. This aspect makes this analogy restrictive and valid only for specific cases. With an objective of making this valid for a wide range of Prandtl number ($0.6 < \text{Pr} < 60$), an analogy has been suggested by Reynolds in the following form.

$$\boxed{\text{St}\,\text{Pr}^{\frac{2}{3}} = \frac{C_f}{2}}$$

Note that, when $\text{Pr} = 1$, this reduces to the Reynolds analogy. Another interesting aspect is that if we assume that the transports of mass, momentum and energy are identical, that is, when the mass, momentum and energy transports are equal, the momentum, thermal, and concentration boundary layers would coincide. It is important to realize that though this kind of assumption results in an interesting situation of making all the three boundary layers identical, it is only an imaginary or hypothetical situation, and not a practical case.

## Example 5.2

Air at 1 atm and 300 K flows over a flat plate of length 1 m. If the boundary layer thickness at the end of the plate is 12 mm, determine the flow velocity, assuming the flow to be laminar.

## Solution

Given, $p = 101325$ Pa, $T = 300$ K, $L = 1$ m, $\delta_L = 12$ mm.

The density and viscosity of the flow are

$$\rho = \frac{p}{RT} = \frac{101325}{287 \times 300}$$

$$= 1.177\,\text{kg/m}^3$$

$$\mu = (1.46 \times 10^{-6}) \times \frac{T^{3/2}}{T + 111}$$

$$= (1.46 \times 10^{-6}) \times \frac{300^{3/2}}{300 + 111}$$

$$= 1.85 \times 10^{-5}\,\text{kg/(m s)}$$

For laminar flow over a flat plate, the boundary layer thickness at $x = L$, from the leading edge, given by Equation (5.70), is

$$\frac{\delta_L}{L} = \frac{5}{\sqrt{\text{Re}_L}}$$

Therefore,

$$\text{Re}_L = \frac{5^2 L^2}{\delta_L^2}$$

$$\frac{\rho U L}{\mu} = \frac{5^2 \times 1^2}{0.012^2} = 173611$$

$$U = \frac{173611 \times \mu}{\rho L}$$

$$= \frac{173611 \times 1.85 \times 10^{-5}}{1.177 \times 1}$$

$$= \boxed{2.73\,\text{m/s}}$$

## Example 5.3

Air at 2 m/s, 1 atm and 300 K flows over a heated thin flat plate of length 0.5 m and width 1 m, maintained at 60°C. Determine the rate of heat transfer from the plate to the air.

## Solution

Given, $U = 2$ m/s, $p = 1$ atm, $T = 300$ K, $T_w = 60 + 273 = 333$ K, $L = 0.5$ m, $w = 1$ m.

The film temperature is

$$T_f = \frac{T + T_w}{2} = \frac{300 + 333}{2}$$

$$= \quad 316.5\,\text{K}$$

For air at $T_f = 316.5$ K, from the properties table in the appendix, $k = 0.027$ W/(m °C) and Pr = 0.7.

The viscosity at 316.5 K is

$$\mu = (1.46 \times 10^{-6}) \times \frac{316^{1.5}}{316.5 + 111}$$

$$= 1.92 \times 10^{-5}\,\text{kg/(m s)}$$

The Reynolds number is

$$\text{Re}_L = \frac{\rho U L}{\mu} = \frac{pUL}{RT_f \mu}$$

$$= \frac{101325 \times 2 \times 0.5}{287 \times 316.5 \times (1.92 \times 10^{-5})}$$

$$= 58098$$

For this laminar flow, by Equation (5.69),

$$\text{Nu} = \frac{hL}{k} = 0.332\sqrt{\text{Re}_L}$$

$$= 0.332 \times \sqrt{58098}$$

$$= 80$$

This gives the heat transfer coefficient as

$$h = \frac{80k}{L} = \frac{80 \times 0.027}{0.5}$$

$$= 4.32\,\text{W/(m}^2\ {}^\circ\text{C)}$$

The rate of heat transfer from the top and bottom surfaces of the plate to air is

$$\dot{Q} = h\,(2A)\,(T_w - T_\infty)$$

$$= 2 \times 4.32 \times (2 \times [0.5 \times 1]) \times (333 - 300)$$

$$= \boxed{142.56\,\text{W}}$$

Note that the area is taken as the sum of the top and bottom surfaces to account for the heat loss from both surfaces.

## 5.9 Thermal Boundary Layer for Flow Past a Heated Plate

From fluid dynamics we know that, for an incompressible laminar flow past a flat plate with zero pressure gradient, the boundary layer thickness $\delta$ is (Blassius solution) given by [2]

$$\boxed{\frac{\delta}{x} = \frac{5}{\sqrt{\mathrm{Re}_x}}} \tag{5.70}$$

where $x$ is the distance from the leading edge of the plate.

Thermal boundary layer over a flat plate under the following two situations is of interest in many practical applications.

- Flow past a flat plate which is maintained at uniform constant temperature over its entire surface.
- Flow past a flat plate supplied with uniform heat flux all over its surface.

In both these cases, the thermal and momentum boundary layers grow over the plate, and the thickness of these boundary layers is function of the Prandtl number. For a plate having uniform temperature over its surface, the local heat transfer rate, given by Fourier's law is,

$$\dot{q}_x = -k \left.\frac{\partial T}{\partial x}\right|_{y=0}$$

where $x$ is the distance along the plate and $y$ is the direction normal to the plate. This heat transferred by conduction would be convected away by the flow over the surface, to maintain the surface temperature at a constant value. Therefore,

$$\dot{q}_x = -k \left.\frac{\partial T}{\partial x}\right|_{y=0} = h_x\left(T_s - T_\infty\right)$$

where $T_s$ is the plate surface temperature, $T_\infty$ is the freestream static temperature, $k$ is the thermal conductivity of the plate material, and $h_x$ is the local convection heat transfer coefficient. Thus, $h_x$ becomes

$$h_x = \frac{-k \left.\dfrac{\partial T}{\partial y}\right|_{y=0}}{(T_s - T_\infty)}$$

Now, for the case of flows with Pr << 1, such as liquid metals (Figure 5.9), we can even ignore the momentum boundary layer because it is very thin compared to the thermal boundary layer. Therefore, in the analysis, the $x$-component of velocity inside the thermal boundary layer can be taken as the freestream velocity itself, that is $u(y) = U_\infty$. For this case, neglecting the viscous dissipation term, the energy equation [Equation (5.31)], can be simplified as

$$u\frac{\partial T}{\partial x} + v\frac{\partial T}{\partial y} = \alpha\frac{\partial T^2}{\partial y^2} \tag{5.71}$$

By Equation (5.41), the nondimensional temperature $\theta$ is

$$\theta = \frac{T - T_s}{T_\infty - T_s}$$

where $T_\infty$ is the freestream temperature, $T_s$ is the temperature at the surface of the plate and $T$ is local temperature. In terms of $\theta$, Equation (5.71) can be expressed as

$$u\frac{\partial \theta}{\partial x} + v\frac{\partial \theta}{\partial y} = \alpha\frac{\partial \theta^2}{\partial y^2} \tag{5.72}$$

It is important to note that, the above definition of nondimensional $\theta$ is different from the conventional one namely, $\theta = (T - T_\infty)/(T_s - T_\infty)$. This is intentionally done to make the boundary conditions and the nondimensional temperature profile of the thermal boundary layer idential to the corresponding boundary conditions and the velocity profile of the momentum boundary layer, as shown in Figure 5.10.

For solving Equation (5.72), we need to find the expression of the velocity components $u$ and $v$ by solving the continuity equation [Equation (5.28)], momentum equations [Equations (5.29) and (5.30)], along with the velocity boundary conditions,

$$u = 0, \;\; v = 0, \;\; \text{at } y = 0$$

$$u \to u_\infty, \;\; \text{at } y = \delta(x)$$

But this is a complex analysis to perform. Therefore, it is usual to find an approximate solution for this problem, using the following procedure.

Integrating the energy equation (5.72), with respect to $y$, over the thermal boundary layer (that is, from 0 to $\delta_t$), and eliminating the $y$-component of velocity, using the continuity equation, we can obtain the *energy integral equation* as

$$\frac{d}{dx}\left[\int_0^{\delta_t} u(1-\theta)\,dy\right] = \alpha\left.\frac{\partial \theta}{\partial y}\right|_{y=0} \quad (0 \le y \le \delta_t) \tag{5.73}$$

Now, let us assume a cubic polynomial for the temperature profile as

$$\theta(x, y) = c_0 + c_1 y + c_2 y^2 + c_3 y^3$$

where $c_0, c_1, c_2, c_3$ are constants, and they can be evaluated using the following temperature boundary conditions:

$$\text{at } y = 0, \qquad \theta = 0$$

$$\text{at } y = \delta_t, \qquad \theta = 1$$

$$\text{at } y = 0, \qquad \frac{\partial^2 \theta}{\partial y^2} = 0$$

$$\text{at } y = \delta_t, \qquad \frac{\partial \theta}{\partial y} = 0,$$

leading to the nondimensional temperature profile as

$$\boxed{\theta(x,y) = \frac{3}{2}\left(\frac{y}{\delta_t}\right) - \frac{1}{2}\left(\frac{y}{\delta_t}\right)^3} \tag{5.74}$$

Substituting this $\theta(x,y)$ into Equation (5.72), and solving for thermal boundary layer thickness, we get

$$\delta_t = \sqrt{\frac{8\alpha x}{U_\infty}} \tag{5.75}$$

Also, from Equation (5.74), we get

$$\left.\frac{\partial \theta}{\partial y}\right|_{y=0} = \frac{3}{2\delta_t} \tag{5.76}$$

The local convective heat transfer coefficient $h_x$ can also be expressed in terms of the nondimensional temperature $\theta(x,y)$, as follows. The $h_x$ in terms of temperature $T$ is

$$h_x = \frac{-k \left.\dfrac{\partial T}{\partial y}\right|_{y=0}}{(T_s - T_\infty)}$$

But in terms of $\theta = \dfrac{T - T_\infty}{T_s - T_\infty}$, we have

$$h_x = k \left.\frac{\partial \theta}{\partial y}\right|_{y=0} \tag{5.77}$$

Substituting Equation (5.76) into Equation (5.77), we get

$$\boxed{h_x = \frac{3}{2}\frac{k}{\delta_t}} \tag{5.78}$$

Now, substituting Equation (5.75) into Equation (5.78), we get

$$h_x \;=\; \frac{3k}{2}\sqrt{\frac{U_\infty}{8\alpha x}}$$

$$= \frac{3k}{2\sqrt{8}} \sqrt{\frac{U_\infty}{x} \frac{\rho c_p}{k}}$$

$$= \frac{3k}{2\sqrt{8}} \sqrt{\frac{\rho U_\infty x}{\mu} \frac{\mu c_p}{k} \frac{1}{x^2}}$$

$$= \frac{3}{2\sqrt{8}} \frac{k}{x} \sqrt{(\mathrm{Re})\,(\mathrm{Pr})}$$

or

$$h_x = 0.53 \frac{k}{x} (\mathrm{Pr})^{\frac{1}{2}} (\mathrm{Re})^{\frac{1}{2}} \tag{5.79}$$

Therefore, the local Nusselt number $\mathrm{Nu}_x$ becomes

$$\mathrm{Nu}_x = \frac{h_x x}{k} = 0.53\,(\mathrm{Pr})^{\frac{1}{2}} (\mathrm{Re})^{\frac{1}{2}} \tag{5.80}$$

This result, obtained by the approximate analysis, agrees closely with Pohlhausen's exact solution for the limiting case of $\mathrm{Pr} \to 0$, which gives $\mathrm{Nu}_x$ as

$$\mathrm{Nu}_x = 0.564\,(\mathrm{Pr})^{\frac{1}{2}} (\mathrm{Re})^{\frac{1}{2}} \tag{5.81}$$

It is usual in literature to combine the Re and Pr and express as Peclet number Pe, resulting in

$$\mathrm{Nu}_x = 0.564\,(\mathrm{Pe})^{\frac{1}{2}} \tag{5.82}$$

When the Prandtl number is more than one ($\mathrm{Pr} > 1$), the momentum boundary layer grows faster than the thermal boundary layer over a flat plate with a constant surface temperature $T_s$. For this case, the growth of the momentum and thermal boundary layers will be as illustrated in Figure 5.10. As shown in the figure, both the boundary layers begin to grow from the leading edge of the plate. Therefore, at some distance downstream of the leading edge, the thermal boundary layer will cross the momentum boundary layer, as shown in the figure. To avoid the complexity caused by this cross-over process, it is a usual practice to consider a starting length $x_0$, from which the thermal boundary layer begins to grow as shown in Figure 5.10.

Consider the energy equation [Equation (5.72)]

$$u\frac{\partial \theta}{\partial x} + v\frac{\partial \theta}{\partial y} = \alpha \frac{\partial \theta^2}{\partial y^2}$$

The boundary conditions are

$$\text{at } y = 0,\ \theta = 0$$

$$\text{at } y = \delta_t,\ \theta = 1$$

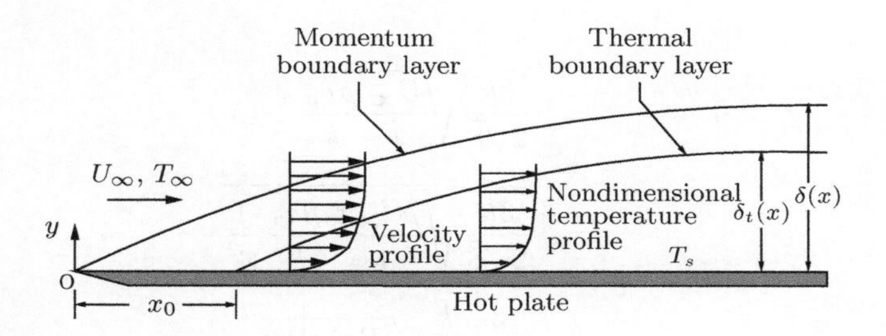


**Figure 5.10**
Momentum and thermal boundary layers over a uniformly heated flat plate, for Pr > 1.

where $\theta$ is the nondimensional temperature.

By integrating the above energy equation, with respect to $y$, for the limits 0 to $\delta_t$, and eliminating the $y$-component of the velocity using continuity equation, we can get the *energy integral equation* [Equation (5.73)] as

$$\frac{d}{dx}\left[\int_0^{\delta_t} u(1-\theta)\,dy\right] = \alpha\left.\frac{\partial \theta}{\partial y}\right|_{y=0} \quad (0 \le y \le \delta_t)$$

There are three unknowns $(u, \delta_t, \theta)$ in this equation. Therefore, two more relations are necessary for solving this equation. The velocity profile and temperature profile can be taken as the additional relations to solve the energy equation.

Assuming the velocity relation as a cubic profile, we can write

$$u(x,y) = c_0 + c_1 y + c_2 y^2 + c_3 y^3 \tag{5.83}$$

where $c_0$, $c_1$, $c_2$ and $c_3$ are constants to be evaluated using the following boundary conditions.

$$\text{at } y = 0, \quad u = 0$$

$$\text{at } y = \delta, \quad u = U_\infty$$

$$\text{at } y = \delta, \quad \frac{\partial u}{\partial y} = 0$$

$$\text{at } y = 0, \quad \frac{\partial u^2}{\partial y^2} = 0$$

Solving, we get

$$\boxed{\frac{u}{U_\infty} = \left(\frac{3}{2}\frac{y}{\delta} - \frac{1}{2}\left(\frac{y}{\delta}\right)^3\right)} \tag{5.84}$$

Also, assuming the temperature relation as a cubic profile, we can write

$$\theta(x,y) = c_0' + c_1' y + c_2' y^2 + c_3' y^3$$

where $c'_0$, $c'_1$, $c'_2$ and $c'_3$ are constants to be evaluated using the following boundary conditions.

$$\begin{aligned} &\text{at } y = 0, && \theta = 0 \\ &\text{at } y = \delta_t, && \theta = 1 \\ &\text{at } y = \delta_t, && \frac{\partial \theta}{\partial y} = 0 \\ &\text{at } y = 0, && \frac{\partial \theta^2}{\partial y^2} = 0 \end{aligned}$$

Solving we get

$$\boxed{\theta(x,y) = \left(\frac{3}{2}\frac{y}{\delta_t} - \frac{1}{2}\left(\frac{y}{\delta_t}\right)^3\right)} \tag{5.85}$$

Note that the procedure and boundary conditions used for obtaining the velocity and temperature profiles are identical. Also, the resulting velocity and temperature profiles are identical.

Substituting the velocity and temperature profiles [Equations (5.84) and (5.85)] into the energy integral Equation (5.73), we get

$$\frac{d}{dx}\left(U_\infty \int_0^{\delta_t} \left[\frac{3}{2}\frac{y}{\delta} - \frac{1}{2}\left(\frac{y}{\delta}\right)^3\right]\left[1 - \frac{3}{2}\frac{y}{\delta_t} + \frac{1}{2}\left(\frac{y}{\delta_t}\right)^3\right] dy\right) = \frac{3}{2}\frac{\alpha}{\delta_t} \tag{5.86}$$

Now, let us define a new variable $\epsilon = \dfrac{\delta_t}{\delta}$. For the present case, Pr $> 1$ and $\epsilon < 1$. Therefore, using this $\epsilon$ in Equation (5.86), and solving, we can get

$$\frac{d}{dx}\left(\delta\left[\frac{3}{20}\epsilon^2 - \frac{3}{280}\epsilon^4\right]\right) = \frac{3}{2}\frac{\alpha}{\delta \epsilon U_\infty} \tag{5.87}$$

Since $\epsilon < 1$, the term $\dfrac{3}{280}\epsilon^4$ can be neglected in comparison with $\dfrac{3}{20}\epsilon^2$. Thus, the above equation reduces to

$$\boxed{\delta\epsilon \frac{d}{dx}\left(\delta\epsilon^2\right) = \frac{10\alpha}{U_\infty}} \tag{5.88}$$

Differentiating this with respect to $x$, and rearranging the terms, we get

$$\begin{aligned} \delta\,\epsilon\,\delta\,\frac{d\epsilon^2}{dx} + \delta\,\epsilon\,\epsilon^2\,\frac{d\delta}{dx} &= \frac{10\alpha}{U_\infty} \\ \delta^2\epsilon\,2\epsilon\,\frac{d\epsilon}{dx} + \delta\epsilon^3\,\frac{d\delta}{dx} &= \frac{10\alpha}{U_\infty} \\ \delta^2\,2\epsilon^2\,\frac{d\epsilon}{dx} + \left(\delta\,\frac{d\delta}{dx}\right)\epsilon^3 &= \frac{10\alpha}{U_\infty} \\ \frac{2}{3}\delta^2\,\frac{d\epsilon^3}{dx} + \left(\delta\,\frac{d\delta}{dx}\right)\epsilon^3 &= \frac{10\alpha}{U_\infty} \end{aligned}$$

or

$$\boxed{\frac{2}{3}\delta^2\frac{d\epsilon^3}{dx}+\left(\delta\frac{d\delta}{dx}\right)\epsilon^3=\frac{10\alpha}{U_\infty}} \tag{5.89}$$

For velocity boundary layer thickness $\delta$ we can use the relation Equation (5.90a), obtained as follows.

Integrating the $x$-momentum equation (5.29), from $y = 0$ to $y = \delta$, and eliminating the $y$-component of velocity using the continuity equation, we can express the *momentum integral equation* as

$$\frac{d}{dx}\left[\int_0^\delta (U_\infty - u)\,u\,dy\right] = \nu\frac{\partial u}{\partial y}\bigg|_{y=0} \qquad (0 \le y \le \delta)$$

By Equation (5.84), the velocity profile is

$$\frac{u(x,y)}{U_\infty} = \left[\frac{3}{2}\frac{y}{\delta} - \frac{1}{2}\left(\frac{y}{\delta}\right)^3\right]$$

Substituting this velocity profile into the momentum integral equation, we get

$$U_\infty^2\frac{d}{dx}\left(\int_0^\delta\left[\frac{3}{2}\left(\frac{y}{\delta}\right)-\frac{1}{2}\left(\frac{y}{\delta}\right)^3\right]\left[1-\frac{3}{2}\left(\frac{y}{\delta}\right)+\frac{1}{2}\left(\frac{y}{\delta}\right)^3\right]dy\right) = \nu\,U_\infty\frac{3}{2\delta}$$

Simplifying this, we get

$$\frac{d}{dx}\left[\frac{39}{280}\delta\right] = \frac{3\nu}{2U_\infty\delta}$$

or

$$\delta\,d\delta = \frac{140}{13}\frac{\nu}{U_\infty}\,dx$$

On integrating the left-hand side in the limits from $\delta = 0$ to $\delta = \delta$, and the right-hand side in the limits from $x = 0$ to $x = x$, we get

$$\delta^2(x) = \frac{280}{13}\frac{\nu x}{U_\infty}$$

or

$$\delta(x) = \sqrt{\frac{280}{13}\frac{\nu x}{U_\infty}} \tag{5.90a}$$

that is,

$$\delta(x) = 4.64\,x\,\mathrm{Re}_x^{-\frac{1}{2}} \tag{5.90b}$$

Substituting Equation (5.90a) into Equation (5.89), we get

$$x\frac{d\epsilon^3}{dx} + \frac{3}{4}\epsilon^3 = \frac{39}{56}\frac{\alpha}{\nu} \tag{5.91}$$

This is a first-order ordinary differential equation in $\epsilon^3$. Therefore, the general solution can be expressed as

$$\epsilon^3(x) = c\,x^{-\frac{3}{4}} + \frac{13}{14}\frac{\alpha}{\nu} \tag{5.92}$$

where $c$ is the integration constant to be evaluated using the starting length boundary condition of the thermal boundary layer, which is

$$\text{at } x = x_0, \;\; \delta_t = 0$$

Thus, at $x = x_0$, $\epsilon = 0$. Therefore,

$$\epsilon = \frac{\delta_t}{\delta} = \left(\frac{13}{14}\right)^{\frac{1}{3}} \mathrm{Pr}^{-\frac{1}{3}} \left[1 - \left(\frac{x_0}{x}\right)^{\frac{3}{4}}\right]^{\frac{1}{3}} \tag{5.93}$$

where $\mathrm{Pr} = \dfrac{\nu}{\alpha}$. If we assume that the thermal boundary layer also begins to grow from the leading edge of the plate, we will have $x_0 = 0$. For $x_0 = 0$, Equation (5.93) simplifies to

$$\epsilon = \frac{\delta_t}{\delta} = 0.976\,\mathrm{Pr}^{-\frac{1}{3}}$$

This gives, the thermal boundary layer thickness as

$$\delta_t = 0.976\,\frac{\delta}{\mathrm{Pr}^{\frac{1}{3}}} \tag{5.94}$$

But by Equation (5.90b), the momentum boundary layer thickness is

$$\delta = 4.64\,x\,\mathrm{Re}_x^{-\frac{1}{2}}$$

Substituting this into Equation (5.93), we get the thermal boundary layer thickness as

$$\delta_t = \frac{0.976 \times 4.64\,x}{\mathrm{Re}^{\frac{1}{2}}\,\mathrm{Pr}^{\frac{1}{3}}}$$

or

$$\delta_t = \frac{4.53x}{\mathrm{Re}^{\frac{1}{2}}\,\mathrm{Pr}^{\frac{1}{3}}} \tag{5.95}$$

From the temperature profile relation [Equation (5.85)], we have

$$\theta(x,y) = \left[\frac{3}{2}\frac{y}{\delta_t} - \frac{1}{2}\left(\frac{y}{\delta_t}\right)^3\right]$$

$$\left.\frac{\partial \theta(x,y)}{\partial y}\right|_{y=0} = \frac{3}{2}\frac{1}{\delta_t}$$

The local convective heat transfer coefficient becomes

$$\begin{aligned} h_x &= k\frac{\partial\theta(x,y)}{\partial y}\bigg|_{y=0} \\ &= \frac{3}{2}\frac{k}{\delta_t} \end{aligned}$$

Substituting for $\delta_t$, from Equation (5.93), the local convective heat transfer coefficient reduces to

$$h_x = 0.331\,\frac{k}{x}\,\mathrm{Re}_x^{\frac{1}{2}}\,\mathrm{Pr}^{\frac{1}{3}}$$

This approximate solutions closely matches the exact solution of the problem given by Schlichting [1] as

$$\boxed{h_x = 0.332\,\frac{k}{x}\,\mathrm{Re}_x^{\frac{1}{2}}\,\mathrm{Pr}^{\frac{1}{3}}} \tag{5.96}$$

This relation is valid for $\mathrm{Re}_x < 5 \times 10^5$. Therefore, the local Nusselt number becomes

$$\boxed{\mathrm{Nu}_x = \frac{h_x x}{k} = 0.332\,\mathrm{Re}_x^{\frac{1}{2}}\,\mathrm{Pr}^{\frac{1}{3}}} \tag{5.97}$$

The average convective heat transfer coefficient is

$$\begin{aligned} h_L &= \frac{1}{L}\int_0^L h_x dx \\ &= \frac{1}{L}\int_0^L 0.332\,\frac{k}{x}\,\mathrm{Re}_L^{\frac{1}{2}}\mathrm{Pr}^{\frac{1}{3}}\,dx \end{aligned}$$

Integrating, we get

$$h_L = 0.664\,\frac{k}{L}\,\mathrm{Re}_L^{\frac{1}{2}}\mathrm{Pr}^{\frac{1}{3}} \tag{5.98}$$

Therefore, the average Nusselt number becomes

$$\boxed{\mathrm{Nu}_L = 0.664\,\mathrm{Re}_L^{\frac{1}{2}}\mathrm{Pr}^{\frac{1}{3}}} \tag{5.99}$$

It is interesting to note for a given length, the average Nusselt number is twice the local Nusselt number, that is

$$\boxed{\mathrm{Nu}_L = 2 \times \mathrm{Nu}_{x=L}} \tag{5.100}$$

The friction and heat transfer coefficients for a flat plate can be determined theoretically by solving the mass, momentum, and energy conservation equations. They can also be determined experimentally and expressed by empirical correlations. In both these approaches, it is found that the *average* Nusselt number can be expressed in terms of the Reynolds and Prandtl numbers in the form

$$\mathrm{Nu} = \frac{hL}{k} = c\,\mathrm{Re}_L^m \mathrm{Pr}^n$$

where $c$, $m$, and $n$ are constants and, $L$ is the *length* of the plate in the flow direction.

The fluid temperature in the thermal boundary layer varies from $T_s$, at the surface of the plate, to about $T_\infty$, at the edge of the boundary layer. The fluid properties also vary with temperature, and thus with the position across the boundary layer. In order to properly account for the variation of the properties with temperature, the fluid properties are usually evaluated at the *film temperature*, $T_f$, defined as

$$T_f = \frac{T_s + T_\infty}{2} \tag{5.101}$$

which is the *average* of the surface and freestream temperatures. The fluid properties corresponding to $T_f$ are assumed to remain constant during the entire flow.

The local friction and heat transfer coefficients vary along the surface of the plate as a result of the changes in the velocity and thermal boundary layers in the flow direction. Usually we are interested in the heat transfer and drag force on the *entire* surface of the plate (that is, due to the entire wetted area of the plate), which can be determined using the average heat transfer and friction coefficients. But if the heat flux and drag force at a certain location is required, then the *local* values of heat transfer and friction coefficients should be determined.

The heat transfer and friction coefficients for the entire plate can be determined using the local values in integral form, given below.

$$h = \frac{1}{L}\int_0^L h_x\,dx$$

$$c_f = \frac{1}{L}\int_0^L c_{f,x}\,dx$$

Once $h$ and $c_f$ are known, the heat transfer rate $\dot{Q}$, and the drag or friction force $D$ can be calculated from

$$\dot{Q}_{\mathrm{conv}} = h\,A\,(T_s - T_\infty)$$

and

$$D = \frac{1}{2}\rho U_\infty^2 A_s c_f$$

where $A_s$ is the surface area, $U_\infty$ is the freestream velocity and $\rho$ is the freestream density.

### 5.9.1 Turbulent Flow

The local friction coefficient at location $x$, for *laminar flow* over a flat plate is given by [2]

$$c_{f,x} = \frac{0.664}{\sqrt{\mathrm{Re}_x}} \tag{5.102}$$

The average friction coefficient and the Nusselt number over the entire plate can be expressed as follows.

$$\begin{aligned} c_f &= \frac{1}{L}\int_0^L c_{f,x}\,dx \\ &= \frac{1}{L}\int_0^L \frac{0.664}{\sqrt{\mathrm{Re}_x}}\,dx \\ &= \frac{0.664}{L}\int_0^L \left(\frac{V_\mathrm{m} x}{\nu}\right)^{-1/2} dx \\ &= \frac{0.664}{L}\left(\frac{V_\mathrm{m}}{\nu}\right)^{-1/2}\int_0^L x^{-1/2}dx \\ &= (2\times 0.664)\left(\frac{V_\mathrm{m} L}{\nu}\right)^{-1/2} \end{aligned}$$

This simplifies to

$$\boxed{C_f = \frac{1.328}{\mathrm{Re}_L^{1/2}}} \tag{5.103}$$

where $V_\mathrm{m}$ is the mean velocity.

By Equation (5.99), the Nusselt number is

$$\mathrm{Nu} = \frac{hL}{k} = 0.664\,\mathrm{Re}_L^{1/2}\,\mathrm{Pr}^{1/3}$$

It is important to note that, the skin friction coefficient $C_f$ and Nu given by Equations (5.103) and (5.99) are twice of $C_f$ and Nu given by Equations (5.102) and (5.97), because Equations (5.102) and (5.97) account for heat transfer from only one surface of the plate, whereas Equations (5.103) and (5.99) account for heat transfer from both top and bottom surfaces of the plate. The above relations give the average friction and heat transfer coefficients for the entire plate when the flow is *laminar* over the *entire* plate. These relations are valid for Reynolds number less than $5 \times 10^5$, which is the *critical* Reynolds number and Prandtl number $\mathrm{Pr} \geq 0.6$. It is important to realize that the critical Reynolds number is that at which the flow changes from laminar to turbulent is a gross statement, since the change over from laminar to turbulent nature is taking place over a range of Reynolds number and not at a particular Reynolds number. Therefore, as we saw already, there are two critical Reynolds numbers, namely the *lower* critical Reynolds number and the *upper* critical Reynolds number. The lower critical Reynolds number is that below which the entire flow field is laminar, and the upper critical Reynolds number is that above which the entire flow field is turbulent.

For turbulent flow over a flat plate, the *local* friction coefficient and Nusselt number at location $x$ are given by [1]

$$c_{f,x} = \frac{0.0592}{\mathrm{Re}_x^{1/5}} \tag{5.104}$$

$$\mathrm{Nu}_x = \frac{h_x x}{k} = 0.0296\,\mathrm{Re}_x^{4/5}\,\mathrm{Pr}^{0.43} \tag{5.105}$$

These results are valid for $5 \times 10^5 \leq \mathrm{Re}_x \leq 10^7$ and $0.6 \leq \mathrm{Pr} \leq 60$. The local friction and heat transfer coefficients in turbulent flow are higher than those in laminar flow, because of the active exchange of transverse momentum in the turbulent boundary layer. Note that, both $c_{f,x}$ and $h$ reach their highest values when the flow becomes fully turbulent, and then decrease by a factor $x^{-1/5}$ in the flow direction, as shown in Figure 5.11.

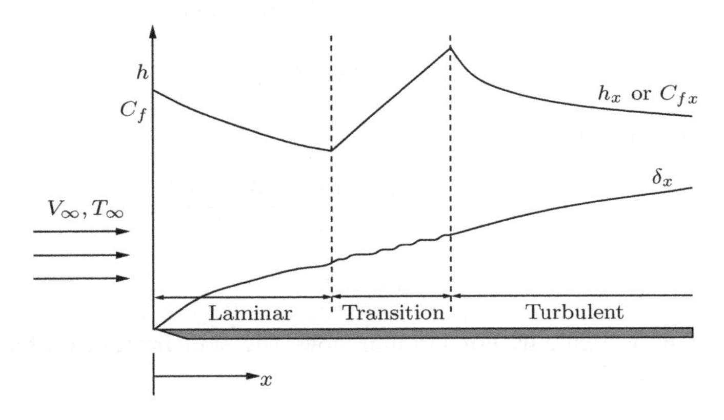


**Figure 5.11**
Variation of local $c_f$ and $h$ for flow over a flat plate.

The average friction coefficient and Nusselt number over the entire plate in turbulent flow are determined, by substituting the above relations into the $c_f$ and Nu relations and integrating, as

$$\boxed{c_f = \frac{0.074}{\mathrm{Re}_L^{1/5}}} \tag{5.106}$$

$$\boxed{\mathrm{Nu} = \frac{hL}{k} = 0.037\,\mathrm{Re}_L^{4/5}\mathrm{Pr}^{0.43}} \tag{5.107}$$

## Example 5.4

Air at 1 atm and 350 K flows over the top surface of a flat plate of length 200 mm and width 1 m. Determine the drag acting over the plate surface when the flow speed is 30 m/s.

## Solution

Given, $p = 101325$ Pa, $T = 350$ K, $L = 0.2$ m, $w = 1$ m, $U = 30$ m/s. For air, the gas constant is $R = 287$ m$^2$/(s$^2$ K).

The density and viscosity of the flow are

$$\begin{aligned} \rho &= \frac{p}{RT} = \frac{101325}{287 \times 350} \\ &= 1.01\,\text{kg/m}^3 \end{aligned}$$

$$\begin{aligned} \mu &= 1.46 \times 10^{-6} \times \left( \frac{350^{3/2}}{350 + 111} \right) \\ &= 2.07 \times 10^{-5}\,\text{kg/(m s)} \end{aligned}$$

The Reynolds number is

$$\begin{aligned} \text{Re}_L &= \frac{\rho U L}{\mu} = \frac{1.01 \times 30 \times 0.2}{2.07 \times 10^{-5}} \\ &= 2.93 \times 10^5 \end{aligned}$$

Thus, the flow is laminar. For laminar flow, the skin friction coefficient, by Equation (5.102), is

$$\begin{aligned} C_f &= \frac{0.664}{\sqrt{\text{Re}_L}} \\ &= \frac{0.664}{\sqrt{2.93 \times 10^5}} \\ &= 0.001227 \end{aligned}$$

The drag acting over the top surface of the plate is

$$\begin{aligned} D &= \frac{1}{2}\rho U^2 S C_f \\ &= \frac{1}{2}\rho U^2 \left(L \times w\right) C_f \end{aligned}$$

$$= \frac{1}{2} \times 1.01 \times 30^2 \times (0.2 \times 1) \times 0.001227$$

$$= \boxed{0.1115\,\mathrm{N}}$$

### 5.9.2 Transition Flow

In some cases, the plate is sufficiently long for the flow to become turbulent, but not long enough to discard the laminar flow region. That is, the flow transition is in progress and not complete. In such a situation, the average $c_f$ and $h$ can be determined by integrating the respective relations over the plate in two parts: the laminar region over the plate in the length range $0 \leq x \leq x_{\mathrm{cri}}$ and the turbulent region for $x_{\mathrm{cri}} \leq x \leq L$, as

$$c_f = \frac{1}{L}\left(\int_0^{x_{\mathrm{cri}}} c_{f,x,\mathrm{lam}}\,dx + \int_{x_{\mathrm{cri}}}^{L} c_{f,x,\mathrm{tur}}\,dx\right)$$

$$h = \frac{1}{L}\left(\int_0^{x_{\mathrm{cri}}} h_{x,\mathrm{lam}}\,dx + \int_{x_{\mathrm{cri}}}^{L} h_{x,\mathrm{tur}}\,dx\right) \tag{5.108}$$

$$h_{x,\mathrm{lam}} = 0.332\left(\frac{k}{x}\right)\left(\frac{U_\infty x}{\nu}\right)^{\frac{1}{2}} \mathrm{Pr}^{\frac{1}{3}} \;\; (0 \leq x \leq x_{\mathrm{cri}}) \tag{5.109}$$

$$h_{x,\mathrm{tur}} = 0.0296\left(\frac{k}{x}\right)\left(\frac{U_\infty x}{\nu}\right)^{0.8} \mathrm{Pr}^{0.43} \;\; (x_{\mathrm{cri}} \leq x \leq L) \tag{5.110}$$

Substituting Equations (5.109) and (5.110) into Equation (5.108), we get the average convective heat transfer coefficient $h$ as

$$h = \frac{1}{L}\left[\int_0^{x_{\mathrm{cri}}} 0.332\left(\frac{k}{x}\right)\left(\frac{U_\infty x}{\nu}\right)^{\frac{1}{2}} \mathrm{Pr}^{\frac{1}{3}}\,dx + \int_{x_{\mathrm{cri}}}^{L} 0.0296\left(\frac{k}{x}\right)\left(\frac{U_\infty x}{\nu}\right)^{0.8} \mathrm{Pr}^{0.43}\,dx\right] \tag{5.111}$$

This simplifies to

$$h = \frac{1}{L}\left[0.332\,k\left(\frac{U_\infty}{\nu}\right)^{0.5} \mathrm{Pr}^{\frac{1}{3}} \int_0^{x_{\mathrm{cri}}} x^{-0.5}\,dx + 0.0296\,k\left(\frac{U_\infty}{\nu}\right)^{0.8} \mathrm{Pr}^{0.43} \int_{x_{\mathrm{cri}}}^{L} x^{-0.2}\,dx\right] \tag{5.112}$$

The average Nusselt number is defined as

$$\mathrm{Nu} = \frac{hL}{k}$$

Therefore, substituting Equation (5.112) and integrating, we get the average Nusselt number as

$$\mathrm{Nu} = 0.037\,\mathrm{Pr}^{0.43}\left(\mathrm{Re}_L^{0.8} - \mathrm{Re}_{x_{\mathrm{cri}}}^{0.8}\right) + 0.664\,\mathrm{Pr}^{\frac{1}{3}}\,\mathrm{Re}_{x_{\mathrm{cri}}}^{0.5}$$

But for flat plate the critical Reynolds number is

$$\mathrm{Re}_{x_{\mathrm{cri}}} = 5 \times 10^5$$

Therefore, the average Nusselt number becomes

$$\mathrm{Nu} = 0.037\,\mathrm{Pr}^{0.43}\left(\mathrm{Re}_L^{0.8} - 36239\right) + 469.5\,\mathrm{Pr}^{\frac{1}{3}}$$

But

$$\mathrm{Pr}^{0.43} \approx \mathrm{Pr}^{\frac{1}{3}}$$

Therefore,

$$\boxed{\mathrm{Nu} = \left[0.037\left(\mathrm{Re}_L^{0.8} - 36239\right) + 469.5\right]\mathrm{Pr}^{\frac{1}{3}}} \tag{5.113}$$

is valid for $\mathrm{Re}_L$ from $5 \times 10^5$ to $5 \times 10^7$ and Pr from 0.6 to 60. The constants in the above two relations will be different for different critical Reynolds number.

The average skin friction coefficient $C_f$ can be expressed [1] as

$$\boxed{c_f = \frac{0.074}{\mathrm{Re}_L^{\frac{1}{5}}} - \frac{1742}{\mathrm{Re}_L}} \tag{5.114}$$

## Example 5.5

Air at 1 atm and 20°C, flows with velocity 35 m/s over the surface of a flat plate which is maintained at 300°C. Calculate the rate at which the heat is transferred per meter width from both sides of the plate over a distance of 0.5 m from the leading edge, treating the flow as laminar.

## Solution

Given, $V = 35$ m/s, $p = 1$ atm, $T_s = 300°\mathrm{C}$, $T_\infty = 20°\mathrm{C}$, $x = 0.5$ m.

The film temperature is

$$\begin{aligned} T_f &= \frac{300 + 20}{2} = 160°\mathrm{C} \\ &= 433\,\mathrm{K} \end{aligned}$$

The density and viscosity at 433 K are

$$\begin{aligned} \rho &= \frac{p}{RT} = \frac{101325}{287 \times 433} \\ &= 0.815\,\mathrm{kg/m^3} \end{aligned}$$

$$\mu = 1.46 \times 10^{-6} \frac{433^{3/2}}{433 + 111}$$

$$= 2.418 \times 10^{-5}$$

Therefore, the Reynolds number becomes

$$\text{Re} = \frac{\rho V x}{\mu}$$

$$= \frac{0.815 \times 35 \times 0.5}{2.418 \times 10^{-5}}$$

$$= 5.9 \times 10^{5}$$

The Reynolds number is more than the critical value. But assuming the flow as laminar, we have by Equation (5.97), the Nusselt number as

$$\text{Nu}_x = 0.332(\text{Re}_x)^{1/2}(\text{Pr})^{1/3}$$

$$= 0.332 \times (5.9 \times 10^{5})^{1/2} \times (0.687)^{1/3}$$

$$= 225$$

since, for air at 433 K, from Table A-4, we have $k = 36.1 \times 10^{-3}$ W/(m K) and $\text{Pr} = 0.687$.

But, by definition,

$$\text{Nu}_x = \frac{hx}{k}$$

Therefore,

$$\frac{h \times 0.5}{36.1 \times 10^{-3}} = 225$$

$$h = \frac{225 \times 0.0361}{0.5}$$

$$= 16.25 \text{ W/(m}^2 \text{ K)}$$

The rate of heat transfer from both sides of the plate over the length of 0.5 m for 1 m width is given by

$$\dot{Q} = h\,(2A)\,(T_s - T_\infty)$$

$$= 16.25 \times (2 \times [0.5 \times 1]) \times (300 - 20)$$

$$= 4550\,\text{W} = \boxed{4.550\,\text{kW}}$$

## Example 5.6

Electrical strip heaters of 50 mm length each embedded all over the surface are used to maintain the surface temperature of a 1-m wide and 0.3-m long flat plate, uniform at 230°C. The power supply to each heater strip is independent of the others. If atmospheric air at 25°C flows at 60 m/s over the plate, what will be the power input to the heaters required to maintain the plate temperature constant and which one among the heaters will receive the maximum electrical power input? Assume $k = 0.0338$ W/(m K) and Pr $= 0.69$, for air.

## Solution

Given, $T = 25$°C, $V = 60$ m/s, $T_w = 230$°C, $L = 50$ mm.

At 25°C the density of the air stream is

$$\rho = \frac{p}{RT} = \frac{101325}{287 \times 298}$$

$$= 1.185\,\text{kg/m}^3$$

since 1 atm = 101325 Pa.

By Sutherland relation, the viscosity coefficient at temperature 25°C is [2]

$$\mu = 1.46 \times 10^{-6}\,\frac{T^{3/2}}{T + 111}$$

$$= 1.46 \times 10^{-6} \times \frac{298^{3/2}}{298 + 111}$$

$$= 1.836 \times 10^{-5}\,\text{kg/(m s)}$$

We know that, the heat transfer will be the maximum in the transition zone. Therefore, the heater located there will require the maximum power input. The Reynolds number based on the first heater length of $L = 50$ mm is

$$\text{Re}_L = \frac{\rho V L}{\mu}$$

$$= \frac{1.185 \times 60 \times 0.05}{1.836 \times 10^{-5}}$$

$$= 1.94 \times 10^5$$

This is less than the flat plate critical Reynolds number of $5 \times 10^5$. Therefore, the flow over the first heater strip is laminar. For transition, the length required for the boundary layer along the plate is

$$\begin{aligned} x_{\text{cri}} &= \frac{\text{Re}_{\text{cri}}\,\mu}{\rho V} \\ &= \frac{(5 \times 10^5)(1.836 \times 10^{-5})}{1.185 \times 60} \\ &= 0.129\,\text{m} \end{aligned}$$

Thus, the heater strip located at a distance of about 0.129 m from the leading edge of the plate will require the maximum power input. Also, the total number of heater strips along the plate length of 0.3 m is 6. The first two heater strips are in the laminar flow regime and the rest 4 are in the transition regime. For the first heater strip,

$$\dot{Q}_{\text{conv1}} = \overline{h}_1 L\, w (T_s - T_\infty)$$

where $\overline{h}_1$ is the average heat transfer coefficient for the first heater strip and it can be determined as follows. By Equation (5.97),

$$\begin{aligned} \text{Nu}_1 &= 0.332\,\text{Re}_1^{1/2}\,\text{Pr}^{1/3} \\ &= 0.332 \times (1.94 \times 10^5)^{1/2} \times (0.69)^{1/3} \\ &= 129.2 \\ \overline{h}_1 &= \frac{\text{Nu}_1\, k}{L_1} = \frac{129.2 \times 0.0338}{0.05} \\ &= 87.34\,\text{W/(m}^2\,\text{K)} \end{aligned}$$

Thus,

$$\begin{aligned} \dot{Q}_{\text{conv1}} &= 87.34 \times (0.05 \times 1)(230 - 25) \\ &= 895.2\,\text{W} \end{aligned}$$

For the second heater strip, $L = 100$ mm, thus,

$$\begin{aligned}
\text{Re}_{L2} &= \frac{1.185 \times 60 \times 0.10}{1.835 \times 10^{-5}} \\
&= 3.87 \times 10^5 \\
\text{Nu}_2 &= 0.332 \times (3.87 \times 10^5)^{1/2} \times (0.69)^{1/3} \\
&= 182.5 \\
\overline{h}_2 &= \frac{182.5 \times 0.0338}{0.1} \\
&= 61.685\,\text{W/(m}^2\,\text{K)} \\
\dot{Q}_{\text{conv2}} &= 61.685 \times (0.05 \times 1)(230 - 25) \\
&= 632.27\,\text{W}
\end{aligned}$$

For the third heater strip, $L = 150$ mm, thus,

$$\text{Re}_{L3} = 5.80 \times 10^5$$

Thus, the flow is a mixed flow consisting of laminar and turbulent zones and the Nusselt number becomes [Equation (5.113)]

$$\begin{aligned}
\text{Nu}_3 &= (0.037\,\text{Re}_{L3}^{4/5} - 871) \times (0.69)^{1/3} \\
&= 564.56
\end{aligned}$$

Hence,

$$\begin{aligned}
\overline{h}_3 &= \frac{564.56 \times 0.0338}{0.15} \\
&= 127.21\,\text{W/(m}^2\,\text{K)} \\
\dot{Q}_{\text{conv3}} &= 127.21 \times (0.05 \times 1) \times (230 - 25) \\
&= 1303.9\,\text{W}
\end{aligned}$$

For the fourth heater strip, $L = 200$ mm, thus,

$$\text{Re}_{L4} = 7.745 \times 10^5$$

$$\begin{aligned}
\text{Nu}_4 &= 911.855 \\
\overline{h}_4 &= \frac{911.855 \times 0.0338}{0.2} \\
&= 154.103\,\text{W}/(\text{m}^2\ \text{K}) \\
\dot{Q}_{\text{conv4}} &= 154.103 \times (0.05 \times 1) \times (230 - 25) \\
&= 1579.6\,\text{W}
\end{aligned}$$

For the fifth heater strip, $L = 250$ mm, thus,

$$\begin{aligned}
\text{Re}_{L5} &= 9.68 \times 10^5 \\
\text{Nu}_5 &= 1240.28 \\
\overline{h}_5 &= \frac{1240.28 \times 0.0338}{0.25} \\
&= 167.686\,\text{W}/(\text{m}^2\ \text{K}) \\
\dot{Q}_{\text{conv5}} &= 167.686 \times (0.05 \times 1) \times (230 - 25) \\
&= 1718.78\,\text{W}
\end{aligned}$$

For the sixth heater strip, $L = 300$ mm, thus,

$$\begin{aligned}
\text{Re}_{L6} &= 11.616 \times 10^5 \\
\text{Nu}_6 &= 1556.07 \\
\overline{h}_6 &= \frac{1556.07 \times 0.0338}{0.30} \\
&= 175.32\,\text{W}/(\text{m}^2\ \text{K}) \\
\dot{Q}_{\text{conv6}} &= 175.32 \times (0.05 \times 1) \times (230 - 25) \\
&= 1797.03\,\text{W}
\end{aligned}$$

It is seen that, the sixth strip at the end of the plate experiences the largest heat transfer and hence requires the maximum electrical input of $\boxed{1797.03\,\mathrm{W}}$. The power requirement will come down if the plate length is such that the local Reynolds number is more than $10^7$, causing the flow to become fully turbulent.

## 5.10 Free Convection

In the preceding section, we considered convection heat transfer in fluid flows that originate from an *external forcing* condition. For example, fluid motion may be induced by a fan or a pump, or it may result from the motion of a solid through the fluid. When a temperature gradient is present during such motion, *forced convection* heat transfer will occur.

However, convection heat transfer also occur in situations where there is no *forced velocity*. Such situations are referred to as *free* or *natural convection*, and they originate when a *body force* acts on a fluid in which there is *density gradient*. The net effect is a *buoyancy force*, which induces free convection currents. In most cases, the density gradients is due to a temperature gradient, and the body force is due to the gravitational field. For free convection boundary layer on a hot vertical plate, as shown in Figure 5.12, the Nusselt number can be obtained by carrying out the order-of-magnitude analysis on the governing equations and the boundary conditions.

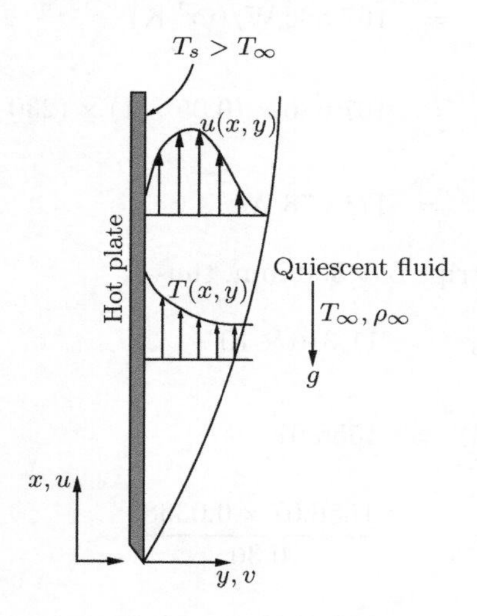


**Figure 5.12**
Boundary layer development on a heated vertical plate.

For the problem of boundary layer development on a heated vertical plate, the governing equations are the following.

The continuity equation is

$$\frac{\partial u}{\partial x} + \frac{\partial v}{\partial y} = 0 \tag{5.115}$$

The momentum equation is

$$u\frac{\partial u}{\partial x} + v\frac{\partial u}{\partial y} = g\beta\left(T - T_\infty\right) + \nu\frac{\partial^2 u}{\partial x^2} \tag{5.116}$$

The energy equation is

$$u\frac{\partial T}{\partial x} + v\frac{\partial T}{\partial y} = \alpha\frac{\partial^2 T}{\partial x^2} \tag{5.117}$$

The boundary conditions are

$$\text{at} \quad y = 0, \quad u = v = 0$$

$$\text{as} \quad y \to \infty, \quad u, v \to 0$$

and

$$\text{at} \quad y = 0, \quad T = T_s$$

$$\text{as} \quad y \to \infty, \quad T \to T_\infty$$

Even though the heat transfer problem considered here appears simple, its analysis is very much involved. Indeed, a thorough knowledge of boundary layer is required to understand this problem. But in our analysis here, without going into the depth of flow physics, let us approach the problem with the integral method used in the study of thermal boundary layer associated with the flow past a heated horizontal plate in Section 5.9. To further simplify the analysis, let us assume the Prandtl number of the flow is approximately unity, rendering the thicknesses of the momentum and thermal boundary layers to become identically equal ($\delta = \delta_t$).

The momentum integral equation for this problem can be obtained by integrating Equation (5.116) with respect to $y$ in the limit from 0 to $\delta$, that is from the wall to the edge of the boundary layer, making use of the continuity equation to eliminate the velocity component $v$, and utilizing the boundary conditions for the velocity component $u$. With this process, Equation (5.116) results in

$$\frac{d}{dx}\left(\int_0^\delta u^2 dy\right) = -\nu\left.\frac{\partial u}{\partial y}\right|_{y=0} + g\beta\int_0^\delta \left(T - T_\infty\right) dy \tag{5.118}$$

This is the *momentum integral equation* for boundary layer development on a heated vertical plate.

Similarly, integrating Equation (5.117) with respect to $y$ in the limit from 0 to $\delta_t$, that is, from the wall to the edge of the thermal boundary layer, making use of the continuity equation to eliminate the velocity component $v$, and utilizing the boundary conditions for temperature, the *energy integral equation* for boundary layer development on a heated vertical plate can be obtained as

$$\frac{d}{dx}\left(\int_0^{\delta_t} u\,[T - T_\infty]\,dy\right) = -\alpha \left.\frac{\partial T}{\partial y}\right|_{y=0} \tag{5.119}$$

The upper limit $\delta_t$ of integration in Equation (5.119) is assumed to be equal to $\delta$. From Equations (5.118) and (5.119), it is seen that these equations are coupled because $u$ and $T$ appear in both. Therefore, they should be solved simultaneously, by choosing suitable velocity and temperature profiles for the velocity and thermal boundary layers over the vertical plate, illustrated in Figure 5.12. Polynomial profiles can be used to represent the velocity and temperature distributions.

The temperature profile may be represented by a second-degree polynomial in the form

$$T(x, y) = c_0 + c_1 y + c_2 y^2$$

where $c_0$, $c_1$, $c_2$ are coefficients which can be evaluated with the following boundary conditions.

$$\text{at } y = 0, \quad T = T_w$$

$$\text{at } y = \delta, \quad T = T_\infty \text{ and } \frac{\partial T}{\partial y} = 0$$

The resulting temperature profile is

$$\boxed{\frac{T(x, y) - T_\infty}{T_w - T_\infty} = \left(1 - \frac{y}{\delta}\right)^2} \tag{5.120}$$

Note that, $\delta$ is used in this expression, instead of $\delta_t$, because $\delta = \delta_t$.

It is interesting to note that, as we did in the analysis of forced convection over a heated horizontal flat plate (Figure 5.10) in Section 5.9, no attempt is made to introduce a dimensionless temperature $\theta$ in the analysis of free convection boundary layer on a hot vertical plate (Figure 5.12). This is because, in the forced convection problem considered, introduction of $\theta$, as in Equation (5.41), rendered the thermal boundary conditions and the dimensionless temperature profile to become identical to the velocity boundary conditions and the velocity profile. But for the problem of free convection on the hot vertical plate, because of the peculiar shape of the velocity profile there is no such possibility. However, to simplify the analysis of free convection the Prandtl number is assumed to be unity, rendering the thickness of the momentum and thermal boundary layers to become identical ($\delta = \delta_t$).

For this problem, the velocity profile in Figure 5.12 shows that, $u = 0$ at the wall as well as at the edge of the boundary layer. Also, the maximum velocity is inside the boundary layer. This velocity profile may be represented by a cubical profile in the form

$$u(x,y) = u_\infty(c_0 + c_1 y + c_2 y^2 + c_3 y^3)$$

where $u_\infty$ is a reference velocity which is a function of $x$ and $c_0$, $c_1$, $c_2$, $c_3$ are constants. To determine these four constants, we need four conditions. The three conditions for the boundary layer are

$$\text{at } y = 0, \quad u = 0$$

$$\text{at } y = \delta, \quad u = 0 \text{ and } \frac{\partial u}{\partial y} = 0$$

The fourth condition is obtained by evaluating the momentum equation (5.116) with the condition at the wall ($y = 0$); $u = v = 0$ and $T = T_w$. Thus, we get

$$\text{at } y = 0, \quad \frac{\partial^2 u}{\partial y^2} = -\frac{g\beta}{\nu}(T_w - T_\infty)$$

With these four conditions, the velocity profile becomes

$$\boxed{u(x,y) = \left[u_\infty \frac{g\beta\delta^2 (T_w - T_\infty)}{4\nu}\right] \frac{y}{\delta}\left(1 - \frac{y}{\delta}\right)^2} \qquad (5.121)$$

It can be shown that, the maximum velocity in the boundary layer occurs at a distance of $y = \delta/3$.

Substitution of the temperature profile given by Equation (5.120) and the velocity profile given by Equation (5.121) into the momentum and energy integral equations (5.118) and (5.119), results in

$$\frac{1}{105}\frac{d}{dx}\left(u_\infty^2 \delta\right) = \frac{1}{3} g\beta (T_w - T_\infty)\,\delta - \nu \frac{u_\infty}{\delta} \qquad (5.122)$$

$$\frac{1}{30}(T_w - T_\infty)\frac{d}{dx}\left(u_\infty^2 \delta\right) = 2\alpha \frac{T_w - T_\infty}{\delta} \qquad (5.123)$$

To solve Equations (5.122) and (5.123) let us assume that, $u_\infty(x)$ and $\delta(x)$ depend on $x$ in the form

$$u_\infty(x) = c_1 x^m \qquad (5.124)$$

$$\delta(x) = c_2\, x^n \qquad (5.125)$$

where $c_1$, $c_2$, $m$, $n$ are coefficients. Substitution of Equations (5.124) and (5.125) into Equations (5.122) and (5.123) results in

$$\frac{(2m+n)c_1^2 c_2}{105} x^{2m+n-1} = \frac{1}{3} g\beta (T_w - T_\infty)\, c_2\, x^n - \frac{\nu c_1}{c_2} x^{m-n} \qquad (5.126)$$

$$\frac{(m+n)c_1c_2}{30}x^{m+n-1} = \frac{2\alpha}{c_2}x^{-n} \tag{5.127}$$

To render these equations invariant with $x$, let us equate the exponents of $x$ on both sides of Equations (5.126) and (5.127), resulting in

$$2m + n - 1 = n = m - n$$

$$m + n - 1 = -n$$

Solving, we get

$$m = \frac{1}{2} \text{ and } n = \frac{1}{4}$$

Using these $m$ and $n$ in Equations (5.126) and (5.127), the variable $x$ is cancelled and a simultaneous solution of the resulting equations gives the coefficients $c_1$ and $c_2$ as

$$c_1 = 5.17\nu\left(\frac{20}{21}+\frac{\alpha}{\nu}\right)^{-\frac{1}{2}}\left[\frac{g\beta\,(T_w - T_\infty)}{\nu^2}\right]^{\frac{1}{2}}$$

$$c_2 = 3.93\left(\frac{20}{21}+\frac{\alpha}{\nu}\right)^{\frac{1}{4}}\left[\frac{g\beta\,(T_w - T_\infty)}{\nu^2}\right]^{-\frac{1}{4}}\left(\frac{\alpha}{\nu}\right)^{-\frac{1}{2}}$$

Substituting these $m$, $n$, $c_1$ and $c_2$ into Equation (5.125), we get the boundary layer thickness as

$$\delta(x) = 3.93\,(0.95 + \mathrm{Pr})^{\frac{1}{4}}\left[\frac{g\beta\,(T_w - T_\infty)}{\nu^2}\right]^{-\frac{1}{4}}\mathrm{Pr}^{-\frac{1}{2}}x^{\frac{1}{4}} \tag{5.128}$$

But we know that,

$$\frac{g\beta\,(T_w - T_\infty)\,x^3}{\nu^2}$$

is the Grashof number Gr. Thus,

$$\boxed{\frac{\delta(x)}{x} = 3.93\mathrm{Pr}^{-\frac{1}{2}}\,(0.95+\mathrm{Pr})^{\frac{1}{4}}\,(0.95+\mathrm{Pr})^{\frac{1}{4}}\,\mathrm{Gr}_x^{-\frac{1}{4}}} \tag{5.129}$$

Once $\delta(x)$ is known, the temperature profile $T(x, y)$ can be determined from Equation (5.120), and the local Nusselt number can be obtained as follows.

### 5.10.1 Local Nusselt Number

The local Nusselt number is defined as

$$\mathrm{Nu}_x \equiv \frac{h_x x}{k}$$

By Newton's law of cooling,

$$\dot{q}_w = h_x (T_w - T_\infty)$$

$$h_x = \frac{\dot{q}_w}{T_w - T_\infty}$$

Therefore,

$$\mathrm{Nu}_x = \frac{\dot{q}_w}{(T_w - T_\infty)} \frac{x}{k}$$

But by Fourier's law, we have

$$\dot{q}_w = -k \left(\frac{\partial T}{\partial y}\right)\bigg|_{y=0}$$

Therefore,

$$\mathrm{Nu}_x = \frac{-k \left(\frac{\partial T}{\partial y}\right)\bigg|_{y=0}}{(T_w - T_\infty)} \frac{x}{k}$$

or

$$\mathrm{Nu}_x = -\frac{x}{(T_w - T_\infty)} \left(\frac{\partial T}{\partial y}\right)\bigg|_{y=0} \tag{5.130}$$

The temperature gradient is determined from Equation (5.120), as

$$\left(\frac{\partial T}{\partial y}\right)\bigg|_{y=0} = -2\frac{(T_w - T_\infty)}{\delta}$$

Substituting this into Equation (5.130), we get the local Nusselt number as

$$\mathrm{Nu}_x = 2\,\frac{x}{\delta} \tag{5.130$a$}$$

Now, substituting the expression for $\delta$, given by Equation (5.129), we get

$$\mathrm{Nu}_x = 0.508\,\mathrm{Pr}^{\frac{1}{2}}\,(0.95 + \mathrm{Pr})^{-\frac{1}{4}}\,\mathrm{Gr}_x{}^{\frac{1}{4}}$$

or

$$\boxed{\mathrm{Nu}_x = \frac{0.51\,\mathrm{Pr}^{\frac{1}{2}}}{(0.95 + \mathrm{Pr})^{\frac{1}{4}}}\,\mathrm{Gr}_x{}^{\frac{1}{4}}} \tag{5.131}$$

This equation gives the variation of the local heat transfer coefficient along the vertical plate, maintained at a constant temperature of $T_s$ in a quiescent fluid environment.

In an alternate form, the local Nusselt number can be expressed as

$$\boxed{\mathrm{Nu}_x = 0.51\,\mathrm{Ra}_x^{\frac{1}{4}} \left(\frac{\mathrm{Pr}}{0.95 + \mathrm{Pr}}\right)^{\frac{1}{4}}} \tag{5.132}$$

where $\mathrm{Ra}_x = (\mathrm{Gr}_x \mathrm{Pr})$ is known as the *Rayleigh number.*

### 5.10.2 Free Convection Correlations

Even though the mechanism of free or natural convection is well known, the complexities of fluid motion makes it difficult to obtain simple analytical relations for heat transfer by solving the governing equations of motion and energy. Some analytical solutions exist for natural convection, but such solutions lack generality since they are obtained for simple geometries under some simplified assumptions. Therefore, excepting some simple cases, heat transfer relations for free convection are based on experimental studies. Although there are numerous such correlations of varying complexities and accuracy available in the literature for any given geometry, only the *simpler* ones are given below for the following reasons.

- The accuracy of simpler relations is usually within the range of uncertainty associated with a problem.
- In order to keep the emphasis on the physics of the problem rather than formula manipulation.

The simple empirical correlations for the *average* Nusselt number Nu in natural convection are of the form

$$\boxed{\mathrm{Nu} = \frac{h\delta}{k} = c\,(\mathrm{Gr\,Pr})^n = c\,\mathrm{Ra}^n} \tag{5.133}$$

where Ra is the Rayleigh number, which is the product of the Grashof and Prandtl numbers.

$$\boxed{\mathrm{Ra} = \mathrm{Gr\,Pr} = \frac{g\beta(T_s - T_\infty)\delta^3}{\nu^2}\,\mathrm{Pr}} \tag{5.134}$$

The values of the constant $c$ and index $n$ in Equation (5.133) depend on the geometry of the surface and the laminar or turbulent nature of the *flow regime*, which is characterized by the range of the Reynolds number. The value of $n$ is usually $1/4$ for laminar flow and $1/3$ for turbulent flow. The constant $c$ is normally less than 1, and in literature it is shown that, $c$ can be taken as 0.59 for laminar flow and 0.10 for turbulent flow.

Note that the Nusselt number in Equation (5.133) is the average or mean Nusselt number. This is because usually in practical applications, the mean Nu over a distance from $x = 0$ to $x = L$ along a plate is of interest. Therefore, in the expression for the average Nusselt number, the average convective heat transfer coefficient, defined as

$$h = \frac{1}{L}\int_0^L h_x dx$$

is used, where $h_x$ is the local convective heat transfer coefficient. From Equation (5.131), we get $h_x \propto x^{-\frac{1}{4}}$. Therefore,

$$\begin{aligned} h &= \frac{1}{L}\int_0^L x^{-\frac{1}{4}}\,dx \\ &= \frac{1}{4}\left[\frac{x^{-\frac{1}{4}}+1}{-\frac{1}{4}+1}\right]_0^L \\ &= \frac{1}{L}\frac{4}{3}\left[L^{\frac{3}{4}}\right] = \frac{4}{3}\left[L^{-\frac{1}{4}}\right] \\ &= \frac{4}{3}\Big[h_x\Big]_{x=L} \end{aligned}$$

Thus, the average Nusselt number Nu becomes,

$$\begin{aligned} \text{Nu} &= \frac{hL}{k} = \frac{L}{k}\frac{4}{3}\Big[h_x\Big]_{x=L} \\ &= \frac{4}{3}\left[\frac{h_x x}{k}\right]_{x=L} \end{aligned}$$

or

$$\boxed{\text{Nu} = \frac{4}{3}\left[\text{Nu}_x\right]_{x=L}} \qquad (5.135)$$

This expression for average Nusselt number is valid only for the case where $h_x \propto x^{-\frac{1}{4}}$. To make this more general, let us assume the local value of average convective heat transfer coefficient $h_x$ depends on $x$ in the form

$$h_x = c\,x^{-n}$$

For this $h_x$ the average convective heat transfer coefficient $h$ over $x = 0$ to $x = L$ becomes

$$\begin{aligned} h &= \frac{c}{L}\int_0^L x^{-n}\,dx = \frac{c}{1-n}L^{-n} \\ &= \frac{1}{1-n}\Big[h_x\Big]_{x=L} \end{aligned}$$

Thus, the average Nusselt number becomes

$$\boxed{\text{Nu} = \frac{1}{1-n}\left[\text{Nu}_x\right]_{x=L}} \qquad (5.136)$$

Now, using the average Nusselt number given by Equation (5.135), the corresponding average value of the local Nusselt number given by Equation (5.131) becomes

$$\begin{aligned} \text{Nu} &= \frac{4}{3}\left[\frac{0.51\,\text{Pr}^{\frac{1}{2}}}{(0.95+\text{Pr})^{\frac{1}{4}}}\,\text{Gr}_x^{\frac{1}{4}}\right] \\ &= \frac{0.68\,\text{Pr}^{\frac{1}{2}}}{(0.95+\text{Pr})^{\frac{1}{4}}}\,\text{Gr}_x^{\frac{1}{4}} \end{aligned}$$

where $\text{Nu} = \dfrac{hL}{k}$. But the product of Gr and Pr can be expressed as Rayleigh number, that is,

$$\text{Ra}_L = \text{Gr}_L\text{Pr}$$

Thus, the above expression for Nu becomes

$$\text{Nu} = 0.68\ (\text{Pr}\text{Gr}_L)^{\frac{1}{4}}\left(\frac{0.68\,\text{Pr}}{0.95+\text{Pr}}\right)^{\frac{1}{4}}$$

or

$$\boxed{\text{Nu} = 0.68\ (\text{Ra}_L)^{\frac{1}{4}}\left(\frac{0.68\,\text{Pr}}{0.95+\text{Pr}}\right)^{\frac{1}{4}}} \tag{5.137}$$

The local and average Nusselt number given by Equations (5.131) and (5.137) are valid for laminar free convection on a vertical plate maintained at a uniform temperature. The fluid properties are evaluated at the film temperature $T_f = (T_w + T_\infty)/2$.

*Note:* It is essential to note that, for the problem under consideration, the transition of the flow from laminar state to turbulent state is observed to be taking place for $\text{Ra}_x$ from $10^8$ to $10^9$ [3]. Hence, Equations (5.131) and (5.137) are valid only for $\text{Ra}_L < 10^9$.

## 5.11 Free Convection on Vertical Planes and Cylinders

Owing to the complications associated with the nature of convection, it is difficult to analyze the convection process theoretically. Therefore, experimental data are often used to develop reliable heat transfer relations. Some such well-known empirical correlations for free convection for a vertical surface in laminar and turbulent flows, with uniform wall temperature or uniform wall heat flux boundary conditions, are given here.

### 5.11.1 Uniform Wall Temperature

For uniform wall temperature, McAdams [4] correlated the average Nusselt number as

$$\text{Nu} = c\,(\text{Gr}_L\text{Pr})^n = c\,\text{Ra}_L^n \tag{5.138}$$

where $L$ is the length of the vertical plate and $c$ and $n$ are constants. For laminar flows in the range of $(\mathrm{Gr}_L\mathrm{Pr})$ from $10^4$ to $10^9$, $c = 0.59$ and $n = 1/4$. For turbulent flows in the range of $(\mathrm{Gr}_L\mathrm{Pr})$ from $10^9$ to $10^{13}$, $c = 0.1$ and $n = 1/3$. The physical properties of the flow are calculated at $T_f = (T_w + T_\infty)/2$.

Churchill and Chu [5] developed two correlations for free convection on a vertical plate under isothermal surface conditions. For laminar flow, with $10^{-1} < \mathrm{Ra}_L < 10^9$, the average Nusselt number is given by

$$\mathrm{Nu} = 0.68 + \frac{0.67\,\mathrm{Ra}_L^{\frac{1}{4}}}{\left[1 + (0.492/\mathrm{Pr})^{\frac{9}{16}}\right]^{\frac{4}{9}}} \tag{5.139}$$

This is valid for all values of Pr.

The correlation which is valid for both laminar and turbulent flows, in the range of $\mathrm{Ra}_L$ from $10^{-1}$ to $10^{12}$, is

$$\sqrt{\mathrm{Nu}} = 0.825 + \frac{0.387\,\mathrm{Ra}_L^{\frac{1}{6}}}{\left[1 + (0.492/\mathrm{Pr})^{\frac{9}{16}}\right]^{\frac{8}{27}}} \tag{5.140}$$

The physical properties of the flow are calculated at $T_f$. Among Equations (5.139) and (5.140), Equation (5.139) yields better results for laminar flows.

### 5.11.2 Uniform Wall Heat Flux

Based on the experimental studies with air [6] and with water [7], the following correlations for local Nusselt number under uniform wall heat flux have been proposed.

For laminar flows in the range of $(\mathrm{Gr}_x^*\mathrm{Pr})$ from $10^5$ to $10^{11}$,

$$\mathrm{Nu}_x = 0.60\,(\mathrm{Gr}_x^*\mathrm{Pr})^{\frac{1}{5}} \tag{5.141}$$

For turbulent flows in the range of $(\mathrm{Gr}_x^*\mathrm{Pr})$ from $2 \times 10^{13}$ to $10^{16}$,

$$\mathrm{Nu}_x = 0.568\,(\mathrm{Gr}_x^*\mathrm{Pr})^{0.22} \tag{5.142}$$

where $\mathrm{Gr}_x^*$ is the modified Grashof number, defined as

$$\mathrm{Gr}_x^* = \mathrm{Gr}_x\mathrm{Nu}_x = \frac{g\beta\,(T_w - T_\infty)\,x^3}{\nu^2}\,\frac{q_w\,x}{T_w - T_\infty} = \frac{g\beta\,q_w\,x^4}{k\,\nu^2} \tag{5.143}$$

and the local Nusselt number is

$$\mathrm{Nu}_x = \frac{x\,h_x}{k}$$

and $q_w$ is the constant wall heat flux.

To determine the average Nusselt number Nu, it is necessary to establish the dependence fo $h_x$ on $x$ in the form $h_x = cx^{-n}$. From Equations (5.141) and (5.142), we have

$$h_x \sim \frac{1}{x}\left(\mathrm{Gr}_x^*\right)^{0.2} \sim \frac{1}{x}(x^4)^{0.2} \sim x^{-0.2}$$

$$h_x \sim \frac{1}{x}\left(\mathrm{Gr}_x^*\right)^{0.22} \sim \frac{1}{x}(x^4)^{0.22} \sim x^{-0.12}$$

Therefore, the average Nusselt number Nu, for Equations (5.141) and (5.142), respectively, becomes

$$\mathrm{Nu} = \frac{1}{1-0.2}\left[\mathrm{Nu}_x\right]_{x=L} = 1.25\left[\mathrm{Nu}_x\right]_{x=L} \qquad (5.141a)$$

$$\mathrm{Nu} = \frac{1}{1-0.12}\left[\mathrm{Nu}_x\right]_{x=L} = 1.136\left[\mathrm{Nu}_x\right]_{x=L} \qquad (5.142a)$$

All physical properties are evaluated at the film temperature $T_f$.

At this stage, it is interesting to note that the Nu correlation for uniform wall temperature given by Equation (5.139) may also be used for uniform wall heat transfer flux condition, provided the right-hand side of Equation (5.139) is expressed in terms of the modified Grashof number $\mathrm{Gr}_L^*$. This can be done by noting,

$$\mathrm{Ra}_L = \mathrm{Gr}_L\mathrm{Pr} \quad \text{and} \quad \mathrm{Gr}_L^* = \mathrm{Gr}_L\mathrm{Nu}$$

That is,

$$\mathrm{Ra}_L = \frac{\mathrm{Gr}_L^*\mathrm{Pr}}{\mathrm{Nu}}$$

Using this in Equation (5.139) and rearranging, we get

$$\mathrm{Nu}^{\frac{1}{4}}\left(\mathrm{Nu} - 0.68\right) = \frac{0.67\left(\mathrm{Gr}_L^*\mathrm{Pr}\right)^{\frac{1}{4}}}{\left[1 + (0.492/\mathrm{Pr})^{\frac{9}{16}}\right]^{\frac{4}{9}}} \qquad (5.144)$$

## Example 5.7

A flat plate of length 1 m, maintained at a constant temperature of 350 K is placed vertically in still atmospheric air at 1 atm and 300 K. (a) Determine the thickness of the thermal boundary layer at the trailing edge of the plate, caused by the free convection heat transfer from the plate to the air. (b) If air flows over the plate with a velocity of 8 m/s, will the forced convection boundary layer thickness at trailing edge be greater or smaller than the free convection boundary layer, in still air?

## Solution

(a) Given, $p = 1$ atm, $T_s = 350$ K, $T_\infty = 300$ K, $L = 1$ m.

The film temperature is

$$T_f = \frac{(T_s + T_\infty)}{2} = \frac{(350 + 300)}{2}$$

$$= 325\,\mathrm{K}$$

For air at 325 K, from Table A-4, we have Pr = 0.7.

The density and viscosity are

$$\rho = \frac{p}{RT} = \frac{101325}{287 \times 325}$$

$$= 1.086\,\mathrm{kg/m^3}$$

$$\mu = (1.46 \times 10^{-6}) \times \frac{325^{1.5}}{325 + 111}$$

$$= 1.96 \times 10^{-5}\,\mathrm{kg/(m\ s)}$$

Also,

$$\beta = \frac{1}{T_f} = \frac{1}{325} = 0.00308\ 1/\mathrm{K}$$

The Grashof number is

$$\mathrm{Gr}_L = \frac{g\beta\,(T_s - T_\infty)\,L^3}{\nu^2}$$

$$= \frac{9.81 \times 0.00308 \times (350 - 300) \times 1^3}{\left(\frac{1.96 \times 10^{-5}}{1.086}\right)^2}$$

$$= 4.64 \times 10^9$$

The Rayleigh number is

$$\mathrm{Ra}_L = \mathrm{Gr}_L\,\mathrm{Pr} = (4.64 \times 10^9) \times 0.7$$

$$= 3.25 \times 10^9$$

This is only slightly greater than the critical Ra of $10^9$. Therefore, the flow may be taken as laminar. By Equation (5.138), the Nusselt number is

$$\mathrm{Nu} = c\,\mathrm{Ra}^n$$

where, $c = 0.59$ and $n = 1/4$, for laminar flow. Thus,

$$\begin{aligned}\text{Nu} &= 0.59 \times (3.25 \times 10^9)^{1/4} \\ &= 140.87\end{aligned}$$

By Equation (5.130a), the Nusselt number, based on plate length is

$$\text{Nu} = 2\,\frac{L}{\delta}$$

where $\delta$ is the boundary layer thickness at the trailing edge of the plate. Thus,

$$\begin{aligned}\delta &= \frac{2L}{\text{Nu}} = \frac{2 \times 1}{140.87} \\ &= 0.0142\,\text{m} \\ &= \boxed{14.2\,\text{mm}}\end{aligned}$$

(b) Given $U = 8$ m/s, therefore, the Reynolds number based on $L$ is

$$\begin{aligned}\text{Re}_L &= \frac{\rho U L}{\mu} = \frac{1.086 \times 8 \times 1}{1.96 \times 10^{-5}} \\ &= 4.43 \times 10^5\end{aligned}$$

The Reynolds number is less than the critical value of $5 \times 10^5$, thus the boundary layer is laminar. For this laminar boundary layer, the thickness is

$$\begin{aligned}\delta &= \frac{5L}{\sqrt{\text{Re}_L}} = \frac{5 \times 1}{\sqrt{4.43 \times 10^5}} \\ &= 0.00751\,\text{m} \\ &= \boxed{7.51\,\text{mm}}\end{aligned}$$

The forced convection boundary layer thickness is less than the free convection boundary layer thickness.

## Example 5.8

A plate of length 0.9 m and width 0.5 m, maintained at a uniform temperature of 500 K is suspended vertically in still air at 300 K and 1 atm. Determine (a) the average heat transfer coefficient for the entire plate and (b) the temperature profile at the trailing edge of the plate.

## Solution

(a) Given, $T_s = 500$ K, $T_\infty = 300$ K, $L = 0.9$ m, $w = 0.5$ m.

The film temperature is

$$T_f = \frac{500 + 300}{2} = 400\,\mathrm{K}$$

For air at 400 K, from Table A-4, Pr = 0.689, $k = 0.03365$ W/(m °C), and

$$\beta = 1/400 = 0.0025\,1/\mathrm{K}$$

The density and viscosity are

$$\rho = \frac{p}{RT} = \frac{101325}{287 \times 400} = 0.883\,\mathrm{kg/m^3}$$

$$\mu = (1.46 \times 10^{-6}) \times \frac{400^{1.5}}{400 + 111} = 2.29 \times 10^{-5}\,\mathrm{kg/(m\ s)}$$

The Grashof number is

$$\mathrm{Gr}_L = \frac{g\beta\,(T_s - T_\infty)\,L^3}{\nu^2} = \frac{9.81 \times 0.0025 \times (500 - 300) \times 0.9^3}{[(2.29 \times 10^{-5})/0.883]^2} = 5.32 \times 10^9$$

This is greater than the critical value of $10^9$. Therefore, by Equation (5.140), the average Nusselt number is

$$\sqrt{\mathrm{Nu}} = 0.825 + \frac{0.387\,\mathrm{Ra}_L^{\frac{1}{6}}}{\left[1 + (0.492/\mathrm{Pr})^{\frac{9}{16}}\right]^{\frac{8}{27}}}$$

where

$$\mathrm{Ra}_L = (\mathrm{Gr}_L\mathrm{Pr}) = 5.32 \times 10^9 \times 0.689 = 3.66 \times 10^9$$

Therefore,

$$\sqrt{\text{Nu}} = 0.825 + \frac{0.387 \times (3.66 \times 10^9)^{1/6}}{\left[1 + (0.492/0.689)^{\frac{9}{16}}\right]^{\frac{8}{27}}}$$

$$= 13.53$$

Therefore,

$$\sqrt{\frac{hL}{k}} = 13.53$$

$$\begin{aligned} h &= \frac{13.53^2 k}{L} \\ &= \frac{13.53^2 \times 0.03365}{0.9} \\ &= \boxed{6.84\,\text{W/(m}^2\,{}^\circ\text{C)}} \end{aligned}$$

(b) For this flow, by Equation (5.133), we have the Nusselt number in terms of the boundary layer thickness as

$$\text{Nu} = \frac{h\delta}{k} = c\,(\text{Gr}\,\text{Pr})^n = c\,\text{Ra}^n$$

where $c = 1/3$ and $n = 0.1$. Thus,

$$\begin{aligned} \frac{h\delta}{k} &= \frac{1}{3} \times (3.66 \times 10^9)^{0.1} \\ &= 3.015 \end{aligned}$$

$$\begin{aligned} \delta &= \frac{3.015 \times 0.03365}{6.84} \\ &= 0.0148\,\text{m} \end{aligned}$$

The temperature profile given by Equation (5.120) is

$$\begin{aligned} \frac{T(x,y) - T_\infty}{T_w - T_\infty} &= \left(1 - \frac{y}{\delta}\right)^2 \\ T(x,y) - T_\infty &= 200\left(1 - \frac{y}{\delta}\right)^2 \\ T(x,y) &= \boxed{200\left(1 - \frac{y}{0.0148}\right)^2 + 300} \end{aligned}$$

Note that, at $y = 0$, we get $T = T_w = 500$ K, and at the outer edge of the boundary layer, that is, at $y = \delta$, the temperature given by the above profile is $T = T_\infty = 300$ K. Thus, the temperature profile satisfies the conditions at the wall and the freestream.

## 5.12 Free Convection on a Horizontal Plate

In the case of free convection on a vertical plate, the gravity effect was identical on the top and bottom surfaces of the plate. But for the case of a horizontal plate, the gravity effect on the free convection will have a different effect depending on whether the surface of the plate over which the free convection taking place faces upward or downward. Thus, the average Nusselt number for free convection on a horizontal plate depends on whether the surface is facing up or down, and the plate surface is warmer or cooler than the surrounding fluid.

For a horizontal plate with uniform wall temperature the mean Nusselt number for free convection, correlated by McAdams [4], is

$$\boxed{\mathrm{Nu} = c\,(\mathrm{Gr}_L\,\mathrm{Pr})^n} \tag{5.145}$$

The coefficient $c$ and the exponent $n$ in Equation (5.145) are listed in Table 5.2.

**Table 5.2** Values of $c$ and $n$ for free convection on a horizontal plate with uniform surface temperature

| Plate orientation | Range of $(\mathrm{Gr}_L\,\mathrm{Pr})$ | $c$ | $n$ | Flow regime |
|---|---|---|---|---|
| Hot-side up | $10^5 - 2\times10^7$ | 0.54 | $\frac{1}{4}$ | Laminar |
| | $2\times10^7 - 3\times10^{10}$ | 0.14 | $\frac{1}{3}$ | Turbulent |
| Hot-side down | $3\times10^5 - 3\times10^{10}$ | 0.27 | $\frac{1}{4}$ | Laminar |

For a horizontal plate with *uniform wall heat flux* with horizontal surface facing up, the correlation for Nusselt number has been proposed by Fujii and Imura [8], as

$$\mathrm{Nu} = 0.13\,(\mathrm{Gr}_L\,\mathrm{Pr})^{\frac{1}{3}} \quad \text{for } (\mathrm{Gr}_L\,\mathrm{Pr}) < 2\times10^8 \tag{5.146}$$

$$\mathrm{Nu} = 0.16\,(\mathrm{Gr}_L\,\mathrm{Pr})^{\frac{1}{3}} \quad \text{for } 5\times10^8 < (\mathrm{Gr}_L\,\mathrm{Pr}) < 10^{11} \tag{5.147}$$

**Table 5.2a** Values of $c$ and $n$ for free convection on a horizontal plate with uniform surface temperature

| Geometry | Range of $(\mathrm{Gr}_L\,\mathrm{Pr})$ | $c$ | $n$ |
|---|---|---|---|
| Vert. planes | $10^4$-$10^9$ | 0.59 | $\frac{1}{4}$ |
| and cylinders | $10^9$-$10^{13}$ | 0.021 | $\frac{2}{5}$ |
| | $10^9$-$10^{13}$ | 0.10 | $\frac{1}{3}$ |
| Hori. cylinders | $10^4$-$10^9$ | 0.53 | $\frac{1}{4}$ |
| | $10^9$-$10^{12}$ | 0.13 | $\frac{1}{3}$ |

For a horizontal plate with heated surface facing down, the correlation for Nusselt number is

$$\mathrm{Nu} = 0.58\,(\mathrm{Gr}_L\,\mathrm{Pr})^{\frac{1}{5}} \quad \text{for } 10^6 < (\mathrm{Gr}_L\,\mathrm{Pr}) < 10^{11} \tag{5.148}$$

The physical properties in the above relation for Nu [Equations (5.146) to (5.148)] should be evaluated at a mean temperature $T_m$, defined as

$$T_m = T_w - 0.25\,(T_w - T_\infty)$$

and the thermal expansion coefficient $\beta$ should be evaluated at the film temperature $T_f = (T_w + T_\infty)/2$.

It is essential to note that the characteristic dimension $L$ to be used in Equation (5.145) is the length of a side for a *square surface*, the mean of the two sides for a *rectangular surface*, and 0.9 times the diameter for a circular disk.

The constants $c$ and $n$ Equation (5.145) have been evaluated by a large number of researchers. Values of $c$ and $n$ for specified range of (Gr Pr) for different shapes, listed by Holman [29], are listed in Table 5.2a.

**Unsymmetrical planforms**

When the body considered is not of square, rectangle, or circular shape, the characteristic dimension for the body can be calculated from

$$L = \frac{A}{P}$$

where $A$ is the surface area and $P$ is the wetted perimeter of the body.

### Number of solids

Heat transfer from a number of plates, cylinders, spheres, and blocks, the average heat transfer coefficient can be correlated using Equation (5.145), with $c = 0.60$ and $n = \frac{1}{4}$ for $(\text{Gr}_L \, \text{Pr})$ in the range from $10^4$ to $10^9$ [29]. The characteristic dimension for rectangular block can be calculated from

$$\frac{1}{L} = \frac{1}{L_h} + \frac{1}{L_v}$$

where $L_h$ and $L_v$ are the horizontal and vertical dimensions, respectively. If is essential to note that, Equation (5.145) with $c = 0.60$ and $n = \frac{1}{4}$ should be used only in the absence of specific data for the particular shape of the body considered.

## Example 5.9

A horizontal rod of length 1.3 m and diameter 25 mm, maintained at a constant surface temperature of 60°C is submerged in water at 20°C. Determine the heat loss from the rod by free convection.

## Solution

Given, $T_s = 60°\text{C}$, $T_\infty = 20°\text{C}$, $L = 1.3$ m, $D = 0.025$ m.

The film temperature is

$$T_f = \frac{60 + 20}{2} = 40°\text{C}$$

For water at 40°C, from Table A-8, $k = 0.635$ W/(m °C), $\text{Pr} = 4.5$, $c_p = 4.174$ kJ/(kg °C), $\mu = 6.3 \times 10^{-4}$ kg/(m s), $\rho = 991$ kg/m$^3$. The thermal expansion coefficient is

$$\begin{aligned} \beta &= \frac{1}{T_f} = \frac{1}{40 + 273} \\ &= 0.0032 \text{ 1/K} \end{aligned}$$

The Grashof number is

$$\begin{aligned} \text{Gr}_L &= \frac{g\beta \left(T_s - T_\infty\right) L^3}{\nu^2} \\ &= \frac{9.81 \times 0.0032 \times (60 - 20) \times 1.3^3}{[(6.3 \times 10^{-4})/991]^2} \\ &= 6.83 \times 10^{12} \end{aligned}$$

Thus, the flow is turbulent.

$$(\text{Gr}_L\,\text{Pr}) = (6.83 \times 10^{12}) \times 4.5$$

$$= 30.735 \times 10^{12}$$

For this value of $(\text{Gr}_L\,\text{Pr})$ there is no data listed in Table 5.2. Therefore, taking the values of $c = 0.14$ and $n = \frac{1}{3}$ for this turbulent flow, we can use Equation (5.145) for estimating the approximate value of the convection coefficient. Thus,

$$\text{Nu} = 0.14(\text{Gr}_L\,\text{Pr})^{\frac{1}{3}}$$

$$= 0.14 \times (30.735 \times 10^{12})^{\frac{1}{3}}$$

$$= 4385.4$$

Therefore,

$$h = \frac{\text{Nu}k}{L} = \frac{4385.4 \times 0.635}{1.3}$$

$$= 2142\,\text{W}/(\text{m}^2\,^\circ\text{C})$$

The heat loss from the rod is

$$\dot{Q} = hA\,(T_s - T_\infty)$$

$$= h\,(\pi DL)\,(T_s - T_\infty)$$

$$= 2142 \times (\pi \times 0.025 \times 1.3) \times (60 - 20)$$

$$= \boxed{8748\,\text{W}}$$

### 5.12.1 Effects of Turbulence

Free convection boundary layer can be laminar or turbulent. Free convection flows originate from a thermal instability, that is, warmer fluid moves vertically upwards relative to cooler fluid. But in forced convection, hydrodynamic instabilities may also arise. That is, disturbances in the flow may be amplified, leading to transition from laminar to turbulent flow. The transition in a free convection boundary layer depends on the relative magnitude of the buoyancy and viscous forces in the fluid. It is customary to correlate the occurrence of transition in terms of Rayleigh number. For vertical plates the critical Rayleigh number is

$$\boxed{\text{Ra}_{\text{cri}} = (\text{Gr}_x\,\text{Pr}) = \frac{g\beta\,(T_s - T_\infty)}{\nu\alpha}\,x^2 \approx 10^9}$$

where $x$ is the length from the leading edge of the plate.

## 5.13 Free Convection from Inclined Surfaces

Free convection over heated plates in water, at various angles of inclination $\theta$, was studied experimentally by Fujii and Imura [8]. The inclination is considered positive when the heated surface faces downward, as shown in Figure 5.13.

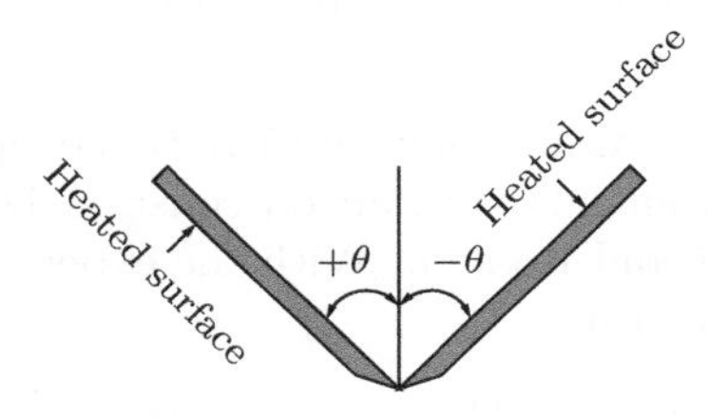


**Figure 5.13**
Sign convention for inclination angle.

For an inclined plate with *heated surface facing downward* ($\theta$ positive), with approximately constant heat flux, the following correlation was obtained for the average Nusselt number

$$\text{Nu}_e = 0.56\,(\text{Gr}_e\;\text{Pr}_e\;\cos\,\theta)^{\frac{1}{4}} \tag{5.149}$$

Note that, the Nusselt number, Grashof number and Prandtl number are expressed with subscript $e$ to emphasize that the properties used in this relation are estimated at a reference temperature $T_e$, defined in Equation (5.150). However, the characteristic length $L$ used continues to be the geometrical length of the inclined surface. Equation (5.149) is valid for $+\theta < 88^\circ$ and $10^5 < (\text{Gr}_e\text{Pr}_e\cos\,\theta) < 10^{11}$. All the properties except $\beta$ in this relation are evaluated at a reference temperature $T_e$, defined as

$$T_e = T_w - 0.25\,(T_w - T_\infty) \tag{5.150}$$

where $T_w$ is the mean wall temperature and $T_\infty$ is the freestream temperature. The value of $\beta$ is determined at a temperature of $T_\infty + 0.25\,(T_w - T_\infty)$. For almost horizontal plates, with heated surface facing downward, in the range of $\theta$ from $88^\circ$ to $90^\circ$, an additional relation for Nu was obtained as

$$\text{Nu}_e = 0.58\,(\text{Gr}_e\text{Pr}_e\;\cos\,\theta)^{\frac{1}{5}} \tag{5.151}$$

This is valid for $10^6 < (\mathrm{Gr}_e \mathrm{Pr}_e \cos\theta) < 10^{11}$.

The empirical correlations become complicated when the heated surface of the inclined plate faces upward. For $\theta$ from $-15°$ to $-75°$, the correlation developed for this orientation of the inclined plate is

$$\mathrm{Nu}_e = 0.14\left[(\mathrm{Gr}_e\mathrm{Pr}_e)^{\frac{1}{3}} - (\mathrm{Gr}_c\mathrm{Pr}_e)^{\frac{1}{3}}\right] + 0.56\,(\mathrm{Gr}_e\mathrm{Pr}_e\cos\theta)^{\frac{1}{4}} \qquad (5.152)$$

This is valid for $10^5 < (\mathrm{Gr}_e \mathrm{Pr}_e \cos\theta) < 10^{11}$. In this relation, $\mathrm{Gr}_c$ is the critical Grashof number, indicating when the Nusselt number starts to separate from the laminar relation Equation (5.149). The critical Grashof number $\mathrm{Gr}_c$ for some values of $\theta$ are listed below.

| $\theta$ | $-15°$ | $-30°$ | $-60°$ | $-75°$ |
|---|---|---|---|---|
| $\mathrm{Gr}_c$ | $5 \times 10^9$ | $2 \times 10^9$ | $10^8$ | $10^6$ |

For $\mathrm{Gr}_e < \mathrm{Gr}_c$ the first term in Equation (5.152) is dropped out.

Experimental measurements with air on constant heat flux surfaces have been conducted by Vliet and Ross [9]. With the experimental date, they correlated the Nusselt number as

$$\mathrm{Nu}_x = \frac{hx}{k} = 0.60\,(\mathrm{Gr}_x^*\mathrm{Pr})^{\frac{1}{5}}$$

This is valid for $10^5 < \mathrm{Gr}_x^* < 10^{11}$. This correlation can be used for laminar region, by replacing $\mathrm{Gr}_x^*$ with $(\mathrm{Gr}_x^* \cos\theta)$, for both upward- and downward-facing heated surfaces.

For turbulent regions, with $(\mathrm{Gr}_x^*\mathrm{Pr})$ in the range from $10^{10}$ to $10^{15}$, they developed the empirical relation as

$$\mathrm{Nu}_x = 0.17\,(\mathrm{Gr}_x^*\mathrm{Pr})^{\frac{1}{4}} \qquad (5.153)$$

where the modified Grashof number $\mathrm{Gr}_x^* = (\mathrm{Gr}_x\mathrm{Nu}_x)$ in Equation (5.153) is the same as that for the vertical plate when the heated surface of the inclined plate faces upward. When the heated surface of the inclined plate faces downward, the $\mathrm{Gr}_x^*$ is replaced by $\mathrm{Gr}_x^* \cos^2\theta$.

### 5.13.1 Free Convection on Inclined Cylinders

For inclined cylinders, the laminar heat transfer under constant heat flux conditions may be calculated using the following correlation proposed by Al-Arabi and Salman [10].

$$\mathrm{Nu}_x = \left[0.6 - 0.488\,(\sin\theta)^{1.03}\right](\mathrm{Gr}_L\mathrm{Pr})^{\frac{1}{4}+\frac{1}{12}(\sin\theta)^{1.75}} \qquad (5.154)$$

where $\theta$ is the angle the cylinder makes with the vertical. All properties except $\beta$ are evaluated at the film temperature, and $\beta$ is calculated at ambient temperature. Equation (5.154) is valid for $(\mathrm{Gr}_L\mathrm{Pr}) < 2 \times 10^8$.

Note: It is essential to note that these correlations are only approximate and the uncertainty associated with the results obtained with these correlations can be as high as 20 percent.

### 5.13.2 Free Convection on a Sphere

For an isothermal sphere of diameter $D$ in a fluid, such as air, having Prandtl number close to unity, the average Nusselt number for free convection has been correlated by Yuge [11], as

$$\text{Nu} = \frac{hD}{k} = 2 + 0.43\,\text{Ra}_D^{\frac{1}{4}} \tag{5.155}$$

This is valid for $1 < \text{Ra}_D < 10^5$.

Based on their experimental data, for free convection on an isothermal sphere in water, Amato and Tien [12] proposed the correlation for average Nusselt number as

$$\text{Nu} = 2 + 0.50\,\text{Ra}_D^{\frac{1}{4}} \tag{5.156}$$

for $3 \times 10^5 < \text{Ra}_D < 8 \times 10^8$ and $10 \leq \text{Nu} \leq 90$. The Rayleigh number is based on the sphere diameter, and is defined as

$$\begin{aligned} \text{Ra}_D &= \text{Gr}_D\text{Pr} \\ &= \frac{g\beta D^3 \left(T_w - T_\infty\right)}{\nu^2}\,\text{Pr} \end{aligned}$$

All the properties are evaluated at the film temperature $T_f$.

Note that, as $\text{Ra}_D \to 0$, Equations (5.155) and (5.156) result in $\text{Nu} \to 2$, which is *the limiting value* for heat conduction from an isothermal sphere in an infinite medium.

### 5.13.3 Simplified Equations for Air

Simplified equations for the convective heat transfer coefficient for heat transfer from various surfaces to air, at atmospheric pressure and moderate temperatures, are given in Table 5.3.

Note: In Table 5.3, the subscript $f$ for Gr and Pr refers to properties calculated at $T_f$. These relations may be extended to higher or lower pressures by multiplying by $\left(\frac{p}{101.32}\right)^{\frac{1}{2}}$, for laminar flows, and by $\left(\frac{p}{101.32}\right)^{\frac{2}{3}}$, for turbulent flows, where $p$ is pressure in kPa. It is essential to note that these simplified relations are only approximate.

## 5.14 Free Convection in Enclosed Spaces

Free convection in enclosed spaces finds application is many problems of practical importance, such as fluid in wall cavities, between window glasses, in the

**Table 5.3** Simplified relations for free convection from various surfaces to air at atmospheric pressure.

| Surface | Laminar $10^5 < (\mathrm{Gr}_f \mathrm{Pr}_f) < 10^9$ | Turbulent $(\mathrm{Gr}_f \mathrm{Pr}_f) > 10^9$ |
|---|---|---|
| Vertical plate or cylinder of length $L$ | $h = 1.42 \left(\frac{\Delta T}{L}\right)^{\frac{1}{4}}$ | $h = 0.95\,(\Delta T)^{\frac{1}{3}}$ |
| Horizontal cylinder of diameter $D$ | $h = 1.32 \left(\frac{\Delta T}{D}\right)^{\frac{1}{4}}$ | $h = 1.24\,(\Delta T)^{\frac{1}{3}}$ |
| Horizontal plate of length $L$ with heated face upward | $h = 1.32 \left(\frac{\Delta T}{L}\right)^{\frac{1}{4}}$ | $h = 1.43\,(\Delta T)^{\frac{1}{3}}$ |
| Horizontal plate of length $L$ with heated face downward | $h = 0.61 \left(\frac{\Delta T}{L^2}\right)^{\frac{1}{5}}$ | |

annulus between concentric cylinders and spheres. To gain an insight into the physics of the nature of the free convection in such enclosed spaces, let us examine the motion of a fluid contained in the space enclosed between two large horizontal plates, shown in Figure 5.14(a). Let the lower and upper plates be at uniform temperatures $T_h$ and $T_c$, respectively, and $T_h > T_c$. The temperature difference $\Delta T = (T_h - T_c)$ imposed on the fluid would cause heat transfer in the upward direction, establishing a temperature profile.

The fluid layers of higher temperature will lie below the lower temperature fluid layers. When the temperature difference $(T_h - T_c)$ is sufficiently small, the viscous forces will overcome the buoyancy forces and the fluid would remain motionless. At this state the heat transfer is solely by conduction. Thus, we have

$$k\frac{(T_h - T_c)}{\delta} \equiv h\,(T_h - T_c)$$

or

$$\frac{h\delta}{k} = \mathrm{Nu}_\delta = 1$$

where $\delta$ is the thickness of the fluid layer.

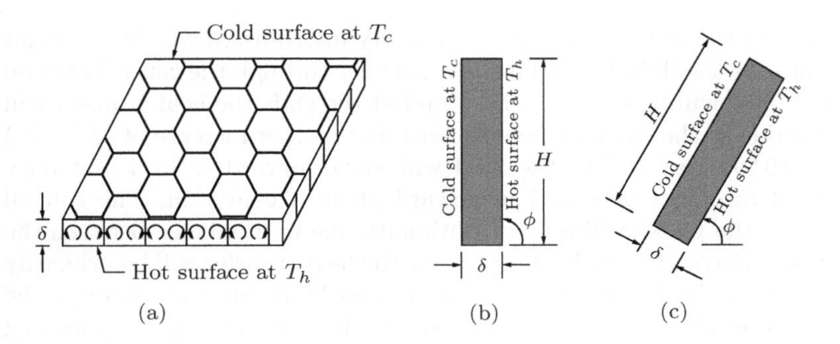


**Figure 5.14**
Free convection in enclosed spaces: (a) horizontal layer, (b) vertical layer, (c) inclined layer.

When $(T_h - T_c)$ is increased beyond a value such that the buoyancy overcomes the viscous forces, convection motion would be initiated in the fluid. It has been verified by both theory and experiments that for a horizontal enclosure, as in Figure 5.14(a), when $(T_h - T_c)$ is increased beyond a value corresponding to the critical Rayleigh number of 1708, the horizontal fluid layer will become unstable. That is, for

$$\text{Ra}_\delta = \text{Gr}_\delta \text{Pr} = \frac{g\beta\,(T_h - T_c)\,\delta^3}{\nu^2}\,\text{Pr} > 1708 \tag{5.157}$$

the fluid layer will become unstable, giving rise to the formation of flow pattern in the form of hexagonal cells, as shown in Figure 5.14(a). These cells are called *Benard cells*, in honor of Benard, who was the first to observe this in the year 1900.

For further increase of $(T_h - T_c)$, such that $\text{Ra}_\delta > 50{,}000$, turbulent free convection will set in, destroying the regular cellular flow pattern. Thus, heating of a horizontal layer of fluid from below can result in three regimes namely, *conduction*, the *cellular convection*, and the *turbulent free convection* regimes, depending on the magnitude of $(T_h - T_c)$. The Nusselt number associated with these regimes are of interest in engineering applications requiring prediction of heat transfer across the fluid layer.

For a fluid layer contained between two vertical plates maintained at constant temperatures of $T_h$ and $T_c$, as illustrated in Figure 5.14(b), for small values of $(T_h - T_c)$ the heat transferred across the fluid layer will be by pure conduction. For such a conduction process,

$$\text{Nu}_\delta = \frac{h\delta}{k} = 1$$

With increase of $(T_h - T_c)$, a circulatory motion of the fluid, leading to Benard cells will result. In these cells, the circulatory flow will be symmetric with respect to the cell center, and the velocity distribution will be invariant with height over the fluid layer. The heat transfer through the central portion of the fluid layer will be by conduction, and at the ends the heat transfer will be complex due to the complex flow pattern. For further increase of $(T_h - T_c)$ leading to $10^4 < \text{Ra}_\delta < 10^5$, the fluid will move upward as in a boundary layer flow along a hot wall and downward along a cold wall. The central layer between the plates will remain stationary due to a balance between the buoyancy and viscous forces. In this process, the heat transfer will be primarily by convection in the boundary layer region, and by conduction through the central region of the layer which is stationary. Increase of $(T_h - T_c)$ leading to $\text{Ra}_\delta > 10^5$ will result in the formation of vertical row of vortices with their axis of rotation in the horizontal direction. The number of these vortices will increase with increase of Grashof number. When the Grashof number increases beyond 50,000, the fluid motion will become turbulent. For vertical layers, the aspect ratio $H/\delta$ is an important parameter governing the end effects and the transition from one regime to another.

When the fluid layer is inclined, as shown in Figure 5.14(c), with the cold plate above the hot plate, for Grashof number above the critical value, free convection will take place in the fluid layer. But this free convection process is more complicated than that in the horizontal and vertical layers.

For fluid contained in the annulus between two horizontal cylinders or two concentric spheres, as shown in Figure 5.15, when the temperature difference $(T_i - T_o)$ is increased to a level leading to critical Grashof number, free convection will begin.

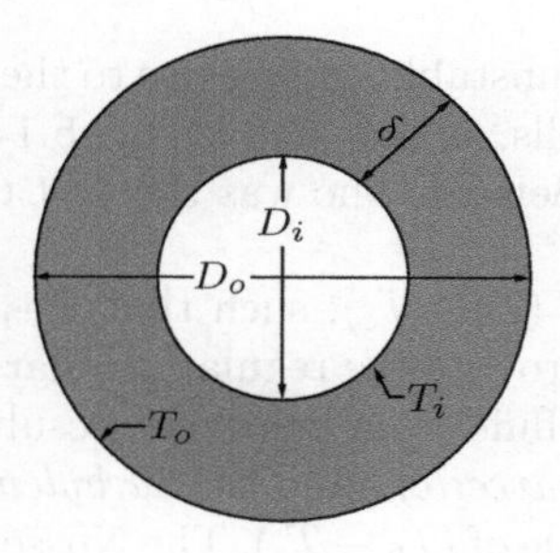


**Figure 5.15**
Free convection in horizontal cylindrical or spherical annulus.

### 5.14.1 Correlation for Free Convection in Enclosed Spaces

Even though numerous studies aiming at the determination of the onset of free convection and the heat transfer coefficient associated with that are reported in open literature [11], heat transfer correlation covering all ranges of parameters are not yet available.

It was seen that the mechanism of heat transfer in free convection in an enclosed space is complicated and hence, prediction of heat transfer associated with such process is not easy. However, simple correlations may prove to be useful for approximate estimation of the free convection coefficient. One such simple correlation for free convection heat transfer coefficient for enclosures is

$$\mathrm{Nu}_\delta = c\,(\mathrm{Ra}_\delta)^n \left(\frac{H}{\delta}\right)^m \tag{5.158}$$

where $\delta$ is the thickness of the fluid layer enclosed by flat surfaces, $\delta = (D_o - D_i)/2$ for fluid layer enclosed in an annulus, and $H$ is the height of the fluid layer. The constants $c$, $n$ and $m$ in Equation (5.158) are listed for some specific configurations in Table 5.4.

Once the Nusselt number is computed, the average heat transfer coefficient $h_m$ can be determined from

$$h_m = \mathrm{Nu}_\delta \frac{k}{\delta}$$

Using this $h_m$, the total heat transfer rate $\dot{Q}$ across the fluid layer can be calculated using

$$\dot{Q} = A_m h_m \left(T_h - T_c\right)$$

where $A_m$ is equal to the wall area $A_w$ for the plane layer, for cylindrical annulus $A_m$ is defined as $A_m = \left(A_o - A_i\right) / \ln\left(A_o/A_i\right)$, and for spherical annulus $A_m = \sqrt{A_o A_i}$.

## 5.14.2 Vertical Layer

A more accurate correlation for the vertical layer of fluid shown in Figure 5.14(b) was proposed by El Sherbiny et al. [14], for air in the range of $H/\delta$ from 5 to 110, and $\mathrm{Ra}_\delta$ from $10^2$ to $2 \times 10^7$, as

$$\mathrm{Nu}_{90^\circ} = \left[\mathrm{Nu}_1, \mathrm{Nu}_2, \mathrm{Nu}_3\right]_{\mathrm{max}} \tag{5.159}$$

where

$$\begin{aligned}
\mathrm{Nu}_1 &= 0.0605\,\mathrm{Ra}^{\frac{1}{3}} \\
\mathrm{Nu}_2 &= \left(1 + \left[\frac{0.104\,\mathrm{Ra}^{0.293}}{1 + (6310/\mathrm{Ra})^{1.36}}\right]^3\right)^{\frac{1}{3}} \\
\mathrm{Nu}_3 &= 0.243\left(\frac{\mathrm{Ra}}{H/\delta}\right)^{0.272}
\end{aligned}$$

The subscript "max" in Equation (5.159) implies that the maximum among the above three Nusselt number should be selected as the appropriate value.

**Table 5.4** Coefficient $c$ and indices $m$ and $n$ in Equation (5.158) [11]

| Geometry | mode | fluid | Pr | $Ra_\delta$ | $\frac{H}{\delta}$ | $c$ | $n$ | $m$ |
|---|---|---|---|---|---|---|---|---|
| Vertical layer | isothermal | gas | - | $< 2000$ | - | 1 | 0 | 0 |
| Vertical cylindrical | isothermal | gas | 0.5-2 | $2 \times 10^3$-$2 \times 10^5$ | 11-42 | 0.197 | 1/4 | $-1/9$ |
| annulus | isothermal | gas | 0.5-2 | $2 \times 10^5$-$10^7$ | 11-42 | 0.073 | 1/3 | $-1/9$ |
| | isothermal or | liquid | $1$-$2 \times 10^4$ | $10^4$-$10^7$ | 10-40 | $0.42\, Pr^{0.012}$ | 1/4 | $-0.3$ |
| | uniform heat flux | liquid | 1-20 | $10^6$-$10^9$ | 1-40 | 0.046 | 1/3 | 0 |
| Horizontal layer | isothermal | gas | - | $< 1700$ | - | 1 | 0 | 0 |
| | bottom hot | gas | 0.5-2 | $1.7 \times 10^3$-$6 \times 10^3$ | - | 0.059 | 0.6 | 0 |
| | | gas | 0.5-2 | $7 \times 10^3$-$3.2 \times 10^5$ | - | 0.212 | 1/4 | 0 |
| | | gas | 0.5-2 | $> 3.2 \times 10^5$ | - | 0.061 | 1/3 | 0 |
| | | liquid | - | $< 1700$ | - | 1 | 0 | 0 |
| | | liquid | 1-5000 | $1.7 \times 10^3$-$6 \times 10^3$ | - | 0.012 | 0.6 | 0 |
| | | liquid | 1-5000 | $6 \times 10^3$-$3.7 \times 10^4$ | - | 0.375 | 0.2 | 0 |
| | | liquid | 1-20 | $3.7 \times 10^4$-$10^8$ | - | 0.13 | 0.3 | 0 |
| | | liquid | 1-20 | $> 10^8$ | - | 0.057 | 1/3 | 0 |
| Horizontal | isothermal | gas | 1-5000 | $6 \times 10^3$-$10^6$ | - | 0.11 | 0.29 | 0 |
| Cylindrical annulus | isothermal | or liquid | 1-5000 | $10^6$-$10^8$ | - | 0.40 | 0.20 | 0 |
| Spherical annulus | isothermal | gas or liquid | 0.7-4000 | $10^2$-$10^9$ | - | 0.228 | 0.226 | 0 |

The Nusselt number can also be defined as

$$\begin{aligned} \mathrm{Nu}_{90^\circ} &= \frac{h\delta}{k} \\ &= \frac{\dot{q}\delta}{k\,(T_h - T_c)} \end{aligned}$$

because $\dot{q} = h\,(T_h - T_c)$. All physical properties in the above relations have to be evaluated at the film temperature $T_f = (T_h + T_c)\,/2$.

## Example 5.10

The vertical space between a double-pane window of height 0.8 m and width 1.6 m is filled with air at a pressure of 1 atm. If the space is 20 mm thick, determine the rate of heat loss through the window when the surface temperatures of the window panes are 15°C and 3°C.

## Solution

Given, $\delta = 0.02$ m, $H = 0.8$ m, $w = 1.6$ m, $T_h = 15$°C, $T_c = 3$°C, $p = 101325$ Pa.

The film temperature is

$$\begin{aligned} T_f &= \frac{T_h + T_c}{2} \\ &= \frac{15+3}{2} = 9^\circ\mathrm{C} \\ &= 9 + 273 \\ &= 282\,\mathrm{K} \end{aligned}$$

At 282 K, from Table A-4, $k = 0.0248$ W/(m °C), Pr = 0.712.

The density, viscosity and $\beta$ are

$$\begin{aligned} \rho &= \frac{p}{RT} = \frac{101325}{287 \times 282} \\ &= 1.252\,\mathrm{kg/m^3} \end{aligned}$$

$$\begin{aligned} \mu &= (1.46 \times 10^{-6}) \times \left(\frac{282^{3/2}}{282 + 111}\right) \\ &= 1.76 \times 10^{-5}\,\mathrm{kg/(m\ s)} \end{aligned}$$

$$\beta = \frac{1}{T_f} = \frac{1}{282}$$

$$= 0.00355 \, 1/\mathrm{K}$$

The Grashof number is

$$\mathrm{Gr}_\delta = \frac{g\beta\,(T_h - T_c)\,\delta^3}{\nu^2}$$

$$= \frac{9.81 \times 0.00355 \times (15 - 3) \times 0.02^3}{[(1.76 \times 10^{-5})/1.252]^2}$$

$$= 1.692 \times 10^4$$

Thus, the Rayleigh number is

$$\mathrm{Ra}_\delta = (\mathrm{Gr}_\delta \mathrm{Pr}) = 1.692 \times 10^4 \times 0.712$$

$$= 1.2 \times 10^4$$

By Equation (5.159), we have

$$\mathrm{Nu}_1 = 0.0605\,\mathrm{Ra}_\delta^{\frac{1}{3}}$$

$$= 0.0605 \times 12000^{\frac{1}{3}}$$

$$= 1.385$$

$$\mathrm{Nu}_2 = \left(1 + \left[\frac{0.104\mathrm{Ra}_\delta^{0.293}}{1 + (6310/\mathrm{Ra}_\delta)^{1.36}}\right]^3\right)^{\frac{1}{3}}$$

$$= \left(1 + \left[\frac{0.104 \times 12000^{0.293}}{1 + (6310/12000)^{1.36}}\right]^3\right)^{\frac{1}{3}}$$

$$= 1.361$$

$$\mathrm{Nu}_3 = 0.242\left(\frac{\mathrm{Ra}_\delta}{(H/\delta)}\right)^{0.272}$$

$$= 0.242\left(\frac{12000}{(0.8/0.02)}\right)^{0.272}$$

$$= 1.142$$

It is found that, $\mathrm{Nu}_1 = 1.385$ is the largest. Thus,

$$\mathrm{Nu}_{90^\circ} = 1.385 = \frac{h\delta}{k}$$

$$h = \frac{1.385 \times 0.0248}{0.02}$$

$$= 1.717\,\mathrm{W/(m^2\ ^\circ C)}$$

The rate of heat loss from the window is

$$\dot{Q} = hA(T_h - T_c)$$

$$= 1.717 \times (0.8 \times 1.6) \times (15 - 3)$$

$$= \boxed{26.37\,\mathrm{W}}$$

### 5.14.3 Inclined Layer

For inclined layer of air, as shown in Figure 5.15(c), with $\phi = 60^\circ$, the correlation for Nusselt number proposed by El Sherbiny et al. [14] is

$$\mathrm{Nu}_{60^\circ} = [\mathrm{Nu}_1, \mathrm{Nu}_2]_{\max} \tag{5.160}$$

where

$$\mathrm{Nu}_1 = \left(1 + \left[\frac{0.093\,\mathrm{Ra}_\delta^{0.314}}{1 + \left[0.5/\left(1 + \mathrm{Ra}_\delta/3160\right)^{20.6}\right]^{0.1}}\right]^7\right)^{\frac{1}{7}}$$

$$\mathrm{Nu}_2 = \left(0.104 + \frac{0.175}{H/\delta}\right)\mathrm{Ra}_\delta^{0.283}$$

The greater of $\mathrm{Nu}_1$ and $\mathrm{Nu}_2$ should be taken as the Nusselt number $\mathrm{Nu}_{60^\circ}$.

By definition, the Nusselt number becomes

$$\mathrm{Nu}_{60^\circ} = \frac{\dot{q}\delta}{k\left(T_h - T_c\right)}$$

Using a straight-line interpolation between Equations (5.159) and (5.160), the Nusselt number for $\phi$ in the range from 60° to 90° can be expressed as

$$\mathrm{Nu}_\phi = \frac{(90^\circ - \phi^\circ)\mathrm{Nu}_{60^\circ} + (\phi^\circ - 60^\circ)\mathrm{Nu}_{90^\circ}}{30^\circ} \tag{5.161}$$

**Inclined Layer with $0 \leq \phi \leq 60°$**

For inclination angle in the range $0 \leq \phi \leq 60°$, Hollands et al. [15] based on their experimental results, correlated the Nusselt number as

$$\mathrm{Nu}_\phi = 1+1.44\left[1-\frac{1708}{\mathrm{Ra}_\delta \cos\phi}\right]^*\left[1-\frac{1708\,(\sin 1.8\phi)^{1.6}}{\mathrm{Ra}_\delta \cos\phi}\right]+\left[\left(\frac{\mathrm{Ra}_\delta \cos\phi}{5830}\right)^{\frac{1}{3}}-1\right]^* \tag{5.162}$$

In this relation, if the quantity in the square bracket with superscript * is negative, it should be set equal to zero. This relation is valid for $\mathrm{Ra}_\delta$ in the range from 0 to $10^5$.

**Horizontal Layer**

For horizontal layer, $\phi = 0°$, therefore, Equation (5.162), simplifies to

$$\mathrm{Nu}_{\phi=0°} = 1+1.44\left[1-\frac{1708}{\mathrm{Ra}_\delta}\right]^*+\left[\left(\frac{\mathrm{Ra}_\delta}{5803}\right)^{\frac{1}{3}}-1\right]^* \tag{5.163}$$

## 5.14.4 Horizontal Cylindrical Annulus

For a cylindrical annulus of length $H$, shown in Figure 5.15, the heat transfer rate was proposed by Raithby and Hollands [16] as,

$$\dot{Q} = \frac{2\pi k_{\text{eff}} H}{\ln(D_o/D_i)}\,(T_i - T_o) \tag{5.164}$$

where

$$\begin{aligned}
\frac{k_{\text{eff}}}{k} &= 0.386\left(\frac{\mathrm{Pr}}{0.861+\mathrm{Pr}}\right)^{\frac{1}{4}}\left(\mathrm{Ra}_{cy}^*\right)^{\frac{1}{4}} \\
\left(\mathrm{Ra}_{cy}^*\right)^{\frac{1}{4}} &= \frac{\ln(D_o/D_i)}{\delta^{\frac{3}{4}}\left(D_i^{-\frac{3}{5}}+D_o^{-\frac{3}{5}}\right)^{\frac{5}{4}}}\mathrm{Ra}_\delta^{\frac{1}{4}} \\
\mathrm{Ra}_\delta &= \frac{g\beta\,(T_i - T_o)\,\delta^3}{\nu^2}\,\mathrm{Pr} \\
\delta &= \frac{1}{2}\,(D_i - D_o)
\end{aligned}$$

This correlation is valid for $10^2 < \mathrm{Ra}_{cy}^* < 10^7$.

## Example 5.11

A horizontal pipe of inside diameter 0.1 m and outside diameter 0.6 m carries steam at 165°C. If the surrounding atmosphere is at 23°C, estimate the heat loss per unit length of the pipe, treating the steam as air.

## Solution

Given, $T_i = 165°\text{C}$, $T_o = 23°\text{C}$, $D_i = 0.1$ m, $D_o = 0.6$ m.

The film temperature is

$$T_f = \frac{(T_s + T_\infty)}{2} = \frac{165 + 23}{2}$$
$$= 94°\text{C}$$

For air at 94°C, from properties table, $k = 0.0313$ W/(m °C), $\nu = 22.8 \times 10^{-6}$ m$^2$/s, $\alpha = 32.8 \times 10^{-6}$ m$^2$/s, Pr $= 0.7$.

$$\beta = \frac{1}{T_f} = \frac{1}{94 + 273}$$
$$= 0.002725 \text{ 1/K}$$

$$\delta = \frac{D_o - D_i}{2} = \frac{0.6 - 0.1}{2}$$
$$= 0.25\,\text{m}$$

The Grashof and Rayleigh numbers are

$$\text{Gr}_\delta = \frac{g\beta\,(T_i - T_o)\,\delta^3}{\nu^2}$$
$$= \frac{9.81 \times 0.002725 \times (165 - 23) \times 0.25^3}{(22.8 \times 10^{-6})^2}$$
$$= 1.14 \times 10^8$$
$$\text{Ra}_\delta = \text{Gr}_\delta\text{Pr} = (1.14 \times 10^8) \times 0.7$$
$$= 7.98 \times 10^7$$

Therefore

$$\left(\text{Ra}^*_{cy}\right)^{\frac{1}{4}} = \frac{\ln\,(D_o/D_i)}{\delta^{\frac{3}{4}}\left(D_i^{-\frac{3}{5}} + D_o^{-\frac{3}{5}}\right)^{\frac{5}{4}}}\text{Ra}_\delta^{\frac{1}{4}}$$
$$= \frac{\ln\,(0.6/0.1)}{0.25^{3/4}\left(0.1^{-\frac{3}{5}} + 0.6^{-\frac{3}{5}}\right)^{\frac{5}{4}}} \times (7.98 \times 10^7)^{\frac{1}{4}}$$

$$= 59$$

$$\frac{k_{\text{eff}}}{k} = 0.386 \left( \frac{\text{Pr}}{0.861 + \text{Pr}} \right)^{\frac{1}{4}} \left(\text{Ra}_{cy}^{*}\right)^{\frac{1}{4}}$$

$$= 0.386 \left( \frac{0.7}{0.861 + 0.7} \right)^{\frac{1}{4}} \times 59^{\frac{1}{4}}$$

$$= 0.875$$

$$k_{\text{eff}} = 0.875 \times 0.0313$$

$$= 0.0274\,\text{W/(m °C)}$$

By Equation (5.164)

$$\dot{Q} = \frac{2\pi k_{\text{eff}} H}{\ln\left(D_o/D_i\right)} \left(T_i - T_o\right)$$

$$= \frac{2\pi \times 0.0274 \times 1}{\ln\left(0.6/0.1\right)} \times (165 - 23)$$

$$= \boxed{13.64\,\text{W/m}}$$

### 5.14.5 Spherical Annulus

For a fluid contained between two concentric spheres of inner and outer diameters $D_i$ and $D_o$, shown in Figure 5.15, Raithby and Hollands [16] correlated the heat transfer rate $\dot{Q}$ as

$$\dot{Q} = k_{\text{eff}} \frac{\pi D_i D_o}{\delta} \left(T_i - T_o\right) \tag{5.165}$$

where

$$\frac{k_{\text{eff}}}{k} = 0.74 \left( \frac{\text{Pr}}{0.861 + \text{Pr}} \right)^{\frac{1}{4}} \left(\text{Ra}_{sph}^{*}\right)^{\frac{1}{4}}$$

$$\text{Ra}_{sph}^{*} = \frac{\delta^{\frac{1}{4}}}{D_i D_o} \frac{\text{Ra}_{\delta}^{\frac{1}{4}}}{\left(D_i^{-\frac{7}{5}} + D_o^{-\frac{7}{5}}\right)^{\frac{5}{4}}}$$

Equation (5.165) is valid for $\text{Ra}_{sph}^{*}$ in the range from $10^2$ to $10^4$.

## Example 5.12

Air at a pressure of 1 atm is contained in the vertical space between two square plates of side 1 m, separated by a distance of 25 mm. When the plates are maintained at constant temperatures of 120°C and 80°C, determine the free convection heat transfer across the air layer.

## Solution

Given, $H = 1$ m, $\delta = 0.025$ m, $T_h = 120$°C, $T_c = 80$°C.

The film temperature is

$$T_f = \frac{T_h + T_c}{2} = \frac{120 + 80}{2}$$

$$= 100^\circ\text{C} = 373\,\text{K}$$

Therefore,

$$\beta = \frac{1}{T_f} = \frac{1}{373}$$

$$= 0.00268\,1/\text{K}$$

$$\rho = \frac{p}{RT} = \frac{101325}{287 \times 373}$$

$$= 0.946\,\text{kg/m}^3$$

$$\mu = (1.46 \times 10^{-6}) \times \left(\frac{373^{3/2}}{373 + 111}\right)$$

$$= 2.17 \times 10^{-5}\,\text{kg/(m s)}$$

At $T_f = 373$ K, from Table A-4, $k = 0.03$ W/(m °C), Pr $= 0.685$.

The Grashof number is

$$\text{Gr}_\delta = \frac{g\beta\,(T_h - T_c)\,\delta^3}{\nu^2}$$

$$= \frac{9.81 \times 0.00268 \times (120 - 80) \times 0.025^3}{\left[(2.17 \times 10^{-5})/0.946\right]^2}$$

$$= 31228.2$$

The Rayleigh number is

$$\begin{aligned} \mathrm{Ra}_\delta &= (\mathrm{Gr}_\delta \mathrm{Pr}) = 31228.2 \times 0.685 \\ &= 2.14 \times 10^4 \end{aligned}$$

For $\mathrm{Ra}_\delta = 2.14 \times 10^4$, from Table 5.4, we have $c = 0.197$, $n = \frac{1}{4}$, $m = -\frac{1}{9}$.

By Equation (5.158),

$$\begin{aligned} \mathrm{Nu}_\delta &= c\,(\mathrm{Ra}_\delta)^n \left(\frac{H}{\delta}\right)^m \\ &= 0.197 \times (2.14 \times 10^4)^{\frac{1}{4}} \times \left(\frac{1}{0.025}\right)^{-\frac{1}{9}} \\ &= 1.58 \end{aligned}$$

The average convective heat transfer coefficient is

$$\begin{aligned} h_m &= \mathrm{Nu}_\delta \frac{k}{\delta} = 1.58 \times \frac{0.03}{0.025} \\ &= 1.896\,\mathrm{W/(m^2\ ^\circ C)} \end{aligned}$$

The free convection heat transfer across the air layer becomes

$$\begin{aligned} \dot{Q} &= h_m A_m (T_h - T_c) \\ &= 1.896 \times (1 \times 1) \times (120 - 80) \\ &= \boxed{75.84\,\mathrm{W}} \end{aligned}$$

## 5.14.6 Non-Newtonian Fluids

When the shear stress-viscosity relation of the fluid does not obey the simple Newtonian expression for shear stress,

$$\tau_w = \mu \frac{\partial u}{\partial y}$$

the fluid is termed *non-Newtonian* and the free convection correlations presented above do not apply. That is, for non-Newtonian fluids such as high viscous polymers and lubricants the proposed correlations are not valid.

### 5.14.7 Combined Natural and Forced Convection

The presence of a temperature gradient in a fluid in a gravity field always gives rise to natural convection currents, and thus leading to heat transfer by natural convection. Therefore, forced convection is always accompanied by natural convection. Also, we know that the convection heat transfer coefficient, free (natural) or forced, is a strong function of the flow velocity. Heat transfer coefficients encountered in forced convection are much higher than those associated with natural convection because of the high flow velocities associated with forced convection. Therefore, we usually ignore natural convection in a heat transfer analysis that involves forced convection, although we pretty well know that forced convection is always accompanied by natural convection. The error involved in ignoring natural convection is negligible at high velocities, but can become considerable at low velocities associated with forced convection. Therefore, it is desirable to have a criterion to assess the relative magnitude of natural convection in the presence of forced convection.

For a fluid, the parameter $Gr/Re^2$ represents the importance of natural convection relative to forced convection. This can easily be inferred since the convection heat transfer coefficient is a strong function of the Reynolds number Re in forced convection and of the Grashof number Gr in natural convection. It is well established that the natural convection is negligible when $Gr/Re^2 < 0.1$, forced convection is negligible when $Gr/Re^2 > 10$, and both are significant when $0.1 < Gr/Re^2 < 10$.

Natural convection may *augment* or *abate* the forced convection heat transfer, depending on the relative direction of the *buoyancy-induced* and *forced convection motions.* Based on the relative direction, as illustrated in Figure 5.16, the flows are broadly classified as *assisting flow*, *opposing flow*, and *transverse flow.*

• In an assisting flow, the buoyant motion is in the *same* direction of the forced motion. Therefore, the natural convection enhances the heat transfer by assisting forced convection. Upward forced flow over a hot surface shown in Figure 5.16(a) is a typical example for assisting flow.

• In an opposing flow, the buoyant motion is in the *opposite* direction to the fluid motion. Therefore, the natural convection resists the forced convection and *decreases* the heat transfer. Upward forced flow over a cold surface shown in Figure 5.16(b) is an opposing flow.

• In transverse flow, the buoyant motion is *perpendicular* to the forced motion, as shown in Figure 5.16(c). This flow enhances fluid mixing and thus promotes heat transfer.

To determine the heat transfer due to combined natural and forced convection conditions, it would appear that we should add the contributions of natural and forced convection in assisting flow and subtract them in opposing flow. But, practical experience indicates differently. Experimental studies

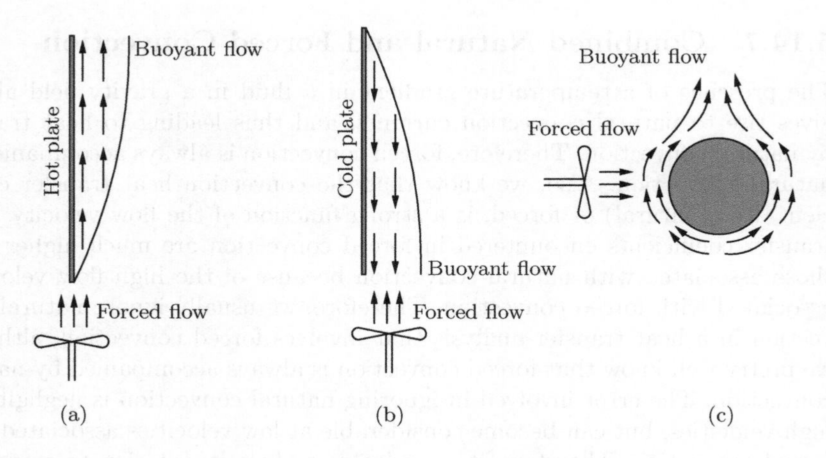


**Figure 5.16**
Illustration of (a) assisting, (b) opposing, and (c) transverse flow.

suggest a correlation of the form

$$\boxed{Nu_{combined} = (Nu^n_{forced} \pm Nu^n_{natural})^{\frac{1}{n}}} \tag{5.166}$$

where $Nu_{forced}$ and $Nu_{natural}$ are determined from the correlations for *pure forced* and *pure natural* convection, respectively. The + sign is for *assisting* and *transverse* flows and the − sign is for *opposing flow*. The exponent $n$ varies between 3 and 4, depending on the geometry involved. It is proved that, $n = 3$ correlates experimental data for vertical surface well. Larger values of $n$ are better suited for horizontal surfaces.

## Example 5.13

The glass cover of a solar collector receives solar radiation at the rate of 700 W/m$^2$. The glass transmits 88 percent of the incident radiation and has an emissivity of 0.9. A collector of size 1.2 m high and 2 m wide is used to heat water which enters the tube attached to the collector plate at a rate of 1 kg/min. The temperature of the glass cover is measured to be 35°C, on a day when the surrounding air temperature is 23°C and wind blows at 30 kmph. The effective temperature for radiation exchange between the glass cover and the open sky is − 40°C. Assuming the black surface collector plate to be perfectly insulated and the only heat loss is through the glass cover, determine (a) the total rate of heat loss from the collector, (b) the collector efficiency, which is the ratio of the amount of heat transferred to the water to the solar energy incident on the collector, and (c) the temperature rise of water as it flows through the collector. Assume Prandtl number to be 0.707.

## Solution

Total radiation energy incident on the glass is

$$\begin{aligned} \dot{Q}_{\text{rad}} &= \dot{q}_{\text{rad}} \times A \\ &= 700 \times (2 \times 1.2) \\ &= 1680\,\text{W} \end{aligned}$$

The glass wall is at 35°C and the sky is at – 40°C. Also, the emissivity of the glass is 0.90. Thus, the heat radiated by the glass to the sky is

$$\begin{aligned} \dot{q}_{\text{rad}} &= \sigma \times 0.1 \times \left(T_s^4 - T_{\text{sky}}^4\right) \\ &= 5.67 \times 10^{-8} \times 0.1 \times (308^4 - 233^4) \\ &= 34.3\,\text{W/m}^2 \end{aligned}$$

Therefore, the total radiation loss from the glass surface area of 2.4 $\text{m}^2$ is

$$\begin{aligned} \dot{Q}_{\text{rad,loss}} &= \dot{q} \times A \\ &= 34.3 \times 2.4 \\ &= 82.3\,\text{W} \end{aligned}$$

Thus, the net radiation received by the glass plate at steady state is

$$\begin{aligned} \dot{Q}_{\text{rad,net}} &= \dot{Q}_{\text{rad}} - \dot{Q}_{\text{rad,loss}} \\ &= 1680 - 82.3 \\ &= 1597.7\,\text{W} \end{aligned}$$

Out of this 1597.7 watts of radiation received, 88 percent is transmitted by the glass plate to the black plate below the glass. Thus, the heat energy received by the plate is

$$\begin{aligned} \dot{Q}_{\text{plate}} &= 1597.7 \times 0.88 \\ &= 1406\,\text{W} \end{aligned}$$

The film temperature is

$$T_f = \frac{T_s + T_\infty}{2} = \frac{35 + 23}{2}$$

$$= 29°\mathrm{C} + 273$$

$$= 302\,\mathrm{K}$$

The density and viscosity of air, at 1 atm and 302 K, are

$$\rho = \frac{101325}{287 \times 302}$$

$$= 1.165\,\mathrm{kg/m^3}$$

$$\mu = 1.46 \times 10^{-6} \frac{302^{3/2}}{302 + 111}$$

$$= 2.97 \times 10^{-5}\,\mathrm{kg/(m\ s)}$$

Air blows at 30 km/h over the glass plate. The corresponding Reynolds number is

$$\mathrm{Re}_L = \frac{\rho V L}{\mu}$$

$$= \frac{1.165 \times (30/3.6) \times 1.2}{2.97 \times 10^{-5}}$$

$$= 3.92 \times 10^5$$

The Reynolds number is less than the critical value of $5 \times 10^5$. Therefore, the flow is laminar. For laminar flow over a flat plate, by Equation (5.97), we have the average Nusselt number as

$$\mathrm{Nu} = \frac{hL}{k} = 0.332\,(\mathrm{Re}_L)^{1/2}\,\mathrm{Pr}^{1/3}$$

$$= 0.332\,(3.92 \times 10^5)^{1/2}\,(0.707)^{1/3}$$

$$= 185.18$$

$$h = \frac{185.18 \times 0.0422}{2}$$

$$= 3.9\,\mathrm{W/(m^2\ °C)}$$

Note that, the Prandtl number is

$$\text{Pr} = \frac{\mu c_p}{k} = \frac{2.97 \times 10^{-5} \times 1005}{k} = 0.707$$

This gives $k = 0.0422$.

(a) The heat loss from the glass plate, by forced convection is

$$\begin{aligned} \dot{Q}_{\text{conv}} &= h\,A\,(T_s - T_\infty) \\ &= 3.9 \times 2.4 \times (35 - 23) \\ &= \boxed{112.32\,\text{W}} \end{aligned}$$

(b) The collector efficiency is

$$\begin{aligned} \eta &= \frac{\text{Heat transferred to the water}}{\text{Solar energy incident on the glass plate}} \\ &= \frac{1406 - 112.32}{1597.7} \\ &= \boxed{0.81} \end{aligned}$$

(c) The heat transferred to the water is

$$\dot{Q}_{\text{water}} = \dot{m}\,c\,\Delta T = \frac{1}{60} \times 4178 \times \Delta T$$

Therefore, the temperature rise of water becomes

$$\begin{aligned} \Delta T &= \frac{(1406 - 112.32) \times 60}{4178} \\ &= \boxed{18.6^\circ\text{C}} \end{aligned}$$

since $c = 4178$ J/(kg °C) for water.

## Example 5.14

A train coach of length 8 m and width 2.8 m travels at 70 kmph. The top surface of the coach is absorbing solar radiation at the rate of 200 W/m$^2$, and the temperature of the ambient air is 30°C. Assuming the roof of the coach to be perfectly insulated and the radiation heat exchange with the surroundings to be small relative to the convective heat transfer, determine the equilibrium temperature of the top surface. Take $k$ for the top surface to be 0.026 W/(m °C).

## Solution

The top surface of the coach can be assumed to be a flat plate. Therefore, the problem is essentially a flat plate exposed to air flow and solar heating.

The density and viscosity of air at 30°C are

$$\rho = \frac{p}{RT}$$

$$= \frac{101325}{287 \times 303} = 1.165\,\text{kg/m}^3$$

$$\mu_{30^\circ\text{C}} = 1.46 \times 10^{-6} \frac{T^{3/2}}{T + 111}$$

$$= 1.46 \times 10^{-6} \times \frac{(303)^{3/2}}{303 + 111}$$

$$= 1.86 \times 10^{-5}\,\text{kg/(m s)}$$

The Reynolds and Prandtl numbers of the flow are

$$\text{Re}_L = \frac{\rho V L}{\mu} = \frac{1.165 \times (70/3.6) \times 8}{1.86 \times 10^{-5}}$$

$$= 97.4 \times 10^5$$

$$\text{Pr} = \frac{\mu c_p}{k} = \frac{1.86 \times 10^{-5} \times 1005}{0.026}$$

$$= 0.719$$

The Reynolds number is greater than the flat plate critical Reynolds number of $5 \times 10^5$ and thus the flow is turbulent. For turbulent flow over a flat plate, by Equation (5.107), we have

$$\text{Nu} = \frac{hL}{k} = 0.037\,(\text{Re}_L)^{4/5}\,\text{Pr}^{0.43}$$

$$= 0.037 \times (97.4 \times 10^5)^{4/5} \times (0.719)^{0.43}$$

$$= 12515$$

Therefore, the film transfer coefficient becomes

$$h = \frac{12515 \times 0.026}{8} = 40.67\,\text{W/(m}^2\,^\circ\text{C)}$$

The surface of the plate receives 200 W/m$^2$ by radiation. Therefore, at steady state, the surface temperature $T_s$ will assume a constant value and the heat received by the plate by radiation and the heat removed from the plate by the forced convection will balance each other. Thus,

$$\begin{aligned}
\dot{q}_{\text{rad}} &= \dot{q}_{\text{conv}} \\
200 &= h\,(T_s - T_\infty) \\
T_s &= \frac{200}{h} + T_\infty \\
&= \frac{200}{40.67} + 30 \\
&= \boxed{34.92^\circ\text{C}}
\end{aligned}$$

The top surface will be at an equilibrium temperature of 34.92°C.

## 5.15 Vortex behind a Circular Cylinder

One of the extensively studied vortex field is that behind a circular cylinder. The wake behind the cylinder is strongly influenced by the Reynolds number based on the cylinder diameter [2,17]. At very low Reynolds numbers (based on the velocity, density, viscosity of the freestream and cylinder diameter), of the order of unity, the flow behaves as if it were purely viscous (inertia effects negligible) and the boundary layers effectively extend to infinity, as illustrated in Figure 5.17(a).

At Reynolds numbers slightly higher than unity, true boundary layers form and remain laminar over the whole surface. Separation occurs on either side near the rear of the cylinder, and a narrow turbulent wake develops. With further increase of the Reynolds number, in the range from 10 to 60, the laminar separation points on either side of the rear of the streamwise diameter move rapidly outward and forward to points near the opposite ends of a transverse diameter, as shown in Figure 5.17(b). This results in a corresponding increase in wake width and consequent form drag.

At some stage, for a value of Reynolds number somewhere between 60 and 140, a pair of symmetrical vortices will begin to develop on either side of the center line behind the laminar separation points, as shown in Figure 5.17(b). These vortices will grow with time (at a given Reynolds number), continuously stretching downstream until a stage is reached when they become unsymmetrical and the system breaks down, one vortex becomes detached and moves away downstream. In practice, perfect symmetry and absolute smoothness is not possible and this causes the vortices to be of different sizes and shed differentially. This makes the vortex formation and shedding become

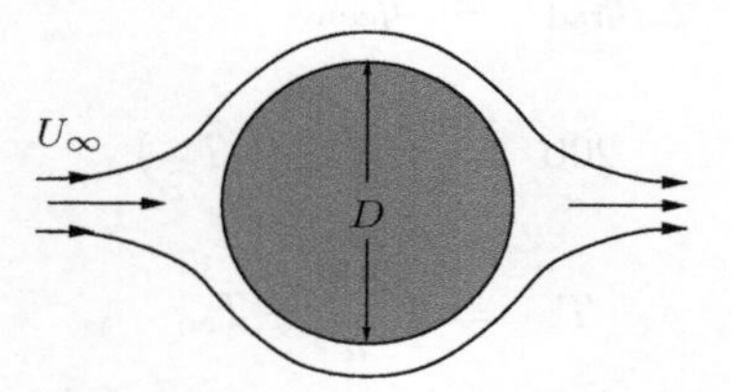


(a) Re < 4, Flow without separation

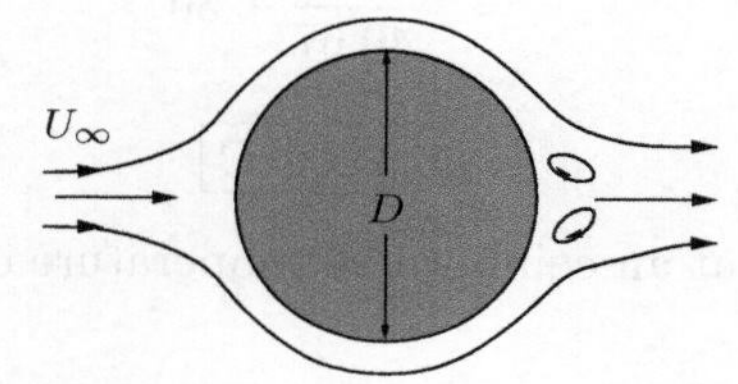


(b) 4 < Re < 60, Pair of vortices at the base

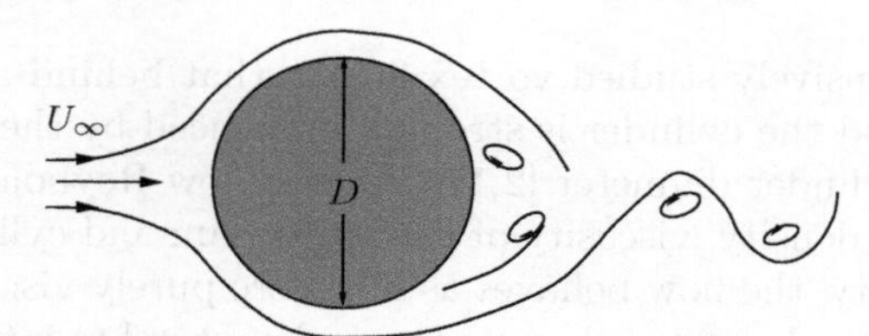


(c) 60 < Re < 5000, Periodic vortex shedding

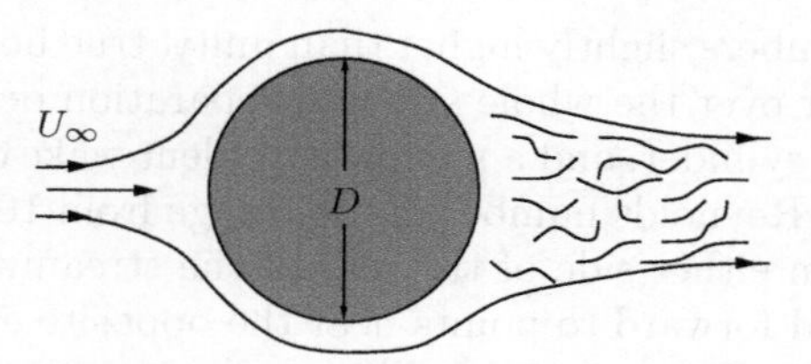


(d) Re > 5000, Wake formation

**Figure 5.17**
Flow around circular cylinder at different Reynolds number range.

alternative, with a frequency for a given shape and initial conditions. The alternative vortex motion behind the cylinder is the popular Karman vortex street, as shown in Figure 5.17(c). The subsequent wake motion, which is typically seen for Reynolds number above 5000, is oscillatory in character. This motion was investigated by Theodor von Karman in the first decade of the $20^{th}$ century. He showed that a stable system of vortices will be shed alternately from the laminar separation point on either side of the cylinder. Thus, a standing vortex will be generated in the region behind the separation point on one side, while a corresponding vortex on the other side will break away from the cylinder and move downstream in the wake. When the attached vortex reaches a particular strength, it will in turn break away and a new vortex will begin to develop again.

### 5.15.1 Drag Coefficient

For a cylinder of diameter $d$ and length $L$ in a uniform flow of velocity $U_\infty$, the drag force $F_D$ acting over the length $L$ of the cylinder is given by

$$F_D = \frac{1}{2}\rho U_\infty^2 S\, C_D$$

where $C_D$ is the drag coefficient and $S$ is the projected area of the cylinder, normal to the flow direction, which is $S = L \times d$.

The drag coefficient $C_D$ of a sphere is greatly influenced by the changes in flow velocity. The $C_D$ for a sphere decreases with increasing air speed, since the earlier transition to turbulent flow results in reduced wake behind the sphere. This action decreases the form or pressure drag, resulting in lower total drag coefficient. The decrease of drag coefficient is rapid in a range of speed in which both the drag coefficient and drag go down. The Reynolds number at which the transition occurs at a given point on the sphere is a function of the turbulence already present in the freestream flow, and hence the drag coefficient of the sphere can be used to measure turbulence of the freestream flow.

The drag coefficient may be calculated from the relation,

$$C_D = \frac{D}{\frac{1}{2}\rho U_\infty^2 \left(\dfrac{\pi d^2}{4}\right)}$$

where $d$ is the sphere diameter. Note that, here $d$ is used as the diameter, instead of $D$, since it is conventional to use uppercase $D$ to represent the drag force.

The variation of cylinder drag coefficient with Reynolds number Re is shown in Figure 5.18. At Reynolds number around $3.5 \times 10^5$, due to boundary layer transition from laminar to turbulent nature, the separation point moves towards the rear end of the cylinder. This shift of the separation point reduces the wake size, leading the reduced drag, as shown in Figure 5.18.

From the above discussion it is evident that when the Reynolds number is negligibly small, the drag is due to the viscous forces only, since the boundary layer remains attached to the cylinder. When the Reynolds number is sufficiently high, causing wake formation, the drag is due to skin friction caused by the viscosity, and the low pressure at the base of the cylinder caused by the flow separation. At Reynolds number of the order of $10^5$, the drag is mainly due to the turbulent eddies in the wake. These results are valid only for air with constant Prandtl number.

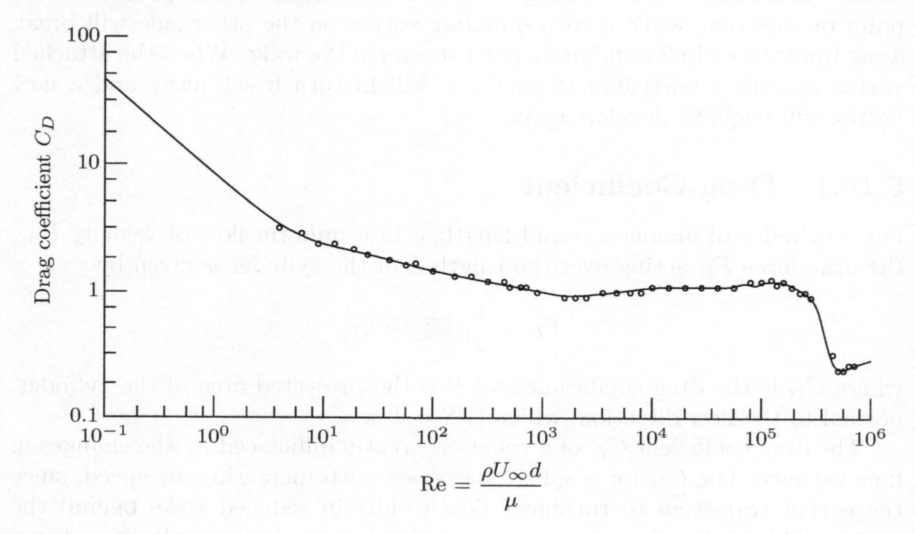


**Figure 5.18**
Cylinder drag coefficient variation with Reynolds number [22].

Several investigators developed correlations to include the effects of Prandtl number in order to extend the applicability of the results to fluids other than gases. Whitaker [18] correlated the average heat transfer coefficient $h_m$ for the flow of gases or liquids across a single cylinder as

$$\mathrm{Nu}_m \equiv \frac{h_m D}{k} = (0.4\,\mathrm{Re}^{\frac{1}{2}} + 0.06\,\mathrm{Re}^{\frac{2}{3}})\,\mathrm{Pr}^{0.4}\left(\frac{\mu_\infty}{\mu_w}\right)^{0.25} \tag{5.167}$$

where $D$ is the diameter of the cylinder, $\mu_w$ is the viscosity at the wall, and $\mu_\infty$ is the viscosity in the freestream. This correlation agrees with the experimental data within $\pm 25$ percent in the range of variables

$$40 < \mathrm{Re} < 10^5$$

$$0.67 < \mathrm{Pr} < 300$$

$$0.25 < \left(\frac{\mu_\infty}{\mu_w}\right) < 5.2$$

where the physical properties are evaluated at the freestream temperature except $\mu_w$, which is evaluated at the wall temperature, for gases. For liquids, the properties are evaluated at the film temperature. Note that Equation (5.167) involves two different functional dependences of the Nusselt number on the Reynolds number. The functional dependence $\mathrm{Re}^{1/2}$ characterizes the contribution from the undetached laminar boundary layer region, and $\mathrm{Re}^{2/3}$ characterizes the contribution from the wake region of the cylinder.

A more general correlation is given by Churchill and Berstein [19] for the average heat transfer coefficient $h_m$ for flow across a single cylinder as

$$\mathrm{Nu}_m = 0.3 + \frac{0.62\,\mathrm{Re}^{\frac{1}{2}}\,\mathrm{Pr}^{\frac{1}{3}}}{\left[1+(0.4/\mathrm{Pr})^{\frac{2}{3}}\right]^{\frac{1}{4}}}\left[1+\left(\frac{\mathrm{Re}}{282,000}\right)^{\frac{5}{8}}\right]^{\frac{4}{5}} \tag{5.168}$$

Equation (5.168) is valid for $10^2 < \mathrm{Re} < 10^7$ and Peclet number Pe $(= \mathrm{Re} \times \mathrm{Pr}) > 0.2$. This correlation underestimates most data by about 20 percent in the Reynolds number range from 20,000 to 400,000. To overcome this situation for this range of Reynolds number, the following modified form of Equation (5.168) is recommended.

$$\mathrm{Nu}_m = 0.3 + \frac{0.62\,\mathrm{Re}^{\frac{1}{2}}\,\mathrm{Pr}^{\frac{1}{3}}}{\left[1+(0.4/\mathrm{Pr})^{\frac{2}{3}}\right]^{\frac{1}{4}}}\left[1+\left(\frac{\mathrm{Re}}{282,000}\right)^{\frac{1}{2}}\right] \tag{5.169}$$

In Equations (5.168) and (5.169), all properties are evaluated at the film temperature. These two correlations were based on the experimental data of many researchers, for fluids such as air, water, and liquid sodium with both constant wall temperature and constant wall heat flux.

Nakai and Okazaki [20] proposed the correlation

$$\mathrm{Nu}_m = \left(0.8237 - \ln\,\mathrm{Pe}^{\frac{1}{2}}\right)^{-1} \tag{5.170}$$

This correlation is valid only for Peclet numbers less than 0.2. Note that, Peclet number Pe is the product of Reynolds number and Prandtl numbers $(\mathrm{Pe} = \mathrm{Re} \times \mathrm{Pr})$.

## Example 5.15

A sphere of diameter 25 mm with its surface temperature maintained constant is exposed to an air stream at 1 atm, 250 K and 30 m/s. If the heat loss from the sphere to air is 30 W, determine the temperature of the sphere surface, taking air properties corresponding to 250 K.

## Solution

Given, $D = 25$ mm, $T_\infty = 250$ K, $U = 30$ m/s, $p = 1$ atm, $\dot{Q} = 30$ W. For air at 250 K, $k = 0.022$ W/(m °C), Pr $= 0.722$.

The air density and viscosity are

$$\rho = \frac{p}{RT} = \frac{101325}{287 \times 250}$$

$$= 1.41\,\text{kg/m}^3$$

$$\mu = (1.46 \times 10^{-6}) \times \frac{250^{1.5}}{250 + 111}$$

$$= 1.6 \times 10^{-5}\,\text{kg/(m s)}$$

The Reynolds number is

$$\text{Re} = \frac{\rho U D}{\mu} = \frac{1.41 \times 30 \times 0.025}{1.6 \times 10^{-5}}$$

$$= 66094$$

By Equation (5.169),

$$\text{Nu}_m = 0.3 + \frac{0.62 \times \sqrt{66094} \times 0.722^{1/3}}{\left[1 + (0.4/0.722)^{2/3}\right]^{1/4}} \times \left[1 + \left(\frac{66094}{282000}\right)^{1/2}\right]$$

$$= 186.45$$

The convection heat transfer coefficient is

$$h_m = \frac{\text{Nu}_m k}{D}$$

$$= \frac{186.45 \times 0.022}{0.025}$$

$$= 164\,\text{W/(m}^2\,{}^\circ\text{C)}$$

The heat transfer rate is

$$\dot{Q} = 30 = h_m \times (\pi D^2) \times (T_s - T_\infty)$$

Thus, the surface temperature becomes

$$T_s = \frac{30}{164 \times (\pi \times 0.025^2)} + 250$$

$$= \boxed{343.16\,\text{K}}$$

### 5.15.2 Variation of $h$ around a Cylinder

In the above discussion on the average heat transfer coefficient correlations for flow past cylinders, the focus was on the determination of the average value of $h_m$. But in practical problems, the local heat transfer coefficient $h(\theta)$ is a strong function of the azimuthal angle $\theta$ variation around the cylinder. The value of $h$ is fairly high at the forward stagnation point S with $\theta = 0°$, shown in Figure 5.19.

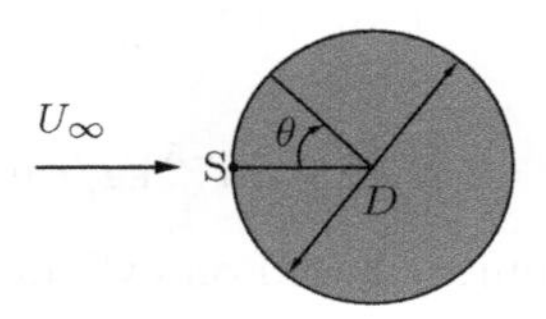


**Figure 5.19**
Flow past a cylinder.

With increase of $\theta$ the heat transfer coefficient decreases continuously as the boundary layer thickens. This decrease continues up to the location where the boundary layer separates from the cylinder wall or the laminar boundary layer transits to turbulent. In the downstream direction, from the separation or transition location, the heat transfer coefficient increases with distance around the cylinder.

## Example 5.16

Air at 1 atm and 300 K flows with a speed of 25 m/s, over a hot circular cylinder of diameter 20 mm, maintained at a constant surface temperature of 360 K. Determine the average heat transfer coefficient, the heat transfer rate per unit length, the average drag coefficient and the drag force per unit length. Assume $k = 0.03$ W/(m °C) and Pr $= 0.7$.

## Solution

Given, $p = 101325$ Pa, $T = 300$ K, $D = 0.02$ m, $T_w = 360$ K, $V = 25$ m/s.

By thermal state equation, the density is

$$\rho = \frac{p}{RT} = \frac{101325}{287 \times 300}$$

$$= 1.177\,\text{kg/m}^3$$

By Sutherland relation, the viscosity is [2]

$$\mu = 1.46 \times 10^{-6} \frac{T^{3/2}}{T + 111}$$

$$= 1.46 \times 10^{-6} \frac{300^{3/2}}{300 + 111}$$

$$= 1.846 \times 10^{-5}\, \text{kg/(m s)}$$

At the wall the temperature is 360 K. The corresponding viscosity is

$$\mu_w = 1.46 \times 10^{-6} \frac{360^{3/2}}{360 + 111}$$

$$= 2.12 \times 10^{-5}\, \text{kg/(m s)}$$

The freestream Reynolds number based on cylinder diameter is

$$\text{Re} = \frac{\rho U_\infty D}{\mu}$$

$$= \frac{1.177 \times 25 \times 0.02}{1.846 \times 10^{-5}}$$

$$= 31880$$

The ratio of $\mu_\infty$ to $\mu_w$ is

$$\frac{\mu_\infty}{\mu_w} = \frac{1.846}{2.12}$$

$$= 0.87$$

The ratio of $\mu_\infty$ to $\mu_w$ is between 0.25 and 5.2. Therefore, by Equation (5.167), the average Nusselt number becomes

$$\text{Nu}_m = \frac{h_m D}{k} = (0.4\,\text{Re}^{0.5} + 0.06\,\text{Re}^{2/3})\,\text{Pr}^{0.4} \left(\frac{\mu_\infty}{\mu_w}\right)^{0.25}$$

$$= (0.4 \times 31880^{0.5} + 0.06 \times 31880^{2/3}) \times 0.7^{0.4} \times \left(\frac{1.846 \times 10^{-5}}{2.12 \times 10^{-5}}\right)^{0.25}$$

$$= 110.3$$

Thus, the average heat transfer coefficient becomes

$$h_m = \frac{\text{Nu}_m k}{D}$$

$$= \frac{110.3 \times 0.03}{0.02}$$

$$= \boxed{165.45\, \text{W/(m}^2\, {}^\circ\text{C)}}$$

The heat transfer rate is

$$\begin{aligned}\dot{Q} &= h_m A \Delta T \\ &= h_m \times (\pi D L) \times (T_w - T_\infty) \\ &= 165.45 \times (\pi \times 0.02 \times 1) \times (360 - 300) \\ &= \boxed{624\,\text{W/m}}\end{aligned}$$

For Re = 31880, from Figure 5.18, the drag coefficient is

$$C_D \approx \boxed{1.5}$$

The drag force acting on the cylinder is

$$\begin{aligned}F_d &= \frac{1}{2}\rho {U_\infty}^2 S C_D \\ &= \frac{1}{2}\rho {U_\infty}^2 (L \times D) C_D \\ &= \frac{1}{2} \times 1.177 \times 25^2 \times (1 \times 0.02) \times 1.5 \\ &= \boxed{11\,\text{N}}\end{aligned}$$

## 5.16 Flow Past a Noncircular Cylinder

Flow past noncircular single cylinders of various geometries have been studied experimentally by Jakob [21]. For flow of gases across long noncircular cylinders, he correlated the average Nusselt number as

$$\text{Nu}_m = \frac{h_m D_e}{k} = c\left(\frac{U_\infty D_e}{\nu}\right)^n \tag{5.171}$$

where $h$ is the average convection coefficient, $U_\infty$ is the freestream velocity, $k$ is the thermal conductivity of the gas, $D_e$ is the characteristic dimension of the noncircular shape, $c$ is a constant which is sensitive to the cross-section, the index $n$ is specific for a given cross-section and $\nu = \mu/\rho$ is the kinematic viscosity. The physical properties of the fluid are evaluated at the arithmetic mean of the freestream and wall temperatures. The constant $c$, index $n$, and the characteristic dimension $D_e$, for selected geometries, are presented in Table 5.5.

**Table 5.5** Values of $c$ and $n$ for a range of Reynolds number, for different geometries [21]

| Geometry | Re range | $n$ | $c$ |
|---|---|---|---|
| $D_e$ | 5000 - 100,000 | 0.588 | 0.222 |
| $D_e$ | 2500 - 15000 | 0.612 | 0.224 |
| $D_e$ | 2500 - 7500 | 0.624 | 0.261 |
| $D_e$ | 5000 - 100,000 | 0.638 | 0.138 |
| $D_e$ | 5000 - 19500 | 0.638 | 0.144 |
| $D_e$ | 5000 - 100,000 | 0.675 | 0.092 |
| $D_e$ | 2500 - 8000 | 0.699 | 0.160 |
| $D_e$ | 4000 - 15000 | 0.731 | 0.205 |
| $D_e$ | 19500 - 100,000 | 0.782 | 0.035 |
| $D_e$ | 3000 - 15000 | 0.804 | 0.085 |

## 5.17 Flow Past a Sphere

The flow field around a sphere is somewhat similar to flow past a cylinder. Therefore, the dependence of the drag and heat transfer coefficients on the Reynolds number for a sphere can be expected to be of the same form as that for a cylinder.

### 5.17.1 Drag Coefficient

The drag force acting on a sphere of diameter $d$, with the average drag coefficient $C_D$, in a flow with freestream velocity $U_\infty$, is given by

$$F_d = \frac{1}{2}\rho {U_\infty}^2 S C_D$$

where $S = \pi d^2/4$ is the projected area of the sphere, normal to the freestream flow. The variation of the average drag coefficient $C_D$ with Reynolds number, for a sphere is shown in Figure 5.20.

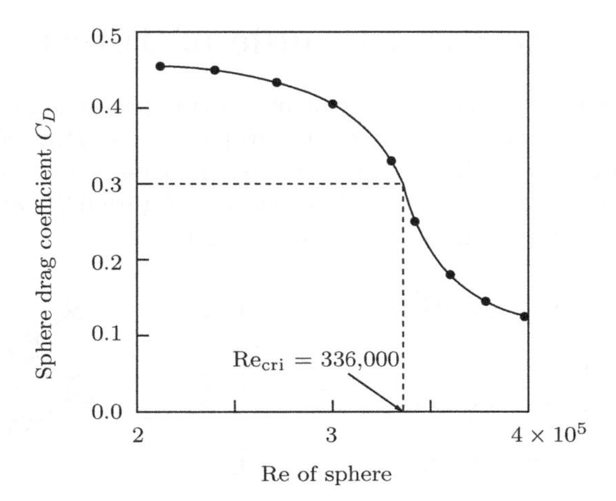


**Figure 5.20**
Sphere drag coefficient variation with Reynolds number [22].

## 5.17.2 Heat Transfer Coefficient

For flow past a sphere of diameter $d$, the simple correlation for average Nusselt number, proposed by McAdams [4], is

$$\text{Nu}_m = \frac{h_m d}{k} = 0.37\,\text{Re}^{0.6} \tag{5.172}$$

where $h_m$ is the mean convection heat transfer coefficient over the entire surface of the sphere and $k$ is the conduction coefficient of the flowing fluid. The properties are calculated at $(T_\infty + T_w)/2$. Equation (5.172) is valid for the Reynolds number range from 17 to 70,000.

A more general correlation for the flow of gases and liquids across a single sphere, reported by Whitaker [18], is of the form

$$\text{Nu}_m = 2 + (0.4\,\text{Re}^{0.5} + 0.06\,\text{Re}^{2/3})\,\text{Pr}^{0.4}\left(\frac{\mu_\infty}{\mu_w}\right)^{0.25} \tag{5.173}$$

This is valid for $3.5 < \text{Re} < 8 \times 10^4$ and $0.7 < \text{Pr} < 380$, $1 < \mu_\infty/\mu_w < 3.2$. All the physical properties are evaluated at the freestream temperature, except $\mu_w$, which is calculated for the wall temperature. For gases, the viscosity correction is neglected but the physical properties are evaluated at the film temperature. When $\text{Re} \to 0$, that is, when there is no flow, the Nusselt number given by Equation (5.173) assumes a limiting value of 2, which represents the steady-state heat conduction from a sphere, at a uniform temperature into the surrounding environment.

## 5.18 Flow across a Bundle of Tubes

Bundle of tubes is a common arrangement in many practical heat exchangers. Therefore, the heat transfer and pressure drop characteristics of tube bundles play a dominant role in the design of heat exchangers. In-line and staggered arrangement of a bundle of tubes, illustrated in Figure 5.21, are some of the commonly used configurations in heat exchangers.

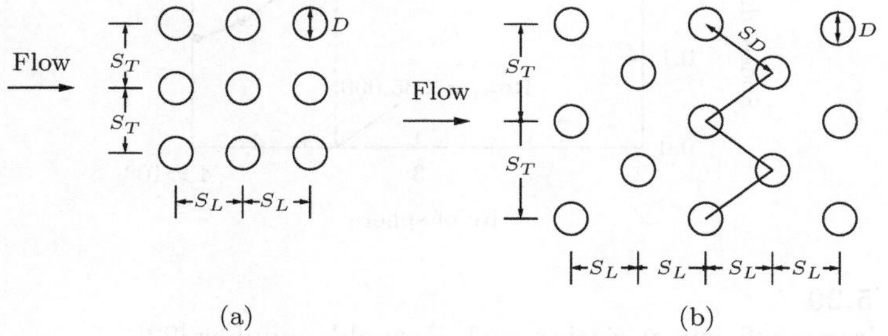

**Figure 5.21**
Tube bundles in (a) in-line arrangement, (b) staggered arrangement.

The tube bundle geometry is characterized by the *transverse pitch* $S_T$ and the *longitudinal pitch* $S_L$, between the tube centers. The *diagonal pitch* $S_D$, between the centers of the tubes in the diagonal row, is also used sometimes for the staggered arrangement. The Reynolds number for flow over the tube bundles is defined with the flow velocity based on the *minimum free-flow area* available for the flow. The minimum area may be taken as either the distance between the tubes in the transverse row or in the diagonal row. Thus, the Reynolds number is

$$\mathrm{Re} = \frac{DG_{\max}}{\mu} \tag{5.174}$$

where $G_{\max}$ is the maximum mass flow velocity, defined as

$$G_{\max} = \rho U_{\max} \tag{5.175}$$

In effect, $G_{\max}$ is the mass flow rate per unit area where the flow velocity is maximum, $D$ is the outside diameter of the tube, $\rho$ is the density of the flow, and $U_{\max}$ is the maximum velocity based on the minimum free-flow area available for fluid flow. If $U_\infty$ is the flow velocity measured at a point in the heat exchanger before the fluid enters the tube bank, then the maximum flow velocity $U_{\max}$ for the in-line arrangement shown in Figure 5.21(a) is determined from

$$U_{\max} = U_\infty \frac{S_T}{S_T - D} = U_\infty \frac{S_T/D}{S_T/D - 1} \tag{5.176}$$

because for the in-line arrangement, $(S_T - D)$ is the minimum free-flow area between the adjacent tubes in a transverse row, per unit length of the tube.

For the staggered arrangement of the tube bank, shown in Figure 5.21(b), the minimum free-flow area may occur between the adjacent tubes in a transverse row or in a diagonal row. When the minimum free-flow area is in a transverse row, $U_{\max}$ should be determined with Equation (5.176), and when the minimum free-flow area is in a diagonal row, it should be determined with

$$U_{\max} = U_\infty \frac{S_T}{2(S_D - D)} = \frac{1}{2} U_\infty \frac{S_T/D}{S_D/D - 1} \tag{5.177}$$

The maximum mass flow rate $G_{\max}$, defined in Equation (5.175), can also be calculated with the relation

$$G_{\max} = \frac{\dot{m}}{A_{\min}} \tag{5.178}$$

where $\dot{m}$ is the total mass flow rate through the tube bundle, in kg/s, and $A_{\min}$ is the total minimum free-flow area.

Because of the complex nature of the flow pattern through the tube bank, it is not possible to predict the heat transfer and pressure loss for flow across the bank, by theoretical approach. Therefore, all the predictions of heat transfer and pressure loss for this problem are based on experimental approach.

Experimentally it has been established that for tube banks with more than 10 rows of tubes in the direction of flow, and the tube length is very large compared to the diameter, the entrance, exit and edge effects are negligible. For such cases, the Nusselt number is a function of the Reynolds number, Prandtl number, $S_T/D$ and $S_L/D$, that is

$$\text{Nu} = f(\text{Re}, \text{Pr}, S_T/D, S_L/D)$$

This functional relation shows that, in addition to Re and Pr, the geometrical arrangement of the tubes, namely in-line and staggered, also influences the Nusselt number.

### 5.18.1 Heat Transfer Correlations

Several investigators studied the heat transfer for air flow over both in-line and staggered tube arrangements, for tube banks with 10 or more transverse rows in the direction of flow. Grimison [23] correlated these data with the relation

$$\frac{h_m D}{k} = c_0 \left( \frac{D G_{\max}}{\mu} \right)^n \tag{5.179}$$

where $D$ is the tube diameter. This relation is valid for air in the Reynolds number range from 2000 to 40,000.

Including the effect of Prandtl number, Equation (5.179) has been generalized to fluids other than air, in the following form

$$\frac{h_m D}{k} = 1.13\, c_0\, \mathrm{Re}^n\, \mathrm{Pr}^{\frac{1}{3}} \tag{5.180}$$

for Reynolds number range from 2000 to 40,000, Pr > 0.7, and number of tube rows more than 10. The Reynolds number in Equation (5.180) is defined as

$$\mathrm{Re} = \frac{DG_{\max}}{\mu}$$

All the physical properties in Equation (5.180) should be evaluated at the mean film temperature. The values of the constant $c_0$ and exponent $n$ in Equation (5.180), for some in-line and staggered arrangement, are listed in Table 5.6.

The effects of the row number on the heat transfer coefficient for a variety of tube arrangements has been studied experimentally by Kays et al. [24]. It was found that, for tube bundles having less than 10 transverse rows in the direction of flow, there was significant reduction in the heat transfer coefficient. Based on the results of their experimental results, the heat transfer coefficient $h_N$ for number of tube rows $N < 10$ could be determined with the relation

$$h_N = c_1 h_{N \geq 10} \tag{5.181}$$

This is valid for $1 \leq N \leq 10$. Table 5.7 lists the values of $c_1$ for both in-line and staggered arrangements of the tubes, with $N$ in the range 1 to 9. The dependence of the results on Reynolds number is only marginal.

Reviewing various studies reported in open literature, Zukauskas [25] proposed the following heat transfer correlations for flow across tube banks.

$$\frac{h_m D}{k} = c_2 \mathrm{Re}^m \mathrm{Pr}^{0.36} \left(\frac{\mathrm{Pr}}{\mathrm{Pr}_w}\right)^n \tag{5.182}$$

where $\mathrm{Pr}_w$ is the Prandtl number evaluated at the wall temperature, and $n = 0$ for gases and $n = 1/4$ for liquids. This relation is valid for $0.7 < \mathrm{Pr} < 500$ and $N \geq 20$. For liquids, the physical properties are evaluated at the bulk mean temperature, because the viscosity correction term is included through the Prandtl number ratio. For gases, the physical properties are evaluated at the film temperature, and the viscosity correction term $\mathrm{Pr}/\mathrm{Pr}_w$ is omitted.

The coefficient $c_2$ and the exponent $m$ in Equation (5.182) were determined by correlating the experimental data for air, water, and oil reported by large number of researchers. Table 5.8 gives the values of $c_2$ and $m$, for some specific cases.

Equation (5.182) correlates the experimental data very well for tube bundles having $N = 20$ or more rows in the direction of flow. For bundles having less than 20 rows ($N < 20$), the Nusselt number can be found from

$$\mathrm{Nu}_N = c_3 \mathrm{Nu}_{N \geq 20} \tag{5.183}$$

**Table 5.6** Constant $c_0$ and exponent $n$ in Equation (5.180) [23]

| | | $S_T/D = 1.25$ | | $S_T/D = 1.50$ | | $S_T/D = 2.0$ | | $S_T/D = 3.0$ | |
|---|---|---|---|---|---|---|---|---|---|
| Arrangement | $S_L/D$ | $c_0$ | $n$ | $c_0$ | $n$ | $c_0$ | $n$ | $c_0$ | $n$ |
| | 0.6 | - | - | - | - | - | - | 0.213 | 0.636 |
| | 0.9 | - | - | - | - | 0.446 | 0.571 | 0.401 | 0.581 |
| | 1.0 | - | - | 0.497 | 0.588 | - | - | - | - |
| | 1.125 | - | - | - | - | 0.478 | 0.565 | 0.518 | 0.560 |
| Staggered | 1.25 | 0.518 | 0.556 | 0.505 | 0.554 | 0.519 | 0.556 | 0.522 | 0.562 |
| | 1.50 | 0.451 | 0.568 | 0.460 | 0.562 | 0.452 | 0.568 | 0.488 | 0.568 |
| | 2.0 | 0.404 | 0.572 | 0.416 | 0.568 | 0.482 | 0.556 | 0.449 | 0.570 |
| | 3.0 | 0.310 | 0.592 | 0.356 | 0.580 | 0.440 | 0.562 | 0.421 | 0.574 |
| | 1.25 | 0.348 | 0.592 | 0.275 | 0.608 | 0.100 | 0.704 | 0.0633 | 0.752 |
| In-line | 1.50 | 0.367 | 0.586 | 0.250 | 0.620 | 0.101 | 0.702 | 0.0678 | 0.744 |
| | 2.0 | 0.418 | 0.570 | 0.299 | 0.602 | 0.229 | 0.632 | 0.198 | 0.648 |
| | 3.0 | 0.290 | 0.601 | 0.357 | 0.584 | 0.374 | 0.581 | 0.286 | 0.608 |

**Table 5.7** Correction factor $c_1$ in Equation (5.181) [24]

| $N$ | 1 | 2 | 3 | 4 | 5 | 6 | 7 | 8 | 9 |
|---|---|---|---|---|---|---|---|---|---|
| In-line | 0.64 | 0.80 | 0.87 | 0.90 | 0.92 | 0.94 | 0.96 | 0.98 | 0.99 |
| Staggered | 0.68 | 0.75 | 0.83 | 0.89 | 0.92 | 0.95 | 0.97 | 0.98 | 0.99 |

**Table 5.8** Values of $c_2$ and $m$ in Equation (5.182) [25]

| Geometry | Re | $c_2$ | $m$ | Remark |
|---|---|---|---|---|
| In-line | $10$ to $10^2$<br>$10^2$ to $10^3$ | 0.8 | 0.40 | Large and moderate longitudinal pitch can be regarded as a single tube |
| | $10^3$ to $2 \times 10^5$ | 0.27 | 0.63 | |
| | $2 \times 10^5$ to $10^6$ | 0.21 | 0.84 | |
| Staggered | $10$ to $10^2$<br>$10^2$ to $10^3$ | 0.9 | 0.40 | About 20% higher than that for single tube |
| | $10^3$ to $2 \times 10^5$ | $0.35\,(S_T/S_N)^{0.2}$ | 0.6 | $S_T/S_L < 2$ |
| | $10^3$ to $2 \times 10^5$ | 0.40 | 0.60 | $S_T/S_L > 2$ |
| | $2 \times 10^5$ to $10^6$ | 0.022 | 0.84 | |

where $c_3$ is the correction factor, which is valid for both in-line and staggered tube arrangements. Variation of $c_3$ with $N$ for in-line and staggered configurations are given in Figure 5.22.

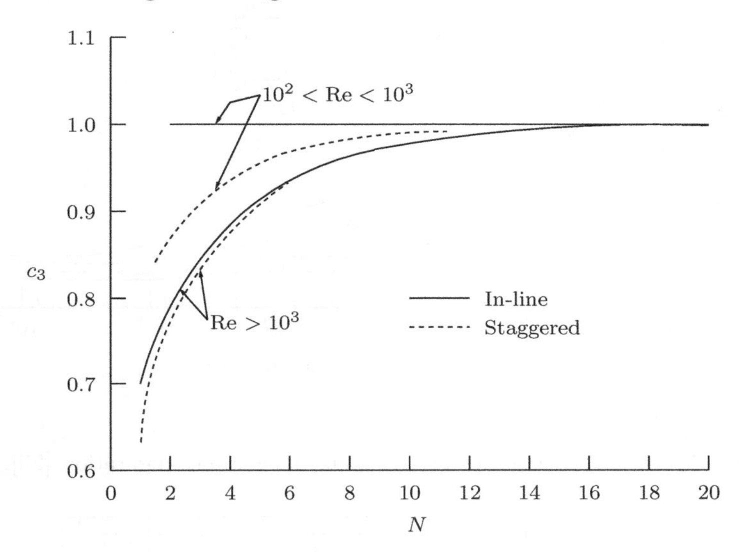


**Figure 5.22**
Variation of $c_3$ with $N$ [25].

## 5.18.2 Pressure Drop Correlations

The pressure drop $\Delta p$ due to fluid friction for flow across tube bank has been correlated by Zukauskas [25] as

$$\Delta p = fZ\frac{NG_{\max}^2}{2\rho} \tag{5.184}$$

where $f$ is the friction factor, $Z$ is the correction factor for effects of tube bundle configuration, $N$ is the number of tube rows in the flow direction, and $G_{\max} = \rho U_{\max}$ is the maximum mass flow velocity. The correction factor $Z = 1$ for square and equilateral triangle arrangements of the tubes.

For in-line arrangement with square configuration, the friction factor variation with Reynolds number, for different $x_L$ values and correction factor $Z$, as a function of $(x_T - 1)/(x_L - 1)$ and Reynolds number are shown in Figures 5.23(a) and (b), respectively, for square arrangement with $Z = 1$. The friction factor $f$ variation with Reynolds number, for equilateral triangular arrangement is shown in Figure 5.24(a). Figure 5.24(b) gives $Z$ variation with $X_T/X_L$.

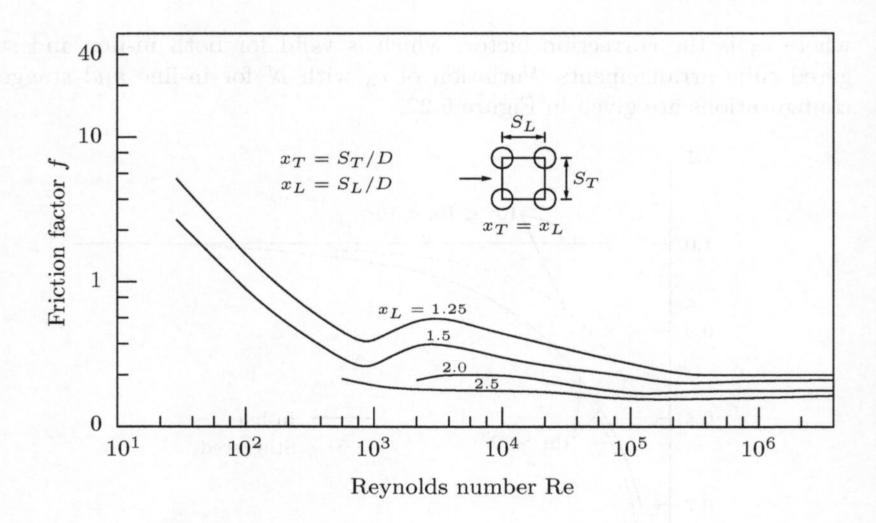


**Figure 5.23a**
Variation of $f$ with Re, for in-line arrangement in square configuration [25].

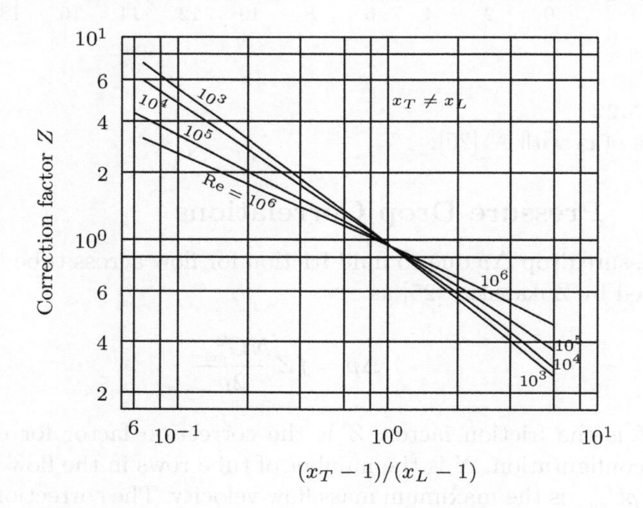


**Figure 5.23b**
Variation of $Z$ with $(x_T - 1)/(x_L - 1)$, for in-line arrangement in square configuration [25].

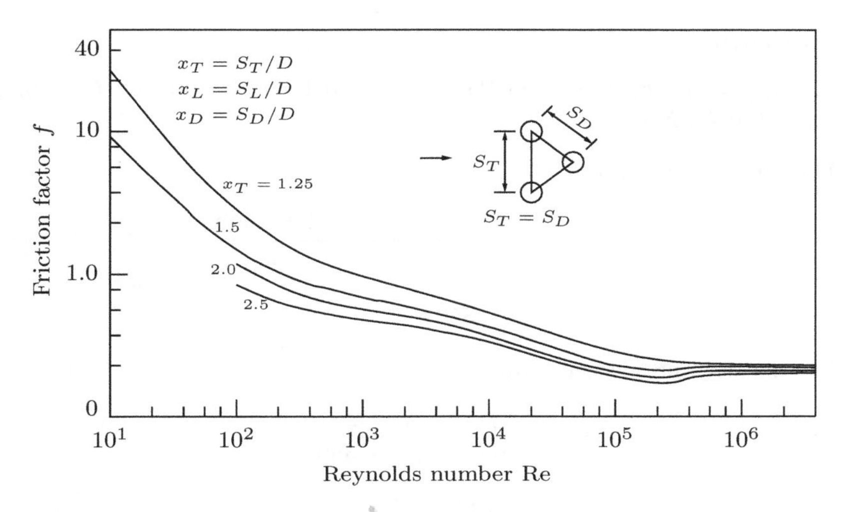


**Figure 5.24a**
Variation of $f$ with Re, for staggered arrangement in triangular configuration [25].

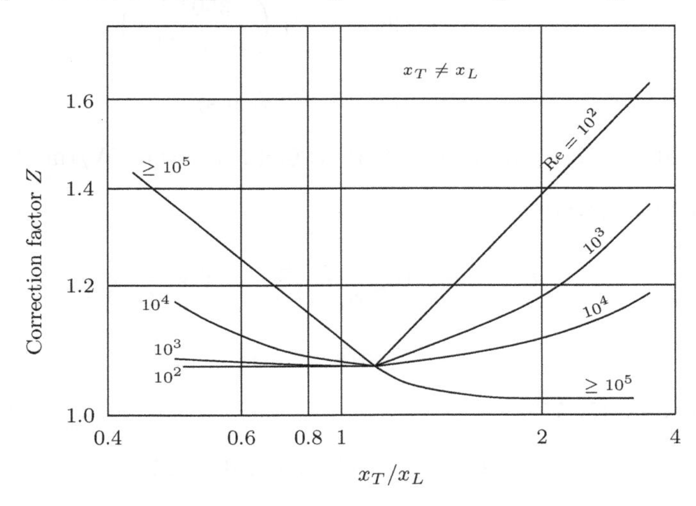


**Figure 5.24b**
Variation of $Z$ with $x_T/x_L$, for staggered arrangement in triangular configuration [25].

## Example 5.17

Air at 1 atm and 300 K flows at 6 m/s across the tubes of a heat exchanger with 15 rows of tubes of outer diameter 20 mm, arranged in-line with equal longitudinal ($S_L/D$) and transverse ($S_T/D$) pitches of 2. Determine the pressure drop across the tube bundle and the average convection coefficient of the flow when the tubes surfaces are at a constant temperature of 400 K.

## Solution

Given, $p = 1\,\text{atm} = 101325$ Pa, $T_\infty = 300$ K, $T_w = 400$ K, $D = 0.02$ m, $U_\infty = 6$ m/s, $N = 15$.

The film temperature is

$$T_f = \frac{T_w + T_\infty}{2} = \frac{400 + 300}{2}$$

$$= 350\,\text{K}$$

The density and viscosity are

$$\rho = \frac{p}{RT} = \frac{101325}{287 \times 350}$$

$$= 1.01\,\text{kg/m}^3$$

$$\mu = (1.46 \times 10^{-6}) \left( \frac{350^{3/2}}{350 + 111} \right)$$

$$= 2.07 \times 10^{-5}\,\text{kg/(m s)}$$

For air at 350 K, from Table A-4, $\Pr = 0.697$, $k = 0.03$ W/(m °C).

By Equation (5.176),

$$U_{\max} = U_\infty \frac{S_T/D}{S_T/D - 1} = 6 \times \frac{2}{2-1}$$

$$= 12\,\text{m/s}$$

By Equation (5.175),

$$G_{\max} = \rho U_{\max} = 1.01 \times 12$$

$$= 12.12\,\text{kg/(m}^2\text{ s)}$$

The Reynolds number, by Equation (5.174), is

$$\text{Re} = \frac{DG_{\max}}{\mu} = \frac{0.02 \times 12.12}{2.07 \times 10^{-5}}$$

$$= 11710$$

For $\text{Re} = 11710$ and $X_L = S_L/D = 2$, from Figure 5.23(a), the friction factor is $f \approx 0.2$.

Also, for $X_L = X_T$, the correction factor $Z = 1$. Therefore, the pressure drop across the tube bundle, given by Equation (5.184), is

$$\begin{aligned}\Delta p &= fZ\,\frac{NG_{\max}^2}{2\rho} \\ &= 0.2 \times 1 \times \left(\frac{15 \times 12.12^2}{2 \times 1.01}\right) \\ &= \boxed{218.16\,\text{Pa}}\end{aligned}$$

For Re $= 11710$, from Table 5.8, $c_2 = 0.27$ and $m = 0.63$.

By Equation (5.182),

$$\frac{h_m D}{k} = c_2 \text{Re}^m \text{Pr}^{0.36} \left(\frac{\text{Pr}}{\text{Pr}_w}\right)^n$$

For gas, $n = 0$. Therefore,

$$\begin{aligned}\frac{h_m D}{k} &= 0.27 \times 11710^{0.63} \times 0.697^{0.36} \\ &= 86.72 \\ h_m &= \frac{86.72\,k}{D} = \frac{86.72 \times 0.03}{0.02} \\ &= \boxed{130.1\,\text{W}/(\text{m}^2\,{}^\circ\text{C})}\end{aligned}$$

### 5.18.3 Liquid Metals

The heat transfer correlations discussed above are valid only for flow of gases with large Reynolds number. They are not valid for fluids, such as liquid metals, having very low Prandtl number. For this kind of flows, based on their experimental data for mercury flowing across staggered tube bundles with 60 to 70 rows deep, and consisting of half an inch diameter tubes arranged in equilateral triangle array having a pitch-to-diameter ratio of 1.375, Hoe et al. [26] and Richards et al. [27] correlated the heat transfer coefficient as

$$\text{Nu}_m = \frac{h_m D}{k} = 4.03 + 0.228\,(\text{Re Pr})^{0.67} \tag{5.185}$$

where Re $= DG_{\max}/\mu$. This relation is valid for Reynolds number range from 20,000 to 80,000. The physical properties are evaluated at the arithmetic average of the bulk fluid temperature and wall surface temperature.

# 5.19 Heat Transfer in High-Speed Flow over a Flat Plate

High-speed subsonic and supersonic flows are encountered in applications such as aircraft, missiles and launch vehicles. For such flows, the effects of compressibility, viscous dissipation, and flow property variation with temperature become important. Analysis of such flows is complex. However, for simple geometries, such as flat plate, the heat transfer rate at a uniform surface temperature $T_w$ can be predicted, by using the low-speed heat transfer coefficient $h_x$ for flow along a flat plate with the temperature difference $(T_w - T_{aw})$. That is, the heat flux $\dot{q}_x$ at the wall of a flat plate in a high-speed flow can be calculated from

$$\dot{q}_x = h_x \left(T_w - T_{aw}\right) \tag{5.186}$$

where $T_w$ is the wall temperature, $T_{aw}$ is the adiabatic wall temperature, and $h_x$ is the local, low-speed heat transfer coefficient.

For using the Equation (5.186) for high-speed heat transfer calculations, the local heat transfer coefficient $h_x$ is obtained from the low-speed heat transfer correlations for flow along a flat plate. For example, for laminar boundary layer flow, $h_x$ is obtained from Equation (5.97) as

$$\mathrm{Nu}_x \equiv \frac{h_x x}{k} = 0.332\,\mathrm{Pr}^{1/3}\,\mathrm{Re}_x^{1/2} \tag{5.187}$$

for $\mathrm{Re}_x < 5 \times 10^5$, that is for laminar flow. Note that, the constant in Equation (5.187) is half of that in Equation (5.98), because only one surface of the plate is considered here.

For turbulent boundary layer flow, from Equation (5.105), we have

$$\frac{h_x x}{k} = 0.0296\,\mathrm{Re}_x^{4/5}\mathrm{Pr}^{0.43}$$

This can be expressed as

$$\begin{aligned}
\frac{h_x}{\rho U c_p}\,\rho U c_p\,\frac{x}{k} &= 0.0296\,\mathrm{Re}_x^{4/5}\mathrm{Pr}^{0.43} \\
\frac{h_x}{\rho U c_p}\,\frac{\rho U x}{\mu}\,\frac{\mu c_p}{k} &= 0.0296\,\mathrm{Re}_x^{4/5}\mathrm{Pr}^{0.43} \\
\frac{h_x}{\rho U c_p}\,\mathrm{Re}_x\,\mathrm{Pr} &= 0.0296\,\mathrm{Re}_x^{\frac{4}{5}}\,\mathrm{Pr}^{\frac{1}{3}} \\
\frac{h_x}{\rho U c_p} &= 0.0296\,\mathrm{Re}_x^{(4/5-1)}\,\mathrm{Pr}^{(0.43-1)}
\end{aligned}$$

That is,

$$\frac{h_x}{\rho U_\infty c_p} = 0.0296\,\mathrm{Pr}^{-0.57}\,\mathrm{Re}_x^{-0.2} \tag{5.188}$$

This is valid for $5 \times 10^5 < \text{Re}_x < 10^7$.

For the Reynolds number range $10^7 < \text{Re}_x < 10^9$ [13],

$$\frac{h_x}{\rho U_\infty c_p} = 0.185\,\text{Pr}^{-\frac{2}{3}}\left(\log \text{Re}_x\right)^{-2.584} \tag{5.189}$$

For Reynolds number $\text{Re}_x$ range from $2 \times 10^5$ to $5 \times 10^5$ [13],

$$\frac{h_x x}{k} = 0.029\,\text{Pr}^{0.43}\,\text{Re}_x^{0.8} \tag{5.190}$$

For an adiabatic wall, by energy equation we have

$$h_0 = h + r\frac{1}{2}U_\infty^2$$

where $h_0$ is the stagnation enthalpy, $h$ is the static enthalpy, $r$ is the *recovery factor*. Assuming the fluid to be perfect, we have $h = c_pT$, therefore,

$$T_{aw} = T_\infty + r\frac{1}{2}\frac{U_\infty^2}{c_p} \tag{5.191}$$

where $T_{aw}$ is the adiabatic wall temperature and $T_\infty$ is the freestream static temperature, $U_\infty$ is the freestream velocity and $c_p$ is the specific heat at constant pressure.

For laminar boundary layer the recovery factor is $r \approx \text{Pr}^{1/2}$ and for turbulent boundary layer, $r \approx \text{Pr}^{1/3}$ [22]. The adiabatic wall temperature is identically equal to the stagnation temperature $T_0$, when $r = 1$.

In high-speed flows, the temperature gradient in the boundary layers are usually large. Therefore, the flow properties vary significantly with temperature. The effects of variation of properties may be approximately included in the heat transfer relations given by Equations (5.187) to (5.190) if the properties of the fluid are evaluated at the following reference temperature [28]

$$T_r = T_\infty + 0.5(T_w - T_\infty) + 0.22(T_{aw} - T_\infty) \tag{5.192}$$

## 5.20 Summary

*Convection* is the mode of heat transfer between a solid and the adjacent fluid that is in motion. The heat transfer that occurs between a solid surface and a moving fluid, when they are at different temperatures is called *convection.* Thus, convection involves the combined effects of conduction and fluid motion. Newton's law of cooling expresses the convection heat transfer as

$$\dot{Q}_{\text{convection}} = hA\left(T_s - T_\infty\right)$$

The convection heat transfer mode comprises of two mechanisms. The energy transfer in convection heat transfer is due to (i) random molecular motion (diffusion), and (ii) the bulk or macroscopic motion of the fluid. The random molecular motion (diffusion) contribution usually dominates near the solid surface where the fluid velocity is low. In fact, at the interface between a solid surface and a fluid, heat is transferred by this mechanism only. The bulk fluid motion contribution originates due to the velocity boundary layer growth as the flow progresses.

Based on the nature of the flow, the convection heat transfer may be classified as

- *Forced* convection, when the flow is caused by some external means, such as a fan, a pump, or atmospheric winds.
- *Free* (or natural) convection, when the flow is induced by buoyancy force in the fluid.

Irrespective of the nature of the convection heat transfer mode, the appropriate heat transfer rate equation is of the form

$$\dot{q} = h\,(T_w - T_\infty)$$

Basically, the velocity boundary layer, the thermal boundary layer and the concentration boundary layer are all convection boundary layers. The velocity boundary layer is of extent $\delta(x)$ and is characterized by the presence of velocity gradients and shear stresses. The thermal boundary layer is of extent $\delta_t\,(x)$ and is characterized by temperature gradients and heat transfer, and the concentration boundary layer is of extent $\delta_c\,(x)$ and is characterized by concentration gradients and species transfer. For engineering applications, the principal manifestations of these three boundary layers are the *surface friction*, *convection heat transfer* and *convection mass transfer*, respectively. The key boundary layer parameters are the friction coefficient $c_f$, the convection heat transfer coefficient $h$ and the mass transfer coefficient $h_m$, respectively.

For fluid flow over any body, there will always exist a velocity boundary layer, and hence surface friction. However, a thermal boundary layer, and hence convection heat transfer will exist only when the body surface and freestream flow are at different temperatures. Similarly, a concentration boundary layer and convection mass transfer will exist only when concentration of a species at the surface differs from its freestream concentration. Situations can arise in which all the three boundary layers are present. In such cases, the boundary layers rarely grow at the same rate, and the values of $\delta$, $\delta_t$, and $\delta_c$ at a given $x$ location are not the same.

Continuity equation can be expressed in differential form as

$$\boxed{\frac{\partial\,(\rho u)}{\partial x} + \frac{\partial\,(\rho v)}{\partial y} = 0}$$

This is a general expression of the *overall* mass conservation requirement, and it must be satisfied at every point in the velocity boundary layer.

The second fundamental law that is pertinent to the velocity boundary layer is Newton's second law of motion. Two kinds of forces may act on the fluid in the boundary layer; the *body* forces, which are proportional to the volume, and the *surface* forces, which are proportional to area. Gravitational, centrifugal, magnetic, and/or electric fields may contribute to the total body force.

The sum of the forces in the $x$-direction is

$$\boxed{\rho\left(u\frac{\partial u}{\partial x}+v\frac{\partial u}{\partial y}\right)=\frac{\partial}{\partial x}\left(\sigma_{xx}-p\right)+\frac{\partial \tau_{yx}}{\partial y}+X}$$

The sum of the forces in the $y$-direction is

$$\boxed{\rho\left(u\frac{\partial v}{\partial x}+v\frac{\partial v}{\partial y}\right)=\frac{\partial \tau_{xy}}{\partial x}+\frac{\partial}{\partial y}\left(\sigma_{yy}-p\right)+Y}$$

The energy conservation requirement may be expressed as

$$-\frac{\partial}{\partial x}\left[\rho u\left(e+\frac{V^2}{2}\right)\right]-\frac{\partial}{\partial y}\left[\rho u\left(e+\frac{V^2}{2}\right)\right]+\frac{\partial}{\partial x}\left(k\frac{\partial T}{\partial x}\right)+\frac{\partial}{\partial y}\left(k\frac{\partial T}{\partial y}\right)$$

$$+\left(Xu+Yv\right)-\frac{\partial}{\partial x}\left(pu\right)-\frac{\partial}{\partial y}\left(pv\right)+\frac{\partial}{\partial x}\left(\sigma_{xx}u+\tau_{xy}v\right)+\frac{\partial}{\partial x}\left(\tau_{yx}u+\sigma_{yy}v\right)+\dot{q}=0$$

This equation is the general form of energy conservation equation for the thermal boundary layer. This represents conservation of *mechanical* and *thermal* energy, and hence it is rarely used for solving heat transfer problems. Instead, a more convenient form, termed the *thermal energy equation*,

$$\rho u\frac{\partial h}{\partial x}+\rho v\frac{\partial h}{\partial y}=\frac{\partial}{\partial x}\left(k\frac{\partial T}{\partial x}\right)+\frac{\partial}{\partial y}\left(k\frac{\partial T}{\partial y}\right)+\left(u\frac{\partial p}{\partial x}+v\frac{\partial p}{\partial y}\right)+\mu\Phi+\dot{q}$$

is used for solving heat transfer problems, where $h$ is the enthalpy per unit mass of mixture, defined as, $h=e+p/\rho$ and $\mu\Phi$ is the *viscous dissipation*, defined as

$$\mu\Phi\equiv\mu\left(\left(\frac{\partial u}{\partial y}+\frac{\partial v}{\partial x}\right)^2+2\left[\left(\frac{\partial u}{\partial x}\right)^2+\left(\frac{\partial v}{\partial y}\right)^2\right]-\frac{2}{3}\left(\frac{\partial u}{\partial x}+\frac{\partial v}{\partial y}\right)^2\right)$$

The boundary layers are very thin; therefore, the following approximations can be applied to them.

- For velocity boundary layer

$$u\gg v$$

$$\frac{\partial u}{\partial y}\gg\frac{\partial u}{\partial x},\ \frac{\partial v}{\partial y},\ \frac{\partial v}{\partial x}$$

- For thermal boundary layer

$$\frac{\partial T}{\partial y} \gg \frac{\partial T}{\partial x}$$

The continuity and the $x$-momentum reduce to

$$\frac{\partial u}{\partial x} + \frac{\partial v}{\partial y} = 0$$

$$u\frac{\partial u}{\partial x} + v\frac{\partial v}{\partial y} = -\frac{1}{\rho}\frac{\partial p}{\partial x} + \nu\frac{\partial^2 u}{\partial y^2}$$

Using the velocity boundary layer approximations, the $y$-momentum reduces to

$$\frac{\partial p}{\partial y} = 0$$

That is, the *pressure does not vary in the direction normal to the surface.*

With the above assumptions and simplifications, the energy equation reduces to

$$u\frac{\partial T}{\partial x} + v\frac{\partial T}{\partial y} = \alpha\frac{\partial^2 T}{\partial y^2} + \frac{\nu}{c_p}\left(\frac{\partial u}{\partial y}\right)^2$$

For solving convection heat transfer problems, we have to consider continuity, momentum, and energy equations simultaneously. Except for very simple situations, it is extremely difficult to get analytical solutions with these equations. However, under simplified conditions it is possible to deal with these equations comfortably. Another way to simplify the equations is to make order of magnitude analysis and neglect few terms in comparison with others.

Volumetric thermal expansion coefficient is given by

$$\beta = -\frac{1}{\rho}\left(\frac{\partial \rho}{\partial T}\right)_p$$

This thermodynamic property of the fluid provides a measure of the amount by which the density changes in response to a change in temperature at constant pressure.

The boundary layer equations are

$$\frac{\partial u^*}{\partial x^*} + \frac{\partial v^*}{\partial y^*} = 0$$

$$u^*\frac{\partial u^*}{\partial x^*} + v^*\frac{\partial u^*}{\partial y^*} = -\frac{1}{\rho}\frac{\partial p^*}{\partial x^*} + \frac{\nu}{UL}\left(\frac{\partial^2 u^*}{\partial x^{*2}} + \frac{\partial^2 u^*}{\partial y^{*2}}\right)$$

$$u^*\frac{\partial v^*}{\partial x^*} + v^*\frac{\partial v^*}{\partial y^*} = -\frac{g\beta L}{v^2}(T_s - T_\infty) + \frac{\nu}{UL}\left(\frac{\partial^2 v^*}{\partial x^{*2}} + \frac{\partial^2 v^*}{\partial y^{*2}}\right)$$

$$u^*\frac{\partial\theta}{\partial x^*}+v^*\frac{\partial\theta}{\partial y^*}=-\frac{U^2}{2c_p\left(T_s-T_\infty\right)}\left[\beta T_\infty+\beta\left(T_s-T_\infty\right)\theta\right]\left[u^*\frac{\partial p^*}{\partial x*}+v^*\frac{\partial p^*}{\partial y^*}\right]$$
$$+\frac{k}{\rho_\infty c_p UL}\left[\frac{\partial^2\theta}{\partial x^{*2}}+\frac{\partial^2\theta}{\partial y^{*2}}\right]+\frac{\mu U}{\rho_\infty c_p UL\left(T_s-T_\infty\right)}\Phi^*$$

where

$$\Phi^*=2\left[\left(\frac{\partial u^*}{\partial x^*}\right)^2+\left(\frac{\partial v^*}{\partial y^*}\right)^2\right]+\left[\frac{\partial u^*}{\partial x^*}+\frac{\partial v^*}{\partial y^*}\right]^2$$

The following groups of parameters, present in the boundary layer equations, are dimensionless groups.

$$\nu/\left(UL\right),\ g\beta L\left(T_s-T_\infty\right)/U^2, k/\left(\rho_\infty c_p UL\right),\ \mu U/\left[\rho_\infty c_p L\left(T_s-T_\infty\right)\right],$$
$$U^2\beta T_\infty/\left[2c_p\left(T_s-T_\infty\right)\right]$$

These equations suggest that the conditions in the velocity and thermal boundary layers depend on the fluid and the material properties, and the length scale $L$.

The group $\frac{g\beta L^3\Delta T}{\nu^2}$ is called Grashof number Gr.

Prandtl number Pr and Eckert number Ec are defined as

$$\mathrm{Pr}=\frac{\mu c_p}{k}=\frac{\nu}{\alpha}$$

and

$$\mathrm{Ec}=\frac{U^2}{c_p\Delta T}$$

In functional form, the dimensionless temperature $\theta$ can be expressed as

$$\theta=\theta\left(x^*,\ y^*,\ \mathrm{Re},\mathrm{Pr},\mathrm{Gr},\mathrm{Ec},\ \beta T_\infty,\ \beta\left(T_s-T_\infty\right)\right)$$

The heat flux at the wall is given by

$$q_{\mathrm{th}}\Big|_{\mathrm{surface}}=-k\left(\frac{\partial T}{\partial n}\right)_{\mathrm{surface}}$$

In nondimensional form, the film transfer coefficient becomes

$$\frac{hL}{k}=-\left(\frac{\partial\theta}{\partial n^*}\right)_{\mathrm{surface}}$$

The group $\frac{hL}{k}$ is known as Nusselt number Nu, and the functional form of $\theta$ can be written as

$$\mathrm{Nu}=\mathrm{Nu}\left[x^*,\ y^*,\ \mathrm{Re},\mathrm{Pr},\mathrm{Gr},\mathrm{Ec},\ \beta T_\infty,\ \beta\left(T_s-T_\infty\right)\right]$$

Reynolds number, $\text{Re} = \dfrac{\rho U L}{\mu}$, may be interpreted as the ratio of inertia to viscous forces.

Prandtl number may be interpreted as a ratio of momentum diffusivity $\nu$ to the thermal diffusivity $\alpha$. The Prandtl number may be viewed as a measure of the relative effectiveness of the momentum and energy transports by diffusion in the velocity and thermal boundary layers, respectively.

Grashof number provides a measure of the ratio of buoyancy forces to viscous forces in the boundary layer. Its role in free convection is identically the same as that of the Reynolds number in forced convection.

Eckert number provides a measure of kinetic energy of the flow relative to the enthalpy difference across the thermal boundary layer. It plays an important role in high speed flows for which the viscous dissipation is significant.

It is important to note that, although similar in form, the Nusselt and Biot numbers differ in both definition and interpretation. The Nusselt number is defined in terms of the thermal conductivity of fluid, whereas the Biot number is based on the thermal conductivity of solid.

The momentum boundary layer thickness $\delta$ is given by

$$\boxed{\frac{\delta}{L} \sim \frac{1}{\sqrt{\text{Re}_L}}}$$

It can be shown that,

$$\boxed{\frac{\delta_t}{L} \sim \frac{1}{(\text{Re}_L \text{Pr})^{1/2}}}$$

The momentum equation of the thermal boundary layer is

$$u^* \frac{\partial \theta}{\partial x^*} + v^* \frac{\partial \theta}{\partial y^*} = \frac{1}{\text{Pr}\,\text{Re}_L} \frac{\partial^2 \theta}{\partial y^{*2}}$$

The ratio of velocity boundary layer thickness to thermal boundary layer thickness is

$$\boxed{\frac{\delta_t}{\delta} = \frac{1}{(\text{Pr})^{1/2}} = \sqrt{\frac{\alpha}{\nu}}}$$

For Pr of order unity, both velocity and thermal boundary layer thickness are of the same order.

The relation between the Nusselt number and skin friction coefficient $c_f$ is

$$\boxed{\text{Nu}_x = \frac{1}{2}\text{Re}_x c_f}$$

This result is usually expressed in the following form with a new nondimensional number, namely the *Stanton number*, St, as

$$\boxed{\text{St} = \frac{1}{2} c_f}$$

The Stanton number is also known as the *modified* Nusselt number. The above relation between St and $c_f$ is known as *Reynolds analogy.*

For a flat plate laminar flow, the skin friction coefficient is given by

$$\boxed{C_f = \frac{0.664}{\sqrt{\text{Re}_x}}}$$

The Reynolds analogy is valid only for laminar flow over a flat plate with zero pressure gradient and Pr = 1.

For incompressible, laminar flow past a flat plate with zero pressure gradient, the boundary layer thickness is given by Blassius as

$$\boxed{\frac{\delta}{x} = \frac{5}{\sqrt{\text{Re}_x}}}$$

For flow with hydrodynamic boundary layer very thin compared to thermal boundary layer, we can show that,

$$\boxed{\frac{\delta_t}{x} = \frac{2}{\sqrt{\text{Re}_x \text{Pr}}}}$$

$$\text{Nu}_x = \frac{hx}{k} = 0.5\,\sqrt{\text{Re}_x \text{Pr}}$$

When a temperature gradient is present during such motion, *forced* convection heat transfer will occur.

For turbulent flow over a flat plate, the *local* friction coefficient and Nusselt number at location $x$ are given by

$$c_{f,x} = \frac{0.0592}{\text{Re}_x^{1/5}}$$

$$\text{Nu}_x = \frac{h_x x}{k} = 0.0296\,\text{Re}_x^{4/5}\,\text{Pr}^{0.43}$$

These results are valid for $5 \times 10^5 \leq \text{Re}_x \leq 10^7$ and $0.6 \leq \text{Pr} \leq 60$. The local friction and heat transfer coefficients are higher in turbulent flow than in laminar flow, because of the active exchange of transverse momentum in the turbulent boundary layer.

In some cases, the plate is sufficiently long for the flow to become turbulent, but not long enough to discard the laminar flow region. That is, the flow transition is in progress and not over. In such a situation, the average $c_f$ and $h$ can be determined by integrating the respective relations over the plate in two parts.

Convection heat transfer will also occur in situations where there is no *forced velocity.* Such situations are referred to as *free* or *natural convection,* and they originate when a *body force* acts on a fluid in which there are *density gradients.* The net effect is a *buoyancy force*, which induces free convection currents.

The Nusselt number for boundary layer on a heated vertical plate is

$$\mathrm{Nu}_x = \frac{0.51\mathrm{Pr}^{1/2}}{(0.95 + \mathrm{Pr})^{1/4}}\,\mathrm{Gr}^{1/4}$$

This equation gives the variation of the local heat transfer coefficient along the vertical plate maintained at a constant temperature of $T_s$ in a quiescent fluid environment.

Excepting some simple cases, heat transfer relations for free convection are based on experimental studies.

The simple empirical correlations for the *average* Nusselt number Nu in natural convection are of the form

$$\mathrm{Nu} = \frac{h\delta}{k} = c\,(\mathrm{Gr}\,\mathrm{Pr})^n = c\,\mathrm{Ra}^n$$

where Ra is the Rayleigh number, which is the product of the Grashof and Prandtl numbers.

$$\mathrm{Ra} = \mathrm{Gr}\,\mathrm{Pr} = \frac{g\beta(T_s - T_\infty)\delta^3}{\nu^2}\,\mathrm{Pr}$$

The values of the constant $c$ and index $n$ depend on the geometry of the surface and the laminar or turbulent nature of the *flow regime*, which is characterized by the range of the Reynolds number. The value of $n$ is usually $1/4$ for laminar flow and $1/3$ for turbulent flow. The constant $c$ is normally less than 1.

Free convection boundary layer can be laminar or turbulent. Free convection flows originate from a thermal instability. But in forced convection, hydrodynamic instabilities may also arise. The transition in a free convection boundary layer depends on the relative magnitude of the buoyancy and viscous forces in the fluid. It is customary to correlate the occurrence of transition in terms of Rayleigh number. For vertical plates the critical Rayleigh number is

$$\mathrm{Ra}_{\mathrm{cri}} = \mathrm{Gr}\,\mathrm{Pr} = \frac{g\beta\,(T_s - T_\infty)}{\nu\alpha}\,x^2 \approx 10^9$$

where $x$ is the length from the leading edge of the plate.

Forced convection is always accompanied by natural convection. Heat transfer coefficients encountered in forced convection are much higher than those associated with natural convection because of the high flow velocities associated with forced convection.

Natural convection is negligible when $\mathrm{Gr}/\mathrm{Re}^2 < 0.1$, forced convection is negligible when $\mathrm{Gr}/\mathrm{Re}^2 > 10$, and both are significant when $0.1 < \mathrm{Gr}/\mathrm{Re}^2 < 10$.

Natural convection may *augment* or *retard* the forced convection heat transfer, depending on the relative direction of the *buoyancy-induced* and *forced convection motions.*

For combined natural and forced convection conditions, experimental studies suggest a correlation of the form

$$\boxed{\mathrm{Nu}_{\mathrm{combined}} = (\mathrm{Nu}^n_{\mathrm{forced}} \pm \mathrm{Nu}^n_{\mathrm{natural}})^{\frac{1}{n}}}$$

where $\mathrm{Nu}_{\mathrm{forced}}$ and $\mathrm{Nu}_{\mathrm{natural}}$ are determined from the correlations for *pure forced* and *pure natural* convection, respectively. The + sign is for *assisting* and *transverse* flows and − sign is for *opposing flow.*

## 5.21 Exercise Problems

5.1 Water at a mean temperature of 22°C flows over a flat plate maintained at 100°C. If the convective heat transfer coefficient is 210 W/(m$^2$ °C), determine the heat transferred per square meter of the plate over a period of 1 hour.

[**Ans.** 58.968 MJ]

5.2 The hot combustion gases of a furnace are separated from the ambient air and its surroundings, which is at 22°C, by a 20-cm-thick brick wall. The brick has a thermal conductivity of 1.22 W/(m K) and surface emissivity of 0.8. Under steady-state conduction, the temperature on the outer surface of the brick wall is measured to be 100°C. Free convection heat transfer to the air adjoining this surface is characterized by a convection coefficient of $h = 20$ W/(m$^2$ K). Determine the temperature on the inner surface of the brick wall.

[**Ans.** 443.48°C]

5.3 A 2.5-mm-thick, 100-mm-wide and 9-cm-long aluminum fin protrudes from a wall. The fin base is maintained at 270°C, and the ambient temperature is 30°C with $h = 10$ W/(m$^2$ K). If the thermal conductivity of aluminum is 200 W/(m K), calculate the heat loss from the fin.

[**Ans.** 39.95 W]

5.4 A mild steel tank of wall thickness 12 mm containing water at 70°C is exposed to atmospheric air 20°C. The thermal conductivity of mild steel is 50 W/(m K). The convection coefficients at the inner and outer surfaces of the tank are 2800 and 11 W/(m$^2$ K), respectively. Calculate (a) the rate of heat loss per unit area of the tank surface, and (b) the temperature of the outer surface of the tank, at the beginning of the heat transfer process.

[**Ans.** (a) 546.45 W/m$^2$, (b) 69.7°C]

5.5 A furnace wall consists of 125-mm-wide refractory bricks and 125-mm-wide insulating bricks, separated by an air gap. The outer surface of the wall is covered with 12-mm-thick plaster. The furnace temperature is 1100°C and the room temperature is 25°C. The heat transfer coefficient from the outer surface of the wall to the air in the room is 17 W/(m$^2$ K), and the thermal resistance of the air gap is 0.16 K/W. The thermal conductivities of the refractory brick, insulating brick and plaster are 1.6, 0.3 and 0.14 W/(m K), respectively. Calculate (a) the rate of heat loss per unit area of the furnace wall surface, (b) the temperature at each interface throughout the wall and (c) the temperature at the outside surface of the wall.

[**Ans.** (a) 1344.9 W, (b) $T_1 = 994.96$°C, $T_2 = 779.77$°C, $T_3 = 219.34$°C, $T_4 = 104.08$°C, (c) $T_o = 104.08$°C]

5.6 A 25-mm-thick thermoplastic sheet, with $k = 5$ W/(m K) and $\alpha = 4\times10^{-7}$ m$^2$/s, is kept in an oven maintained at a uniform temperature of 100°C. What should be the least time of exposure for the sheet to attain a minimum temperature of 75°C everywhere, if its initial temperature is 25°C? For the oven, take $h = 25$ W/(m$^2$ K).

[**Ans.** 169.27 minutes]

5.7 A 50-mm-thick large aluminum plate at 200°C is suddenly exposed to a convection environment at 70°C with a heat transfer coefficient of 525 W/(m$^2$ K). Calculate the temperature at a depth of 12.5 mm from one of the faces of the plate after a lapse of 1 minute. Take $\alpha = 8.4\times10^{-5}$ m$^2$/s, $k = 215$ W/(m K).

[**Ans.** 145.34°C]

5.8 Air at 27°C and 1 atm flows at a speed of 5 m/s over a plate. Calculate the boundary layer thickness at distances of 25 cm and 50 cm from the leading edge of the plate.

[**Ans.** 4.43 mm, 6.27 mm]

5.9 A passive solar house is heated by 50 glass containers each containing 2000 liters of water heated to 80°C, during a day by absorbing solar energy. The house has to be maintained at 22°C, all times during a winter night, for 10 hours. During this period, the average heat loss from the house to outdoor is 50,000 kJ/h. A thermostat-controlled 15 kW back-up electrical resistance heater turns on whenever necessary, to keep the house at 22°C. (a) How long does the electrical heating system need to run that night? (b) How long would the electrical heater has to run that night, if the house had no solar heating? Assume that the mass of 1 liter of water is 1 kg, and the average specific heat of water is 4.2 kJ/(kg °C).

[**Ans.** (a) 14.2 minutes, (b) 9.26 hours]

5.10 An electric iron box is left on board, with its base exposed to the ambient atmospheric air at 20°C. The convection heat transfer coefficient between the iron box base and the surrounding air is 35 W/(m$^2$ °C). If the base has an emissivity of 0.6 and a surface area of 0.02 m$^2$, determine the heat supply required to keep the base at a steady temperature of 674°C.

[**Ans.** 1000 W]

5.11 An electric wire of diameter 4 mm and length 1.5 m extends across a room which is maintained at a uniform temperature of 25°C. Heat is generated in the wire as a result of resistance heating, and the surface temperature of the wire is measured to be 230°C, in steady operation. Also, the voltage drop and electric current through the wire are measured to be 180 volts and 2 amperes, respectively. Neglecting heat transfer by radiation, determine the convection coefficient for the heat transfer between the wire surface and the room air.

[**Ans.** 93.2 W/(m$^2$ °C)]

5.12 A 12 kg of liquid water, initially at 10°C, is to be heated to 90°C in a copper vessel, by burning wood of specific heat 0.65 kJ/(kg °C) whose combustion liberates heat steadily at 200°C. The vessel weighs 1.2 kg and has an average specific heat of 385 J/(kg °C). Neglecting the heat loss from the vessel, determine how long it will take to heat the water. Take specific heat of water as 4.18 kJ/(kg °C), and the mass of wood burned is 50 grams per second.

[**Ans.** 10.38 minutes]

5.13 Hot air at 100°C blows over a flat plate surface area of 12 m$^2$, maintained at a temperature of 25°C. If the rate of heat transfer from air to the plate is 49.5 kW, determine the average convective heat transfer coefficient.

[**Ans.** 55 W/(m$^2$ °C)]

5.14 A 900 kg of water at 21°C is cooled, by adding 70 kg of ice at −4°C. If the melting temperature and heat of fusion of ice at atmospheric pressure are 0°C and 333.7 kJ/kg, taking specific heat of water as 4.2 kJ/(kg °C), determine the final equilibrium temperature of the mixture.

[**Ans.** 13.46°C]

5.15 The temperature of a gas stream at 100°C is to be measured by a thermocouple whose junction can be approximated as a sphere of diameter 1 mm. The properties of the junction are $\rho = 8500$ kg/m$^3$, $k = 35$ W/(m °C), and $c = 320$ J/(kg °C), and the convection heat transfer coefficient between the junction and the gas is $h = 210$ W/(m$^2$ °C). Neglecting the radiation losses from the junction, determine how long it will take for the thermocouple to read 99 percent of the initial temperature difference, if the thermocouple is initially at 20°C.

[**Ans.** 2.9 s]

5.16 A house is maintained at 22°C at all times. The walls of the house are insulated with a material having an $L/k$ value or a thermal resistance of 3.38 (m$^2$ °C)/W. During a winter night, the outside air temperature is 4°C and wind at 50 km/h is blowing parallel to a 3-m-high and 8-m-long wall of the house. If the heat transfer coefficient on the interior surface of the wall is

8 W/(m$^2$ °C), determine the rate of heat loss from that wall of the house. Assume the radiation heat transfer to be negligible.

[**Ans.** [122.28 W]

5.17 A circuit board of length 180 mm, height 120 mm, and thickness 3 mm houses 80 closely spaced logic chips, each dissipating 0.04 W, from one side. The board is impregnated with copper fillings and has an effective thermal conductivity of 16 W/(m °C) across the circuit board. Heat is dissipated from the backside of the board over which, air at 40°C is forced to flow at 6.7 m/s using a fan. Determine the temperatures of the front and back surfaces of the circuit board. Take Prandtl number at 40°C as 0.711.

[**Ans.** 46.25°C on the back surface, 46.53°C on the front surface]

5.18 A thermocouple junction of spherical shape is used to measure temperature in a gas stream. The convection heat transfer coefficient between the junction surface and the gas is 400 W/(m$^2$ K). The thermophysical properties of the junction material are $\rho$ = 8500 kg/m$^2$, $c$ = 400 J/(kg K) and $k$ = 20 W/(m K). (a) Determine the thermocouple junction diameter required to have a response time of 1 second. (b) If the junction is at 25°C and is placed in a gas stream of 200°C, how long will it take for the junction to reach 199°C?

[**Ans.** (a) 0.706 mm, (b) 5.17 s]

5.19 Air at 6 kPa and 300°C flows over a 0.5-m-long plate with a velocity of 10 m/s. Determine the cooling rate per unit width of the plate required to maintain the plate surface at 27°C. Assume $k$ = 0.0364 W/(m K) and Pr = 0.687, for air.

[**Ans.** 458.6 W]

5.20 Air at 1 atm and 120°C flows with a velocity of 3 m/s, over a 3-m-long flat plate. Neglecting the radiation effects, determine the total drag per unit width of the plate.

[**Ans.** 0.054 N]

5.21 Air at 90 kPa and 22°C flows over the top surface of a flat plate of 1-m-width and 5-m-length, at 10 m/s. The plate is maintained at a constant temperature of 130°C. Taking $k$ = 0.03 W/(m °C) and Pr = 0.7, for air, determine the rate of heat transfer from the plate, if the air flows parallel to (a) 5-m-length side, and (b) 1-m-width side.

[**Ans.** (a) 9973.8 W, (b) 6291 W]

5.22 A thin square silicon chip of side 10 mm is fixed on an aluminum plate of same cross-section and 8 mm thickness, with epoxy layer of thickness 0.02 mm. The top surface of the silicon chip, and the bottom surface of the aluminum plate are cooled by air at 25°C and convection heat transfer coefficient $h$ = 100 W/(m$^2$ K). If the chip dissipates $10^4$ W/m$^2$, under steady-state condition, determine the temperature at the chip surface. Assume $k$ = 238 W/(m K) for aluminum, and the conduction to be one-dimensional and the resistance to heat transfer through the epoxy layer is $0.9 \times 10^{-4}$ (m$^2$ K)/W.

[**Ans.** 75.31°C]

5.23 Two large plates, maintained at 20°C, are separated by an oil film of thickness 2 mm. The upper plate moves at a constant speed of 2 m/s, and the bottom plate is held stationary. (a) Obtain the expressions for the velocity and temperature variation across the oil film. (b) Find the maximum temperature in the oil film and the heat flux from the oil to the moving and stationary plates.

[**Ans.** (a) $T = T_0 + \dfrac{\mu U_\infty^2}{2k}\left(\dfrac{y}{h} - \dfrac{y^2}{h^2}\right)$, (b) 22.75°C, 800 W/m$^2$]

5.24 The cooling of fruits and vegetables, in refrigerated air at temperature 4°C, pressure 1 atm and velocity 0.3 m/s, is modeled as a combination of convection, radiation and evaporation of air. The heat transfer coefficient for the combination of convection, radiation and evaporation of air, in the velocity range $0.1 < V < 0.3$ m/s is, determined experimentally and expressed as

$$h = 5\,k\,\frac{\mathrm{Re}^{\frac{1}{3}}}{D}$$

where $k$ is the conduction coefficient of air and $D$ is the characteristic length. (a) Calculate the initial rate of heat transfer from the surface of the fruit, whose area is equivalent to the surface area of a 5-mm-diameter sphere, if the initial temperature of the fruit is 20°C and the thermal conductivity is 0.8 W/(m K). (b) Determine the initial temperature gradient at the surface of the fruit and the Nusselt number corresponding to that.

[**Ans.** (a) 103.75 W, (b) −16520.70 K/m, 51.63]

5.25 The local surface heat transfer coefficient $h_x$ for a rough flat plate is given as

$$h_x(x) = ax^{-0.1}$$

where $a$ is a coefficient with units W/(m$^{1.9}$ K), and $x$ is the distance from the plate leading edge. Develop an expression for the ratio of the average heat transfer coefficient $h$ over the plate length $x$ to the local heat transfer coefficient $h_x$, at $x$.

[**Ans.** $\dfrac{h_{x,\mathrm{ave}}}{h_x(x)} = 1.11$]

5.26 Considering the control volume shown in Figure P5.26, for the steady-state condition with $y$-component of velocity $v = 0$, temperature $T = T(y)$, and density $\rho =$ constant, (a) show that, $u = u(y)$ if $v = 0$ everywhere. (b) Derive the $x$-momentum equation and simplify it with appropriate assumptions. (c) Derive the energy equation and simplify it.

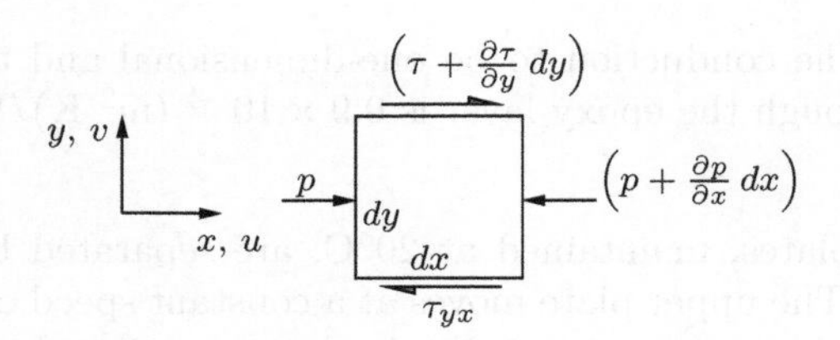


**Figure P5.26**
Control volume.

[**Ans.** (b) $u\frac{\partial u}{\partial x} = -\frac{1}{\rho}\frac{\partial p}{\partial x} + \mu\frac{\partial^2 u}{\partial y^2}$, (c) $\rho\, c\, u\frac{\partial T}{\partial x} + u\,\frac{\partial p}{\partial x} = k\,\frac{\partial T^2}{\partial x^2}\,\mu\left(\frac{\partial u}{\partial y}\right)^2 + \mu\, u\,\frac{\partial^2 u}{\partial y^2}$]

5.27 Air at 20°C and 1 atm, flows over a flat plate at 2 m/s. The thermal boundary layer thickness over the plate is 10 percent larger than the velocity boundary layer thickness. Determine the ratio of thermal boundary layer thickness to velocity boundary layer thickness, if the fluid is water under the same flow conditions.

[**Ans.** 1.69]

5.28 Air stream at 20°C, 1 atm and 2 m/s is used to cool an electronic circuit board. A 4 mm × 4 mm chip is located at 100 mm from the leading edge of the board. The convection heat transfer coefficient is modeled (assuming the air properties at 20°C) with the empirical correlation,

$$\mathrm{Nu}_x = 0.04\,\mathrm{Re}_x^{0.8}\,\mathrm{Pr}^{\frac{1}{3}}$$

Estimate the surface temperature of the chip if it is dissipating 26 mW.

[**Ans.** 107.3°C]

5.29 Air at 1 atm and 300 K flows over a flat plate of length 0.75 m and width 0.5 m. If the flow velocity is 1 m/s, determine the drag force acting on the plate.

[**Ans.** 0.00268 N]

5.30 Water at 20°C flows over a square plate of side 2 m. If the flow velocity is 2.2 m/s, determine the drag force over the plate, assuming the transition Reynolds number as $5 \times 10^5$.

[**Ans.** 59.3 N]

5.31 Air at 1 atm and 2°C flows over a flat plate of length 1.5 m and width 1 m, with a velocity of 20 m/s. If the plate is maintained at a temperature of 50°C, determine the average heat transfer coefficient and the heat transfer rate from the plate to the air.

[**Ans.** 47.42 W/(m$^2$ °C), 6828.5 W]

5.32 A horizontal pipe of diameter 300 mm, maintained at a constant surface temperature of 250°C is passing through a room with air at 1 atm and 15°C. Calculate the free convection heat loss per unit length of the pipe.

[**Ans.** 1470.64 W]

5.33 Air at a pressure of 1 atm is contained in the vertical space between two square plates of side 0.5 m, separated by a distance of 15 mm. When the plates are maintained at constant temperatures of 100°C and 40°C, determine the free convection heat transfer across the air layer.

[**Ans.** 40.2 W]

5.34 A vertical plate of length 2 m and width 0.5 m is maintained at a uniform temperature of 400 K. The left surface of the plate is insulated, and the right face is exposed to still air at 1 atm and 300 K. Determine the rate of heat loss from the plate.

[**Ans.** 573 W]

5.35 A 1-m-long horizontal cylinder of diameter 50 mm, maintained at a uniform temperature of 140°C is exposed to still air at 1 atm and 300 K. Determine the convective heat transfer coefficient associated with the heat transfer from cylinder to air.

[**Ans.** 5.995 W/(m$^2$ °C)]

# References

1. Schlichting, H., *Boundary Layer Theory*, McGraw-Hill Book Company, New York, 1968.
2. Rathakrishnan, E., *Fluid Mechanics—an Introduction*, 2nd ed., Prentice Hall of India, New Delhi, 2007.
3. Eckert, E. R. G. and Soehngen, E., "Interferometric studies on the stability and transition to turbulence of a free convection boundary layer," *Proc. Gen. Discuss. Heat Transfer ASME-IME*, London, 1951.
4. McAdams, W. H., *Heat Transmission*, 3rd ed., McGraw-Hill, New York, 1954.
5. Churchill, S. W. and Chu, H. H. S., "Correlating equations for laminar and turbulent free convection from a vertical plate," *Int. J. Heat Mass Transfer*, Vol. 16, p. 1025, 1953.
6. Vliet, G. C., "Net convection local heat transfer on constant heat flux inclined surfaces," *J. Heat Transfer*, Vol. 91C, pp. 511-516, 1969.
7. Vliet, G. C. and Liu, C. K., "An experimental study of natural convection boundary layers," *J. Heat Transfer*, Vol. 91C, pp. 5117-531, 1969.
8. Fujii, T. and Imura, H., "Natural heat transfer from a plate with arbitrary inclination", *Int. J. Heat and Mass Transfer*, Vol. 15, p. 755, 1972.

9. Vliet, G. C. and Ross, D. C., "Turbulent natural convection on upward and downward facing inclined constant heat flux surfaces," *ASME Pap.* 74-WA/HT-32.
10. Al-Arabi, M. and Salman, Y. K., "Laminar natural convection heat transfer from an inclined cylinder," *Int. J. Heat and Mass Transfer*, Vol. 23, pp. 45-51, 1980.
11. Yuge, T., "Experiments on heat transfer from spheres including natural and forced convections," *J. Heat Transfer*, Vol. 82C, pp. 214-220, 1960.
12. Amato, W. S. and Tien, C., "Free convection heat transfer from isothermal spheres in water," *Int. J. Heat and Mass Transfer*, Vol. 15, pp. 327-339, 1972.
13. Necati Özişik, M., *Heat Transfer a Basic Approach*, McGraw-Hill Book Company, New York, 1985.
14. El Sherbiny, S. M., Raithby, G. D., and Holland, K. G. T., "Heat transfer by natural convection across vertical and inclined air layers," *J. Heat Transfer*, Vol. 104C, pp. 96-102, 1982.
15. Hollands, K. G. T., Unny, T. E., Raithby, G. D., and Konicek, L., "Free convection heat transfer across inclined air layers," *J. Heat Transfer*, Vol. 98C, pp. 189-193, 1976.
16. Raithby, G. D. and Hollands, K. G. T., "A general method of obtaining approximate solution to laminar and turbulent free convection problems," in T.F Irvine and J.P Hartnett (eds.), *Advances in Heat Transfer*, Vol. 11, Academic, New York, pp. 265-315, 1975.
17. Houghton, E. L., and Carruthers, N. B., *Aerodynamics for Engineering Students*, Hodder Arnold, 3rd ed., 1982.
18. Whitaker, S., "Forced convection heat transfer calculations for flow in pipes, past flat plates, single cylinder, and for flow in packed beds and tube bundles," *AIChE J.*, Vol. 18, pp. 361-371, 1972.
19. Churchill, S. W. and Berstein, M., "A correlating equation for forced convection from gases and liquids to a circular cylinder in cross flow," *J. Heat Transfer*, Vol. 99, pp. 300-306, 1977.
20. Nakai, S. and Okazaki, T., "Heat transfer from a horizontal circular wire at small Reynolds and Grashof numbers – 1. Pure convection," *Int. J. Heat and Mass Transfer*, Vol. 18, p. 387, 1975.
21. Jakob, M., *Heat Transfer, Vol. 1*, Wiley, New York, 1949.
22. Rathakrishnan, E., *Instrumentation, Measurements, and Experiments in Fluids*, CRC/Taylor & Francis Group, Boca Raton, Florida, 2007.
23. Grimison, E. D., "Correlation and utilization of new data on flow resistance and heat transfer for cross flow of gases over tube banks," *Trans. ASME*, Vol. 59, pp. 583-594, 1937.
24. Kays, W. M., London, A. L., and Lo, R. K., "Heat transfer and friction characteristics for gas flow normal to tube banks - using a transient test technique," *Trans. ASME*, Vol. 76, p.387, 1954.
25. Zukauskas, A., "Heat transfer from tubes in cross-flow," *Advances in Heat Transfer*, Vol. 8, pp. 93-160, 1972.

26. Hoe, R. J., Dropkin, D. and Dwyer, O. E., "Heat transfer rates to cross flowing mercury in a staggered tube bank, I," *Trans. ASME*, Vol. 79, pp. 899-908, 1957.
27. Richards, C. L., Dwyer, O. E. and Dropkin, D., "Heat transfer rates to cross flowing mercury in a staggered tube bank, II," *ASME-AIChE Heat Transfer Conf.* Paper 57-HT-11, 1957.
28. Eckert, E. R. G., "Engineering relations for heat transfer and friction in high-velocity laminar and turbulent boundary layer flow over surface with constant pressure and temperature," *Trans. ASME*, Vol. 78, pp. 1273-1284, 1956.
29. Holman, J. P., *Heat Transfer*, 5th ed., McGraw-Hill Book Co., New York, 1981.

# Chapter 6

# Radiation Heat Transfer

## 6.1 Introduction

We have seen that in contrast to the mechanism of conduction and convection, where energy transfer through a material medium is involved, heat may also be transferred into regions where perfect vacuum exists. The mechanism in this case is *electromagnetic radiation*, which is propagated as a result of a temperature difference. This mode of energy transfer through electromagnetic radiation is called *thermal radiation*. Thus, thermal radiation is that electromagnetic radiation emitted by a body as a result of its temperature.

## 6.2 Radiation Mechanism

There are many types of electromagnetic radiation, but the thermal radiation is only one. Irrespective of its type, a radiation is propagated at the speed of light $c$ ($3 \times 10^{10}$ cm/s), given by

$$c = \lambda\nu$$

where $\lambda$ is the wavelength and $\nu$ is the frequency. The wavelength is usually expressed in centimeters or angstroms (1 Å$= 10^{-8}$ cm) or micrometers. Thermal radiation lies in the range of wavelength from about 0.1 to 100 $\mu$m, as shown in Figure 6.1, which shows a portion of electromagnetic spectrum.

Thermal radiation propagates in the form of discrete quanta with each quantum having an energy of

$$E = h\nu$$

where $h$ is the Planck's constant, and is equal to $6.625 \times 10^{-34}$ J-s.

To gain an insight into the process of radiation propagation, let us consider each quantum as a particle having mass, momentum and energy, as in the case of the molecules of a gas. Thus, the radiation might be regarded as a *photon*

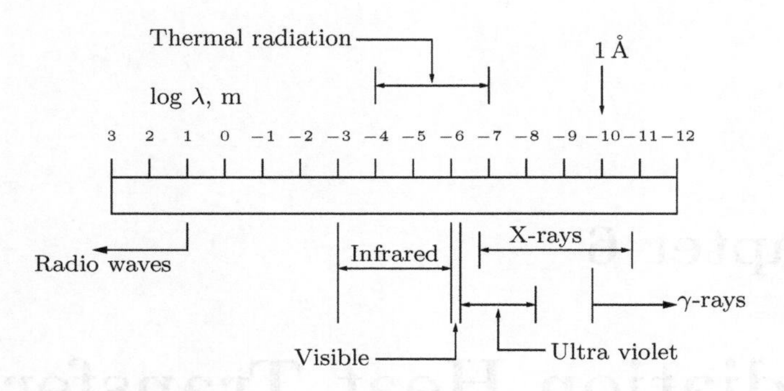


**Figure 6.1**
Electromagnetic spectrum.

*gas* which may flow from one place to another. The expressions for mass, momentum and energy of the particles could be expressed as

$$\text{Mass } m = \frac{h\nu}{c^2}$$

$$\text{Momentum} = \frac{h\nu}{c^2}c = \frac{h\nu}{c}$$

$$\text{Energy } E = mc^2 = h\nu$$

For radiation in such a gas, an expression for the energy density of a radiation $u_\lambda$ per unit volume and per unit wavelength can be expressed, using the principle of quantum statistical thermodynamics [1], as

$$u_\lambda = \frac{8\pi hc}{\lambda^5\left(e^{(hc)/(\lambda kt)} - 1\right)}$$

where $k$ is the Boltzmann constant ($1.38066 \times 10^{-23}$ J/(molecule K)).

Integrating the energy density over all wavelengths, the energy emitted is found to be proportional to the fourth power of absolute temperature.

$$\boxed{E_b = \sigma T^4} \tag{6.1}$$

Equation (6.1) is called *Stefan-Boltzmann law*, $E_b$ is the energy radiated per unit time per unit area, by an ideal radiator, and $\sigma$ is the Stefan-Boltzmann constant, equal to $5.669 \times 10^{-8}$ W/(m$^2$ K$^4$). The subscript $b$ denotes that this is the radiation from a blackbody. This is called *blackbody radiation*, because materials which obey this law would appear black to the eye, owing to the fact that they do not reflect any radiation. Thus, a *blackbody* is that which absorbs all radiation incident on it. The energy $E_b$, radiated by a blackbody, is termed the *emissive power* of a blackbody.

An ideal thermal radiator, also called as *blackbody* will emit energy at a rate proportional to the product of the fourth power of the absolute temperature $(T)$ and its surface area $(A)$. Thus,

$$\dot{Q}_{\text{emitted}} = \sigma A T^4 \tag{6.2}$$

where the proportionality constant $\sigma$ is called the Stefan-Boltzmann constant, and the equation is called the Stefan-Boltzmann law for thermal radiation. This law is valid only for thermal radiation emitted by blackbodies.

According to Stefan-Boltzmann law, the net radiation exchange between two surfaces will be

$$\frac{\dot{Q}_{\text{net exchange}}}{A} \propto \left(T_1^4 - T_2^4\right) \tag{6.2a}$$

We saw that, the blackbody is that which radiates energy according to Stefan-Boltzmann law. Any body whose surface is coated black, such as a piece of metal covered with carbon black can be regarded as a blackbody. Other types of surfaces, like polished or painted metal surface, do not radiate as much energy as the blackbody. However, the total radiation emitted by these surfaces still follows the Stefan-Boltzmann law. Such imperfect black surfaces are termed *gray surfaces.* To account for the gray nature of such surfaces, a factor which relates the radiation of the gray surface to that of an ideal black surface is used in thermal radiation analysis. Such a factor is called *emissivity* $\epsilon$.

## 6.3 Radiation Parameters

The radiant energy leaving a surface varies with direction. Therefore, measures of radiation from a surface are specified with the direction as an important parameter. One of the prime parameters in radiation heat transfer analysis is the *intensity of radiation.* Consider the elemental area $dA_s$ radiating thermal energy, as shown in Figure 6.2.

The *intensity of radiation* $I$, defined as the rate of energy leaving a surface in a given direction per unit projected area (area normal to the direction) per unit solid angle, is given by

$$I(\beta, \theta) = \frac{d\dot{Q}(\beta, \theta)}{dA_p\, d\Omega} \tag{6.3}$$

where $\dot{Q}(\beta, \theta)$ is the rate at which energy leaves the surface $dA_s$, within the solid angle $d\Omega$ and in the direction $(\beta, \theta)$, $dA_p = dA_s \cos\beta$ is the projection of elemental area in the direction normal to $n_2$, $d\Omega = dA_s/R^2$ is the elemental solid angle, $\beta$ is angle measured from normal $n_1$ to the surface $dA_s$ to the

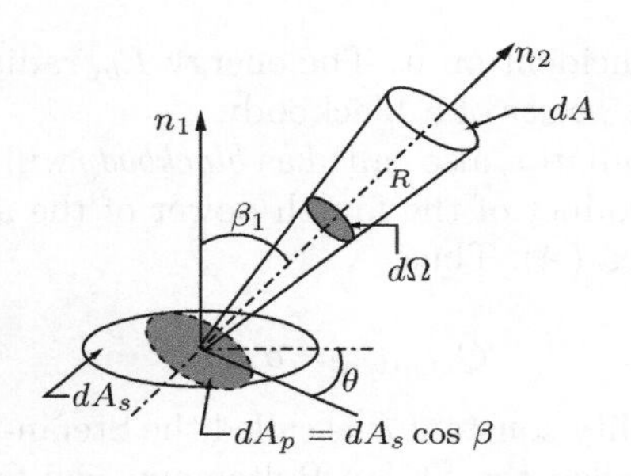


**Figure 6.2**
Radiation from a surface.

direction $n_2$ ($0 \leq \beta \leq \pi/2$), and $\theta$ is the azimuthal angle on the elemental area $dA_s$ ($0 \leq \theta \leq 2\pi$).

The *brightness of a surface from a given direction of view depends on the radiation emitted per unit projected area in that direction.* A surface which is equally bright in all directions is called a *diffuse surface.* That is, the intensity of radiation is independent of direction for such surfaces.

The *emissive power* $e$ of a surface is defined as the ratio of energy emission (in all directions) from the surface per unit surface area, that is,

$$e\,(T_s) = \frac{d\,\dot{Q}_e\,(T_s)}{d\,A_s} \tag{6.4}$$

where $\dot{Q}_e\,(T_s)$ is the rate of energy emitted by the surface. This depends on the surface temperature $T_s$.

The total energy $\dot{Q}$ leaving a surface, given by Equation (6.3), includes the energy emitted $\dot{Q}_e$, the energy reflected $\dot{Q}_r$, and the energy transmitted $\dot{Q}_t$. Therefore, using Equations (6.3) and (6.4), we can write

$$e\,(T_s) = \int_\Omega I_e \cos\beta\, d\Omega \tag{6.5}$$

where $I_e$ is the intensity of emission from the surface.

The relation between the emissive power $e\,(T_s)$ and $I_e$ given by Equation (6.5) can be casted in a compact form as follows.

The solid angle $d\Omega$ in the direction $(\beta, \theta)$ is

$$d\Omega = \sin\beta\, d\beta\, d\theta$$

Substituting this into Equation (6.5), we have

$$e\,(T_s) = \int_\theta \int_\beta I_e \cos\beta \sin\beta\, d\beta\, d\theta$$

Integrating over $0 \leq \theta \leq 2\pi$ and $0 \leq \beta \leq \frac{\pi}{2}$, we get

$$\begin{aligned} e(T_s) &= I_e \Big[\theta\Big]_0^{2\pi} \left[\frac{1}{2}\sin^2\beta\right]_0^{\frac{\pi}{2}} \\ &= I_e \times 2\pi \times \frac{1}{2} \\ &= \boxed{\pi I_e} \end{aligned}$$

A *blackbody* is defined as a surface which absorbs all radiation incident on it. With this definition it is possible to show that, of all surfaces at a given temperature, the blackbody emits the maximum possible radiation and this radiation is diffuse.

By Stefan-Boltzmann law, the emissive power of a blackbody is given by

$$e_b(T_s) = n^2 \sigma T_s^4 \tag{6.6}$$

where $e_b$ is emissive power of the blackbody at temperature $T_s$ $\left(\mathrm{W/m^2}\right)$, $\sigma$ is Stefan-Boltzmann constant, $5.669 \times 10^{-8}$ $\mathrm{W/(m^2\ K^4)}$, and $n$ is the refractive index of the medium bounding the surface, for gaseous medium like air $n \approx 1.0$.

## 6.3.1 Properties of Surfaces

When radiant energy strikes a material surface, part of the radiation is *reflected*, part is *absorbed*, and part is *transmitted*, as shown in Figure 6.3.

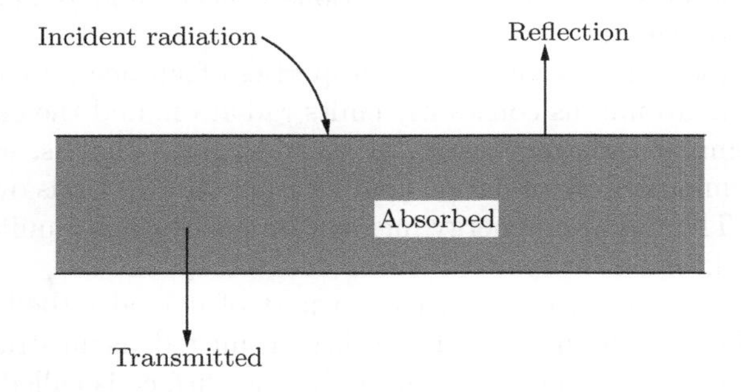


**Figure 6.3**
Radiant energy striking a material surface.

We define the *reflectivity* $\rho$ as the fraction reflected, the *absorptivity* $\alpha$ as the fraction absorbed, and the *transmissivity* $\tau$ as the fraction transmitted. Thus,

$$\boxed{\rho + \alpha + \tau = 1} \tag{6.7}$$

Most solid bodies do not transmit radiation, so that for many applied problems the transmissivity may be taken as zero. Therefore,

$$\rho + \alpha = 1$$

Two types of reflection phenomena are usually encountered when radiation strikes a surface. If the angle of incidence is equal to the angle of reflection, the reflection is called *specular* reflection. When the incident beam is distributed uniformly in all directions after reflection, the reflection is called *diffuse* reflection. The specular and diffuse reflections are illustrated in Figure 6.4.

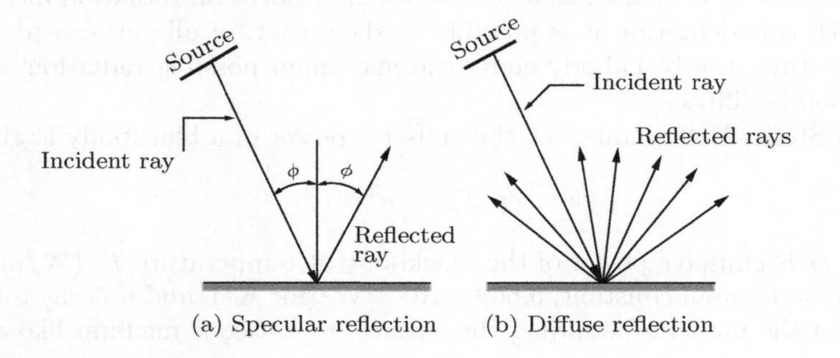


**Figure 6.4**
Specular and diffuse reflections.

The influence of surface roughness on thermal-radiation properties ($\rho$, $\alpha$, $\tau$) of materials is a matter of serious concern and remains as a subject of active research.

From the above discussions on the properties of surfaces, it can be inferred that everything around us constantly emits radiation, and the emissivity represents the emission characteristics of these bodies. That is, every body is constantly bombarded by radiation coming from all directions over a range of wavelengths. The radiation energy incident on a surface per unit surface area per unit time is called *irradiation* and is denoted by $G$.

When a radiation strikes a surface, a part of it is absorbed, part of it is reflected, and the remaining part, if any, is transmitted, as illustrated in Figure 6.3. The fraction of irradiation absorbed by the surface is called absorptivity $\alpha$, the fraction reflected is termed reflectivity $\rho$, and the fraction transmitted is called transmissivity $\tau$. That is,

$$\alpha = \frac{\text{absorbed radiation}}{\text{incident radiation}} = \frac{G_{\text{absorbed}}}{G}$$

$$\rho = \frac{\text{reflected radiation}}{\text{incident radiation}} = \frac{G_{\text{reflected}}}{G}$$

$$\tau = \frac{\text{transmitted radiation}}{\text{incident radiation}} = \frac{G_{\text{transmitted}}}{G}$$

But, in accordance with the first law of thermodynamics, the sum of absorbed, reflected, and transmitted radiation energy must be equal to the incident radiation, that is,

$$G_{\text{absorbed}} + G_{\text{reflected}} + G_{\text{transmitted}} = G$$

Diving each term by $G$, we get Equation (6.7).

The above definitions are for *total hemispherical* properties, since $G$ represents the radiation energy incident on the surface from all directions and all wavelengths. Thus, $\alpha$, $\rho$, and $\tau$ are the *average* properties of the medium for all directions and all wavelengths. However, these properties can be defined for a specific wavelength and direction. For example, the *spectral* absorptivity, reflectivity, and transmissivity of a surface can be defined as

$$\alpha_\lambda = \frac{G_{\lambda\text{-absorbed}}}{G}$$

$$\rho_\lambda = \frac{G_{\lambda\text{-reflected}}}{G}$$

$$\tau_\lambda = \frac{G_{\lambda\text{-transmitted}}}{G}$$

where $G_\lambda$ is the radiation energy incident at the wavelength $\lambda$, and $G_{\lambda\text{-absorbed}}$, $G_{\lambda\text{-reflected}}$, and $G_{\lambda\text{-transmitted}}$ are the absorbed, reflected, and transmitted portions of it, respectively. Similar definitions can be given for the *directional* properties; $\alpha_\theta, \rho_\theta$ and $\tau_\theta$, in the direction $\theta$, by replacing all occurrences of the subscript $\lambda$ in the above equations of $\alpha_\lambda$, $\rho_\lambda$ and $\tau_\lambda$ with $\theta$.

The average $\alpha$, $\rho$, and $\tau$ of a surface can also be determined in terms of their spectral counterparts as

$$\alpha = \frac{\int_0^\infty \alpha_\lambda G_\lambda \, d\lambda}{\int_0^\infty G_\lambda \, d\lambda}$$

$$\rho = \frac{\int_0^\infty \rho_\lambda G_\lambda \, d\lambda}{\int_0^\infty G_\lambda \, d\lambda}$$

$$\tau = \frac{\int_0^\infty \tau_\lambda G_\lambda \, d\lambda}{\int_0^\infty G_\lambda \, d\lambda}$$

It is essential to note that, among $\alpha$, $\rho$, and $\tau$, the reflectivity $\rho$ alone is *bidirectional* in nature. That is, $\rho$ of a surface depends not only on the direction

**Table 6.1** Rays and their wavelength $\lambda$ range

| Type of ray | Wavelength $\lambda$ ($\mu$m) |
|---|---|
| Cosmic rays | $10^{-10}$ – $10^{-7}$ |
| $\gamma$-rays | $10^{-7}$ – $3 \times 10^{-4}$ |
| X-rays | $10^{-5}$ – $10^{-2}$ |
| Ultraviolet rays | $10^{-2}$ – $4 \times 10^{-1}$ |
| Visible rays | $4 \times 10^{-1}$ – $9 \times 10^{-1}$ |
| Infrared waves | $9 \times 10^{-1}$ – $10^{2}$ |
| Thermal radiation | $10^{-1}$ – $10^{2}$ |
| Microwaves | $10^{2}$ – $2 \times 10^{5}$ |
| Radio waves | $2 \times 10^{5}$ – $10^{10}$ |
| Electrical power waves | $> 10^{10}$ |

of the incident radiation but also on the direction of reflection. In practice, for simplicity, surfaces are assumed to reflect in a perfectly *specular* or *diffuse* manner. Reflection from smooth and polished surfaces approximates specular reflection. In radiation analysis, smoothness is defined relative to wavelength. A surface is regarded *smooth* if the height of the surface roughness is much smaller than the wavelength of the incident radiation.

Although all electromagnetic waves have the same general feature, waves of different wavelengths differ significantly in their behavior. The electromagnetic radiation encountered in practice covers a wide range of wavelengths from less than $10^{-10}$ $\mu$m, for cosmic rays, to more than $10^{10}$ $\mu$m, for electrical power waves. The electromagnetic spectrum also includes gamma rays, x-rays, ultraviolet radiation, visible light, infrared radiation, thermal radiation, microwaves and radio waves. The types of waves and their wavelength $\lambda$ range are listed in Table 6.1.

## Example 6.1

The coating on a plate is cured by exposing to an infrared lamp supplying 2 kW/m$^2$, by irradiation. The plate absorbs 80% of the incident energy and its emissivity is 0.5. The plate is exposed to a large surrounding, and air at 20°C with convection coefficient of $h = 15$ W/(m$^2$ K) flows over the plate surface. At a steady state, what will be the temperature at the surface of the coating?

## Solution

Given, irradiation $G = 2000$ W/m$^2$, $\alpha = 0.8$, $\epsilon = 0.5$, $T_\infty = 20 + 273 = 293$ K.

At steady state, the energy received by the coating should be transferred out to the air by convection and to the surrounding space by radiation. Thus,

$$\dot{q}_{\text{absorbed}} = \dot{q}_{\text{conv}} + \dot{q}_{\text{rad}}$$

$$\alpha\, G = h\,(T_s - T_\infty) + \sigma\,\epsilon\,\left(T_s^4 - T_\infty^4\right)$$

$$0.8 \times 2000 = 15 \times (T_s - 293) + (5.67 \times 10^{-8}) \times 0.5 \times \left(T_s^4 - 293^4\right)$$

$$1600 = 15\,T_s - 4395 + (2.835 \times 10^{-8})T_s^4 - 209$$

$$6204 = 15\,T_s + (2.835 \times 10^{-8})\,T_s^4$$

Solving this by trial and error, we get

$$T_s = 377\,\text{K} = \boxed{104°\text{C}}$$

The surface of the coating will be at 104°C at steady state.

# 6.4 The Greenhouse Effect

It can be noticed that, when a car is left under direct sunlight on a sunny day, the interior of the car gets much warmer than the outside environment. The reason for this is the spectral transmissivity curve of the window glasses and wind-screen, which resembles an arc, as shown in Figure 6.5.

The window glass of thickness encountered in practice transmits over 90 percent of radiation in the visible range, and is practically opaque to radiation in the larger-wavelength infrared region of the electromagnetic spectrum ($\lambda > 3\,\mu$m). Therefore, glass behaves as a transparent window in the wavelength range $0.3\;\mu\text{m} < \lambda < 3\,\mu$m, in which over 90 percent of solar radiation is emitted. On the other hand, the entire radiation emitted by surfaces at room temperature falls in the infrared region. Consequently, glass allows the solar radiation to enter but does not allow infrared radiation from the interior surfaces to leave. This causes a rise in the interior temperature, as a result of energy buildup in the car. This heating effect, due to the non-gray characteristic of the glass (or clear plastic) is known as the *greenhouse effect.*

On the earth, greenhouse effect is experienced on a larger scale. The earth's surface, which warms up during the day as a result of absorption of solar

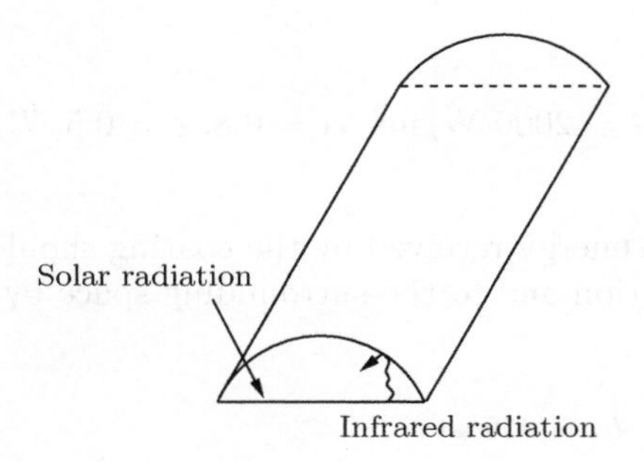


**Figure 6.5**
A greenhouse traps energy by allowing the solar radiation to come in but not allowing the infrared radiation to go out.

energy, cools down at night by radiating its energy into the space as infrared radiation. This is in accordance with Kirchhoff's identity [Equation (6.10)], which is valid for gray surfaces such as earth surface. The combustion products such as carbon-dioxide and water vapor in the atmosphere transmit bulk of the solar radiation but absorb the infrared radiation emitted by the surface of the earth. Thus, there is concern that energy trapped on earth ultimately will cause global warming leading to drastic changes in weather patterns.

In coastal areas with high humidity, there is no drastic change between day and night temperatures. This is because the humidity acts as a barrier on the path of the infrared radiation coming from the earth, causing retardation of the cooling process during night. In places like a desert where the sky is clear, there is a large difference between day and night temperatures because of the absence of any barrier to infrared radiation.

### 6.4.1 Blackbody Radiation

We saw that a body at a temperature above absolute zero emits radiation in all directions over a wide range of wavelengths. The amount of radiation energy emitted from a surface, at a given wavelength, depends on the material of the body, the smoothness of its surface and the surface temperature. Therefore, different bodies may emit different amounts of radiation per unit surface area, even when they are at the same temperature. Therefore, to identify the *maximum* amount of radiation that can be emitted by a surface at a given temperature, it is necessary to define an idealized body, termed *blackbody*, to serve as a standard, against which the radiative properties of real surfaces can be compared.

A *blackbody* is defined as a perfect emitter and absorber of radiation. At a specified temperature and wavelength, no surface can emit more energy than a blackbody. A blackbody absorbs *all* radiation incident on it, regardless of wavelength and direction. Also, a blackbody emits radiation energy uniformly in all directions. That is, a blackbody is a *diffuse* emitter.

The radiation energy emitted by a blackbody, per unit surface area, per unit time is expressed [Equation (6.1)] as

$$\boxed{E_b = \sigma T^4}$$

where $\sigma = 5.67 \times 10^{-8}$ W/(m$^2$ K$^4$) is the Stefan-Boltzmann constant. This relation is known as *Stefan-Boltzmann law*, and $E_b$ is called the *blackbody emissive power.*

The Stefan-Boltzmann law gives the *total* emissive power $E_b$ of a blackbody, which is the sum of the radiation emitted over all wavelengths. But the *spectral emissive power of a blackbody*, defined as the amount of radiation energy emitted by a blackbody at an absolute temperature $T$, per unit time, per unit surface area, and per unit wavelength about the wavelength $\lambda$ will be of interest. For example, the amount of radiation an incandescent light bulb emits in the visible wavelength spectrum will be of more interest than the total amount of radiation that the bulb emits.

The relation for the spectral blackbody emissive power $E_{b\lambda}$ was developed by Max Planck in 1901 in conjunction with his quantum theory. This relation is known as Planck's distribution law, and is expressed as

$$E_{b\lambda}(T) = \frac{c_1}{\lambda^5 \left[\exp\left(c_2/\lambda T\right) - 1\right]} \tag{6.8}$$

where $c_1 = 2\,\pi h c_0^2 = 3.741 \times 10^8\,(\text{W}\,\mu\text{m}^4)/\text{m}^2$ and $c_2 = hc_0/k = 1.439 \times 10^4\,(\mu\text{m} \times \text{K})$. The temperature $T$ is absolute temperature of the surface, $\lambda$ is the wavelength of the radiation emitted, $k = 1.3805 \times 10^{-23}$ J/K is the Boltzmann constant, and $c_0 = 2.998 \times 10^8$ m/s is the speed of light in vacuum. This relation is valid for a surface in a *vacuum* or a *gas*. For other media, it needs to be modified by replacing $c_1$ by $c_1/n^2$, where $n$ is the index of refraction of the medium.

## 6.4.2 Emissive Power

The *emissive power* $e$ of a body is defined as the energy emitted by the body per unit area and per unit time. The ratio of the emissive power of a body to that of a blackbody at the same temperature is defined as the *emissivity* $\epsilon$.

$$\boxed{\epsilon\,(T_s) = \frac{e\,(T_s)}{e_b\,(T_s)}} \tag{6.9}$$

But blackbody is a perfect emitter; therefore, the emissivity of a surface can vary only from 0 to 1.0.

It is important to note that the concept of blackbody is an *idealization*; that is, a perfect blackbody does not exist – all surfaces reflect radiation to some extent.

The properties namely, the reflectivity $\rho$, the absorptivity $\alpha$ and the transmissivity $\tau$ are strictly total properties since they represent ratios of total energies, that is, energy integrated over all wavelengths. These properties when defined for every wavelength are called *spectral properties.* If all the spectral properties of a surface are independent of wavelength, then such a surface is called a *gray surface.* In other words, for a gray surface the spectral and total properties are *identical.*

For the special case of a surface which is gray and diffuse, it can be shown that,

$$\boxed{\alpha = \epsilon\,(T_s)} \tag{6.10}$$

This relation is called *Kirchhoff's identity.*

For nonmetallic surfaces the assumptions *gray* and *diffuse* are reasonably valid, while for metallic surfaces such assumptions lead to erroneous results.

For an opaque gray-diffuse surface, Equations (6.3) and (6.5) can be combined to give

$$\boxed{\rho = 1 - \epsilon} \tag{6.11}$$

since $\tau = 0$.

## Example 6.2

A spherical ball of 10 cm diameter maintained at a constant temperature of 1100 K is suspended in air. Assuming the ball to closely approximate a blackbody, determine (a) the total blackbody emissive power, (b) the total amount of radiation emitted by the ball in 10 minutes, and (c) the spectral blackbody emissive power at a wavelength of 3 $\mu$m.

## Solution

Given, $d = 0.1$ m, $T = 1100$ K, $t = 600$ s, $\lambda = 3$ $\mu$m.

(a) The total blackbody emissive power is given by the Stefan-Boltzmann law, Equation (6.1), as

$$E_b = \sigma T^4 = (5.67 \times 10^{-8}) \times (1100^4)$$

$$= \boxed{83014.5\,\mathrm{W/m^2}}$$

(b) The total amount of radiation emitted from the ball in 10 minutes is given by the product of the blackbody emissive power obtained above and the surface area of the ball and the given time duration.

$$\begin{aligned} Q_{\text{rad}} &= E_b A t \\ &= E_b \times (\pi d^2) \times t \\ &= 83014.5 \times (\pi \times 0.1^2) \times 600 \\ &= 1564786.5\,\text{W-s} \\ &= \boxed{1564.8\,\text{kJ}} \end{aligned}$$

(c) The spectral blackbody emissive power at a wavelength of 3 $\mu$m can be determined from Planck's distribution law, Equation (6.8), as

$$\begin{aligned} E_{b\lambda} &= \frac{c_1}{\lambda^5 \left[\exp\left(\frac{c_2}{\lambda T}\right) - 1\right]} \\ &= \frac{3.741 \times 10^8}{3^5 \left[\exp\left(\frac{1.439 \times 10^4}{3 \times 1100}\right) - 1\right]} \\ &= \boxed{19914.8\,\text{W/(m}^2\,\mu\text{m)}} \end{aligned}$$

## 6.5 The View or Configuration Factor

In radiation heat transfer problems, we wish to obtain a general expression for the energy exchange between surfaces when they are maintained at different temperatures. The problem becomes essentially one of determining the amount of energy which leaves one surface and reaches the other, as shown in Figure 6.6.

Let the surfaces $dA_1$ and $dA_2$ be diffuse, gray and isothermal. The net energy leaving $dA_1$ and arriving at $dA_2$, given by Equation (6.3), is

$$d\dot{Q}_{1\to 2} = I_1 \cos\beta\, d\Omega\, dA_1 \tag{6.12}$$

where $I$ is the intensity of radiation leaving surface $dA_1$ and $d\Omega_1$ is the solid angle subtended by the surfaces $dA_1$ and $dA_2$. The total energy leaving from one side of $dA_1$ is given by

$$\dot{Q}_1 = \int_{2\pi} I_1 \cos\beta\, dA_1\, d\Omega \tag{6.13}$$

This total energy leaving $dA_1$ can be calculated by constructing an imaginary hemisphere of radius $r$ centered at $dA_1$, as shown in Figure 6.6. Noting that,

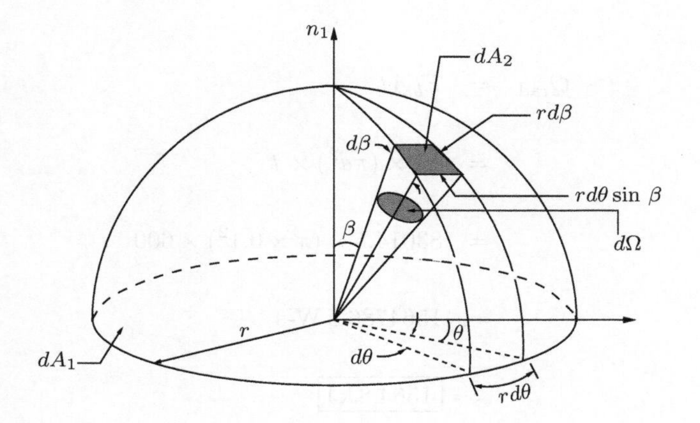


**Figure 6.6**
Radiation exchange between elemental surfaces of area $dA_1$ and $dA_2$.

an identified area on a sphere is always normal to the radius of the sphere at that area, we can express the solid angle $d\Omega$ as

$$\begin{aligned} d\,\Omega &= \frac{dA_2}{r^2} = \frac{(r\,d\beta)(r\,d\theta \sin\,\beta)}{r^2} \\ &= \sin\,\beta\,d\beta\,d\theta \end{aligned}$$

Since the surface $dA_1$ is diffuse, $I_1$ in Equation (6.13) can be taken out of the integration over the solid angle, to result in

$$\begin{aligned} \dot{Q}_1 &= I_1 dA_1 \int_0^{2\pi}\int_0^{\pi/2} \cos\,\beta \sin\,\beta\,d\beta\,d\theta \\ &= I_1 dA_1 \int_0^{\pi/2} \sin\,\beta\,d(\sin\,\beta) \int_0^{2\pi} d\theta \\ &= I_1 dA_1 \left[\frac{\sin^2\beta}{2}\right]_0^{\pi/2} \left[\theta\right]_0^{2\pi} \\ &= I_1 dA_1 \left[\frac{1}{2}\right] 2\pi \end{aligned}$$

that is,

$$\dot{Q}_1 = \pi\,I_1\,dA_1 \tag{6.14}$$

From Equations (6.12) and (6.14), we can express the fraction of energy leaving surface $dA_1$ and reaching surface $dA_2$, called *view factor*, as

$$F_{1\to 2} = \frac{\dot{Q}_{1\to 2}}{\dot{Q}_1} = \frac{\cos\beta\, d\Omega}{\pi} \tag{6.15}$$

But,

$$d\Omega = \frac{dA_2 \cos\theta}{r^2}$$

Therefore, we have

$$\boxed{F_{1\to 2} = \frac{\cos\beta\,\cos\theta\, dA_2}{\pi\, r^2}} \tag{6.16}$$

$F_{1\to 2}$ is called the *configuration* or *view factor* between the infinitesimal areas considered. Multiplying Equation (6.16) by the area $dA_1$, we get

$$\boxed{dA_1\, F_{1\to 2} = \frac{\cos\beta\,\cos\theta\, dA_1\, dA_2}{\pi\, r^2}} \tag{6.17}$$

From the symmetry of the right-hand side of Equation (6.17), we can write

$$\boxed{dA_1\, F_{1\to 2} = dA_2\, F_{2\to 1}} \tag{6.18}$$

This is called *reciprocity relation.*

In reciprocity relation, $F_{1\to 2}$ is the fraction of energy leaving surface 1 and reaching surface 2, and $F_{2\to 1}$ is the fraction of energy leaving surface 2 and reaching surface 1. Therefore, the radiation shape factor, in general, may be defined as $F_{m\to n}$, the fraction of energy leaving surface $m$ and reaching surface $n$. The radiation shape factor is also called a *view factor*, *angle factor*, and *configuration factor.*

The amount of energy leaving surface 1 and arriving at surface 2 is

$$E_{b1} A_1 F_{1\to 2}$$

and the amount of energy leaving surface 2 and arriving at surface 1 is

$$E_{b2} A_2 F_{2\to 1}$$

If both surfaces are at the same temperature, there can be no heat exchange between them, that is, $Q_{1\to 2} = 0$. Also, $E_{b1} = E_{b2}$, therefore,

$$A_1 F_{1\to 2} = A_2 F_{2\to 1}$$

The net heat exchange becomes

$$\begin{aligned} Q_{1\to 2} &= A_1 F_{1\to 2}\,(E_{b1} - E_{b2}) \\ &= A_2 F_{2\to 1}\,(E_{b1} - E_{b2}) \end{aligned}$$

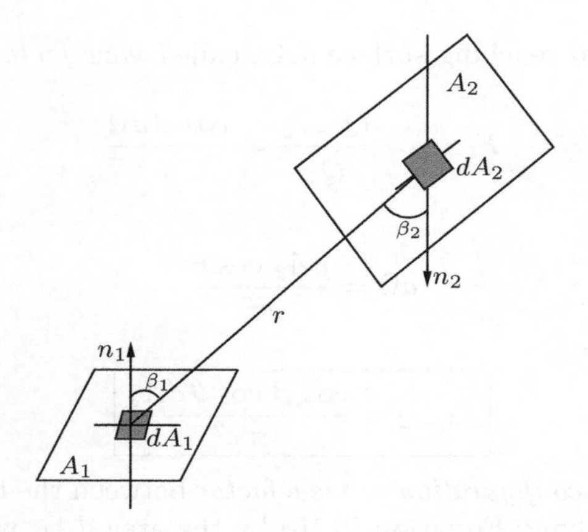


**Figure 6.7**
Energy exchange between two areas.

To determine the energy exchange between two simple surfaces, let us consider two surfaces of elemental area $dA_1$ and $dA_2$, as shown in Figure 6.7.

The elemental area on surface $A_2$, normal to direction $r$, $dA_n$ is given by

$$dA_n = \cos\beta_2 dA_2$$

Therefore, the amount of energy leaving $dA_1$ and arriving at $dA_2$ is given by

$$d\dot{Q}_{1\to 2} = E_{b1}\cos\beta_1\cos\beta_2\frac{dA_1 dA_2}{\pi r^2}$$

Similarly, the energy leaving $dA_2$ and arriving at $dA_1$ is given by

$$d\dot{Q}_{2\to 1} = E_{b2}\cos\beta_2\cos\beta_1\frac{dA_2 dA_1}{\pi r^2}$$

Thus, the net energy exchange between surfaces 1 and 2 becomes,

$$\boxed{d\dot{Q}_{\text{net},2\to 1} = (E_{b1} - E_{b2})\int_{A_2}\int_{A_1}\cos\beta_1\cos\beta_2\frac{dA_1 dA_2}{\pi r^2}} \quad (6.19)$$

This integration gives either $A_1F_{1\to 2}$ or $A_2F_{2\to 1}$. To evaluate this integral, the specific geometries of surfaces $A_1$ and $A_2$ must be known. To gain an understanding about the evaluation, let us work out a simple problem and then present the results for more complicated geometries in tabular form.

Let us consider the radiation from a small area $dA_1$ parallel to a flat disk of area $A_2$, as shown in Figure 6.8.

Let the disk $A_2$ be of circular shape of radius $x$. Thus,

$$dA_2 = 2\pi x dx$$

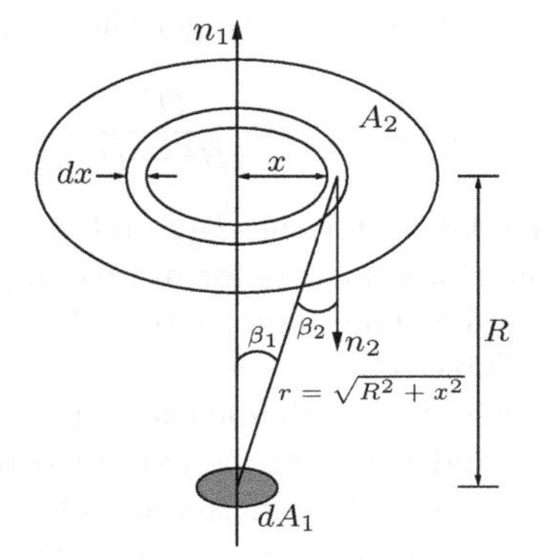


**Figure 6.8**
Radiation exchange between an elemental area and a large disk.

For this arrangement of $dA_1$ and $A_2$, $\beta_1 = \beta_2$. Therefore, Equation (6.19) becomes,

$$dA_1 F_{dA_1 \to A_2} = dA_1 \int_{A_2} \cos^2 \beta_1 \frac{2\pi x}{\pi r^2} dx$$

where $r = \sqrt{R^2 + x^2}$ and $\cos \beta_1 = \dfrac{R}{\sqrt{R^2 + x^2}}$. Therefore,

$$dA_1 F_{dA_1 \to A_2} = dA_1 \int_0^{D/2} \frac{2R^2 x}{(R^2 + x^2)^2} dx$$

where $D$ is the diameter of disk $A_2$.

This integral may be evaluated by substitution. Let $R^2 + x^2 = t$. Therefore, $2xdx = dt$. Substituting this, we get

$$dA_1 F_{dA_1 \to A_2} = dA_1 \int_0^{D/2} R^2 \frac{dt}{t^2}$$

Integrating, we obtain,

$$\begin{aligned}
dA_1 F_{dA_1 \to A_2} &= dA_1 \left[\frac{-R^2}{t}\right]_0^{D/2} = dA_1 \left[\frac{-R^2}{R^2 + x^2}\right]_0^{D/2} \\
&= dA_1 \left[\frac{-R^2}{R^2 + D^2/4} + 1\right] = dA_1 \left[\frac{-4R^2}{4R^2 + D^2} + 1\right] \\
&= dA_1 \left[\frac{-4R^2 + 4R^2 + D^2}{4R^2 + D^2}\right] = dA_1 \left[\frac{D^2}{4R^2 + D^2}\right]
\end{aligned}$$

Canceling $dA_1$, being a constant area, we get the shape factor as

$$F_{dA_1 \to A_2} = \frac{D^2}{4R^2 + D^2}$$

This is the shape factor relation for the $dA_1$ and $A_2$, shown in Figure 6.8.

The calculation of the shape factors for more complex geometries are presented in references 3, 5 and 6. The shape factor relations for some well-defined geometries are listed in Table 6.2.

The configuration factor $F_{1\to 2}$, defined for the energy exchange between two elemental areas $dA_1$ and $dA_2$, can be extended to define the energy exchange between two finite areas $A_1$ and $A_2$, as follows.

By definition, the configuration factor, namely the fraction of energy leaving surface $dA_1$ and reaching surface $dA_2$, given by Equation (6.15), is

$$F_{1\to 2} = \frac{\dot{Q}_{1\to 2}}{\dot{Q}_1} \tag{6.20}$$

where $\dot{Q}_{1\to 2}$ is the net energy leaving $dA_1$ and reaching surface $dA_2$, and $\dot{Q}_1$ is the total energy leaving $dA_1$. By Equation (6.14),

$$\dot{Q}_1 = \pi I_1 dA_1$$

**Table 6.2** Shape factor relations for some specific geometries

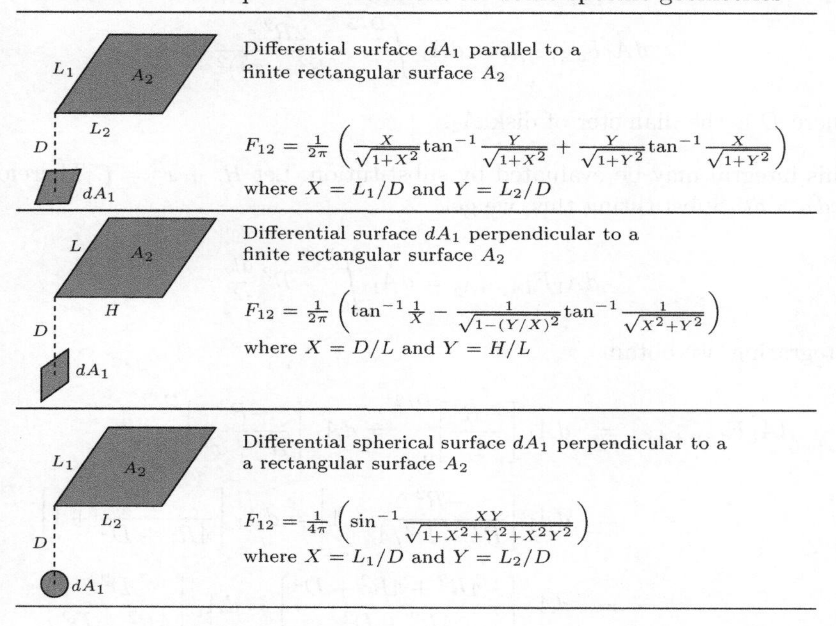


| | |
|---|---|
| Differential surface $dA_1$ parallel to a finite rectangular surface $A_2$ | $F_{12} = \frac{1}{2\pi}\left(\frac{X}{\sqrt{1+X^2}}\tan^{-1}\frac{Y}{\sqrt{1+X^2}} + \frac{Y}{\sqrt{1+Y^2}}\tan^{-1}\frac{X}{\sqrt{1+Y^2}}\right)$ where $X = L_1/D$ and $Y = L_2/D$ |
| Differential surface $dA_1$ perpendicular to a finite rectangular surface $A_2$ | $F_{12} = \frac{1}{2\pi}\left(\tan^{-1}\frac{1}{X} - \frac{1}{\sqrt{1-(Y/X)^2}}\tan^{-1}\frac{1}{\sqrt{X^2+Y^2}}\right)$ where $X = D/L$ and $Y = H/L$ |
| Differential spherical surface $dA_1$ perpendicular to a a rectangular surface $A_2$ | $F_{12} = \frac{1}{4\pi}\left(\sin^{-1}\frac{XY}{\sqrt{1+X^2+Y^2+X^2Y^2}}\right)$ where $X = L_1/D$ and $Y = L_2/D$ |

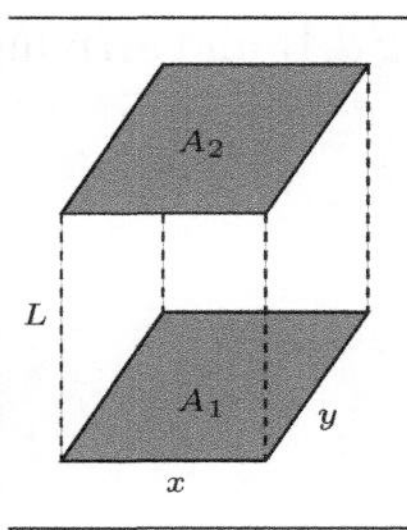


Rectangular surfaces $A_1$ and $A_2$ aligned parallel

$$F_{12} = \frac{2}{\pi XY}\Bigg\{ \ln\left(\frac{(1+X^2)(1+Y^2)}{1+X^2+Y^2}\right)^{1/2} + X(1+Y^2)^{1/2}\tan^{-1}\frac{X^2}{(1+Y^2)^{1/2}} + Y(1+X^2)^{1/2}\tan^{-1}\frac{Y}{(1+X^2)^{1/2}} - X\tan^{-1}X - Y\tan^{-1}Y\Bigg\}$$

where $X = x/L$ and $Y = y/L$

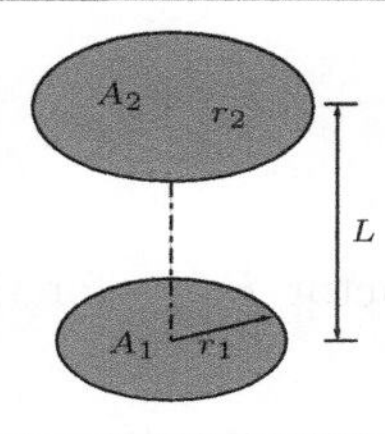


Coaxial parallel disks

$$F_{12} = \frac{1}{2}\left[S - (S - 4(r_2/r_1)^2)^{1/2}\right]$$

$$S = 1 + \frac{1+R_2^2}{R_1^2}$$

where $R_1 = r_1/L$ and $R_2 = r_2/L$

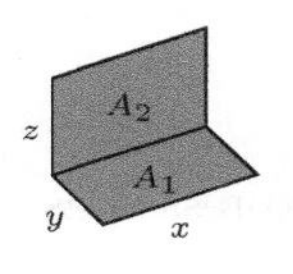


Rectangular plates $A_1$ and $A_2$, at 90 degrees with a common edge

$$F_{12} = \frac{1}{\pi W}\Bigg(W\tan^{-1}\frac{1}{W} + H\tan^{-1}\frac{1}{H} - (H^2+W^2)^{1/2}\tan^{-1}\frac{1}{(H^2+W^2)^{1/2}} + \frac{1}{4}\ln\left\{\frac{(1+W^2)(1+H^2)}{1+W^2+H^2}\left[\frac{W^2(1+W^2+H^2)}{(1+W^2)(W^2+H^2)}\right]^{W^2} \times \left[\frac{H^2(1+H^2+W^2)}{(1+H^2)(H^2+W^2)}\right]^{H^2}\right\}\Bigg)$$

where $H = z/x$ and $W = y/x$

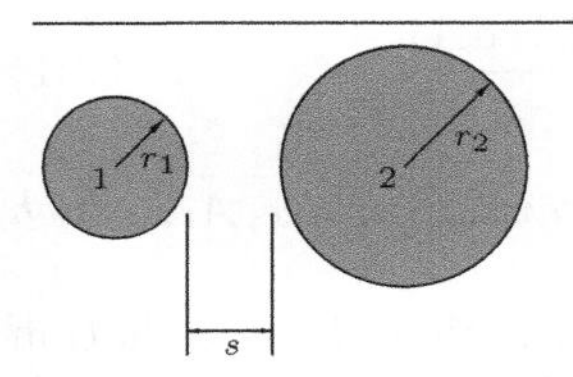


Parallel cylinders of different radii

$$F_{12} = \frac{1}{2\pi}\Big\{\pi + [C^2 - (R+1)^2]^{1/2} - [C^2 - (R-1)^2]^{1/2} + (R-1)\cos^{-1}[(R/C) - (1/C)] - (R+1)\cos^{-1}[(R/C) + (1/C)]\Big\}$$

where $C = 1 + R + S$,
$R = r_2/r_1$ and $S = s/r_1$

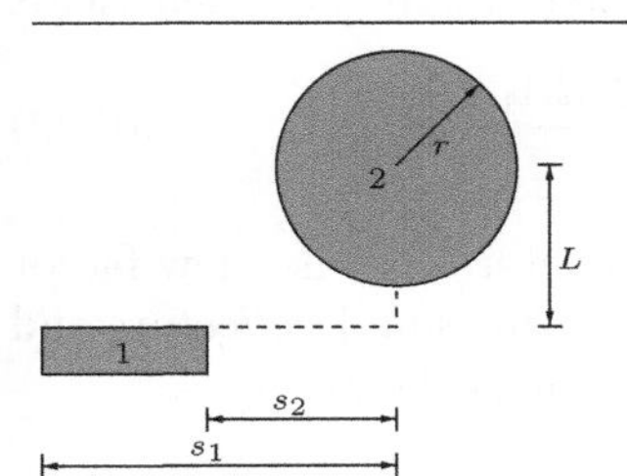


Cylinder and a parallel rectangular surface

$$F_{12} = \frac{r}{s_1 - s_2}\left[\tan^{-1}\left(\frac{s_1}{L}\right) - \tan^{-1}\left(\frac{s_2}{L}\right)\right]$$

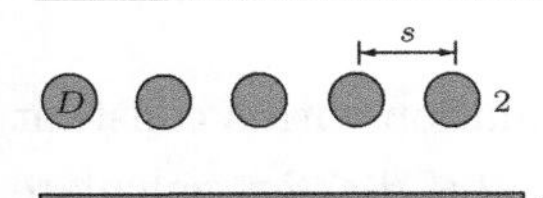


An infinite plane and a row of cylinders ofdiameter $D$

$$F_{12} = 1 - \left[1 - \left(\frac{D}{s}\right)^2\right]^{1/2} + \left(\frac{D}{s}\right)\tan^{-1}\left(\frac{s^2 - D^2}{s^2}\right)^{1/2}$$

Note: It is essential to note that the resulting values of $\sin^{-1}$, $\cos^{-1}$ and $\tan^{-1}$ functions in Table 6.2 are in radians.

Referring to Figure 6.7, we have the amount of energy leaving $dA_1$ and arriving at $dA_2$ as

$$\dot{Q}_{1\to 2} = E_{b1} \cos\beta_1 \cos\beta_2 \frac{dA_1 dA_2}{\pi r^2}$$

But, for a blackbody $E_{bi} = \pi I$; therefore,

$$\dot{Q}_{1\to 2} = \pi I \cos\beta_1 \cos\beta_2 \frac{dA_1 dA_2}{\pi r^2}$$

Substituting this into Equation (6.20), we have

$$F_{1\to 2} = \frac{I \cos\beta_1 \cos\beta_2 \frac{dA_1 dA_2}{\pi r^2}}{I_1 dA_1}$$

Integrating over the areas $A_1$ and $A_2$, we get the view factor $F_{1\to 2}$ for the energy exchange between two finite areas $A_1$ and $A_2$ as

$$F_{1\to 2} = \frac{\int_{A_2}\int_{A_1} \dot{Q}_{1\to 2}\, dA_1\, dA_2}{\int_{A_1} \dot{Q}_1\, dA_1}$$

Using Equations (6.12) and (6.14), Equation (6.15) can be expressed as

$$F_{1\to 2} = \frac{\int_{A_2}\int_{A_1} I_1 \cos\beta_1 \cos\beta_2\, dA_1\, dA_2}{\pi \int_{A_1} I_1\, dA_1}$$

Now, assuming the intensity of radiation leaving surface 1 to be uniform over the entire surface, $F_{1\to 2}$ can be simplified to

$$F_{1\to 2} = \frac{1}{A_1}\int_{A_2}\int_{A_1} \frac{\cos\beta_1 \cos\beta_2\, dA_1\, dA_2}{\pi} \tag{6.21}$$

Equation (6.21) gives the *fraction of the radiation leaving surface* $A_1$, *which is intercepted by surface* $A_2$.

Similarly, the view factor $F_{2\to 1}$ is defined as the fraction of the radiation which leaves $A_2$ and is intercepted by $A_1$. The same development then yields

$$F_{2\to 1} = \frac{1}{A_2}\int_{A_2}\int_{A_1} \frac{\cos\beta_1 \cos\beta_2\, dA_1\, dA_2}{\pi} \tag{6.22}$$

Either of Equation (6.21) or (6.22) may be used to determine the view factor associated with any two surfaces that are *diffuse emitters* and *reflectors* and have *uniform radiosity* (the total radiation leaving the surface).

## Example 6.3

If a blackbody of surface area 25 cm$^2$ is subjected to radiation with a constant intensity of $10^5$ W/m$^2$ over the solid angle $0 \le \beta \le \frac{\pi}{3}$, $0 \le \theta \le 2\pi$, calculate the radiation energy received by the blackbody.

## Solution

Given, $I = 10^5$ W/m$^2$, $A = 25$ cm$^2$.

By Equation (6.13), the radiation energy received by the blackbody is

$$\dot{Q} = \int_{2\pi} IA \cos\beta \, d\Omega$$

where the solid angle $d\Omega$ is

$$d\Omega = \sin\beta \, d\beta \, d\theta$$

Substituting for $d\Omega$, we get

$$\begin{aligned}
\dot{Q} &= IA \int_0^{2\pi} \int_0^{\frac{\pi}{3}} \cos\beta \sin\beta \, d\beta \, d\theta \\
&= IA \Big[\theta\Big]_0^{2\pi} \left[\frac{1}{2}\sin^2\beta\right]_0^{\frac{\pi}{3}} \\
&= IA\,2\pi \left[\frac{1}{2} \times (0.866)^2\right] \\
&= IA\,2\pi \left[\frac{1}{2} \times \frac{3}{4}\right] \\
&= \frac{3}{4}\pi I A \\
&= \frac{3}{4} \times \pi \times 10^5 \times \left(25 \times 10^{-4}\right) \\
&= \boxed{589\,\text{W}}
\end{aligned}$$

## Example 6.4

Find the expression for the view factor $F_{1\rightarrow 2}$ between the differential area $dA_1$ and the large parallel plane of area $A_2$, at a distance $D$ from $dA_1$, shown in Figure E6.4.

## Solution

The view factor, in general, can be determined by integrating Equation (6.21). But this integration becomes tedious for complex geometries. For such shapes usually numerical methods are employed to obtain the view factor. For simple geometries, Equation (6.21) can be integrated to get the analytical expression

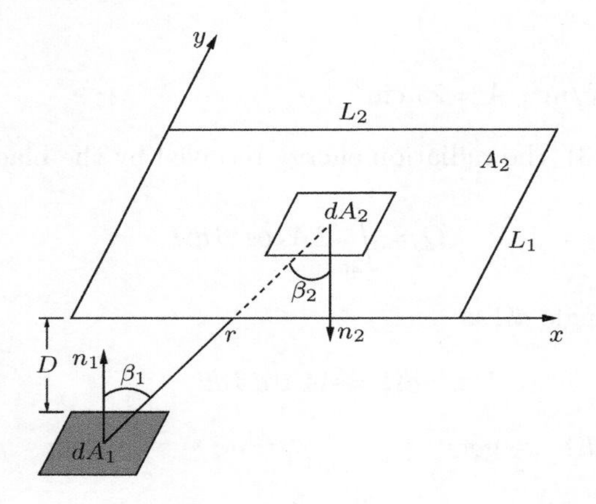


**Figure E6.4**
A small surface placed parallel to a large surface.

for the view factor. For the planes shown in Figure E6.4, the view factor $F_{1\to 2}$ is given by Equation (6.21) as

$$F_{1\to 2} = \frac{1}{dA_1} \int_{dA_1} \int_{A_2} \frac{\cos\beta_1 \cos\beta_2 \, dA_1 \, dA_2}{\pi r^2}$$

But $A_2 >> dA_1$, therefore, the view of $dA_2$ from $dA_1$ can be regarded as independent of the position on $dA_1$. Hence,

$$F_{1\to 2} = \frac{1}{\pi} \int_{A_2} \frac{\cos\beta_1 \cos\beta_2}{r^2} \, dA_2$$

For the parallel planes in Figure E6.4, $\beta_1 = \beta_2$ and $\cos\beta_1 = \cos\beta_2 = \dfrac{D}{r}$, where $r^2 = D^2 + x^2 + y^2$. Therefore, the above expression for view factor becomes

$$F_{1\to 2} = \frac{1}{2\pi} \int_0^{L_1} \int_0^{L_2} \frac{D^2}{(D^2 + x^2 + y^2)^2} \, dx dy$$

On integration, this yields the view factor as

$$F_{1\to 2} = \frac{1}{2\pi} \left[ \frac{L_1}{\sqrt{L_1^2 + D^2}} \tan^{-1}\left( \frac{L_2}{\sqrt{L_1^2 + D^2}} \right) + \frac{L_2}{\sqrt{L_2^2 + D^2}} \tan^{-1}\left( \frac{L_1}{\sqrt{L_2^2 + D^2}} \right) \right]$$

Note that, this relation is the same as that given in Table 6.2.

### 6.5.1 View Factor Relation

From Equations (6.21) and (6.22), it is seen that,

$$\boxed{A_1 F_{1\to 2} = A_2 F_{2\to 1}} \qquad (6.23)$$

This expression is called the *reciprocity relation* between surfaces $A_1$ and $A_2$.

Another important view factor relation pertains to the surfaces of an *enclosure* such as that shown in Figure 6.9.

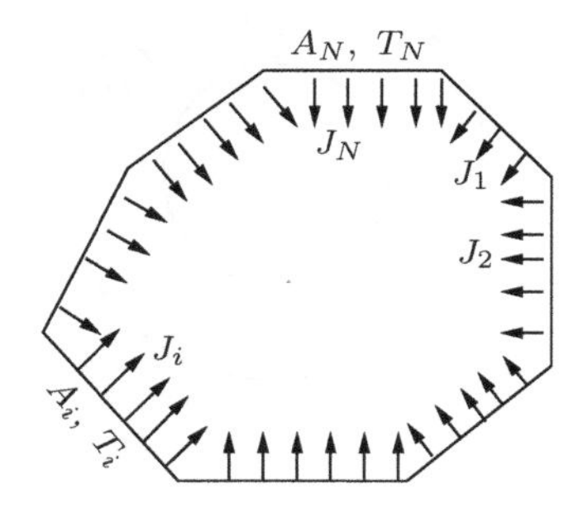


**Figure 6.9**
Radiation exchange in an enclosure.

From the definition of view factor, the *summation rule*

$$\boxed{\sum_{j=1}^{N} F_{i \to j} = 1} \tag{6.24}$$

may be applied to each of the $N$ surfaces in the enclosure. This rule follows from the conservation requirement that all the radiation leaving surface $A_i$ be intercepted by the enclosure surfaces. The term $F_{i \to i}$ appearing in this summation represents the fraction of the radiation that leaves surface $A_i$ and is directly intercepted by $A_i$. If the surface is concave, it sees *itself*, and $F_{i \to i}$ is non-zero. But for a plane or convex surface, $F_{i \to i} = 0$.

To compute radiation exchange in an enclosure of $N$ surfaces, a total number of $N^2$ view factor is needed. However, all the view factors need not be calculated *directly*. A total of $N$ view factors may be obtained from $N$ equations associated with application of the summation rule, Equation (6.24), to each of the surfaces in the enclosure. In addition $N(N-1)/2$ view factors may be obtained from the $N(N-1)/2$ applications of the reciprocity relation, Equation (6.23), which are possible for the enclosure. Thus, only

$$\left[N^2 - N - \frac{N(N-1)}{2}\right] = \frac{N(N-1)}{2}$$

view factors need to be determined directly. For example, in a three-surface enclosure this requirement corresponds to only $3(3-1)/2 = 3$ view factors. The remaining six view factors may be obtained by solving the six equations that result from use of Equations (6.23) and (6.24).

To illustrate the above procedure, consider the two-surface enclosure involving the spherical surfaces shown in Figure 6.10.

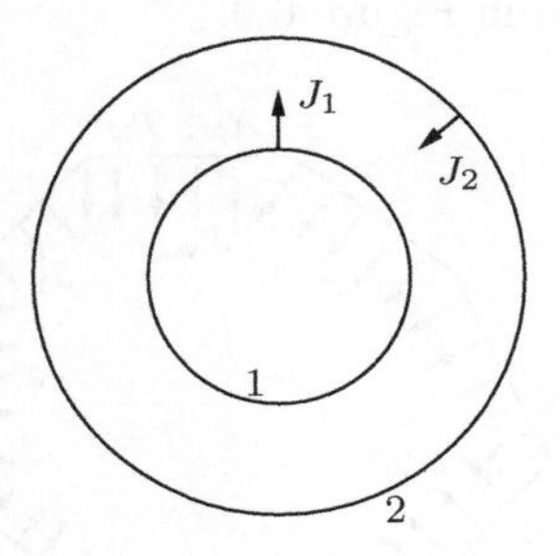


**Figure 6.10**
View factor for the enclosure formed by two spheres.

Although the enclosure is characterized by $N^2 = 4$ view factors, only

$$N(N-1)/2 = 1$$

view factor need to be determined directly. In this case such a determination may be made by *inspection.* Note that all radiation leaving the outer surface of the inner sphere must reach the inner surface of the outer surface, hence it follows that $F_{1\to 2} = 1$. The same may be said of radiation leaving the outer surface, since this surface sees itself. However, from reciprocity relation, Equation (6.23), we have

$$F_{2\to 1} = \left(\frac{A_1}{A_2}\right) F_{1\to 2}$$

But $F_{1\to 2} = 1$; therefore,

$$F_{2\to 1} = \left(\frac{A_1}{A_2}\right)$$

By the summation rule, Equation (6.24), we have

$$F_{1\to 1} + F_{1\to 2} = 1$$

But for the present case, $F_{1\to 1} = 0$. Again by the summation rule,

$$F_{2\to 1} + F_{2\to 2} = 1$$

That is,

$$F_{2\to 2} = 1 - F_{2\to 1}$$

But $F_{2\to 1} = \left(\frac{A_1}{A_2}\right)$, thus,

$$F_{2\to 2} = 1 - \left(\frac{A_1}{A_2}\right)$$

For more complicated geometries, the view factor must be determined by solving the double integral of Equation (6.21).

## 6.6 Blackbody Radiation Exchange

We have seen that, in general, radiation may leave a surface due to both reflection and emission, and on reaching a second surface, it may experience reflection as well as absorption. However, the analysis of radiation exchange can be simplified for surfaces that may be approximated as blackbodies, since there is no reflection from a blackbody. Hence, from a blackbody energy leaves only as a result of emission, and all incident radiation is absorbed.

Examine the radiation exchange between two black surfaces of arbitrary shapes, shown in Figure 6.11.

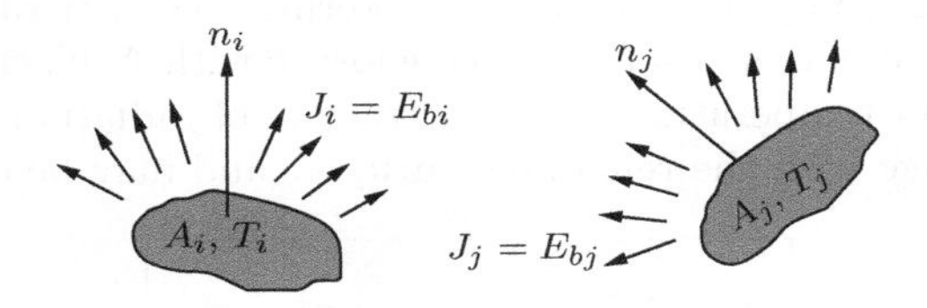


**Figure 6.11**
Radiation exchange between two black surfaces.

Let $\dot{Q}_{i\to j}$ be the rate at which radiation *leaves* surface $A_i$, and is *intercepted* by surface $A_j$. It follows that,

$$\dot{Q}_{i\to j} = (A_i\, J_i)\; F_{i\to j} \tag{6.25}$$

where $J_i$ is the radiation flux leaving surface $A_i$, called *radiosity*. But for a blackbody, radiosity is equal to the emissive power $E_b$; therefore,

$$\dot{Q}_{i\to j} = A_i\, F_{i\to j}\, E_{bi} \tag{6.26}$$

Similarly, the rate at which radiation leaves surface $A_j$, and is intercepted by surface $A_i$ becomes

$$\dot{Q}_{j\to i} = A_j\, F_{j\to i}\, E_{bj} \tag{6.27}$$

The *net radiative exchange* between two surfaces may then be defined as

$$\dot{Q}_{ij} = \dot{Q}_{i\to j} - \dot{Q}_{j\to i}$$

Substituting Equations (6.26) and (6.27), we get

$$\boxed{\dot{Q}_{ij} = A_i\, F_{i\to j}\, E_{bi} - A_j\, F_{j\to i}\, E_{bj}} \tag{6.28}$$

This is the net radiative exchange between surfaces of area $A_i$ and $A_j$, at temperatures $T_i$ and $T_j$.

If the plates are of equal area, Equation (6.28) can be simplified as follows. By Equations (6.1) and (6.23), we have

$$E_{bi} = \sigma T_i^4, \; E_{bj} = \sigma T_j^4, \quad \text{and} \quad F_{i \to j} = F_{j \to i}$$

Substituting these, we get

$$\boxed{\dot{Q}_{ij} = A_i F_{i \to j} \sigma \left(T_i^4 - T_j^4\right)} \tag{6.29}$$

This gives the *net* rate at which radiation leaves surface $i$ as a result of its interaction with surface $j$, which is equal to the *net* rate at which surface $j$ *gains* radiation due to its interaction with surface $i$.

Equation (6.29) may also be used to determine the net radiation transfer from any surface in an *enclosure.* In an enclosure with $N$ black surfaces maintained at different temperatures, the net transfer of radiation from surface $A_i$ is due to exchange with the remaining surfaces, and may be expressed as

$$\boxed{\dot{Q}_i = \sum_{j=1}^{N} A_i F_{i \to j} \sigma \left(T_i^4 - T_j^4\right)} \tag{6.30}$$

## 6.7 Radiation Exchange in an Enclosure

The foregoing results, even though useful, have only limited application, since they are valid only for blackbodies. The blackbody idealization, although closely approximated by some surfaces, can never be achieved precisely. A major difficulty associated with radiation exchange between non-black surfaces is due to surface reflection. Consider the enclosure shown in Figure 6.12. In this enclosure, radiation must experience multiple reflections at all surfaces, with partial absorption occurring at each surface.

Analysis of radiation exchange in an enclosure may be simplified by making the following assumptions.

- Each surface of the enclosure is assumed to be *isothermal,* and be characterized by *uniform radiosity* $J$ and *irradiation* $G$.

- All the surfaces are *opaque, diffuse* and *gray.*

- The medium within the enclosure is *nonparticipating.*

Usually, the problem is one in which the temperature $T_i$ associated with each of the surfaces is known, and the objective is to determine the *net radiative heat flux* $q_i$ from each surface.

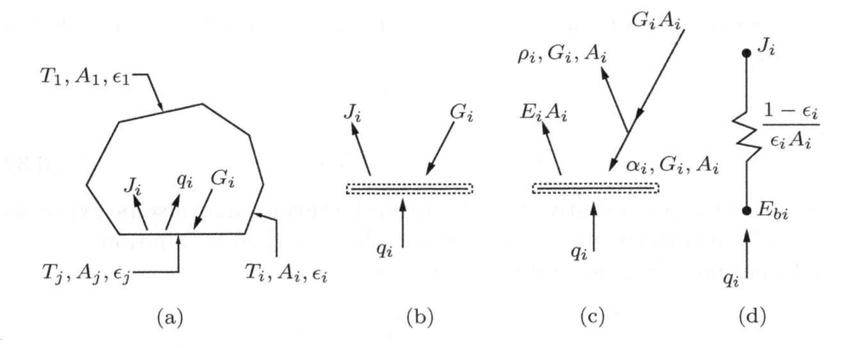


**Figure 6.12**
Radiation exchange in an enclosure of diffuse, gray surfaces with a nonparticipating medium. (a) Schematic of the enclosure, (b) Radiative balance according to Equation (6.29), (c) Radiative balance according to Equation (6.30), (d) Network element representing the net radiation transfer from a surface.

### 6.7.1 Net Radiation Exchange at a Surface

The *net* rate at which radiation leaves surface $i$ represents the net effect of radiative interaction occurring at the surface (Figure 6.12(b)). It is the rate at which energy would have to be transferred to the surface, by other means to maintain it at a constant temperature. It is equal to the difference between the surface radiosity and irradiation and may be expressed as

$$\dot{Q}_i = A_i \, (J_i - G_i) \tag{6.31}$$

where $J_i$ is the radiosity, which represents the radiation flux leaving surface $i$, and $G_i$ is the irradiation, which gives the radiation flux incident of surface $i$.

From Figure 6.12(c) and the definition of the radiosity $J_i$, we can write

$$J_i \equiv E_i + \rho_i \, G_i$$

where $E_i$ and $\rho_i$ are the emissive power and reflectivity of surface $i$, respectively.

Substituting $J_i$ into Equation (6.31), we have

$$\begin{aligned} \dot{Q}_i &= A_i \, (E_i + \rho_i \, G_i - G_i) \\ &= A_i \, (E_i + (\rho_i - 1)G_i) \end{aligned}$$

By Equation (6.7),

$$\alpha_i + \rho_i + \tau_i = 1$$

For the present case of solid surface, the transmissivity $\tau_i$ can be taken as zero. Thus,

$$\rho_i - 1 = -\alpha_i$$

Hence,

$$\dot{Q}_i = A_i \left(E_i - \alpha_i \, G_i\right) \tag{6.32}$$

It is seen that the net radiative transfer from the surface may also be expressed in terms of the surface emissive power and the absorbed irradiation.

By Equation (6.9), the emissive power is

$$E_i = \epsilon \, E_{bi}$$

For an opaque, diffusive, gray surface, $\alpha_1 = \epsilon_i$; therefore,

$$\rho_i = 1 - \epsilon_i$$

The radiosity for surface $i$ is the sum of radiation emitted and radiation reflected by the surface. Thus,

$$\begin{aligned} J_i &= \text{radiation emitted} + \text{radiation reflected} \\ &= E_i + \rho_i G_i \end{aligned}$$

Substituting for $E_i$ and $\rho_i$, the radiosity becomes

$$J_i = \epsilon_i \, E_{bi} + (1 - \epsilon_i) \, G_i \tag{6.33}$$

This gives the irradiation $G_i$ as

$$G_i = \frac{J_i - \epsilon_i \, E_{bi}}{1 - \epsilon_i}$$

Substituting this into Equation (6.31), we get

$$\begin{aligned} \dot{Q}_i &= A_i \left( J_i - \frac{J_i - \epsilon_i \, E_{bi}}{1 - \epsilon_i} \right) \\ &= A_i \left( \frac{(1 - \epsilon_i) \, J_i - J_i + \epsilon_i E_{bi}}{1 - \epsilon_i} \right) \\ &= A_i \left( \frac{(E_{bi} - J_i) \, \epsilon_i}{1 - \epsilon_i} \right) \end{aligned}$$

or

$$\boxed{\dot{Q}_i = \frac{E_{bi} - J_i}{\dfrac{1 - \epsilon_i}{\epsilon_i \, A_i}}} \tag{6.34}$$

This equation gives the net radiative heat transfer rate from a surface. This transfer may be represented by the network element of Figure 6.11(d), with $(E_{bi} - J_i)$ as the *driving potential* and $\left(\dfrac{1-\epsilon_i}{\epsilon_i\, A_i}\right)$ as the *surface radiative resistance.*

It is interesting to note the following:

• For a black surface $i$, the emissivity $\epsilon_i = 1$. This implies that the surface radiative resistance $\left(\dfrac{1-\epsilon_i}{\epsilon_i\, A_i}\right) = 0$. Also, Equation (6.33) reduces to

$$J_i = E_{bi} = \sigma\, T_i^4$$

Thus, for a black surface the radiosity is equal to the blackbody emmisive power of the surface.

• From Equation (6.34) it is seen that the driving potential $(E_{bi} - J_i)$ dictates whether the heat transfer due to radiation will be from the body to the surroundings or vice versa. When the emissive power $E_{bi}$ a body would have, if it were black, exceeds its radiosity $J_i$, the net radiation will be from the body to the surroundings. If $E_{bi}$ is less than $J_i$, the net radiation will be from the surroundings to the body.

## 6.7.2 Radiation Exchange between Surfaces

For estimating the radiation exchange between surface $i$ and other surfaces, the *radiosity* $J_i$ of surface $i$, defined as *the total radiation which leaves surface $i$ per unit surface area, per unit time*, must be known. To determine $J_i$, it is necessary to consider the radiation exchange between the surfaces of the enclosure, illustrated in Figure 6.11.

The *irradiation* $G_i$, defined as the radiation flux incident on surface $i$, can be evaluated from the radiosities of all the surfaces in the enclosure. In particular from the definition of the view factor, it follows that the total rate at which radiation reaches surface $i$ from all surfaces, including $i$, is

$$A_i\, G_i = \sum_{j=1}^{N} A_j\, F_{j\,\rightarrow\, i}\, J_j$$

By the reciprocity relation, Equation (6.18), we have $A_i\, F_{i\,\rightarrow\, j} = A_j\, F_{j\,\rightarrow\, i}$; therefore,

$$A_i\, G_i = \sum_{j=1}^{N} A_i\, F_{i\,\rightarrow\, j}\, J_j$$

Substituting this into Equation (6.31), we get

$$\dot{Q}_i = A_i \left( J_i - \sum_{j=1}^{N} F_{i\,\rightarrow\, j}\, J_j \right)$$

By summation rule, Equation (6.24), we have

$$\sum_{j=1}^{N} F_{i\to j} = 1$$

Therefore, $J_i$ may be expressed as

$$\sum_{j=1}^{N} F_{i\to j}\, J_i$$

Replacing $J_i$ with

$$\sum_{j=1}^{N} F_{i\to j} J_i$$

we get

$$\dot{Q}_i = A_i \left( \sum_{j=1}^{N} F_{i\to j}\, J_i - \sum_{j=1}^{N} F_{i\to j}\, J_j \right)$$

or

$$\boxed{\dot{Q}_i = \sum_{j=1}^{N} A_i F_{i\to j}\, (J_i - J_j) = \sum_{j=1}^{N} \dot{Q}_{ij}} \tag{6.35}$$

That is, the net rate of radiation transfer $\dot{Q}_i$ from surface $i$ is equal to the sum of components $\dot{Q}_{ij}$ related to radiative exchange with the other surfaces. Each component may be represented by a network element, for which $(J_i - J_j)$ is the driving potential, and $(A_i\, F_{i\to j})^{-1}$ is the *space* or *geometrical resistance.*

Combining Equations (6.34) and (6.35) we can write

$$\boxed{\frac{E_{bi} - J_i}{(1-\epsilon_j)/\epsilon_i\, A_i} = \sum_{j=1}^{N} \frac{J_i - J_j}{(A_i\, F_{i\to j})^{-1}}} \tag{6.36}$$

This radiative exchange may be represented as a network as shown in Figure 6.13.

As illustrated in Figure 6.13, Equation (6.36) represents the radiation balance for the *mode* associated with surface $A_i$.

It is essential to note that Equation (6.36) is useful, especially when the surface temperature $T_i$ (hence $E_{bi}$) is known. Even though this situation is typical, it is not the case always. For example, a situation may arise for which the net radiation transfer rate $\dot{Q}_i$ at the surface, rather than the temperature $T_i$, is known. For such cases, Equation (6.35) may be rearranged as

$$\boxed{\dot{Q}_i = \sum_{j=1}^{N} \frac{J_i - J_j}{(A_i\, F_{i\to j})^{-1}}} \tag{6.37}$$

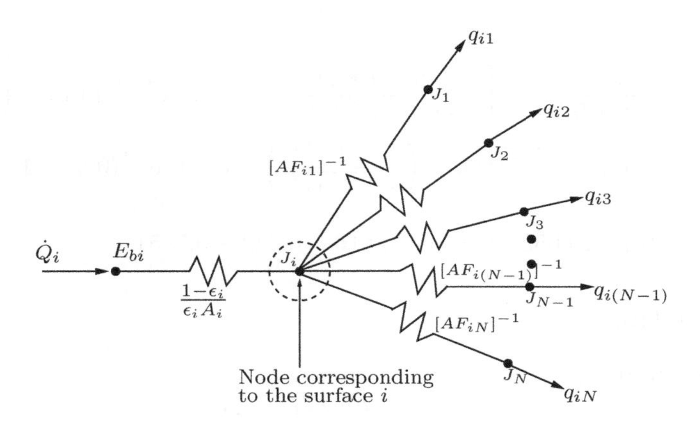


**Figure 6.13**
Network for radiative exchange between surface $i$ and the remaining surfaces in an enclosure.

where $(J_i - J_j)$ is the driving potential and $\dfrac{1}{A_i\, F_{i\to j}}$ is a space or geometrical resistance.

## Example 6.5

Two rectangular black plates of length and width 0.5 m and 1.0 m, respectively, are placed one above the other parallel with a vertical spacing of 1.0 m. The top plate is maintained at 1000°C and the bottom plate is maintained at 500°C. Determine the radiation heat transfer between these two plates.

## Solution

Given, $x = 0.5$ m, $y = 1$ m and $L = 1$ m, $T_1 = 1000$°C, $T_2 = 500$°C.

Therefore,

$$X = x/L = 0.5/1 = 0.5$$

$$Y = y/L = 1/1 = 1$$

From Table 6.2, we have the view factor as

$$F_{12} = \frac{2}{\pi XY}\left\{ \ln\left[\frac{(1+X^2)(1+Y^2)}{1+X^2+Y^2}\right]^{\frac{1}{2}} + X\sqrt{1+Y^2}\tan^{-1}\left(\frac{x}{\sqrt{1+Y^2}}\right) + Y\sqrt{1+X^2}\tan^{-1}\left(\frac{Y}{\sqrt{1+X^2}}\right) - X\tan^{-1}(X) - Y\tan^{-1}(Y)\right\}$$

$$= \frac{2}{\pi \times 1 \times 0.5}\left\{ \ln\left[\frac{(1+0.5^2)\times(1+1)}{1+0.5^2+1^2}\right]^{\frac{1}{2}} + 0.5\sqrt{1+1}\tan^{-1}\left(\frac{0.5}{\sqrt{1+1}}\right) + 1\sqrt{1+0.5^2}\tan^{-1}\left(\frac{1}{\sqrt{1+0.5^2}}\right) - 0.5\tan^{-1}(0.5) - 1\tan^{-1}(1)\right\}$$

$$= 1.273 \times (0.0527 + 0.24 + 0.816 - 0.232 - 0.785)$$

$$= 0.1167$$

By Equation (6.37),

$$\dot{Q}_{12} = A_1 F_{12}(J_1 - J_2)$$

For black surfaces,

$$J = E_b = \sigma T^4$$

Therefore,

$$\dot{Q}_{12} = A_1 F_{12}(E_{b1} - E_{b2}) = A_1 F_{12}\sigma\left(T_1^4 - T_2^4\right)$$

Given, $T_1 = 1273$ K and $T_2 = 773$ K, thus,

$$\dot{Q}_{12} = (5.67 \times 10^{-8}) \times (0.5 \times 1) \times 0.1167 \times (1273^4 - 773^4) = \boxed{7.507\,\text{kW}}$$

### 6.7.3 Two-Surface Enclosure

Two surfaces that exchange radiation only with each other, as shown in Figure 6.14, is the simplest example of an enclosure. Because there are only two surfaces, the net rate of radiation transfer $\dot{Q}_1$ from surface 1 must be equal to the net rate of radiation transfer $-\dot{Q}_2$ to surface 2, and both these quantities must be equal to the net rate at which radiation is exchanged between surfaces 1 and 2. Thus,

$$\dot{Q}_1 = -\dot{Q}_2 = \dot{Q}_{12}$$

From Figure 6.14(b) it is seen that the total resistance to the radiation exchange between surfaces 1 and 2 is made up of two surface resistances and the geometrical resistance. Therefore, the net radiation exchange between the surfaces may be expressed as

$$\dot{Q}_{12} = \dot{Q}_1 = -\dot{Q}_2 = \frac{\sigma\left(T_1^4 - T_2^4\right)}{\frac{1-\epsilon_1}{\epsilon_1 A_1} + \frac{1}{A_1 F_{1\rightarrow 2}} + \frac{1-\epsilon_2}{\epsilon_2 A_2}} \tag{6.38}$$

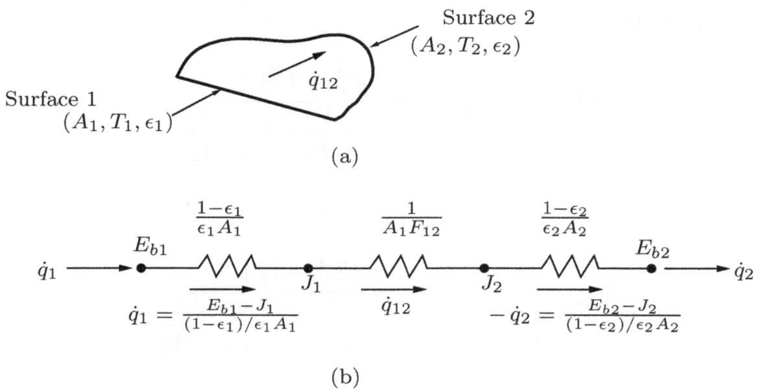

Surface 2
$(A_2, T_2, \epsilon_2)$
$\dot{q}_{12}$
Surface 1
$(A_1, T_1, \epsilon_1)$

(a)

$\frac{1-\epsilon_1}{\epsilon_1 A_1}$ $\frac{1}{A_1 F_{12}}$ $\frac{1-\epsilon_2}{\epsilon_2 A_2}$

$E_{b1}$ $E_{b2}$

$\dot{q}_1$ $J_1$ $J_2$ $\dot{q}_2$

$\dot{q}_1 = \frac{E_{b1} - J_1}{(1-\epsilon_1)/\epsilon_1 A_1}$ $\dot{q}_{12}$ $-\dot{q}_2 = \frac{E_{b2} - J_2}{(1-\epsilon_2)/\epsilon_2 A_2}$

(b)

**Figure 6.14**
Schematic of (a) two-surface enclosure and (b) its network representation.

This result is applicable to any two diffuse, gray surfaces which form an *enclosure.*

## Example 6.6

A small object of surface area $A_1 = 50$ cm$^2$, surface temperature 100°C and emissivity 0.9 is enclosed by a large enclosure of inner surface area $A_2$, as shown in Figure E6.6. If the surface temperature of the enclosure is 60°C, determine the radiation loss from the object to the enclosure.

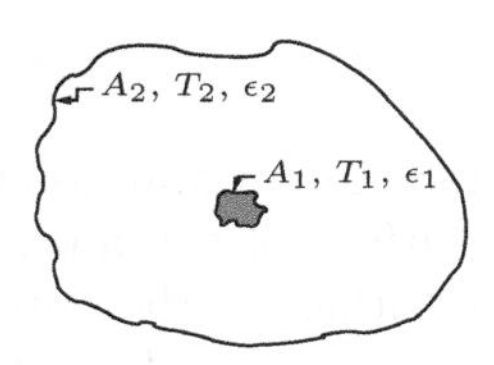

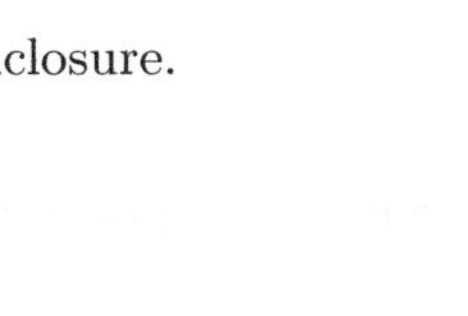

**Figure E6.6**
A small object in a large enclosure.

## Solution

Given, $A_1 = 50$ cm$^2$, $T_1 = 100 + 273 = 373$ K, $T_2 = 60 + 273 = 333$ K, $\epsilon_1 = 0.9$.

By Equation (6.38),

$$\dot{Q}_{12} = \frac{\sigma\left(T_1^4 - T_2^4\right)}{\frac{1-\epsilon_1}{\epsilon_1 A_1} + \frac{1}{A_1 F_{12}} + \frac{1-\epsilon_2}{\epsilon_2 A_2}}$$

or

$$\dot{Q}_{12} = A_1 \frac{\sigma\left(T_1^4 - T_2^4\right)}{\frac{1-\epsilon_1}{\epsilon_1} + \frac{1}{F_{12}} + \frac{1-\epsilon_2}{\epsilon_2}\frac{A_1}{A_2}}$$

Given that, $A_2$ is large. Therefore, $\frac{A_1}{A_2}$ can be neglected. Also, all the radiation leaving surface 1 reaches the inner surface of the enclosure, thus $F_{12} = 1$. Therefore,

$$\begin{aligned} \dot{Q}_{12} &= A_1 \frac{\sigma\left(T_1^4 - T_2^4\right)}{\frac{1-\epsilon_1}{\epsilon_1} + 1} \\ &= A_1 \epsilon_1 \sigma\left(T_1^4 - T_2^4\right) \\ &= (50 \times 10^{-4}) \times 0.9 \times (5.67 \times 10^{-8}) \times (373^4 - 333^4) \\ &= \boxed{1.8\,\text{W}} \end{aligned}$$

## Example 6.7

Two long cylinders of diameters 10 cm and 20 cm are kept concentric (one inside the other). The outer surface of the inner cylinder has $T_1 = 1000$ K and $\epsilon_1 = 0.7$, while the inner surface of the outer cylinder has $T_2 = 500$ K and $\epsilon_2 = 0.3$. Calculate the net heat transfer between the two cylinders per unit length.

## Solution

This is a two-body problem. Since the cylinders are very long, very little radiation leaks out at the ends.

All the radiation leaving the inner cylinder arrives at the outer cylinder, thus $F_{12} = 1.0$. Considering unit length, we have the surface area of the cylinders as

$$A_1 = \pi d_1 = \pi \times 0.1 = 0.314\,\mathrm{m}^2$$

$$A_2 = \pi d_2 = \pi \times 0.2 = 0.628\,\mathrm{m}^2$$

The resistances and emissive powers are

$$\frac{1-\epsilon_1}{\epsilon_1 A_1} = \frac{1-0.7}{0.7 \times 0.314} = 1.365$$

$$\frac{1-\epsilon_2}{\epsilon_2 A_2} = \frac{1-0.3}{0.3 \times 0.628} = 3.715$$

$$\frac{1}{A_1 F_{12}} = \frac{1}{0.314 \times 1.0} = 3.185$$

$$E_{b1} = \sigma T_1^4 = (5.67 \times 10^{-8})(1000^4) = 56700\,\mathrm{W/m}^2$$

$$E_{b2} = \sigma T_2^4 = (5.67 \times 10^{-8})(500^4) = 3543.75\,\mathrm{W/m}^2$$

The total thermal resistance is

$$R_{\mathrm{th}} = 1.365 + 3.715 + 3.185$$

$$= 8.265\,1/\mathrm{m}^2$$

The net heat transfer between the two cylinders per unit length is

$$\dot{Q} = \frac{E_{b1} - E_{b2}}{R_{\mathrm{th}}}$$

$$= \frac{56700 - 3543.75}{8.265}$$

$$= \boxed{6.43\,\mathrm{kW}}$$

## Example 6.8

Two parallel plates of size 1 m $\times$ 2 m are spaced 1 m apart. One plate is maintained at 1500 K and the other at 1000 K. The emissivities of the plates are 0.3 and 0.7, respectively. The plates are placed in a large room whose walls are at 300 K. The plates exchange heat with each other and with the room, but only the plate surfaces facing each other are to be considered in the analysis. (a) Calculate the net transfer of heat from each plate and to the room, assuming the shape factors as $F_{12} = F_{21} = 0.285$. (b) Check whether the shape factors assumed are correct.

## Solution

(a) This is basically a three-body problem. The radiation network is shown in Figure ES6.8.

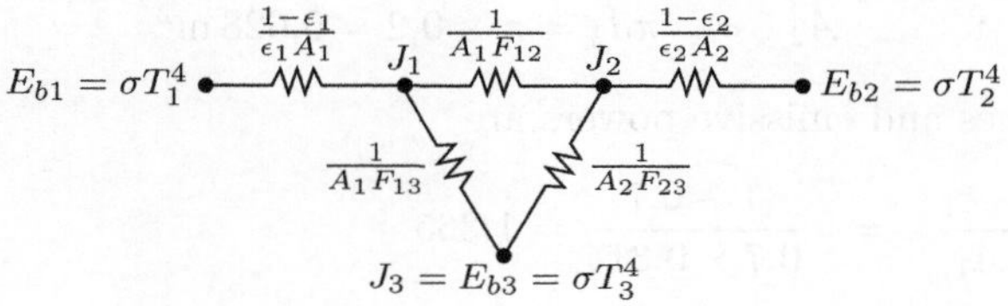

$E_{b1} = \sigma T_1^4$ — $\frac{1-\epsilon_1}{\epsilon_1 A_1}$ — $J_1$ — $\frac{1}{A_1 F_{12}}$ — $J_2$ — $\frac{1-\epsilon_2}{\epsilon_2 A_2}$ — $E_{b2} = \sigma T_2^4$

$\frac{1}{A_1 F_{13}}$ $\frac{1}{A_2 F_{23}}$

$J_3 = E_{b3} = \sigma T_3^4$

**Figure E6.8**
Radiation network.

Given, $T_1 = 1500$ K, $A_1 = 2$ m$^2$, $\epsilon_1 = 0.3$, $T_2 = 1000$ K, $\epsilon_2 = 0.7$ and $T_3 = 300$ K.

Because the area of the room $A_3$ is very large, the resistance $\dfrac{1-\epsilon_3}{\epsilon_3 A_3}$ may be taken as zero. Thus, $E_{b3} = J_3$. Also,

$$F_{12} = F_{21} = 0.285$$

Therefore, by summation rule, we have

$$F_{13} = 1 - F_{12} = 0.715$$

$$F_{23} = 1 - F_{21} = 0.715$$

The resistances in the network are calculated as

$$\frac{1-\epsilon_1}{\epsilon_1 A_1} = \frac{1-0.3}{0.3 \times 2} = 1.167$$

$$\frac{1-\epsilon_2}{\epsilon_2 A_2} = \frac{1-0.7}{0.7 \times 2} = 0.214$$

$$\frac{1}{A_1 F_{12}} = \frac{1}{2 \times 0.285} = 1.754$$

$$\frac{1}{A_1 F_{13}} = \frac{1}{2 \times 0.715} = 0.699$$

$$\frac{1}{A_2 F_{23}} = \frac{1}{2 \times 0.715} = 0.699$$

To calculate the heat at each surface, we should determine the radiosities $J_1$ and $J_2$. The network is solved by setting the sum of the heat currents entering nodes $J_1$ and $J_2$ to zero.

$$\text{Node } J_1: \quad \frac{E_{b1} - J_1}{1.167} + \frac{J_2 - J_1}{1.754} + \frac{E_{b3} - J_1}{0.699} = 0 \qquad \text{(i)}$$

$$\text{Node } J_2: \quad \frac{J_1 - J_2}{1.754} + \frac{E_{b3} - J_1}{0.699} + \frac{E_{b2} - J_2}{0.214} = 0 \qquad \text{(ii)}$$

Also,

$$E_{b1} = \sigma T_1^4 = (5.67 \times 10^{-8})(1500)^4 = 287.04\,\text{kW/m}^2$$

$$E_{b2} = \sigma T_2^4 = (5.67 \times 10^{-8})(1000)^4 = 56.700\,\text{kW/m}^2$$

$$E_{b3} = \sigma T_3^4 = J_3 = (5.67 \times 10^{-8})(300)^4 = 0.4593\,\text{kW/m}^2$$

With these values, Equations (i) and (ii) become

$$\frac{J_2}{1.754} - J_1\left(\frac{1}{1.167} + \frac{1}{1.754} + \frac{1}{0.699}\right) + \frac{287.04}{1.167} + \frac{0.4593}{0.699} = 0$$

$$\frac{J_1}{1.754} - J_2\left(\frac{1}{1.754} + \frac{1}{1.699} + \frac{1}{0.214}\right) + \frac{0.4593}{0.699} + \frac{56.7}{0.214} = 0$$

Solving, we get

$$J_1 = 96\,\text{kW/m}^2$$

$$J_2 = 47.4\,\text{kW/m}^2$$

The total heat lost by plate 1 is

$$\dot{Q}_1 = \frac{E_{b1} - J_1}{(1 - \epsilon_1)/\epsilon_1 A_1} = \frac{287.04 - 96}{1.167}$$

$$= \boxed{163.7\,\text{kW}}$$

Similarly, the total heat lost by plate 2 is

$$\dot{Q}_2 = \frac{E_{b2} - J_2}{(1 - \epsilon_2)/\epsilon_2 A_2} = \frac{56.7 - 47.4}{0.214}$$

$$= \boxed{43.5\,\text{kW}}$$

The total heat acquired by the room is

$$\dot{Q}_3 = \frac{J_1 - J_3}{1/A_1 F_{13}} + \frac{J_2 - J_3}{1/A_2 F_{23}}$$

$$= \frac{96 - 0.4593}{0.699} + \frac{47.4 - 0.4593}{0.699}$$

$$= \boxed{203.84\,\text{kW}}$$

(b) Given, $x = 1$ m, $y = 2$ m and $L = 1$ m. Therefore,

$$X = x/L = 1, \; Y = y/L = 2$$

From Table 6.2, we have

$$\begin{aligned} F_{12} &= \frac{2}{2\,\pi}\left[\ln\left(\frac{2\times 5}{1+1+4}\right)^{1/2} + \sqrt{5}\tan^{-1}\frac{1}{\sqrt{5}} \right. \\ &\qquad \left. + 2\sqrt{2}\tan^{-1}\frac{2}{\sqrt{2}} - \tan^{-1}1 - 2\tan^{-1}2\right] \\ &= \frac{1}{\pi}(0.255 + 0.94 + 2.7 - 0.785 - 2.2) \\ &= 0.2897 \end{aligned}$$

Thus, the assumed values of $F_{12}$ and $F_{21}$ are close to the actual value.

## Example 6.9

The surfaces of the cylindrical furnace of radius 1 m and height 1 m, shown in Figure EP6.9, are maintained at constant temperatures. The side surface of the furnace closely approximates a blackbody. Determine the net rate of radiation heat transfer at each surface at steady-state, if the view factor $F_{12} = 0.38$.

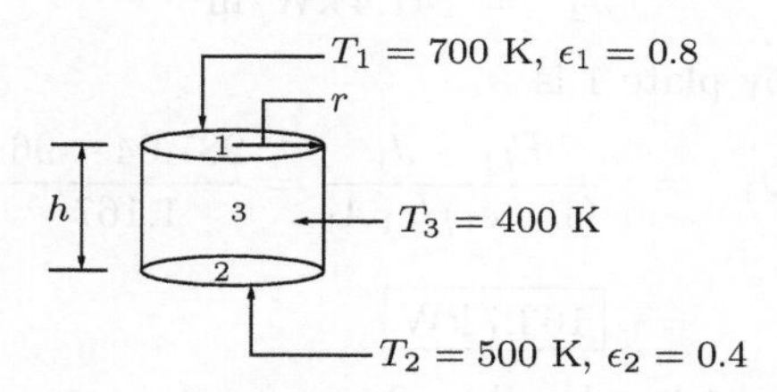


**Figure EP6.9**
Cylindrical furnace.

## Solution

Given, $r = 1$ m, $h = 1$ m, $F_{12} = 0.38$.

The area of the surfaces are

$$A_1 = A_2 = \pi\,r^2 = \pi \times 1$$

$$= \quad 3.14\,\text{m}^2$$

$$A_3 = 2\pi rh = 2\pi \times 1 \times 1$$

$$= \quad 6.28\,\text{m}^2$$

By summation rule,

$$F_{11} + F_{12} + F_{13} = 1$$

The base surface is flat and thus, $F_{11} = 0$. Therefore,

$$F_{13} = 1 - F_{11} - F_{12}$$

$$= 1 - 0 - 0.38 = 0.62$$

The top and bottom surfaces are symmetric about the side surface; therefore, $F_{21} = F_{12} = 0.38$ and $F_{23} = F_{32} = 0.62$.

The view factor $F_{31}$ can be determined from the reciprocity rule, Equation (6.23),

$$A_1F_{13} = A_3F_{31}$$

$$F_{31} = \frac{A_1}{A_3}F_{13} = \frac{3.14}{6.28} \times 0.62$$

$$= 0.31$$

Also, $F_{13} = F_{31} = 0.31$ due to symmetry.

The radiation network associated with the problem is as shown in Figure ES6.9.

The thermal resistances are

$$R_1 = \frac{1-\epsilon_1}{A_1\epsilon_1} = \frac{1-0.8}{3.14 \times 0.8} = 0.08$$

$$R_2 = \frac{1-\epsilon_2}{A_2\epsilon_2} = \frac{1-0.4}{3.14 \times 0.4} = 0.48$$

$$R_3 = \frac{1-\epsilon_3}{A_3\epsilon_3} = 0 \quad \text{(since 3 is a blackbody)}$$

$$R_{12} = \frac{1}{A_1F_{12}} = \frac{1}{3.14 \times 0.38} = 0.84$$

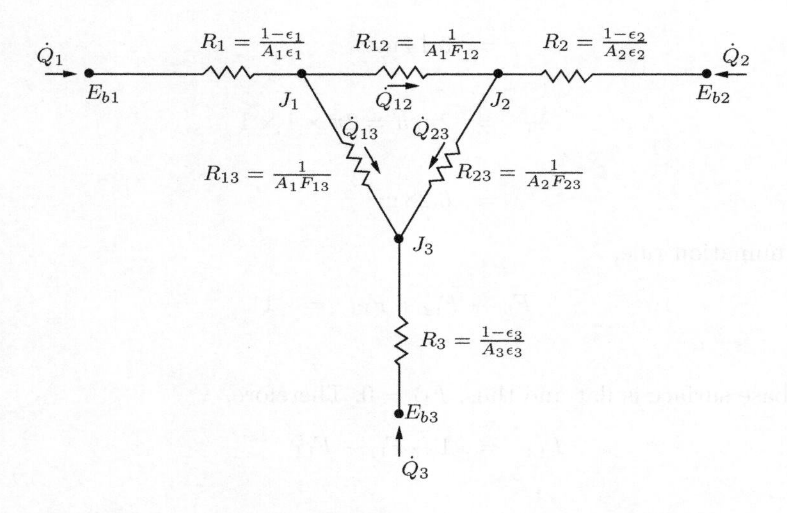


**Figure ES6.9**
Radiation network.

$$R_{13} = \frac{1}{A_1F_{13}} = \frac{1}{3.14 \times 0.62} = 0.51$$

$$R_{23} = \frac{1}{A_2F_{23}} = \frac{1}{3.14 \times 0.62} = 0.51$$

The emissive power of the surfaces are

$$E_{b1} = \sigma T_1^4 = 5.67 \times 10^{-8} \times 700^4 = 13614\,\mathrm{W/m^2}$$

$$E_{b2} = \sigma T_2^4 = 5.67 \times 10^{-8} \times 500^4 = 3544\,\mathrm{W/m^2}$$

$$E_{b3} = \sigma T_3^4 = 5.67 \times 10^{-8} \times 400^4 = 1452\,\mathrm{W/m^2}$$

To calculate the heat flow at each surface, we should determine the radiosities $J_1$, $J_2$ and $J_3$. The network is solved by setting the sum of the heat currents entering the nodes $J_1$ and $J_2$ to zero.

$$\text{Node } J_1: \quad \frac{E_{b1} - J_1}{0.08} + \frac{J_2 - J_1}{0.84} + \frac{E_{b3} - J_1}{0.51} = 0$$

$$\text{Node } J_2: \quad \frac{J_1 - J_2}{0.84} + \frac{E_{b2} - J_2}{0.48} + \frac{E_{b3} - J_2}{0.51} = 0$$

But, $E_{b3} = J_3$. Thus,

$$\frac{13614 - J_1}{0.08} + \frac{J_2 - J_1}{0.84} + \frac{1452 - J_1}{0.51} = 0$$

$$\frac{J_1 - J_2}{0.08} + \frac{3544 - J_2}{0.48} + \frac{1452 - J_1}{0.51} = 0$$

Solving, we get $J_1 = 11418$ W/m$^2$ and $J_2 = 4562$ W/m$^2$.

The net rates of radiation heat transfer at the three surfaces are

$$\dot{Q}_1 = \frac{E_{b1} - J_1}{(1 - \epsilon_1)/(\epsilon_1 A_1)} = \frac{13614 - 11418}{0.08}$$

$$= \boxed{27450 \text{ W}}$$

$$\dot{Q}_2 = \frac{E_{b2} - J_2}{(1 - \epsilon_2)/(\epsilon_2 A_2)} = \frac{3544 - 4562}{0.48}$$

$$= \boxed{-2121 \text{ W}}$$

$$\dot{Q}_3 = \frac{J_3 - J_1}{1/(A_1 F_{13})} + \frac{J_3 - J_2}{1/(A_2 F_{23})}$$

$$= \frac{1452 - 11418}{0.51} + \frac{1452 - 4562}{0.51}$$

$$= \boxed{-25639 \text{ W}}$$

Note that the direction of net radiation heat transfer is from the top surface to the base and side surfaces. The algebraic sum of these quantities must be zero, that is,

$$\dot{Q}_1 + \dot{Q}_2 + \dot{Q}_3 = 27450 - 2121 - 25639 = -310 \text{ W}$$

We get the sum as $-310$ W instead of zero. This is due to the accumulation of truncation error in the calculation at different steps.

## 6.8 Radiation Shields

Radiation shields are surfaces meant for reducing the net radiation transfer between two surfaces. Radiation shields are constructed from low emissivity (high reflectivity) materials. Consider the radiation shield, surface 3, placed between two large, parallel planes, as shown in Figure 6.15.

The net rate of radiation transfer between surfaces 1 and 2, without the radiation shield, given by Equation (6.38), is

$$\dot{Q}_{12} = \frac{\sigma\left(T_1^4 - T_2^4\right)}{\dfrac{1-\epsilon_1}{\epsilon_1 A_1} + \dfrac{1}{A_1 F_{1\to 2}} + \dfrac{1-\epsilon_2}{\epsilon_2 A_2}}$$

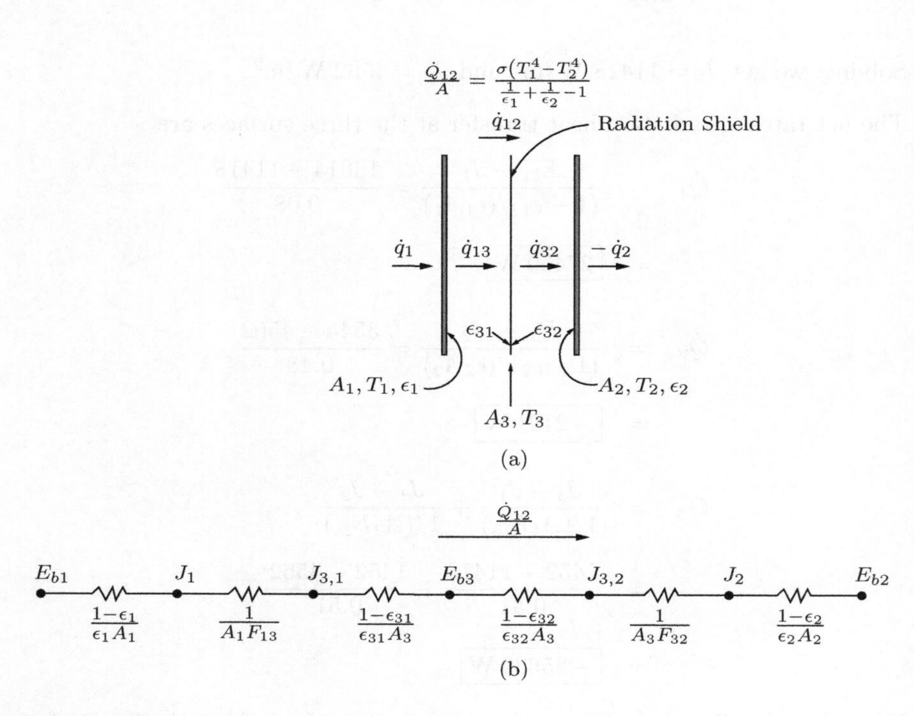


**Figure 6.15**
Radiation exchange between two large parallel planes, with a radiation shield placed in between them.

But, $A_1 = A_2 = A$ and $F_{1\to 2} = 1$. Therefore,

$$\begin{aligned}
\dot{Q}_{12} &= \frac{\sigma\left(T_1^4 - T_2^4\right)}{\dfrac{1-\epsilon_1}{\epsilon_1 A} + \dfrac{1}{A} + \dfrac{1-\epsilon_2}{\epsilon_2 A}} \\
&= \frac{\sigma A\left(T_1^4 - T_2^4\right)}{\dfrac{1}{\epsilon_1} + \dfrac{1}{\epsilon_2} - 1}
\end{aligned}$$

or

$$\frac{\dot{Q}_{12}}{A} = \frac{\sigma\left(T_1^4 - T_2^4\right)}{\dfrac{1}{\epsilon_1} + \dfrac{1}{\epsilon_2} - 1}$$

But with the radiation shield, additional resistances are present, as shown in Figure 6.15(b), and the heat transfer rate is reduced. It is important to note that the emissivity associated with one side of the shield ($\epsilon_{31}$) may differ from that associated with the opposite side ($\epsilon_{32}$) and the radiosities will also differ. Considering all the resistances, we have the total resistance as

$$R_{\text{th,total}} = \frac{1-\epsilon_1}{A_1\epsilon_1} + \frac{1}{A_1 F_{1\rightarrow 3}} + \frac{1-\epsilon_{31}}{A_3\epsilon_{31}} + \frac{1-\epsilon_{32}}{A_3\epsilon_{32}} + \frac{1}{A_3 F_{3\rightarrow 2}} + \frac{1-\epsilon_2}{A_2\epsilon_2}$$

Assuming that, $A_1 = A_2 = A_3 = A$, we have $F_{1\rightarrow 3} = F_{3\rightarrow 2} = 1$, therefore,

$$\begin{aligned} R_{\text{th,total}} &= \frac{1-\epsilon_1}{A\,\epsilon_1} + \frac{1}{A} + \frac{1-\epsilon_{31}}{A\,\epsilon_{31}} + \frac{1-\epsilon_{32}}{A\,\epsilon_{32}} + \frac{1}{A} + \frac{1-\epsilon_2}{A\,\epsilon_2} \\ &= \frac{1}{A}\left(\frac{1}{\epsilon_1} + \frac{1-\epsilon_{31}}{\epsilon_{31}} + \frac{1-\epsilon_{32}}{\epsilon_{32}} + \frac{1}{\epsilon_2}\right) \end{aligned}$$

Hence, the rate of heat transfer between surfaces 1 and 2, with a radiation shield in between them becomes

$$\dot{Q}_{12} = \frac{\sigma A\left(T_1^4 - T_2^4\right)}{R_{\text{th,total}}}$$

or

$$\boxed{\frac{\dot{Q}_{12}}{A} = \frac{\sigma\left(T_1^4 - T_2^4\right)}{\dfrac{1}{\epsilon_1} + \dfrac{1}{\epsilon_2} + \dfrac{1-\epsilon_{31}}{\epsilon_{31}} + \dfrac{1-\epsilon_{32}}{\epsilon_{32}}}} \tag{6.39}$$

The above procedure may be extended to deal with multiple radiation shields. For the special case of multiple radiation shields (for $N$ shields) with all shields having equal emissivity, it can be shown that,

$$(\dot{q}_{12})_N = \frac{1}{N+1}\,(\dot{q}_{12})_0 \tag{6.40}$$

where $(\dot{q}_{12})_0$ is the radiation transfer rate with no shield.

Note: The expression in Equation (6.40) is compact because of the assumption that the surfaces and the radiation shield are of equal area. If the surfaces are of different area, then the resulting expression for the heat transfer between the plates will not be compact.

## Example 6.10

Two large parallel planes with emissivities 0.4 and 0.7 exchange heat. Find the percentage reduction in heat transfer, when a polished aluminum radiation shield ($\epsilon = 0.04$) is placed between them.

## Solution

The heat transfer without the shield is

$$\begin{aligned}\frac{\dot{Q}_{12}}{A} &= \frac{\sigma\left(T_1^4 - T_2^4\right)}{\dfrac{1}{\epsilon_1} + \dfrac{1}{\epsilon_2} - 1} \\ &= \frac{\sigma\left(T_1^4 - T_2^4\right)}{\dfrac{1}{0.4} + \dfrac{1}{0.7} - 1} \\ &= 0.341\sigma\left(T_1^4 - T_2^4\right)\end{aligned}$$

The radiation network for the problem with the shield in place is shown in the Figure ES6.10.

$q/A$

$E_{b1}$ — $(1-\epsilon_1)/\epsilon_1$ — $J_1$ — $1/F_{13}$ — $J_{31}$ — $(1-\epsilon_{31})/\epsilon_{31}$ — $E_{b3}$ — $(1-\epsilon_{32})/\epsilon_{32}$ — $J_{32}$ — $1/F_{32}$ — $J_2$ — $(1-\epsilon_2)/\epsilon_2$ — $E_{b2}$

**Figure ES6.10**
Radiation network.

$$\begin{aligned}\frac{1-\epsilon_1}{\epsilon_1} &= \frac{1-0.4}{0.4} \\ &= 1.5\end{aligned}$$

$$\begin{aligned}\frac{1-\epsilon_{31}}{\epsilon_{31}} &= \frac{1-0.04}{0.04} \\ &= 24\end{aligned}$$

$$\begin{aligned}\frac{1-\epsilon_{32}}{\epsilon_{32}} &= \frac{1-0.04}{0.04} \\ &= 24\end{aligned}$$

$$\frac{1}{F_{13}} = 1$$

$$\frac{1}{F_{23}} = 1$$

$$\frac{1-\epsilon_2}{\epsilon_2} = \frac{1-0.7}{0.7} = 0.43$$

The total resistance is

$$1.5 + (2 \times 24) + (2 \times 1) + 0.43 = 51.93$$

Therefore, the heat transfer with radiation shield becomes

$$\begin{aligned}\left(\dot{Q}_{12}/A\right)_{\text{shield}} &= \frac{\sigma\left(T_1^4 - T_2^4\right)}{51.93} \\ &= 0.01926\,\sigma\left(T_1^4 - T_2^4\right)\end{aligned}$$

Thus, the reduction in heat transfer is

$$\begin{aligned}\frac{\left(\dot{Q}_{12}/A\right)_{\text{shield}} - \left(\dot{Q}_{12}/A\right)}{\left(\dot{Q}_{12}/A\right)} &= \frac{0.341 - 0.01926}{0.341} \\ &= \boxed{94.35\,\%}\end{aligned}$$

### 6.8.1 Radiating Surface

*Radiating surface* is an idealized surface characterized by zero net radiation transfer ($\dot{Q}_i = 0$). This assumption is common to many industrial applications. Radiating surface may only be approximated to such real surfaces that are well insulated on one side, and for which convection effects may be neglected on the opposite (radiation) side. Let us consider surface $i$, illustrated in Figure 6.11. The net radiation flux $q_i$ leaving surface $i$ is given by

$$q_i = J_i - G_i$$

where $J_i$ is the radiation flux leaving surface $i$, termed radiosity and $G_i$ is the radiation flux incident on surface $i$, known as irradiation.

Now, let $\epsilon_i$ and $\rho_i$ be the emissivity and reflectivity of surface $i$. Then the radiosity of surface $i$ becomes

$$J_i = \epsilon_i E_{bi} + \rho_i G_i$$

where $E_{bi}$ is the emissive power of a surface identical to $i$. Assuming that the radiation properties—the reflection $\rho_i$, transmission $\tau_i$ and absorption $\alpha_i$—are uniform and independent of direction, and the surface is opaque and gray, we have

$$\alpha_i + \rho_i + \tau_i = 1$$

But for opaque surface, $\tau = 0$, and for gray surface, $\alpha = \epsilon$, thus,

$$\rho_i = 1 - \alpha_i = 1 - \epsilon_i$$

Therefore, the radiosity becomes

$$J_i = \epsilon_i E_{bi} + (1 - \epsilon_i) G_i$$

Substituting this, we get the net radiation flux $q_i$ leaving surface $i$, as

$$q_i = \epsilon_i E_{bi} + (1 - \epsilon_i) G_i - G_i$$

For the radiating surface with $\dot{q}_i = 0$, this gives

$$G_i = E_{bi}$$

With this, the radiosity becomes

$$J_i = E_{bi}$$

Hence, for a radiating surface, we have

$$\boxed{G_i = J_i = E_{bi}}$$

This shows that if the radiosity of a radiating surface is known, its temperature is readily determined. In an enclosure, the equilibrium temperature of a radiating surface is determined by its interaction with the other surfaces. The equilibrium temperature is independent of the emissivity of the reradiating surface.

Consider the three-surface enclosure with surface $R$ as the reradiating surface, and the corresponding network shown in Figure 6.16. Surface $R$ is assumed to be well insulated, and convection effects are assumed to be negligible. Hence, with $\dot{Q}_R = 0$, the net radiation transfer from surface 1 must be equal to the net radiation transfer to surface 2. The network is a simple series-parallel arrangement, and from its analysis, we get

$$\dot{Q}_1 = -\dot{Q}_2 = \frac{E_{bi} - E_{b2}}{\dfrac{1-\epsilon_1}{\epsilon_1 A_1} + \dfrac{1}{A_1 F_{12} + [(1/A_1 F_{1R}) + (1/A_2 F_{2R})]^{-1}} + \dfrac{1-\epsilon_2}{\epsilon_2 A_2}} \tag{6.41}$$

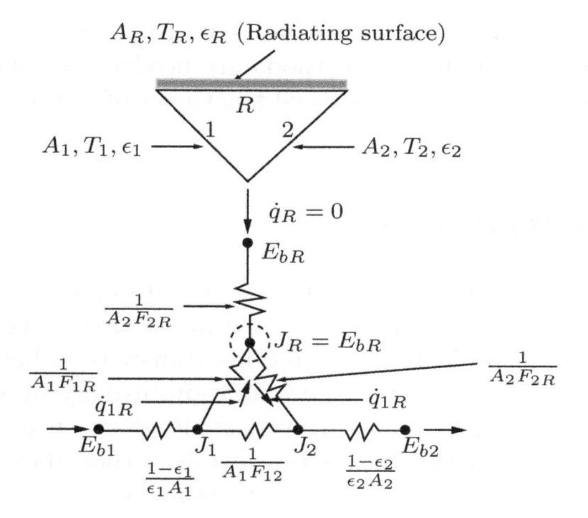


**Figure 6.16**
A three-surface enclosure with one surface radiating.

Knowing $\dot{Q}_1 = -\dot{Q}_2$, Equation (6.30) may be applied to surfaces 1 and 2 to determine their radiosities $J_1$ and $J_2$. Once $J_1$ and $J_2$ and the geometrical resistances are known, the radiosity of the radiating surface $J_R$ may be determined from the requirement that

$$\frac{J_1 - J_R}{\dfrac{1}{A_1\, F_{1R}}} - \frac{J_R - J_2}{\dfrac{1}{A_2\, F_{2R}}} = 0 \tag{6.42}$$

The temperature of the reradiating surface may be determined from the requirement that $\sigma\, T_R^4 = J_R$.

## 6.9 Limitations

So far, we have been discussing the means for predicting radiation exchange between surfaces. In our discussions, although we have developed methods to predict radiation exchange, it is important to realize the inherent limitations of these methods. To begin with, we have considered *isothermal, opaque, gray* surfaces that *emit* and *reflect diffusely*, and that are characterized by *uniform* surface *radiosity* and *irradiation.* For enclosures we have also considered the medium that separates the surfaces to be *nonparticipating*; that is, it neither absorbs nor scatters the surface radiation, and it emits no radiation.

Even though we made so many assumptions in their development, the foregoing conditions, and the related equation may often be used to obtain reliable first estimates and, in some cases, highly accurate results for radiation transfer in an enclosure. However, sometimes the assumptions are grossly inappropriate and more refined prediction methods are needed. For such methods one may refer to books devoted to advanced treatment of radiation transfer, such as reference 3.

## 6.10 Gas Radiation

Radiation exchange between a gas and a heat transfer surface is considerably more complex than the situations discussed in the preceding sections. Unlike most solid bodies, gases in many situations transport radiation even at low temperatures. Gases absorb and emit radiation, usually in a certain narrow wavelength bands. At low temperatures, some gases such as nitrogen and oxygen transmit heat, without radiating, while carbon dioxide, water vapor and hydrocarbon gases radiate to an appreciable extent.

Let us consider the gas layer, shown in Figure 6.17. With reference to this layer, let a monochromatic beam with radiation spectral intensity of $I_\lambda$ and thickness $dx$ propagate through the gas layer. While traveling through the gas, the intensity of the monochromatic beam will be attenuated as a result of absorption by the gas layer and augmented as a result of emission of radiation owing to the temperature of the gas layer.

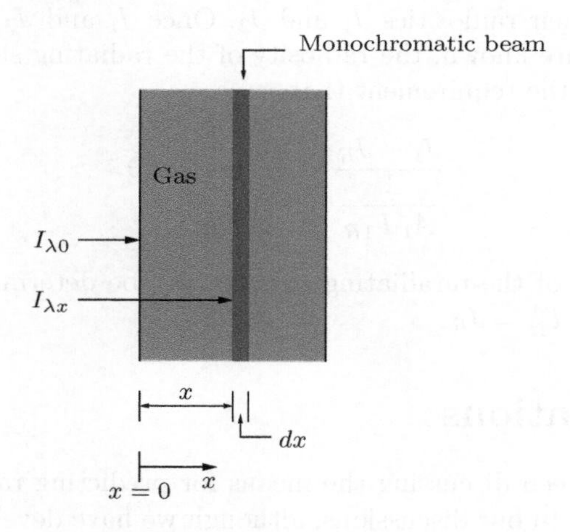


**Figure 6.17**
Absorption in a gas layer.

The decrease of radiation intensity, resulting from absorption in the layer is assumed to be proportional to the thickness of the layer and the intensity of radiation at that point. Thus,

$$dI_\lambda = -a_\lambda I_\lambda dx \tag{6.43}$$

where the proportionality constant $a_\lambda$ is called the *monochromatic absorption coefficient.* Integrating from the edge ($x = 0$) of the gas layer, where the spectral intensity of the beam is $I_{\lambda 0}$, to the edge of the beam ($x = x$), where the spectral intensity of the beam is $I_{\lambda x}$, we get

$$\begin{aligned}
\int_{I_{\lambda 0}}^{I_{\lambda x}} dI_\lambda &= \int_0^x -a_\lambda I_\lambda dx \\
\int_{I_{\lambda 0}}^{I_{\lambda x}} \frac{dI_\lambda}{I_\lambda} &= \int_0^x -a_\lambda dx \\
\Big[\ln I_\lambda\Big]_{I_{\lambda 0}}^{I_{\lambda x}} &= -a_\lambda \Big[x\Big]_0^x \\
\ln\left(\frac{I_{\lambda x}}{I_{\lambda 0}}\right) &= -a_\lambda\, x
\end{aligned}$$

or

$$\frac{I_{\lambda x}}{I_{\lambda 0}} = e^{-a_\lambda x} \tag{6.44}$$

Equation (6.44) is called *Beer's law.* This represents the familiar exponential-decay formula experienced in many types of radiation analysis dealing with absorption. In accordance with our discussions in Section 6.3, the monochromatic transmissivity will be given by

$$\tau_\lambda = e^{-a_\lambda x} \tag{6.45}$$

If the gas is nonreflecting, then, by Equation (6.7),

$$\tau_\lambda + \alpha_\lambda = 1$$

and

$$\alpha_\lambda = 1 - e^{-a_\lambda x} \tag{6.46}$$

Gases frequently absorb only narrow wavelength bands. Analysis of gas-radiation problems are quite complicated, and for such calculations references 2-4 may be consulted. For practical calculations, reference 2 presents simplified procedures to determine the emittance of water vapor and carbon dioxide gas.

## 6.11 Radiation from Real Surfaces

Radiation from actual surfaces differs significantly from the ideal surfaces assumed in our discussions. For example, real surfaces are not perfectly diffuse, and hence the intensity of emitted radiation is not constant over all directions. The radiation emission characteristics of materials which are electrical conductors differs significantly from that of materials which are not conductors of electricity. Conductors such as copper emit more energy in a direction having large azimuth angle. As a result of this basic difference in the behavior of conductors and nonconductors, an electric conducting material, say a sphere of copper will appear bright around the rim since more energy is emitted at large angles. A sphere made up of a nonconducting material will have the opposite behavior, and will appear bright at the center and dark at the edge.

The reflection and absorption of thermal radiation from real surfaces are strongly influenced by the surface of the object and the surrounding environment in which it is present. There properties are dependent on the direction and wavelength of the incident radiation. But the distribution of the intensity of radiation with wavelength may be a complicated function of the temperature and surface characteristics of the object and the characteristics of the surroundings. Let the incident radiation on a surface per unit time, per unit area and per unit wavelength be $\epsilon_\lambda$. Then the absorptivity $\alpha$ will be given by the ratio of the energy absorbed to the energy incident on the surface,

$$\alpha = \frac{\int_0^\infty \alpha_\lambda G_\lambda \, d\lambda}{\int_0^\infty G_\lambda \, d\lambda}$$

For a gray body with $G_\lambda = \epsilon =$ constant, the absorptivity can be evaluated easily. Also, it can be shown that, Kirchhoff's law [Equation (6.10)] for monochromatic radiation may be expressed as

$$\epsilon_\lambda = \alpha_\lambda$$

For a *gray body*, the monochromatic emissivity $\epsilon_\lambda$ is independent of wavelength. Therefore, for a gray body $\alpha_\lambda =$ constant, and the above expression for absorptivity implies that the absorptivity is a constant which is independent of the wavelength distribution and incident radiation. Furthermore, since the absorptivity $\alpha$ and emissivity $\epsilon$ are constant over all wavelengths for a gray body, they must be independent of the temperature also. It is essential to note that, the real surfaces are not always *gray* in nature, hence assuming gray body behavior for real surfaces might result in significant error in the radiation analysis.

## 6.12 Solar Radiation

Solar radiation is a form of thermal radiation having a particular wavelength distribution. The intensity of solar radiation is strongly influenced by the atmospheric condition, period of the year, and the angle of incidence of the rays from the sun to the surface of the earth. At the outer limit of the atmosphere, the total solar irradiation $G$ is 1395 W/m$^2$, when the earth is at its mean distance from the sun. This value is called the *solar constant*. It is essential to note that all the energy expressed by the solar constant does not reach the surface of the earth, because the carbon dioxide and water vapor in the atmosphere will absorb considerable portion of the solar energy passing through the atmosphere. Further, the solar radiation incident on the earth's surface is also influenced by the dust and other pollutants present in the atmosphere.

The solar radiation reaching the surface of the earth does not behave like the radiation from an ideal graybody, even though outside the atmosphere the distribution of energy follows more of an ideal pattern. To determine an equivalent blackbody temperature for the solar radiation, the wavelength of about 0.5 $\mu$m, at which the solar flux is found to be the maximum, may be employed. With this consideration the equivalent solar temperature for the thermal radiation is found to be about 5800 K [1].

Solar radiation analysis would be comfortable if all metals exhibit graybody behavior. But, because the solar radiation is concentrated at short wavelengths, unlike much longer wavelengths for most earthbound thermal radiation, a particular material may exhibit entirely different absorption and transmission properties for the two types of radiation. The typical example for this behavior is the greenhouse effect. Usually a glass transmits radiation at wavelengths below 2 $\mu$m, thus it transmits a large portion of solar radiation incident on it. However, the glass is opaque to radiation with long wavelengths above 3 or 4 $\mu$m. All the low-temperature radiation emitted by the objects in a greenhouse is of such a long wavelength character and hence remains trapped in the greenhouse. Thus, the glass allows much more radiation to come in than can escape, thereby producing the familiar heating effect. The solar radiation absorbed by objects in the greenhouse must eventually be dissipated to the surroundings by convection from the outer wall surface of the greenhouse.

In many instances, the total absorptivity for solar radiation can be different from the absorptivity for blackbody radiation at some moderate temperature. The absorptivities of solar and low-temperature radiations for some surfaces are compared in Table 6.3.

## 6.13 Radiation Properties of the Environment

In our discussions in this chapter, we saw that the major portion of the solar energy is concentrated in the short-wavelength region. However, real surfaces may exhibit substantially different absorption properties for solar radiation than that of the long-wavelength *earthbound* radiation. The direct solar radiation incident on a horizontal surface per unit area, per unit time is called *insolation*. The insolation, represented with symbol $I$, is used to describe the intensity of solar radiation in the science of meteorology and hydrology. In meteorology, the intensity of solar radiation is given unit "langley". Ly, which is 1 cal/cm$^2$. Insolation and radiation intensity are often expressed in langleys per unit time, for example, the Stefan-Boltzmann constant would be

$$\sigma = 0.826 \times 10^{-10}\,\text{Ly}/(\text{min K}^4)$$

Radiation heat transfer environment is dictated by the absorption, scattering, and reflection properties of the atmosphere and natural surfaces. The

**Table 6.3** Absorptivities of solar and low-temperature (around 25°C) radiations [7]

| Surface | Solar | Low-temperature |
|---|---|---|
| Polished aluminum | 0.15 | 0.04 |
| Polished copper | 0.18 | 0.03 |
| Tarnished copper | 0.65 | 0.75 |
| Cast iron | 0.94 | 0.21 |
| Polished Stainless steel No. 301 | 0.37 | 0.60 |
| White marble | 0.46 | 0.95 |
| Asphalt | 0.90 | 0.90 |
| Brick (red) | 0.75 | 0.93 |
| Gravel | 0.29 | 0.85 |
| Flat black lacquer | 0.96 | 0.95 |
| White paints and pigments | 0.12 - 0.16 | 0.90 - 0.95 |

scattering that occurs in the atmosphere is of two types, namely the *molecular scattering* and *particulate scattering*. Molecular scattering is caused by the interaction of radiation with individual molecules, and the particulate scattering is due to the interaction of radiation with the many types of particles suspended in the air. The blue color of the sky is the result of the scattering of the violet (short) wavelength by the air molecules.

Water droplets, dust, and smog are the important types of particle scattering centers. The scattering process is governed by the size of the particle in comparison with the wavelength of radiation. Maximum scattering is found to occur when the wavelength and particle size are equal and decrease progressively for longer wavelengths. For wavelengths smaller than the particle size, the radiation tends to be reflected.

The reflection properties of surfaces are described by defining a parameter termed *albedo A*, defined as

$$A = \frac{\text{reflected energy}}{\text{incident energy}} \tag{6.47}$$

The atmosphere absorbs radiation selectively in narrow-wavelength bands. The earth's atmosphere acts like a greenhouse, trapping the incident solar radiation and provides energy and warmth we require.

The absorption and scattering of radiation may be described with Beer's law [Equation (6.44)]

$$\frac{I_{\lambda x}}{I_{\lambda 0}} = e^{-a_\lambda x}$$

where $a_\lambda$ is the monochromatic absorption coefficient and $x$ is the thickness of the layer absorbing the radiation.

In the study of atmospheric problems, it is usual to assume that the absorption and scattering processes can be imposed on each other and may be expressed in the form of Beer's law over all wavelengths. The appropriate coefficients are defined as $a_{ms}$, the average molecular scattering coefficient over all wavelengths, $a_{ps}$, the average particulate coefficient over all wavelengths, and $a$, the average absorption coefficient over all wavelengths. The total attenuation coefficient over all wavelengths becomes

$$a_t = a_{ms} + a_{ps} + a$$

The radiation insolation at the earth's surface can be expressed as

$$\frac{I_c}{I_o} = e^{-a_t\, m} = e^{-n a_{ms}\, m} \tag{6.48}$$

where $I_c$ is the insolation for the direct, cloudless-sky at earth's surface, and $I_o$ is the insolation at the outer limits of earth's atmosphere, $m$ is the relative thickness of the air mass and $n$ is defined as the turbidity factor of the air, defined as

$$n = \frac{a_t}{a_{ms}} \tag{6.49}$$

The molecular scattering coefficient for air at atmospheric pressure is given as [8]

$$a_{ms} = 0.128 - 0.054 \log m \tag{6.50}$$

The relative thickness of the air mass is calculated as the cosecant of the solar altitude $\alpha$. The turbidity factor is thus a convenient means of specifying atmospheric purity and clarity. The turbidity ranges from about 2 for clear air to 4 or 5 for smoggy, industrial environment.

The insolation at the outer edge of the atmosphere is given by

$$I_o = E_{bo} \sin \alpha \tag{6.51}$$

where $\alpha$ is the angle the solar rays make with the horizontal and $E_{bo}$ is the solar constant equal to 1395 W/m$^2$.

## Example 6.11

In a smoggy atmosphere, the direct, cloudless-sky insolation for a solar altitude angle of 75° is 800 W/m$^2$. Determine the turbidity of this atmosphere.

## Solution

Given, $I_c = 800\ \text{W/m}^2$, $\alpha = 75°$.

The insolation at the outer edge of the atmosphere, by Equation (6.51), is

$$\begin{aligned} I_o &= E_{bo} \sin\alpha = 1395 \times \sin 75° \\ &= 1347.5\,\text{W/m}^2 \end{aligned}$$

The relative thickness of air mass is

$$m = \text{cosec}\, 75 = 1.035$$

The molecular scattering for air at atmospheric pressure, given by Equation (6.50), is

$$\begin{aligned} a_{ms} &= 0.128 - 0.054 \log m \\ &= 0.128 - 0.054 \times \log(1.035) \\ &= 0.1272 \end{aligned}$$

By Equation (6.48), the insolation for the cloudless-sky is

$$I_c = I_o e^{(-n a_{ms}\, m)}$$

Therefore,

$$\begin{aligned} 800 &= 1347.5\, e^{(-n \times 0.1272 \times 1.035)} \\ e^{(-0.1316\, n)} &= \frac{800}{1347.5} = 0.5937 \\ -0.1316\, n &= \ln(0.5937) = -0.5214 \\ n &= \frac{0.5214}{0.1316} \\ &= \boxed{3.962} \end{aligned}$$

The turbidity factor of the given atmosphere is 3.962.

## 6.14 Radiation Absorption in Water

The radiation absorption in natural bodies of water influences the evaporation rates and the dispersion of water vapor in the atmosphere. Due to these, the absorption of radiation in water bodies becomes important. Experimental measurements show that the solar radiation is absorbed rapidly in the top layers of water followed by an approximately exponential decay with depth in the water. The incident radiation follows a variation of [9]

$$\frac{I_z}{I_s} = (1-\beta)\, e^{-az} \tag{6.52}$$

where $I_s$ is the intensity at the top surface and $I_z$ is the intensity at a depth $z$, $\beta$ is the fraction of energy absorbed at the surface, and $a$ is an absorption or extinction coefficient of the material. The parameter $\beta$ is a measure of the long-wavelength content for solar radiation, since the shorter wavelengths easily penetrate the water. This coefficient $\beta$ is about 0.4 for water in lakes, and is assumed to be independent of time. The extinction coefficient $a$ is found to vary considerably. For example $a$ is found to be 0.16 $m^{-1}$ for extremely clear water [10], and 0.89 $m^{-1}$ for turbid water [11].

## Example 6.12

On a clear day with a solar altitude of 90°, if the rate of heat-generation due to solar radiation at a depth of 500 mm from the surface of a lake water is 235 $W/m^3$, determine the extinction coefficient.

## Solution

Given, $\dot{q} = 235$ $W/m^3$, $\alpha = 90°$, $z = 0.5$ m.

The heat generation $\dot{q}$ is the difference between the incoming and outgoing radiation heat flux. Therefore,

$$\begin{aligned} \dot{q}_z &= \frac{I_z - I_{(z+\Delta z)}}{\Delta z} \\ &= -\frac{\Delta I_z}{\Delta z} \end{aligned}$$

In the limiting case,

$$\dot{q}_z = -\frac{dI_z}{dz}$$

From Equation (6.52),

$$\frac{dI_z}{dz} = -I_s a\,(1-\beta)\, e^{-az}$$

Therefore,

$$\dot{q} = I_s a\,(1-\beta)\, e^{-az}$$

The insolation at the surface of the lake water, by Equation (6.51), is

$$I_o = E_{bo} \sin \alpha = 1395\,\mathrm{W/m^2}$$

The relative thickness of air mass is

$$\begin{aligned} m &= \operatorname{cosec} \alpha = \operatorname{cosec} 90 \\ &= 1.0 \end{aligned}$$

The molecular scattering coefficient for air at atmospheric pressure, by Equation (6.50), is

$$\begin{aligned} a_{ms} &= 0.128 - 0.054 \log m \\ &= 0.128 - 0.054 \log(1) \\ &= 0.128 \end{aligned}$$

For clear sky, the turbidity $n = 2$. For lake water, the long-wavelength constant $\beta = 0.4$. Therefore, $I_c$, by Equation (6.48), becomes

$$\begin{aligned} I_c &= I_o e^{-n a_{ms} m} \\ &= 1395\, e^{-2 \times 0.128 \times 1} \\ &= 1079.93\,\mathrm{W/m^2} \end{aligned}$$

For the present problem, $I_c = I_s$, thus,

$$\begin{aligned} \dot{q} &= 235 = 1079.93 \times a \times (1 - 0.4) \times e^{-a \times 0.5} \\ a\, e^{-0.5\, a} &= \frac{235}{1079.93 \times (1 - 0.4)} \\ &= 0.363 \end{aligned}$$

Solving, we get the extinction coefficient $a$ as $\boxed{0.456\,\mathrm{m^{-1}}}$.

## 6.15 Radiation Effect on Temperature Measurement

Let us examine the thermometer placed in a gas stream, as shown in Figure 6.18.

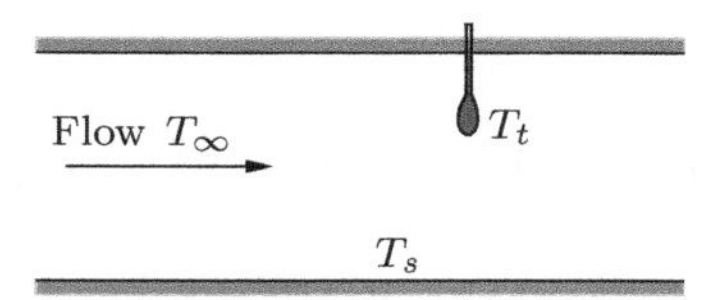


**Figure 6.18**
Thermometer in a gas stream.

The temperature of the gas stream can be determined by applying the overall energy balance for the thermometer bulb. Let $T_\infty$ be the temperature of the gas stream, $T_s$ be the effective temperature of surrounding causing radiation, and $T_t$ be the temperature indicated by the thermometer. When $T_\infty$ greater than $T_s$, energy will be transferred by convection to the thermometer, and then radiated to the surroundings. Thus the energy balance becomes

$$hA\left(T_\infty - T_s\right) = \sigma A \epsilon \left(T_t^4 - T_s^4\right) \tag{6.53}$$

where $h$ is the convection coefficient of the gas, $A$ is the surface area of the thermometer bulb and $\epsilon$ is its emissivity. In this energy balance relation, the surroundings is assumed to be either very large or black so that the radiation relation

$$\dot{Q} = \sigma A \epsilon \left(T_t^4 - T_s^4\right)$$

can be applied.

It is essential to note that the temperature indicated by the thermometer is not the actual temperature of the gas stream but some radiation-convection equilibrium temperature. Significant error can result if the temperature measurement in this energy balance is not properly accounted. More often radiation shields are used to minimize this error.

## Example 6.13

Hot air flows through a duct with a constant wall temperature of 35°C. A thermometer inserted into the air stream reads 150°C. If the emissivity of the thermometer bulb is 0.8 and the convection coefficient of air is 250 W/(m$^2$ °C), calculate the actual temperature of the air stream.

## Solution

Given, $T_s = 35 + 273 = 308$ K, $T_t = 150 + 273 = 423$ K, $\epsilon = 0.8$, $h = 250$ W/(m$^2$ °C).

By Equation (6.53),

$$hA\left(T_\infty - T_s\right) = \sigma A \epsilon \left(T_t^4 - T_s^4\right)$$

Therefore,

$$
\begin{aligned}
250\,(T_\infty - 423) &= 0.8 \times (5.67 \times 10^{-8}) \times \left(423^4 - 308^4\right) \\
&= 1044 \\
T_\infty &= \frac{1044}{250} + 423 \\
&= \boxed{427.18\,\mathrm{K}}
\end{aligned}
$$

## 6.16 Radiation Heat Transfer Coefficient

In our discussions on convection in Chapter 5, the convection heat transfer coefficient $h$ was defined, without considering the radiation effect, by

$$\dot{Q}_{\text{conv.}} = hA\,(T_w - T_\infty)$$

where $A$ is the surface area, $T_w$ is the wall temperature and $T_\infty$ is the freestream temperature. But radiation heat transfer problems are often closely associated with convection problems. Because of this, in the analysis, dealing with both convection and radiation becomes essential to account for both the effects. Therefore, defining a radiation heat transfer coefficient $h_r$ (involving convection and the associated radiation) will be of high value. We can define $h_r$ as

$$\dot{Q}_{\text{rad.}} = h_r A\,(T_1 - T_2)$$

where $T_1$ and $T_2$ are the temperatures of the two bodies exchanging heat by radiation. The total heat transfer is then the sum of the convection and radiation,

$$\dot{Q} = (h_c + h_r)A_1\,(T_w - T_\infty) \tag{6.54}$$

assuming that the second-radiation-exchange surface is an enclosure and is at the same temperature as the fluid. For example, the heat loss by free convection and radiation from a hot steam pipe passing through a room could be calculated from Equation (6.54).

In many situations, the convection heat transfer coefficient is not strongly dependent on temperature. But this is not so with radiation heat transfer coefficient. The value of $h_r$ corresponding to Equation (6.38), can be determined from

$$\frac{\dot{Q}}{A_1} = \frac{\sigma\left(T_1^4 - T_2^4\right)}{\dfrac{1}{\epsilon_1} + \dfrac{A_1}{A_2}\left(\dfrac{1-\epsilon_2}{\epsilon_2}\right)} = h_r\,(T_1 - T_2)$$

This gives the radiation heat transfer coefficient as

$$\boxed{h_r = \frac{\sigma\left(T_1^2 + T_2^2\right)\left(T_1 + T_2\right)}{\dfrac{1}{\epsilon_1} + \dfrac{A_1}{A_2}\left(\dfrac{1-\epsilon_2}{\epsilon_2}\right)}} \tag{6.55}$$

## 6.17 Summary

Thermal radiation is that electromagnetic radiation emitted by a body as a result of its temperature.

There are many types of electromagnetic radiation, but the thermal radiation is only one. Irrespective of its type, a radiation is propagated at the speed of light $c$ ($3 \times 10^{10}$ cm/s), given by

$$c = \lambda\nu$$

Thermal radiation propagates in the form of discrete quanta with each quantum having an energy of

$$E = h\nu$$

where $h$ is the Planck's constant and is equal to $6.625 \times 10^{-34}$ J-s.

The radiation might be regarded as a *photon gas* which may flow from one place to another. The expressions for mass, momentum and energy of the particles could be expressed as

$$\begin{aligned}
\text{Mass } m &= \frac{h\nu}{c^2} \\
\text{Momentum} &= \frac{h\nu}{c^2}c = \frac{h\nu}{c} \\
\text{Energy } E &= mc^2 = h\nu
\end{aligned}$$

For radiation in such a gas, an expression for the energy density of a radiation $u_\lambda$ per unit volume and per unit wavelength can be expressed, using the principle of quantum statistical thermodynamics [1], as

$$u_\lambda = \frac{8\pi hc}{\lambda^5\left(e^{(hc)/(\lambda kt)} - 1\right)}$$

where $k$ is the Boltzmann constant ($1.38066 \times 10^{-23}$ J/(molecule K)).

The energy emitted is found to be proportional to the fourth power of absolute temperature.

$$E_b = \sigma T^4$$

This is called Stefan-Boltzmann law, $E_b$ is the energy radiated per unit time and per unit area by an ideal radiator, and $\sigma$ is the Stefan-Boltzmann constant, equal to $5.669 \times 10^{-8}$ W/(m$^2$ K$^4$). The energy $E_b$, radiated by a blackbody, is termed the *emissive power* of a blackbody.

The energy emitted by an ideal thermal radiator, also called the *blackbody* is given by

$$\dot{Q}_{\text{emitted}} = \sigma A T^4$$

According to Stefan-Boltzmann law, the net radiation exchange between two surfaces becomes

$$\frac{\dot{Q}_{\text{net exchange}}}{A} \propto \left(T_1^4 - T_2^4\right)$$

Imperfect black surfaces are termed *gray surfaces.* A factor which relates the radiation of the gray surface to that of an ideal black surface is called *emissivity* $\epsilon$.

The *intensity of radiation* $I$ is defined as the rate of energy leaving a surface in a given direction per unit projected area (area perpendicular to the direction of outward normal) per unit solid angle, or

$$I\left(\beta,\theta\right) = \frac{d\,\dot{Q}\left(\beta,\theta\right)}{dA_p\,d\Omega}$$

The *brightness of a surface from a given direction of view depends on the radiation emitted per unit projected area in that direction.* A surface which is equally bright in all directions is called a *diffuse surface.*

The *emissive power* $e$ of a surface is defined as the ratio of energy emission (in all directions) from the surface per unit surface area, that is,

$$e\left(T_s\right) = \frac{d\,\dot{Q}_e\left(T_s\right)}{d\,A_s}$$

The relation between the emissive power $e\left(T_s\right)$ and $I_e$ is given by

$$e\left(T_s\right) = \boxed{\pi\,I_e}$$

A *blackbody* is defined as a surface which absorbs all radiation incident on it.

By Stefan-Boltzmann law, the emissive power of a blackbody is given by

$$e_b\left(T_s\right) = n^2\,\sigma\,T_s^4$$

When radiant energy strikes a material surface, part of the radiation is *reflected*, part is *absorbed*, and part is *transmitted.* We define the *reflectivity* $\rho$ as the fraction reflected, the *absorptivity* $\alpha$ as the fraction absorbed, and the *transmissivity* $\tau$ as the fraction transmitted. Thus,

$$\boxed{\rho + \alpha + \tau = 1}$$

Two types of reflection phenomena are usually encountered when radiation strikes a surface. If the angle of incidence is equal to the angle of reflection, the reflection is called *specular* reflection. When the incident beam is distributed uniformly in all directions after reflection, the reflection is called *diffuse* reflection.

The radiation energy incident on a surface per unit surface area per unit time is called *irradiation* $G$.

The heating effect, due to the non-gray characteristic of the glass (or clear plastic) is known as the *greenhouse effect.*

A *blackbody* is defined as a perfect emitter and absorber of radiation. At a specified temperature and wavelength, no surface can emit more energy than a blackbody. A blackbody absorbs *all* radiation incident on it, regardless of wavelength and direction. Also, a blackbody emits radiation energy uniformly in all directions. The radiation energy emitted by a blackbody per unit surface area per unit time is

$$\boxed{E_b = \sigma T^4}$$

$E_b$ is called the *blackbody emissive power.*

The relation for the spectral blackbody emissive power $E_{b\lambda}$ was developed by Max Planck in 1901, in conjunction with his quantum theory. This relation is known as Planck's distribution law and is expressed as

$$E_{b\lambda}(T) = \frac{c_1}{\lambda^5 \left[\exp\left(c_2/\lambda T\right) - 1\right]}$$

The *emissive power* $e$ of a body is defined as the energy emitted by the body per unit area and per unit time. The ratio of the emissive power of a body to that of a blackbody at the same temperature is defined as the *emissivity* $\epsilon$.

$$\boxed{\epsilon\left(T_s\right) = \frac{e\left(T_s\right)}{e_b\left(T_s\right)}}$$

For the special case of a surface which is gray and diffuse it can be shown that,

$$\boxed{\alpha = \epsilon\left(T_s\right)}$$

This relation is called *Kirchhoff's identity.*

In radiation heat transfer problems the energy exchange between surfaces when they are maintained at different temperatures depends on the parameter known as view factor.

*View factor* is the fraction of energy leaving surface $dA_1$ and reaching surface $dA_2$, given by

$$\boxed{F_{1\rightarrow 2} = \frac{\cos\beta\,\cos\theta\,dA_2}{\pi\,r^2}}$$

An important view factor relation is

$$\boxed{A_1\,F_{1\rightarrow 2} = A_2\,F_{2\rightarrow 1}}$$

This expression is called the *reciprocity relation* between surfaces $A_1$ and $A_2$.

The *summation rule*

$$\boxed{\sum_{j=1}^{N} F_{i\rightarrow j} = 1}$$

may be applied to each of the $N$ surfaces in the enclosure. This rule follows from the conservation requirement that all the radiation leaving surface $A_i$ be intercepted by the enclosure surfaces.

The *net radiative exchange* between two surfaces is

$$\boxed{\dot{Q}_{ij} = A_i\,F_{i\rightarrow j}\,\sigma\left(T_i^4 - T_j^4\right)}$$

This gives the *net* rate at which radiation leaves surface $i$ as a result of its interaction with surface $j$, which is equal to the *net* rate at which surface $j$ *gains* radiation due to its interaction with surface $i$.

Analysis of radiation exchange in an enclosure may be simplified by making the following assumptions:

- Each surface of the enclosure is assumed to be *isothermal* and characterized by *uniform radiosity* and *irradiation.*
- All the surfaces are *opaque, diffuse* and *gray.*
- The medium within the enclosure is *nonparticipating.*

The radiative heat transfer rate from a surface is given by

$$\boxed{\dot{Q}_i = \frac{E_{bi} - J_i}{\dfrac{1-\epsilon_i}{\epsilon_i\,A_i}}}$$

where $(E_{bi} - J_i)$ as the *driving potential* and $\left(\dfrac{(1-\epsilon_i)}{\epsilon_i\,A_i}\right)$ as the *surface radiative resistance.*

The total rate at which radiation reaches surface $i$ from all surfaces, including $i$, is

$$\boxed{\dot{Q}_i = \sum_{j=1}^{N} A_i F_{i\rightarrow j}\,\left(J_i - J_j\right) = \sum_{j=1}^{N} \dot{Q}_{ij}}$$

That is, the net rate of radiation transfer $\dot{Q}_i$ from surface $i$ is equal to the sum of components $\dot{Q}_{ij}$ related to radiative exchange with the other surfaces. This may also be expressed as

$$\boxed{\frac{E_{bi} - J_i}{(1 - \epsilon_j)/\epsilon_i A_i} = \sum_{j=1}^{N} \frac{J_i - J_j}{(A_i A_{i \to j})^{-1}}}$$

In a two-surface enclosure, the net rate of radiation transfer $\dot{Q}_1$ from surface 1 must be equal to the net rate of radiation transfer $-\dot{Q}_2$ to surface 2. Thus,

$$\dot{Q}_1 = -\dot{Q}_2 = \dot{Q}_{12}$$

Radiation shields are surfaces meant for reducing the net radiation transfer between two surfaces. A radiation shield essentially introduces an additional thermal resistance, leading to the reduction of heat transfer. Radiation shields are constructed from low emissivity (high reflectivity) materials.

*Radiating surface* is an idealized surface characterized by zero net radiation transfer ($\dot{q}_i = 0$). For the case with $\dot{q}_i = 0$,

$$G_i = J_i = E_{bi}$$

Hence, if the radiosity of a radiating surface is known, its temperature is readily determined.

Radiation exchange between a gas and a heat transfer surface is complex. Unlike most solid bodies, gases in many situations transport radiation. Gases absorb and emit radiation, usually in a certain narrow wavelength bands.

Some gases, such as nitrogen and oxygen transmit heat, without radiating, at low temperatures, while carbon dioxide, water vapor and hydrocarbon gases radiate to an appreciable extent.

The relation

$$\frac{I_{\lambda x}}{I_{\lambda 0}} = e^{-a_\lambda x}$$

is called *Beer's law.* This represents the familiar exponential-decay formula experienced in many types of radiation analyses dealing with absorption.

Radiation from actual surfaces differs significantly from the ideal surfaces. For example, real surfaces are not perfectly diffuse, and hence the intensity of emitted radiation is not constant over all directions.

The reflection and absorption of thermal radiation from real surfaces are strongly influenced by the surface characteristics of the object and the surrounding environment in which it is present. These properties are dependent on the direction and wavelength of the incident radiation.

Solar radiation is a form a thermal radiation having a particular wavelength distribution. At the outer limit of the atmosphere the total solar irradiation $G$ is 1395 $W/m^2$, when the earth is at its mean distance from the sun. This value is called the *solar constant.*

The solar radiation reaching the surface of the earth does not behave like the radiation from an ideal gray body, even though outside the atmosphere the distribution of energy follows more of an ideal pattern. In many instances, the total absorptivity for solar radiation can be different from the absorptivity for blackbody radiation at some moderate temperature.

Real surfaces may exhibit substantially different absorption properties for solar radiation than that of the long-wavelength *earthbound* radiation. The direct solar radiation incident of a horizontal surface per unit area and per unit time is called *insolation.*

Radiation heat transfer environment is dictated by the absorption, scattering, and reflection properties of the atmosphere and natural surfaces. The scattering occurs in the atmosphere is of two types, namely the *molecular scattering* and *particulate scattering.*

The reflection properties of surfaces are described by defining a parameter termed *albedo* $A$, defined as

$$A = \frac{\text{reflected energy}}{\text{incident energy}}$$

The atmosphere absorbs radiation selectively in narrow-wavelength bands. The earth's atmosphere acts like a greenhouse, trapping the incident solar radiation and provides energy and warmth we require. The absorption and scattering of radiation may be described with Beer's law

$$\frac{I_{\lambda x}}{I_{\lambda 0}} = e^{-a_\lambda x}$$

where $a_\lambda$ is the monochromatic absorption coefficient and $x$ is the thickness of the layer absorbing the radiation.

The insolation at the outer edge of the atmosphere is given by

$$I_o = E_{bo} \sin\alpha$$

where $\alpha$ is the angle of the solar rays with the horizontal and $E_{bo}$ is the solar constant (1395 W/m$^2$).

The radiation absorption in natural bodies of water influences the evaporation rates and the dispersion of water vapor in the atmosphere. The solar radiation is absorbed rapidly in the top layers of water, followed by an approximately exponential decay with depth in the water.

A radiation heat transfer coefficient $h_r$ (involving convection and the associated radiation) can be defined as

$$\dot{Q}_{\text{rad.}} = h_r A\,(T_1 - T_2)$$

where $T_1$ and $T_2$ are the temperatures of the two bodies exchanging heat by radiation. This gives the radiation heat transfer coefficient as

$$\boxed{h_r = \frac{\sigma\left(T_1^2 + T_2^2\right)(T_1 + T_2)}{\dfrac{1}{\epsilon_1} + A_1\left(\dfrac{1-\epsilon_2}{A_2\epsilon_2}\right)}}$$

## 6.18 Exercise Problems

6.1 Consider a person of exposed surface area of 1.6 $m^2$, emissivity 0.75, and surface temperature 37°C standing in a large room with its walls at a uniform surface temperature. Determine the rate of heat loss from that person by radiation, if the walls are at (a) 35°C and (b) 25°C.

[**Ans.** (a) 16.06 W, (b) 91.79 W]

6.2 A furnace cavity, in the form of a cylinder of 100 mm diameter and 200 mm length, is open at one end to a large surrounding environment at 30°C. The sides and bottom of the cavity are maintained at constant temperatures of 1200°C and 1500°C, respectively, by electrical heating. All the surfaces are perfectly insulated. Assuming the sides and bottom as blackbodies, determine the power required to maintain the sides at constant temperatures. Assume the view factor between the bottom and the open top to be 0.06.

[**Ans.** 4396.8 W]

6.3 Two large parallel plates maintained at uniform temperatures of $T_1 = 1000$ K and $T_2 = 700$ K, have emissivities $\epsilon_1 = 0.3$ and $\epsilon_2 = 0.8$, respectively. Determine the net rate of radiation heat transfer between the two surfaces per unit surface area of the plate.

[**Ans.** 12024 W/$m^2$]

6.4 The surfaces of a cubical furnace of side 5 m, shown in Figure P6.4, are maintained at uniform temperatures. If the surfaces closely approximate black surface, determine (a) the net rate of radiation heat transfer between the base and the side surfaces, (b) the net rate of radiation heat transfer between the base and top surfaces, and (c) the net rate of radiation heat transfer from the base surface. Take the view factor $F_{12} = 0.2$.

[**Ans.** (a) 987033.6 W, (b) – 304365.6 W, (c) 682668 W]

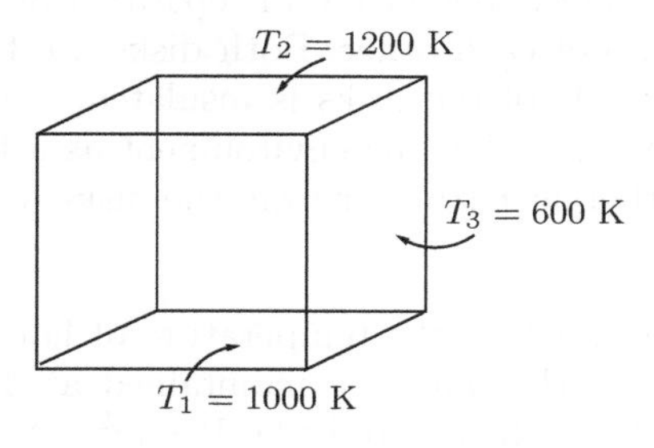


**Figure P6.4**
A cubical furnace.

6.5 A flat-plate solar collector with no cover plate has a selective absorber surface of emissivity 0.1 and solar absorptivity 0.95. At a given time of a day the absorber surface temperature $T_s = 120°C$ when the solar irradiation is 750 W/m$^2$, the effective sky temperature is $-10°C$, and the ambient air temperature $T_\infty = 30°C$. The convection heat transfer coefficient closely approximates

$$\overline{h} = 0.22\,(T_s - T_\infty)^{4/3}\ \mathrm{W/(m^2\ K)}$$

Calculate the useful heat that can be removed from the collector and the efficiency of the collector. Assume the absorptivity of the collector for irradiation from the surrounding to be 0.1.

[**Ans.** 515.64 W/m$^2$, 0.724]

6.6 An enclosure is made up of 15 surfaces. How many view factors does this geometry involve? How many of these view factors can be determined using the reciprocity and summation rules?

[**Ans.** 225, 120]

6.7 A hemispherical furnace has a flat circular base of diameter $D$. Determine the view factors from the dome of this furnace to its base.

[**Ans.** 0.25]

6.8 Find the view factors from the base of a cube to each of the other 5 sides.

[**Ans.** 0.2]

6.9 A hemispherical furnace of diameter 5 m has a flat base. The dome of the furnace closely approximates black surface and the base has an emissivity of 0.7. The dome and the base are maintained at uniform temperatures of 1000 K and 400 K, respectively. Determine the net rate of radiation heat transfer from the dome to the base surface, during steady operation.

[**Ans.** 759.4 kW]

6.10 Two parallel disks of diameter 0.6 m separated by a distance 0.6 m are located directly on top of each other. Both disks are black and maintained at 1000 K. The back side of the disks is insulated, and the disks are in an environment at 300 K. Treating the environment as a blackbody, determine the net rate of radiation heat transfer from the disks to the environment.

[**Ans.** 15901.7 W]

6.11 A thermocouple measures the temperature of hot air flowing in a duct as 600 K. The walls of the duct are maintained at 450 K. Assuming the convection heat transfer coefficient to be 80 W/(m$^2$ °C), determine the actual temperature of the air if the emissivity of the thermocouple junction is 0.7.

[**Ans.** 644 K]

6.12 A thermocouple measures the temperature of hot air flowing in a duct as 850 K, when the duct walls are maintained at 500 K. If the actual temperature of the hot air is 1000 K, determine the convection heat transfer coefficient of the air assuming the emissivity of the thermocouple junction is 0.6.

[**Ans.** 104.21 W/(m$^2$ °C)]

6.13 A thin aluminum sheet with an emissivity of 0.1 on both sides is placed between two large parallel plates maintained at uniform temperatures of 600 K and 300 K, as shown in Figure P6.13. The emissivity of the plates are 0.3 and 0.6. Determine the net rate of radiation heat transfer between the two plates per unit surface area. What will be the percentage increase in heat transfer if the shield is removed?

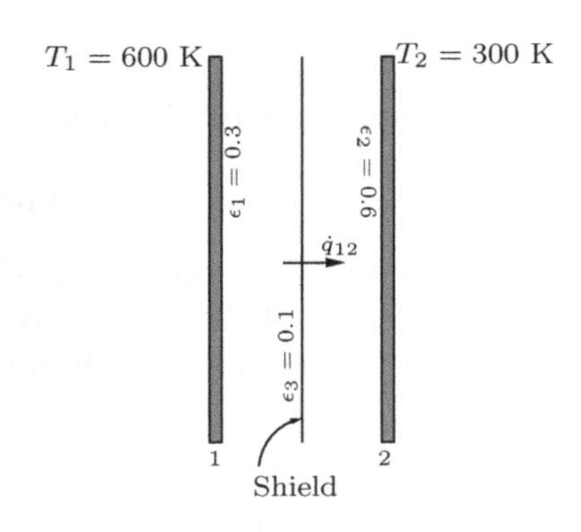


**Figure P6.13**
Radiation shield.

[**Ans.** 299.5 W, 491.8%]

6.14 The top and side surfaces of a cubical furnace of side 3 m closely approximate black surfaces. The temperature of these surfaces is 1200 K. If the bottom surface maintained at 800 K transfers 594.4 kW of heat by radiation, determine the emissivity of the bottom surface.

[**Ans.** 0.7]

6.15 The inner surface of a hot spherical tank is maintained at a constant temperature of 400°C. If a hole of diameter 10 mm is made at the tank wall, determine the radiation energy escaping through this hole to the surroundings.

[**Ans.** 0.913 W]

6.16 For the hemisphere and the plane shown in Figure P6.16, determine the view factors $F_{11}$, $F_{12}$, and $F_{21}$.

[**Ans.** 1, $\frac{1}{2}$, $\frac{1}{2}$]

6.17 A thermometer bulb of emissivity 0.9 in an air stream indicates 32°C, when the actual temperature of the stream is 40°C. If the temperature of the

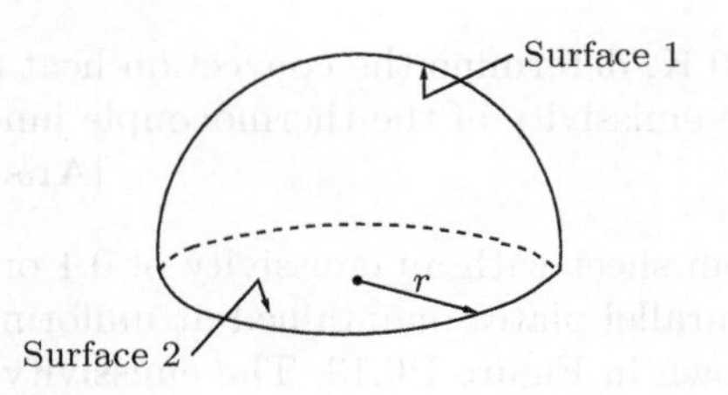


**Figure P6.16**
A hemisphere covering a plane.

atmosphere surrounding the air stream is 10°C, calculate the convection heat transfer coefficient of the air stream.

[**Ans.** 14.29 W/(m$^2$ °C)]

6.18 A thin layer of water exposed to clear sky at −20°C. The ambient air is at 0°C, and the convection coefficient of the air is 5 W/(m$^2$ °C). Assuming the emissivity of water as 0.9, determine the equilibrium temperature of the water layer.

[**Ans.** 265 K]

6.19 A water layer at 6°C is exposed to clear sky on a day with ambient air temperature at 10°C. If the emissivity and convection coefficient of water, respectively, are 0.95 and 10 W/(m$^2$ °C), determine the temperature of the sky.

[**Ans.** 270 K]

6.20 Find the view factors from the base of cube, shown in Figure P6.20, to its five surfaces.

[**Ans.** $F_{12} = F_{13} = F_{14} = F_{15} = F_{16} = 0.2$]

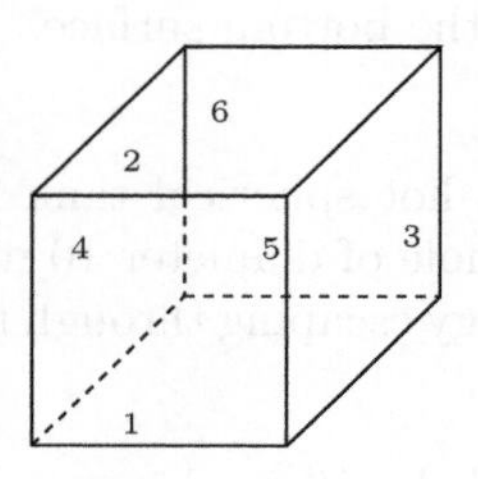


**Figure P6.20**
A cube.

6.21 Two large plates kept parallel are maintained at 800 K and 600 K. If both plates are of emissivity 0.8, calculate the net radiative heat transfer between the plates.

[**Ans.** 10584 $W/m^2$]

6.22 A radiation shield is placed between two large plates of emissivities $\epsilon_1 = 0.5$ and $\epsilon_2 = 0.8$. The emissivity of the radiation shield is $\epsilon_{31} = 0.1$ for the side facing plate 1, and $\epsilon_{32} = 0.05$ for the side facing plate 2. The temperature of plate 1 is 800 K and the net radiative heat transfer per unit area is 500 $W/m^2$. Determine the temperature of plate 2.

[**Ans.** 605 K]

6.23 A blackbody of surface area 2 $cm^2$ emits radiation at 875 K. Calculate the radiation energy emitted into a solid angle subtended by $0 \leq \theta \leq 2\pi$ and $0 \leq \beta \leq \frac{\pi}{6}$.

[**Ans.** 1.66 W]

# References

1. Holman, J. P., *Thermodynamics*, 3rd ed., McGraw-Hill Book Co., New York, 1980.

2. Hottel, H. C. and Sarofim, A. F., *Radiative Transfer*, McGraw-Hill Book Co., New York, 1967.

3. Siegel, R. and Howel, J. R., *Thermal Radiation Heat Transfer*, 2nd ed., McGraw-Hill Book Co., New York, 1980.

4. Eckert, E. R. G. and Drake, R. M., *Analysis of Heat and Mass Transfer*, McGraw-Hill Book Co., New York, 1972.

5. Mackey, C. O., Wright Jr., L. T., Clark, R. E. and Gay, N. R., *Radiant Heating and Cooling, Part I*, Cornell University, *Eng. Exp. Stn. Bull.*, Vol. 32, 1943.

6. Hamilton, D.C. and Morgan, W. R., *Radiant Interchange Configuration Factors*, NACA Tech. Note 2836, 1952.

7. Gubareff, G. G., Janssen, J. E. and Torborg, R. H., *Thermal Radiation Properties Survey*, 2nd ed. Minneapolis Honeywell Regulator Co., Minneapolis, 1960.

8. Eagleson, P. S., *Dynamic Hydrology*, McGraw-Hill Book Co., New York, 1970.

9. Dake, J. M. K. and Harleman, D. R. F., "Thermal Stratification in Lakes: Analytical and Laboratory Studies," *Res.*, Vol. 5, No. 2, p. 484, April 1969.

10. Goldman, C. R. and Carter, C. R., "An Investigation by Rapid Carbon-14 Bioassay of Factors Affecting the Cultural Entrophication of Lake Tahoe, California," *J. Water Pollut. Control Fed.*, p. 1044, July 1965.

11. Bechmann, R. W. and Goldman, C. R., "Hypolimnetic Heating in Castle Lake, California," *Limnol. Oceanog.*, Vol. 10, p. 2, April 1965.

12. Rathakrishnan, E., ***Instrumentation, Measurements, and Experiments in Fluids***, Taylor & Francis, CRC Press, Boca Raton, FL, 2007.

# Chapter 7

# Mass Transfer

## 7.1 Introduction

Mass transfer can result from several different phenomena. There is a mass transfer associated with convection in that mass is transported from one place to another in the flow system. This type of mass transfer occurs on a macroscopic level and is usually treated in the science of *fluid dynamics*. When a mixture of gases or liquids is contained in a container such that there exists a concentration gradient of one or more of the constituents across the system, there will be a mass transfer on a microscopic level as a result of diffusion from the regions of high concentration to the regions of low concentration. In our discussions in this book, we are primarily concerned with some of the simple relations which may be used to calculate mass diffusion.

## 7.2 Mass Transfer Process

Mass diffusion and turbulent fluctuations are the two primary causes of mass transfer. That is, mass transfer may occur not only on a molecular basis, but also in turbulent flows. In turbulent flows, mass diffusion will take place at a faster rate compared to the mass diffusion in a stationary system or in a laminar flow. This is because of the rapid-eddy mixing processes caused by turbulence, as in the case of heat transfer and viscous dissipation, where the turbulent mixing process enhances the transports of energy (heat) and momentum.

Although it is not intended to discuss in this introductory text, it will be beneficial to note that mass diffusion can also occur as a result of temperature gradient. Such a mass transfer caused by a temperature gradient is termed *thermal diffusion*. Similarly, a concentration gradient can cause molecular motion and the associated friction because of the viscous effects, leading to a temperature gradient and a consequent heat transfer. These two effects are referred to as *coupled phenomena* and may be treated by the method of irreversible thermodynamics.

In our discussions in this book, *concentration gradient is the driving potential* for mass transfer. The concentrations of species in a multi-component system may be expressed by their *mass density* $\rho_i$, defined as the mass of species $i$ per unit volume of the mixture and *molar density* $\rho_i^*$, defined as the number of moles of species $i$ per unit volume of the mixture, for analyzing the diffusion in liquids or solids and gases, respectively.

The relation between the mass density and molar density is

$$\boxed{\rho_i^* = \frac{\rho_i}{M_i}} \tag{7.1}$$

where $M_i$ is the molecular weight of the species $i$.

Sometimes, it is useful to express the concentration of a species as the fraction of the total mass or total number of moles rather than the absolute measures, such as the mass density $\rho_i$ or the molar density $\rho_i^*$.

The *mass fraction* $m_i$ is the ratio of the mass of species $i$ to the total mass of the mixture. It may be expressed as

$$m_i = \frac{\rho_i}{\rho} \tag{7.2}$$

where $\rho_i$ is the mass density (mass/volume) of species $i$ and $\rho$ is the mass density of the mixture, given by

$$\rho = \Sigma\, \rho_i$$

The *mole fraction* $m_i^*$ is the ratio of the number of moles of species $i$ to the total number of moles of the mixture, given by

$$m_i^* = \frac{\rho_i^*}{\rho^*} \tag{7.2a}$$

where $\rho_i^*$ is the molar density (molar mass/molar volume) of species $i$ and $\rho^*$ is the molar density of the mixture.

*Note:* It is important to note that, the Molar mass $M$ is the mass of one mole of a substance. The molecular mass of a substance is the mass of one molecule of that substance, in unified atomic mass unit(s) u[1] (equal to 1/12 the mass of one atom of the isotope carbon-12[2]). This is numerically equivalent to the relative molecular mass ($M_r$) of a molecule, frequently referred to by the term *molecular weight*, which is the ratio of the mass of that molecule to 1/12 of the mass of carbon-12 and is a dimensionless number. For example, the molecular weight of ammonia, $NH_3$, is 17 and the molecular weight of oxygen gas, $O_2$, is 32.

Molecular mass differs from more common measurements of the mass of chemicals, such as molar mass, by taking into account the isotopic composition of a molecule rather than the average isotopic distribution of many molecules. As a result, molecular mass is a more precise number than molar mass; however, it is more accurate to use molar mass on bulk samples. This means that molar mass is appropriate most of the time except when dealing with single molecules.

It may be expressed that,

$$\rho^* = \Sigma\, \rho_i^*$$

From the above definitions, we can express the mass and mole fractions as

$$\boxed{\Sigma\, m_i = 1} \tag{7.3}$$

$$\boxed{\Sigma\, m_i^* = 1} \tag{7.4}$$

The *mean molecular weight* $M$ of the mixture can be defined as

$$M = \frac{\rho}{\rho^*}$$

Thus,

$$\begin{aligned} M &= \frac{\rho}{\rho^*} \\ &= \frac{\Sigma\, \rho_i}{\rho^*} \\ &= \frac{\Sigma\, M_i \rho^*}{\rho^*} \end{aligned}$$

That is

$$M = \Sigma\, m_i^* M_i \tag{7.5}$$

In a multi-species mixture the various constituent species may move at different velocities. Let $V_i$ be the velocity of $i^{\text{th}}$ species relative to a stationary frame of reference. Now, we can define two average velocities as follows.

The *mass average velocity* $V$ is defined as

$$\begin{aligned} V &= \frac{\Sigma\, \rho_i V_i}{\Sigma\, \rho_i} = \frac{\Sigma\, \rho_i V_i}{\rho} \\ &= \Sigma \left(\frac{\rho_i}{\rho}\right) V_i \end{aligned}$$

That is,

$$V = \boxed{\Sigma\, m_i V_i} \tag{7.6}$$

The *molar average velocity* $V^*$ is defined as

$$V^* = \frac{\Sigma \rho_i^* V_i^*}{\Sigma \rho_i^*} = \frac{\Sigma \rho_i^* V_i^*}{\rho^*}$$

$$= \Sigma \left( \frac{\rho_i^*}{\rho^*} \right) V_i^*$$

That is,

$$V^* = \boxed{\Sigma m_i^* V_i^*} \tag{7.7}$$

Now, the velocity of a species with respect to the mass average velocity, namely the *mass diffusion velocity* is given by

$$\text{Mass diffusion velocity} = V_i - V$$

and the velocity of a species with respect to the molar average velocity, namely the *molar diffusion velocity* is given by

$$\text{Molar diffusion velocity} = V_i - V^*$$

With the above definitions, the mass flux $n_i$, and the molar flux $n_i^*$, across a stationary surface in space can be expressed as

$$n_i = \rho_i V_i \tag{7.8}$$

and

$$n_i^* = \rho_i^* V_i^* \tag{7.9}$$

Across a surface moving with mass average velocity $V$, the mass diffusion flux $j_i$, becomes

$$j_i = \rho_i \left( V_i - V \right) \tag{7.10}$$

Across a surface moving with molar average velocity $V^*$, the molar diffusion flux $j_i^*$, becomes

$$j_i^* = \rho_i^* \left( V_i - V^* \right) \tag{7.11}$$

The mass diffusion flux $j_i$, which is the amount of species $i$ transported per unit area perpendicular to the direction of transport and per unit time, is expressed as kg/(m$^2$ s). The molar diffusion flux $j_i^*$, which is the number of moles of species $i$ transported per unit area perpendicular to the direction of transport and per unit time, is expressed as kmol/(m$^2$ s).

The mass fraction $m_i$ and and mass flux $n_i$ can be related as follows.

$$\begin{aligned} n_i &= \rho_i V_i = \rho_i (V + V_i - V) \\ &= \rho_i V + \rho_i (V_i - V) \\ &= \rho_i V + j_i \end{aligned}$$

That is,

$$n_i = m_i (\rho V) + j_i \tag{7.12}$$

The mass flux $n_i$ is also given by

$$\Sigma n_i = \Sigma \rho_i V_i = \rho V$$

Substituting this into Equation (7.12), we get

$$\boxed{n_i = m_i (\Sigma n_i) + j_i} \tag{7.12a}$$

Similarly, the molar flux $n_i^*$ becomes

$$\begin{aligned} n_i^* &= \rho_i^* V_i = \rho_i^* (V^* + V_i - V^*) \\ &= \rho_i^* V^* + \rho_i^* (V_i - V^*) \\ &= \rho_i^* V^* + j_i^* \end{aligned}$$

That is,

$$n_i^* = m_i^* (\rho^* V^*) + j_i^* \tag{7.13}$$

The molar flux $n_i^*$ is also given by

$$\Sigma n_i^* = \Sigma \rho_i^* V_i = \rho^* V^*$$

Substituting this into Equation (7.13), we get

$$\boxed{n_i^* = m_i^* (\Sigma n_i^*) + j_i^*} \tag{7.13a}$$

The flux quantities in Equations (7.12a) and (7.13a) are vectors.

## Example 7.1

Oxygen gas stored in a tank is at a pressure of 6 bar and temperature 30°C. Determine the mass and molar densities of oxygen in the tank.

## Solution

Given, $p = 6$ bar $= 6 \times 10^5$ Pa, $T = 30 + 273 = 303$ K. Molecular weight of oxygen is 32.

At the given pressure and temperature oxygen can be treated as a perfect gas. By thermal equation of state,

$$p = \rho RT$$

where $R$ is the gas constant of oxygen, given by

$$R = \frac{R_u}{M}$$

where $R_u$ is the universal gas constant and is equal to 8314 $m^2/(s^2\ K)$. Thus,

$$R = \frac{8314}{32} = 259.8\,m^2/(s^2\ K)$$

Substituting for $p$, $T$, and $R$, we get the mass density $\rho$ as

$$\rho = \frac{6 \times 10^5}{259.8 \times 303}$$

$$= \boxed{7.62\,kg/m^3}$$

By Equation (7.1), the molar density $\rho^*$ is

$$\rho^* = \frac{\rho}{M} = \frac{7.62}{32}$$

$$= \boxed{0.238\,kmol/m^3}$$

## Example 7.2

Dry air at standard atmospheric state has 78 percent of nitrogen, 21 percent of oxygen, and 1 percent of argon and other gases. (a) Determine the mass fractions of nitrogen and oxygen in air. (b) Verify the mass fraction summation relation given by Equation (7.3).

## Solution

Given the molar fractions of $N_2$, $O_2$ and Ar are 0.78, 0.21 and 0.01, respectively. Their molecular weights are $m_{N_2} = 28$, $m_{O_2} = 32$ and $m_{Ar} = 39.95$.

Therefore, the molar mass of air becomes

$$M_{air} = 0.78 \times 28 + 0.21 \times 32 + 0.01 \times 39.95$$

$$= 28.96$$

(a) Thus, the mass fractions of $N_2$ and $O_2$ are

$$m_{\mathrm{N_2}} = 0.78 \times \frac{28}{28.96} = \boxed{0.754}$$

$$m_{\mathrm{O_2}} = 0.21 \times \frac{32}{28.96} = \boxed{0.232}$$

(b) Now, treating the argon and other gases as one unit equivalent to argon, we have the mass fraction of 1 percent of the constituent gases as

$$m_{\mathrm{Ar}} = 0.01 \times \frac{39.95}{28.96} = 0.0138$$

Therefore,

$$\Sigma m = m_{\mathrm{N_2}} + m_{\mathrm{O_2}} + m_{\mathrm{Ar}} = 0.754 + 0.232 + 0.0138 \approx \boxed{1}$$

That is, the summation relation given by Equation (7.3) is satisfied.

## 7.3 Fick's Law of Diffusion

Let us consider a system of dry air contained between two large parallel plates, at a distance $2b$ apart, as shown in Figure 7.1. The bottom plate is kept wet with water and the upper plate is kept dry.

If the temperature and total pressure at the lower plate are constant, the concentration gradient of water vapor is also a constant (since the molar fraction is equal to the partial-pressure fraction). The concentration of water vapor at the upper plate can be taken to be effectively zero. That is, there is a concentration gradient between the plates. If the temperature of the system is far below the boiling temperature of water, the concentration of water vapor in air will be very small. Hence, only a small concentration of water vapor (referred with subscript $w$) diffuse through large concentration of still air (species $A$). Thus, the molar diffusion flux of water vapor into air is given by

$$j_w^* \propto \rho^* \frac{m_{w0}^*}{2b}$$

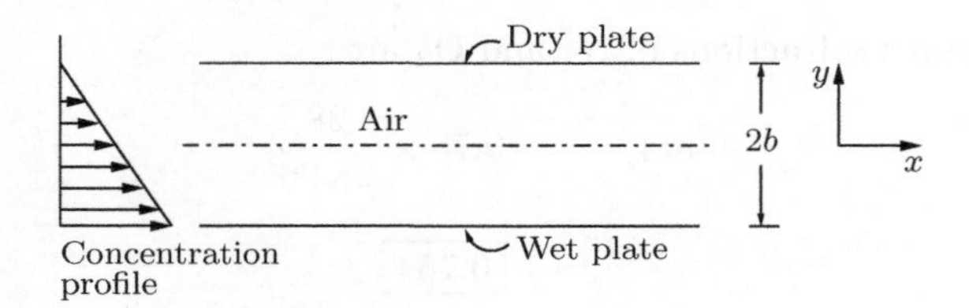


**Figure 7.1**
Air between wet and dry parallel plates.

$$= \rho^* D_{wA} \frac{m_{w0}^*}{2b}$$

where $\rho^*$ is the mixture molar-density, $m_{w0}^*$ is the molar fraction of water at the lower plate (it is zero at the upper plate) and $D_{wA}$ is the proportionality constant.

For the diffusion of species $A$ into species $B$, the molecular flux relation, in the differential form, can be written as

$$\boxed{j_A^* = -\rho^* D_{AB} \frac{\partial m_A^*}{\partial y}} \tag{7.14}$$

where the constant $D_{AB}$ is called the *binary diffusivity* of $A$ in $B$, and the minus sign indicates that the mass diffusion is from higher-concentration region to lower-concentration region. This relation is known as *Fick's first law of diffusion.* The dimensions of binary diffusivity $D_{AB}$ are length$^2$/time (m$^2$/s), as in the case of the thermal diffusivity $\alpha$ and kinematic viscosity $\nu$.

For mass diffusion in a three-dimensional system, the molecular diffusion flux $j_A^*$ and mass diffusion flux $j_A$ become

$$j_A^* = -\rho^* D_{AB} \nabla m_A^* \tag{7.15}$$

and

$$j_A = -\rho D_{AB} \nabla m_A \tag{7.16}$$

where the del operator $\nabla$ is

$$\nabla \equiv i\frac{\partial}{\partial x} + j\frac{\partial}{\partial y} + k\frac{\partial}{\partial z}$$

The binary diffusivity $D_{AB}$ in the mass and molar diffusion flux relations, given by Equations (7.15) and (7.16), are the same.

For binary diffusion across stationary surfaces, substituting Equation (7.15) into Equation (7.12) we get the mass flux as

$$n_A = \rho_A V - \rho D_{AB} \nabla m_A \tag{7.17}$$

For a binary mixture, by Equation (7.2), we have

$$m_A = \frac{\rho_A}{\rho}$$

Therefore,

$$\rho_A = \rho m_A$$

Hence,

$$\begin{aligned} \rho_A V &= m_A \rho V \\ &= m_A (n_A + n_B) \end{aligned}$$

Because by Equation (7.8), $\rho V = n$ and for a binary mixture, $n = n_A + n_B$.
Substituting this expression for $\rho_A V$ into Equation (7.17), we get

$$n_A = m_A (n_A + n_B) - \rho D_{AB} \bigtriangledown m_A \tag{7.17a}$$

Similarly, substituting Equation (7.16) into Equation (7.13), we get the molar flux as

$$n_A^* = \rho_A^* V^* - \rho^* D_{AB} \bigtriangledown m_A^* \tag{7.18}$$

But

$$\rho_A^* V^* = m_A^* (n_A^* + n_B^*)$$

Therefore,

$$n_A^* = m_A^* (n_A^* + n_B^*) - \rho^* D_{AB} \bigtriangledown m_A^* \tag{7.18a}$$

These relations are useful for solving problems involving mass transfer.

## 7.4 Species Conservation Equation

Examine a two-dimensional binary diffusion process. Let $A$ and $B$ be the two species having distributed mass sources with strength $\dot{S}_{m,A}$ and $\dot{S}_{m,B}$ per unit volume. Consider an infinitesimal control volume, of dimensions $\delta x$, $\delta y$ along $x$- and $y$-directions, and unity along $z$-direction, as shown in Figure 7.2.

For conservation of mass,

Rate of change of mass of a species contained in the control volume = (Rate of production of that species within the control volume) − (Efflux of that species across the control surface)

The mass of species $A$ contained in the control volume is

$$\text{density} \times \text{volume} = \rho_A \times (\delta x \delta y)$$

where $\rho_A$ is the density of species $A$.

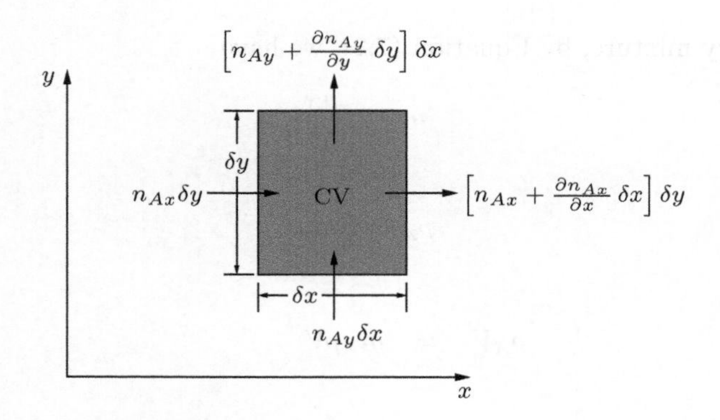


**Figure 7.2**
Control volume (CV) for a two-dimensional binary diffusion.

The time rate of change of the mass of species $A$ contained in the control volume is

$$\frac{\partial \rho_A}{\partial t}\,\delta x \delta y$$

where $\rho_A$ is the mass density of species $A$.

The net mass efflux of species $A$ through the control surfaces of the control volume considered is

Mass flux flowing out of CV − Mass flux flowing into the CV

that is,

$$\left(\left[n_{Ax} + \frac{\partial n_{Ax}}{\partial x}\delta x\right]\delta y + \left[n_{Ay} + \frac{\partial n_{Ay}}{\partial y}\delta y\right]\delta x\right)_{\text{out}} - \left(n_{Ax}\delta y + n_{Ay}\delta x\right)_{\text{in}}$$

This simplifies to

$$\left[\frac{\partial n_{Ax}}{\partial x} + \frac{\partial n_{Ay}}{\partial y}\right]\delta x \delta y$$

where $n_{Ax}$ and $n_{Ay}$ are the $x$ and $y$ components of mass flux $n_A$ of species $A$.

The production of species $A$, namely the generation of species A by a mass source, within the control volume is given by

$$\dot{S}_{m,A}\,\delta x \delta y$$

where $\dot{S}_{m,A}$ is the strength of the mass generating source with units kg/(s m$^3$).

Therefore, the conservation of mass of species $A$ results in

$$\frac{\partial \rho_A}{\partial t} + \frac{\partial n_{Ax}}{\partial x} + \frac{\partial n_{Ay}}{\partial y} = \dot{S}_{m,A}$$

or

$$\boxed{\frac{\partial \rho_A}{\partial t} + \nabla . n_{Ax} = \dot{S}_{m,A}} \tag{7.19}$$

Similarly, for species $B$ the mass conservation equation is

$$\boxed{\frac{\partial \rho_B}{\partial t} + \nabla . n_{Bx} = \dot{S}_{m,B}} \tag{7.19a}$$

For a binary mixture consisting of species $A$ and $B$, we have

$$\begin{aligned} \rho &= \rho_A + \rho_B \\ \rho V &= n_A + n_B \end{aligned}$$

Also,

$$\dot{S}_{m,A} + \dot{S}_{m,B} = 0$$

since the rate of production of one species should be equal to the rate of disappearance of the other, to maintain the overall mass balance.

Combining the above relations, we get the continuity equation for the mixture as

$$\boxed{\frac{\partial \rho}{\partial t} + \nabla \cdot (\rho V) = 0} \tag{7.20}$$

The continuity equation for a specific species is called the *species continuity equation.*

We can obtain the alternate form of the species continuity equation in terms of molecular density as

$$\frac{\partial \rho_A^*}{\partial t} + \nabla . n_A^* = \dot{S}_{m,A}^* \tag{7.20a}$$

where $\dot{S}_{m,A}^*$ is the rate of production of species $A$ in kmol/(m$^3$ s).

Substituting the expressions for $n_A$ and $n_A^*$, given by Equations (7.17) and (7.18) into Equations (7.19) and (7.19b), we get the following forms of mass conservation equation.

$$\frac{\partial \rho_A}{\partial t} + \nabla . (\rho_A V) = \nabla . (\rho D_{AB} \nabla m_A) + \dot{S}_{m,A}$$

Replacing $\rho_A$ with $(\rho m_A)$, this becomes

$$\frac{\partial \rho m_A}{\partial t} + \nabla . (\rho m_A V) = \nabla . (\rho D_{AB} \nabla m_A) + \dot{S}_{m,A} \tag{7.21}$$

and

$$\frac{\partial \rho_A^*}{\partial t} + \nabla . (\rho_A^* V) = \nabla . (\rho^* D_{AB} \nabla m_A^*) + \dot{S}_{m,A}^*$$

or

$$\frac{\partial \rho^* m_A^*}{\partial t} + \nabla.(\rho^* m_A^* V) = \nabla.(\rho^* D_{AB} \nabla m_A^*) + \dot{S}_{m,A}^* \tag{7.22}$$

Equations (7.21) and (7.22) are special cases corresponding to constant mass density $\rho$ and constant molar density $\rho^*$, respectively.

In dilute solutions of liquids or solids, the mass density $\rho$ of the mixture (also termed solution) can be approximately taken as a constant. For such a case of constant $\rho$, if the value for the binary diffusion coefficient $D_{AB}$ is also assumed to be a constant, Equation (7.21) reduces to

$$\boxed{\frac{\partial m_A}{\partial t} + \nabla.(m_A V) = D_{AB} \nabla^2 m_A + \frac{\dot{S}_{m,A}}{\rho}} \tag{7.23}$$

For gas mixtures of constant pressure and temperature, the molar density $\rho^*$ is constant. Thus, for constant pressure and temperature gas-diffusion (with constant value of $D_{AB}$), we have

$$\boxed{\frac{\partial m_A^*}{\partial t} + \nabla.(m_A^* V^*) = D_{AB} \nabla^2 m_A^* + \frac{\dot{S}_{m,A}^*}{\rho^*}} \tag{7.24}$$

The mass and molar conservations given by Equations (7.23) and (7.24) are known as *Fick's second law*.

The boundary conditions depend on the nature of the problems. In general, we encounter four types of boundary conditions. They are the following:

1. Specified concentration at the boundaries, such as

$$\text{at} \quad x = 0, \qquad m_A = m_{A0}$$

2. For impermeable surfaces,

$$\text{at} \quad n = 0, \qquad \frac{\partial m_A}{\partial n} = 0$$

where $n$ is the direction normal to the surface.

3. Specified wall flux, for example,

$$-\rho D_{AB} \frac{\partial m_A}{\partial x} = j_{A,w}$$

4. A specified mass transfer coefficient $h_m$ at the wall. The mass transfer coefficient $h_m$ is defined as

$$j_A = h_m (m_{A,w} - m_{A,f}) \tag{7.25}$$

where subscript $w$ represents wall condition and subscript $f$ represents condition far away from the wall. Thus, the convective mass transfer coefficient

$h_m$ has the units of kg/(m$^2$ s). In terms of molar quantities, the mass transfer coefficient $h_m^*$ is given by

$$j_A^* = h_m^* \left(m_{A,w}^* - m_{A,f}^*\right) \tag{7.26}$$

The units of $h_m^*$ is kmol/(m$^2$ s).

By boundary condition 3,

$$j_{A,w} = -\rho D_{AB} \frac{\partial m_A}{\partial x}$$

and

$$j_{A,w}^* = -\rho^* D_{AB} \frac{\partial m_A^*}{\partial x}$$

Substituting these into Equations (7.25) and (7.26), we get

$$\begin{aligned} -\rho D_{AB} \left(\frac{\partial m_A}{\partial x}\right)_w &= h_m \left(m_{A,w} - m_{A,f}\right) \\ -\left(\frac{\partial m_A}{\partial x}\right)_w &= \frac{h_m}{\rho D_{AB}} \left(m_{A,w} - m_{A,f}\right) \end{aligned}$$

and

$$\begin{aligned} -\rho^* D_{AB} \left(\frac{\partial m_A^*}{\partial x}\right)_w &= h_m^* \left(m_{A,w}^* - m_{A,f}^*\right) \\ -\left(\frac{\partial m_A^*}{\partial x}\right)_w &= \frac{h_m^*}{\rho^* D_{AB}} \left(m_{A,w}^* - m_{A,f}^*\right) \end{aligned}$$

## 7.5 Steady-State Diffusion in Dilute Solutions in Stationary Media

For diffusion of a small quantity of gas $A$ through a stationary gas $B$, or for diffusion of component $A$ in a dilute solution of a liquid or solid $B$, the mass or molar average velocities are small. Further, since the mass fraction $m_A$ or molar-fraction $m_A^*$ are also small, the terms $\nabla.(m_A V)$ or $\nabla.(m_A^* V^*)$ in Equations (7.23) and (7.24) are negligible compared to other terms.

For steady-state condition, the species concentration equations [Equations (7.23) and (7.24)] reduce to

$$\begin{aligned} \nabla^2 m_A &= 0 \\ \nabla^2 m_A^* &= 0 \end{aligned}$$

For one-dimensional diffusion of a single species $A$, such as that shown in Figure 7.3, the species concentration equation is

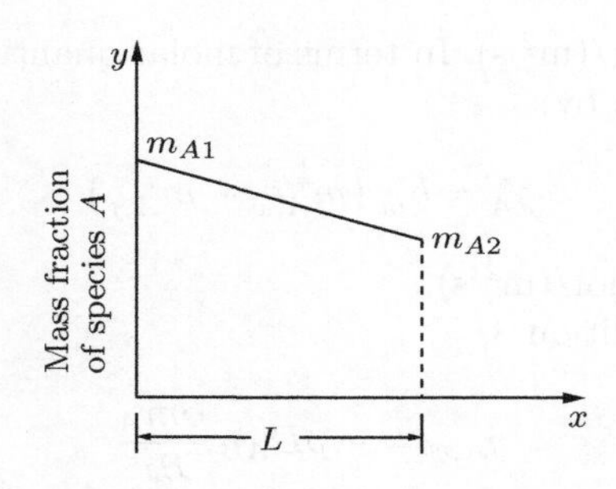


**Figure 7.3**
Concentration profile.

$$\nabla^2 m_A = 0$$

The boundary conditions are

$$\text{at} \quad x = 0, \quad m_A = m_{A1}$$

and

$$\text{at} \quad x = L, \quad m_A = m_{A2}$$

Integrating the governing equation, we get

$$m_A = c_1 x + c_2$$

where $c_1$ and $c_2$ and integration constants. Substituting the boundary conditions, we get

$$m_{A1} = c_2$$

$$m_{A2} = c_1 L + c_2$$

Solving we get the constants as

$$c_2 = m_{A1}$$

$$c_1 L = m_{A2} - c_2$$

$$c_1 = \frac{m_{A2} - m_{A1}}{L}$$

Substituting for $c_1$ and $c_2$, into the above relation for $m_A$, we get

$$m_A = (m_{A2} - m_{A1}) \frac{x}{L} + m_{A1}$$

For this one-dimensional diffusion of single species $A$, the mass flux given by Equation (7.16), is

$$\begin{aligned} n_A &= j_A = -\rho\, D_{AB}\,\frac{\partial m_A}{\partial x} \\ &= -\rho\, D_{AB}\,\frac{(m_{A2} - m_{A1})}{L} \end{aligned}$$

The mass flow current $\dot{Q}_{m,A}$ is defined as the rate of mass flow passing through per unit area and per unit time. In other words, the mass flow current is the product of mass flux and cross-sectional area of the passage. That is,

$$\boxed{\dot{Q}_{m,A} = -n_A A}$$

The negative sign in this equation indicates that the mass diffusion is in the direction opposite to the unit normal to the surface area.

Substituting for $n_A$, we get the mass flow current as

$$\boxed{\dot{Q}_{m,A} = \frac{\rho\, A\, D_{AB}}{L}\,(m_{A2} - m_{A1})} \qquad (7.27)$$

This can be expressed as

$$\dot{Q}_{m,A} = \frac{(m_{A2} - m_{A1})}{\dfrac{L}{\rho\, A\, D_{AB}}}$$

By definition

$$\boxed{\dot{Q}_{m,A} = \frac{\text{mass flow potential}}{\text{resistance}}}$$

Therefore, the resistance to mass transfer becomes

$$\boxed{R_{m,A} = \frac{L}{\rho\, D_{AB}\, A}} \qquad (7.28)$$

Following the same procedure as in convection heat transfer analysis in Chapter 5, the mass flow current or diffusion of fluid film on a solid surface can be expressed as

$$\dot{Q}_{m,A} = h_m A\,(m_{A2} - m_{A1})$$

or

$$\dot{Q}_{m,A} = \frac{(m_{A2} - m_{A1})}{\dfrac{1}{h_m A}}$$

where $h_m$ is the convection mass transfer coefficient. The resistance to the diffusion of fluid film on a solid surface becomes

$$\boxed{R_f = \frac{1}{h_m\, A}} \tag{7.29}$$

We can find the mass diffusion flux across a number of slabs in series in a manner similar to that used in heat transfer analysis.

Similarly, the mass diffusion resistance of a cylindrical shell of inner radius $r_1$ and outer radius $r_2$, and length $L$ will be

$$R_{m,A} = \frac{\ln\left(\dfrac{r_2}{r_1}\right)}{2\pi L\,(\rho\, D_{AB})} \tag{7.30}$$

For a spherical shell, the mass diffusion resistance becomes

$$R_{m,A} = \frac{\dfrac{1}{r_1} - \dfrac{1}{r_2}}{4\pi\,(\rho\, D_{AB})} \tag{7.31}$$

where $r_1$ and $r_2$ are the inner and outer radius of the spherical shell, respectively. This resistance is maximum when $r_2 \to \infty$.

The minimum value of film mass-transfer coefficient for a spherical shell is given by

$$\boxed{h_m = \frac{\rho\, D_{AB}}{r_1}}$$

## 7.6 Transient Diffusion in Dilute Solution in Stationary Media

Examine the problem of unsteady diffusion of species $A$ across an infinite slab of material $B$ as shown in Figure 7.4.

The species concentration equation [Equation (7.23)] for this diffusion process, in the absence of mass source or mass sink (that is, with $\dot{S}_{m,A} = 0$) becomes

$$\boxed{\frac{\partial m_A}{\partial t} = D_{AB}\, \nabla^2\, m_A} \tag{7.32}$$

where the mass diffusivity $D_{AB}$ is assumed to be constant, and the Laplacian $\nabla^2$ for a three-dimensional case is

$$\nabla^2 \equiv \frac{\partial^2}{\partial x^2} + \frac{\partial^2}{\partial y^2} + \frac{\partial^2}{\partial z^2}$$

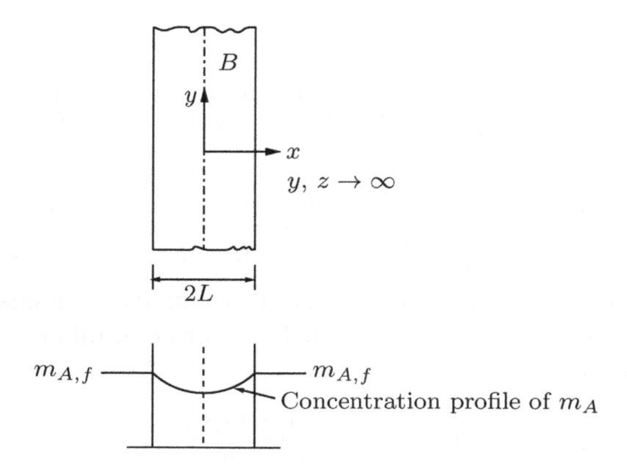


**Figure 7.4**
Unsteady diffusion across an infinite slab.

This reduces to

$$\nabla^2 = \frac{\partial^2}{\partial x^2}$$

for the mass diffusion across an infinite slab considered, because the slab is of finite thickness $2L$ along the $x$-direction, and extends to infinity along the $y$- and $z$-directions, thus rendering the mass transport as essentially a one-dimensional diffusion along the $x$-direction.

Let the initial mass fraction or concentration of species $A$ be $m_{A,0}$, and uniform throughout the slab. Let at time $t = 0$, the slab is exposed to a fluid mixture containing $A$, with concentration $m_{A,f}$. Thus,

$$\left. m_A \right|_{x=\pm L} = m_{A,0}, \text{ for } t < 0$$

$$\left. m_A \right|_{x=\pm L} = m_{A,f}, \text{ for } t \geq 0$$

That is, the slab with uniform mass fraction $m_{A,0}$ of species $A$ is suddenly subjected to a concentration gradient by exposing it to an environment with species $A$ concentration of $m_{A,f}$ at $t = 0$. It is important to note that, $t < 0$ implies the period before exposing the slab to an environment with mass fraction of species $A$ as $m_{A,f}$ at $t = 0$. In other words, $t < 0$ should not be taken as time in the negative scale.

Analogous to the boundary conditions given by Newton's law of cooling for a surface subjected to a temperature gradient, the boundary conditions for the present problem of one-dimensional mass diffusion through an infinite slab, subjected to a concentration gradient of $(m_{A,f} - m_{A,0})$, can be expressed as

$$\left.\frac{\partial m_A}{\partial x}\right|_{x=L} = -\frac{h_m}{\rho D_{AB}}\left(m_A\Big|_{x=L} - m_{A,f}\right)$$

and

$$\left.\frac{\partial m_A}{\partial x}\right|_{x=-L} = \frac{h_m}{\rho D_{AB}}\left(m_A\Big|_{x=-L} - m_{A,f}\right)$$

Now, let us define a dimensionless mass fraction $\theta_m$, dimensionless length $x^*$, dimensionless time $t^*$ (or mass transfer Fourier number $\mathrm{Fo}_m$), and mass transfer Biot number $\mathrm{Bi}_m$ as

$$\theta_m = \frac{m_A - m_{A,f}}{m_{A,0} - m_{A,f}}$$

$$x^* = \frac{x}{L}$$

$$t^* = \frac{tD_{AB}}{L^2} = \mathrm{Fo}_m$$

$$\mathrm{Bi}_m = \frac{h_m}{\rho D_{AB}}$$

With the above-defined dimensionless parameters, we can cast the governing equation, for unsteady diffusion of species $A$ across an infinite slab $B$ of width $2L$, shown in Figure 7.4, as follows.

Differentiating $\theta_m$ with respect to time $t$, we get

$$\frac{\partial \theta_m}{\partial t} = \frac{1}{(m_{A,0} - m_{A,f})}\frac{\partial m_A}{\partial t}$$

Therefore,

$$\begin{aligned}\frac{\partial m_A}{\partial t} &= (m_{A,0} - m_{A,f})\frac{\partial \theta_m}{\partial t} \\ &= (m_{A,0} - m_{A,f})\frac{\partial \theta_m}{\partial\left(\frac{tD_{AB}}{L^2}\right)}\frac{D_{AB}}{L^2} \\ &= (m_{A,0} - m_{A,f})\frac{\partial \theta_m}{\partial t^*}\frac{D_{AB}}{L^2}\end{aligned}$$

Substituting this expression of $\frac{\partial m_A}{\partial t}$ into the governing Equation (7.32), we get

$$\frac{D_{AB}}{L^2}(m_{A,0} - m_{A,f})\frac{\partial \theta_m}{\partial t^*} = D_{AB}\frac{\partial^2 m_A}{\partial x^2}$$

or

$$\frac{\partial \theta_m}{\partial t^*} = \frac{L^2}{(m_{A,0} - m_{A,f})}\frac{\partial^2 m_A}{\partial x^2} \tag{7.32a}$$

Now, Differentiating $\theta_m$ twice, with respect to $x$, we get

$$\frac{\partial \theta_m}{\partial x} = \frac{1}{(m_{A,0} - m_{A,f})} \frac{\partial m_A}{\partial x}$$

$$\frac{\partial^2 \theta_m}{\partial x^2} = \frac{1}{(m_{A,0} - m_{A,f})} \frac{\partial^2 m_A}{\partial x^2}$$

This gives

$$\frac{\partial^2 m_A}{\partial x^2} = (m_{A,0} - m_{A,f}) \frac{\partial^2 \theta_m}{\partial x^2}$$

Replacing $\frac{\partial^2 m_A}{\partial x^2}$ on the right-hand side of the above governing Equation (7.32a), we get

$$\frac{\partial \theta_m}{\partial t^*} = L^2 \frac{\partial^2 m_A}{\partial x^2}$$

or

$$\frac{\partial \theta_m}{\partial t^*} = \frac{\partial^2 m_A}{\partial (x^2/L^2)}$$

$$= \frac{\partial^2 m_A}{\partial x^{*2}}$$

Thus, in terms of the dimensionless parameters $\theta_m$, $t^*$ and $x^{*2}$ the governing equation becomes

$$\boxed{\frac{\partial \theta_m}{\partial t^*} = \frac{\partial^2 \theta_m}{\partial x^{*2}}} \tag{7.33}$$

In terms of the dimensionless parameters, the boundary conditions are expressed as follows.

$$\left.\frac{\partial m_A}{\partial x}\right|_{x=L} = -\frac{h_m}{\rho D_{AB}} (m_{A,0} - m_{A,f})$$

$$(m_{A,0} - m_{A,f}) \frac{\partial \theta_m}{\partial x} = -\frac{h_m}{\rho D_{AB}} \left( m_A\Big|_{x=L} - m_{A,f} \right)$$

$$\frac{\partial \theta_m}{\partial x} = -\frac{h_m}{\rho D_{AB}} \frac{\left( m_A\Big|_{x=L} - m_{A,f} \right)}{(m_{A,0} - m_{A,f})}$$

But at $x = L$, $x^* = x/L = L/L = 1$, and at $x = -L$, $x^* = x/L = -L/L = -1$. Therefore, the boundary conditions in terms of $x^*$ become

$$\frac{\partial \theta_m}{\partial x^*} = -\mathrm{Bi}_m \theta_m, \quad \text{at} \quad x^* = 1$$

$$\frac{\partial \theta_m}{\partial x^*} = \mathrm{Bi}_m \theta_m, \quad \text{at} \quad x^* = -1$$

Therefore, in the functional form, $\theta_m$ becomes

$$\boxed{\theta_m = \theta_m(x^*, t^*, \mathrm{Bi}_m)} \tag{7.33a}$$

## 7.7 Diffusion in a Semi-Infinite Slab

Consider the problem of mass diffusion of species $A$ in a semi-infinite slab of material $B$, shown in Figure 7.5, in which the concentration of species $A$ is very small initially, and there is no generation of species $A$ within the slab.

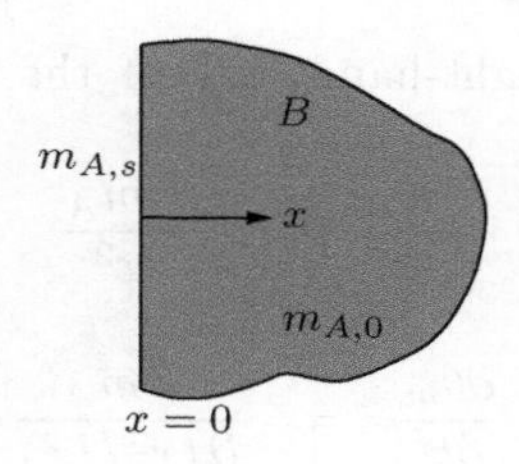


**Figure 7.5**
Diffusion through a semi-infinite slab.

Given that the initial concentration of species $A$ in the slab $B$, $m_{A,0}$ is very small and $\dot{S}_{m,A} = 0$. For small values of $m_A$, the governing Equation (7.23) becomes

$$\boxed{\frac{\partial m_A}{\partial t} = D_{AB}\,\nabla^2\, m_A} \tag{7.34}$$

At time $t = 0$ the concentration of species $A$ at the surface of the slab at $x = 0$ is raised to $m_{A,s}$ and held constant. With $m_{A,s} > m_{A,0}$, species $A$ will diffuse into the semi-infinite slab, and the concentration profile with time would appear as shown in Figure 7.6.

The initial and boundary conditions for the governing equation are the following.

$$\begin{aligned} m_A &= m_{A,0}, \quad \text{for } t < 0 \\ m_A &= m_{A,s}, \quad \text{for } x = 0,\ t \geq 0 \\ m_A &\to m_{A,0}, \quad \text{as } x \to \infty \text{ for all } t \end{aligned}$$

Note: It is essential to note that, here, $t < 0$ does not mean that the time is negative. The time less than zero here refers to the period before $t = 0$, at which a concentration gradient was imposed at the surface at $x = 0$ of the semi-infinite slab of material $B$, in which the initial concentration of species $A$ was $m_{A,0}$.

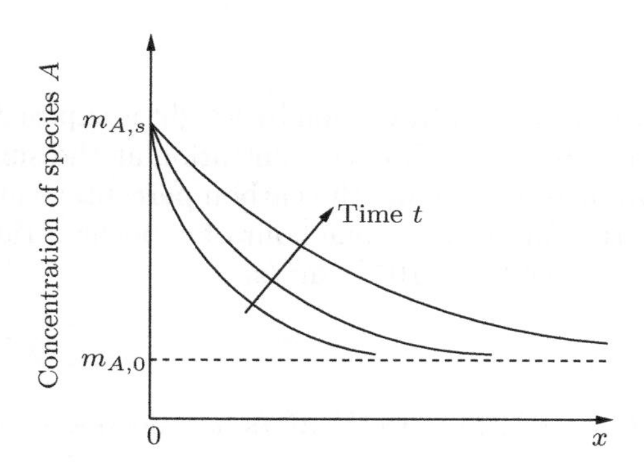


**Figure 7.6**
Concentration profile in a semi-infinite slab as a function of time.

Again, proceeding as in the corresponding unsteady heat conduction analysis in Chapter 4, we can define the dimensionless mass fraction as

$$\theta_m = \frac{m_A - m_{A,0}}{m_{A,s} - m_{A,0}}$$

In terms of $\theta_m$, the governing Equation (7.34) becomes

$$\frac{\partial \theta_m}{\partial t} = D_{AB} \frac{\partial^2 \theta_m}{\partial x^2} \tag{7.35}$$

The initial and the boundary conditions become

$$\begin{aligned} \theta_m(x,t) &= 0, \quad \text{for} \quad t < 0 \\ \theta_m(0,t) &= 1, \quad \text{for} \quad t \geq 0 \\ \theta_m(x,t) &= 0, \quad \text{as} \quad x \to \infty \end{aligned}$$

Now, analogous to $\eta = \dfrac{x}{\sqrt{4\alpha t}}$, defined in unsteady heat transfer analysis in Chapter 4, let us define a parameter $\eta = \dfrac{x}{\sqrt{4D_{AB}t}}$. The solution for Equation (7.35) becomes

$$\boxed{\theta = \frac{2}{\sqrt{\pi}} \int_0^{\eta} e^{-\eta^2} d\eta = \text{erfc}\,(\eta)} \tag{7.36}$$

where

$$\text{erfc}\,(\eta) = 1 - \text{erf}\,(\eta)$$

is the complimentary error function.

The penetration depth of mass diffusion is given by

$$\boxed{x = 1.8\,\sqrt{4\,D_{AB}\,t}} \tag{7.37}$$

## Example 7.3

A big mild steel piece having 0.2% carbon by weight is exposed to a carburizing atmosphere and thereby the carbon concentration at the surface is increased to 0.7% and maintained. Determine the carbon percentage at (a) 0.1 mm and (b) 0.4 mm below the surface after one hour of exposure. Take the diffusivity $D$ of carbon in mild steel to be $10^{-11}$ m$^2$/s.

## Solution

Given, $C_0 = 2\%$, $C_s = 7\%$, $D = 10^{-11}$ m$^2$/s, $t = 3600$ s.

Treating the problem as unsteady diffusion in a semi-infinite solid, by Equation (7.36), we have

$$\theta = \text{erfc}\,(\eta)$$

where

$$\theta = \frac{C - C_0}{C_s - C_0}$$

and

$$\eta = \frac{x}{\sqrt{4D_{AB}t}}$$

(a) Therefore, at 0.1 mm from the surface, after 1 hour, the value for $\eta$ is

$$\begin{aligned} \eta &= \frac{10^{-4}}{\sqrt{4 \times 10^{-11} \times 3600}} \\ &= 0.264 \end{aligned}$$

For $\eta = 0.264$, from error function table (Table 4.1), we have

$$\text{erf}\,(0.264) = 0.29$$

Therefore, the complimentary error function becomes

$$\begin{aligned} \text{erfc}\,(\eta) &= 1 - \text{erf}\,(0.264) \\ &= 1 - 0.29 \\ &= 0.71 \end{aligned}$$

Thus,

$$\begin{aligned} \theta &= 0.71 = \frac{C - 0.002}{0.007 - 0.002} \\ C &= 0.71 \times (0.007 - 0.002) + 0.002 \\ &= \boxed{0.555 \text{ percent}} \end{aligned}$$

(b) At 0.4 mm from the surface after 1 hour, the value of $\eta$ is

$$\eta = \frac{0.4 \times 10^{-3}}{\sqrt{4 \times 10^{-11} \times 3600}}$$
$$= 1.054$$

For $\eta = 1.054$, from error function table (Table 4.1), we have

$$\text{erf}\,(1.054) = 0.864$$

Therefore, the complimentary error function becomes

$$\text{erfc}\,(\eta) = 1 - \text{erf}\,(1.054)$$
$$= 1 - 0.864$$
$$= 0.136$$

Thus,

$$\theta = 0.136 = \frac{C - 0.002}{0.007 - 0.002}$$
$$C = 0.136 \times (0.007 - 0.002) + 0.002$$
$$= \boxed{0.268\ \text{percent}}$$

## 7.8 Diffusion in Nondilute Gases

In a nondilute solution the bulk velocity cannot be ignored. Therefore, the convection effects must be considered in the analysis. Consider one-dimensional diffusion of a gas $A$ into a stationary gas $B$ in a passage, as shown in Figure 7.7.
Let the molar fraction of $A$ at $x = 0$ be $m^*_{A,0}$, and that at $x = L$ be $m^*_{A,L}$, both of them may not be negligible compared to one another.

In the absence of any (homogeneous) chemical reaction inside the passage that is, with $\dot{S}^*_{m,A} = 0$, shown in Figure 7.7, the species concentration Equation (7.20a) becomes

$$\frac{\partial \rho^*}{\partial t} + \nabla . n^*_A = 0$$

For a steady-state diffusion, this reduces to

$$\frac{dn^*_A}{dx} + \frac{dn^*_B}{dx} = 0$$

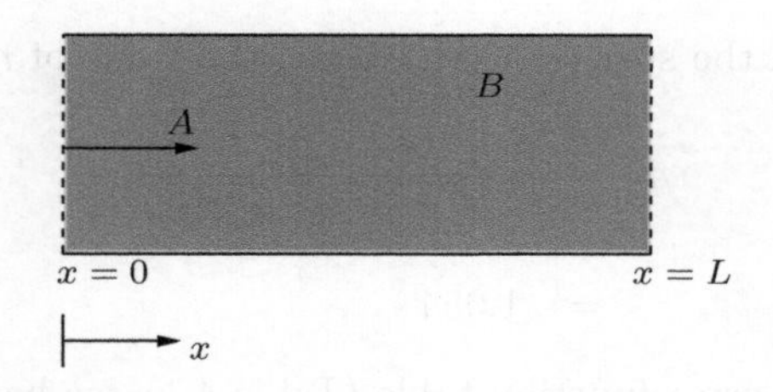


**Figure 7.7**
One-dimensional diffusion of a gas into another gas which is stationary.

Gas $B$ is stationary and hence $n_B^* = 0$ everywhere in the system. Therefore, the above equation simplifies to

$$\frac{dn_A^*}{dx} = 0$$

That is,

$$n_A^* = \text{constant}$$

The molar flux relation for this diffusion process, by Equation (7.13), is

$$n_A^* = m_A^* \left(n_A^* + n_B^*\right) + j_A^*$$

By Fick's law of diffusion, Equation (7.14), we have the molar diffusion flux $j_A^*$ as

$$\mathrm{j}_A^* = -\rho^* D_{AB} \frac{\partial m_A^*}{\partial x}$$

Substituting this into the above relation for $n_A^*$, we get

$$n_A^* = m_A^*(n_A^* + n_B^*) - \rho^* D_{AB} \frac{dm_A^*}{dx}$$

But $n_B^* = 0$, therefore,

$$\boxed{n_A^* = -\frac{\rho^* D_{AB} \dfrac{dm_A^*}{dx}}{1 - m_A^*}} \tag{7.38}$$

The boundary conditions are

$$\text{at } x = 0, \qquad m_A^*(0) = m_{A,0}^*$$
$$\text{at } x = L, \qquad m_A^*(L) = m_{A,L}^*$$

Equation (7.38), being first order, requires only one boundary condition. Therefore, the second boundary condition may be used to evaluate $n_A^*$.

Equation (7.38) gives

$$n_A^* = -\frac{\rho^* D_{AB} \dfrac{dm_A^*}{dx}}{1 - m_A^*}$$

$$\frac{dm_A^*}{1 - m_A^*} = -\frac{n_A^*}{\rho^* D_{AB}} dx$$

Integrating between the limits $x = 0$ and $x = L$, we get

$$\int_{m_{A,0}^*}^{m_{A,L}^*} \frac{dm_A^*}{1 - m_A^*} = -\int_0^L \frac{n_A^*}{\rho^* D_{AB}} dx$$

$$\Big[\ln (1 - m_A^*)\Big]_{m_{A,0}^*}^{m_{A,L}^*} = -\frac{n_A^*}{\rho^* D_{AB}} \Big[x\Big]_0^L$$

$$\ln \left(\frac{1 - m_{A,L}^*}{1 - m_{A,0}^*}\right) = -\frac{n_A^* L}{\rho^* D_{AB}}$$

That is,

$$\frac{1 - m_{A,L}^*}{1 - m_{A,0}^*} = \exp\left(-\frac{n_A^* L}{\rho^* D_{AB}}\right) \tag{7.39}$$

For any $x$, this becomes

$$\frac{1 - m_{A,x}^*}{1 - m_{A,0}^*} = \exp\left(-\frac{n_A^* x}{\rho^* D_{AB}}\right) \tag{7.39a}$$

Now, dividing Equation (7.39a) by Equation (7.39), we have

$$\frac{\left(\dfrac{1 - m_{A,x}^*}{1 - m_{A,0}^*}\right)}{\left(\dfrac{1 - m_{A,L}^*}{1 - m_{A,0}^*}\right)} = \frac{\exp\left(-\dfrac{n_A^* x}{\rho^* D_{AB}}\right)}{\exp\left(-\dfrac{n_A^* L}{\rho^* D_{AB}}\right)}$$

$$= \exp\left(-\frac{n_A^*}{\rho^* D_{AB}}[x - L]\right)$$

or

$$\boxed{\frac{1 - m_{A,x}^*}{1 - m_{A,0}^*} = \left(\frac{1 - m_{A,L}^*}{1 - m_{A,0}^*}\right) \exp\left(-\frac{n_A^*}{\rho^* D_{AB}}[x - L]\right)} \tag{7.40}$$

Rearranging Equation (7.39), we have

$$-\frac{n_A^* L}{\rho^* D_{AB}} = \ln\left(\frac{1-m_{A,L}^*}{1-m_{A,0}^*}\right)$$

$$n_A^* = -\frac{\rho^* D_{AB}}{L}\ln\left(\frac{1-m_{A,L}^*}{1-m_{A,0}^*}\right)$$

where

$$m_{A,L}^* = \frac{\rho_{A,L}^*}{\rho^*}$$

Also,

$$m_{B,L}^* = \frac{\rho_{B,L}^*}{\rho^*}$$

Therefore,

$$\begin{aligned} m_{A,L}^* + m_{B,L}^* &= \frac{\rho_{A,L}^*}{\rho^*} + \frac{\rho_{B,L}^*}{\rho^*} \\ &= \frac{\rho_{A,L}^* + \rho_{B,L}^*}{\rho^*} \\ &= 1 \end{aligned}$$

Thus,

$$1 - m_{A,L}^* = m_{B,L}^*$$

Therefore,

$$n_A^* = -\frac{\rho^* D_{AB}}{L}\ln\left(\frac{1-m_{A,L}^*}{1-m_{A,0}^*}\right) = -\frac{\rho^* D_{AB}}{L}\ln\frac{m_{B,L}^*}{m_{B,0}^*} \tag{7.41}$$

Equation (7.40) can be expressed as

$$\frac{m_{B,x}^*}{m_{B,0}^*} = \exp\left(\frac{-n_A^*(x-L)}{\rho^* D_{AB}}\right)\left(\frac{m_{B,L}^*}{m_{B,0}^*}\right) \tag{7.42}$$

By Equation (7.13),

$$n_A^* = m_A^*\left(n_A^* + n_B^*\right) + j_A^*$$
$$n_B^* = m_B^*\left(n_A^* + n_B^*\right) + j_B^*$$

But $n_B^* = 0$, therefore, the above relations for $n_A^*$ and $n_B^*$ simplify to

$$j_B^* = -\, m_B^* n_A^*$$

and

$$j_A^* = n_A^*(1 - m_A^*) = m_B^* n_A^*$$

Thus,

$$j_A^* + j_B^* = 0$$

since total molar flux across a surface moving with the molar average velocity should be zero. In other words, the *total molar flux across a surface moving with the molar average velocity is zero.*

Note that even though gas $B$ is stationary, $j_B^*$ is non-zero. Thus, gas $B$ is diffusing into gas $A$ at the same rate as gas $A$ diffuses into $B$ although $B$ is stationary.

## 7.9 Convective Mass Transfer

The mass transfer associated with significant bulk velocity is termed *convective mass transfer.* That is, in a convective mass transfer the species involved in the mass diffusion will move with an appreciable speed. Examine the diffusion of species $A$, with velocity $V_\infty$, density $\rho_f$ and mole fraction $\dot{m}_{A,f}^*$, through species $B$ zone of length $L$, shown in Figure 7.8.

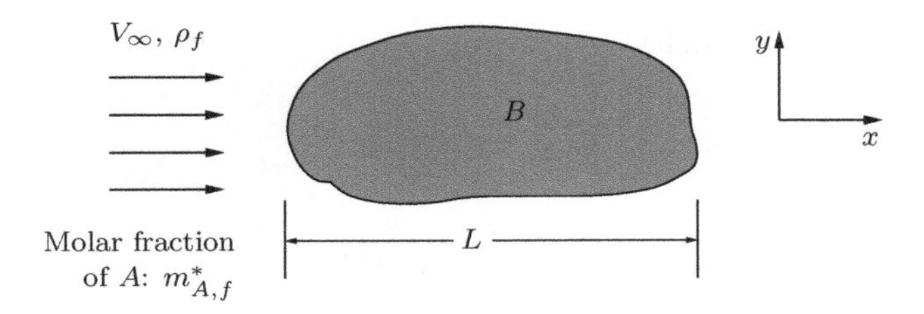


**Figure 7.8**
Motion of species $A$ over a zone.

When convection is present, the continuity and momentum equations governing the motion of the mixture should be considered along with species conservation equation. Thus, the governing equations for the convective motion for a steady flow are the following. For steady, incompressible, two-dimensional flow, the continuity equation [Equation (5.32)] is

$$\frac{\partial V_x}{\partial x} + \frac{\partial V_y}{\partial y} = 0$$

The $x$ and $y$ momentum equations [Equations (5.33) and (5.34)] are

$$V_x \frac{\partial V_x}{\partial x} + V_y \frac{\partial V_x}{\partial y} = -\frac{1}{\rho_f}\frac{\partial p}{\partial x} + \frac{\mu}{\rho_f}\left(\frac{\partial^2 V_x}{\partial x^2} + \frac{\partial^2 V_x}{\partial y^2}\right)$$

$$V_x \frac{\partial V_y}{\partial x} + V_y \frac{\partial V_y}{\partial y} = -g\left(\frac{\rho}{\rho_f} - 1\right) + \frac{\mu}{\rho_f}\left(\frac{\partial^2 V_y}{\partial x^2} + \frac{\partial^2 V_y}{\partial y^2}\right)$$

The equation governing the diffusion of species $A$, given by Equation (7.34) is

$$V_x \frac{\partial m_A}{\partial x} + V_y \frac{\partial m_A}{\partial y} = D_{AB}\left[\frac{\partial^2 m_A}{\partial x^2} + \frac{\partial^2 m_A}{\partial y^2}\right]$$

where $V_x$ and $V_y$ are the components of flow velocity in the $x$ and $y$ directions, respectively, $p$ is the pressure, $\rho$ is the mixture density, $\rho_f$ is the density of species $A$ in the freestream (that is, upstream of the zone of species $B$), $g$ is the gravitational acceleration, and $D_{AB}$ is the mass diffusivity.

Let us define the nondimensional lengths along $x$ and $y$, respectively, as

$$x^* = \frac{x}{L}$$

$$y^* = \frac{y}{L}$$

the nondimensional velocity and pressure as

$$V^* = \frac{V}{V_\infty}$$

$$p^* = \frac{p}{\frac{1}{2}\rho_f V_\infty^2}$$

and the dimensionless mass fraction of species $A$ at any instant of time and any location as

$$\theta_{m,A} = \frac{m_A - m_{A,f}}{m_{A,w} - m_{A,f}}$$

where $m_{A,f}$ is the mass fraction of species $A$ in the freestream, and $m_{A,w}$ is the mass fraction of species at the wall surface.

In terms of the nondimensional parameters, the governing equations become

$$\frac{\partial V_x^*}{\partial x^*} + \frac{\partial V_y^*}{\partial y^*} = 0$$

$$V_x^* \frac{\partial V_x^*}{\partial x^*} + V_y^* \frac{\partial V_x^*}{\partial y^*} = -\frac{1}{2}\frac{\partial p^*}{\partial x^*} + \frac{1}{\text{Re}}\left(\frac{\partial^2 V_x^*}{\partial x^{*2}} + \frac{\partial^2 V_x^*}{\partial y^{*2}}\right)$$

$$V_x^* \frac{\partial V_y^*}{\partial x^*} + V_y^* \frac{\partial V_y^*}{\partial y^*} = \frac{\text{Gr}_m}{\text{Re}^2}\,\theta_{m,A} + \frac{1}{\text{Re}}\left(\frac{\partial^2 V_y^*}{\partial x^{*2}} + \frac{\partial^2 V_y^*}{\partial y^{*2}}\right)$$

$$V_x^* \frac{\partial \theta_{m,A}}{\partial x^*} + V_y^* \frac{\partial \theta_{m,A}}{\partial y^*} = \frac{1}{\text{Re}\,\text{Sc}} \nabla^{*2} \theta_{m,A}$$

where

$$\text{Gr}_m = -\frac{\beta_m g\, L^3 \left(m_{A,w} - m_{A,f}\right)}{\nu^2}$$

That is,

$$\text{Gr}_m = \frac{g\, L^3 \left(\rho_f - \rho_w\right)}{\nu^2\, \rho_f} \tag{7.43}$$

The dimensionless parameter $\text{Gr}_m$ is the *Grashoff number* and the parameter $\text{Sc} = \frac{\nu}{D_{AB}}$ is called *Schmidt number* (similar to Prandtl number in heat transfer), which is the ratio of momentum diffusivity to mass diffusivity and Re is the Reynolds number. Therefore, in the functional form $\theta_m$ can be expressed as

$$\boxed{\theta_m = \theta_m\left(\text{Re, Sc, Gr}_m\right)} \tag{7.44}$$

For the diffusion of species $A$ into a system containing species $B$, the mass diffusion flux, given by Equation (7.16) is

$$j_w = -\rho D_{AB} \left.\frac{\partial m_A}{\partial y}\right|_{\text{wall}}$$

In the presence of convective motion, the mass diffusion flux is also given by

$$j_w = h_m \left(m_{A,w} - m_{A,f}\right) \tag{7.45}$$

where $h_m$ is the convective mass transfer coefficient. Thus,

$$h_m \left(m_{A,w} - m_{A,f}\right) = -\rho D_{AB} \left.\frac{\partial m_A}{\partial y}\right|_{\text{wall}}$$

We defined $\theta_{m,A}$ as

$$\theta_{m,A} = \frac{m_A - m_{A,f}}{m_{A,w} - m_{A,f}}$$

Therefore,

$$m_A = \left(m_{A,w} - m_{A,f}\right) \theta_{m,A} + m_{A,f}$$

Differentiating with respect to $y$, we get

$$\left.\frac{\partial m_A}{\partial y}\right|_{\text{wall}} = \left(m_{A,w} - m_{A,f}\right) \left.\frac{\partial \theta_{m,A}}{\partial y}\right|_{\text{wall}}$$

Substituting this into the above relation for $h_m$, we get

$$\begin{aligned} h_m &= -\rho D_{AB} \left.\frac{\partial \theta_{m,A}}{\partial y}\right|_{\text{wall}} \\ &= -\rho D_{AB} \left.\frac{\partial \theta_{m,A}}{\partial \left(\frac{y}{L}\right)} \frac{1}{L}\right|_{\text{wall}} \\ &= -\rho D_{AB} \left.\frac{\partial \theta_{m,A}}{\partial y^*} \frac{1}{L}\right|_{\text{wall}} \\ \frac{h_m L}{\rho D_{AB}} &= -\left.\frac{\partial \theta_{m,A}}{\partial y^*}\right|_{\text{wall}} \end{aligned}$$

The dimensionless group on the left-hand side is

$$\frac{h_m L}{\rho D_{AB}} = \text{Nu}_m \tag{7.46}$$

where $\text{Nu}_m$ is the *mass transfer Nusselt number*. The mass transfer Nusselt number is also known as *Sherwood number*.

Thus, the mass transfer Nusselt number becomes

$$\text{Nu}_m = \frac{h_m L}{\rho D_{AB}} = -\left.\frac{\partial \theta_{m,A}}{\partial y^*}\right|_{\text{wall}} \tag{7.47}$$

But $\dfrac{\partial \theta_{m,A}}{\partial y^*}$ depends on Re, Sc and $\text{Gr}_m$, therefore, the $\text{Nu}_m$ also known as Sherwood number can be expressed in functional form as

$$\text{Nu}_m = \text{Nu}_m(\text{Re, Sc, Gr}_m) \tag{7.48}$$

In forced convection, $\text{Gr}_m$ is insignificant and thus $\text{Nu}_m$ depends only on Re and Sc. In free convection, Reynolds number is not important and $\text{Nu}_m$ depends only on Sc and $\text{Gr}_m$.

Once the Nusselt number $\text{Nu}_m$ is known, the mass transfer coefficient $h_m$ can be determined. Using $h_m$ the mass diffusion flux $j_A$ can be calculated. For calculating the mass diffusion flux $n_A$, given by Equation (7.12),

$$n_A = \rho_A V + j_A$$

or

$$n_A = m_A(n_A + n_B) + j_A$$

can be employed.

## 7.10 Counterdiffusion in Gases

In a binary mixture of two gases, say $A$ and $B$, they will diffuse simultaneously in opposite direction to each other. That is, gas $A$ will diffuse through gas $B$, and vice versa. They would diffuse at the molar rate but in opposite directions.

Let us assume that two large tanks containing uniform mixtures of gases $A$ and $B$ at different concentrations are suddenly connected by a small pipe. Also, let us assume that both the tanks are at the same total pressure $p$ and at uniform temperature $T$. Let the gas $A$ diffuses from higher concentration (density) to the lower concentration (density). This diffusion of gas $A$ would cause the gas $B$ to diffuse at the same rate but in the opposite direction through the connecting pipe. When the tanks are sufficiently large, steady-state equimolal countermolal diffusion would take place in the connecting pipe. That is, *the total molar flux with respect to stationary coordinates would be zero.* Thus,

$$n_A^* + n_B^* = 0 \text{ or } n_A^* = -n_B^* \tag{7.49}$$

That is, in a steady-state equimolal counterdiffusion, the molar fluxes of gases $A$ and $B$ relative to stationary coordinates are equal, and in the opposite directions.

If the mixture is considered to be perfect gas, the molar concentrations $\rho_A^*$ and $\rho_B^*$ can be related to the partial pressures $p_A$ and $p_B$ of the gases $A$ and $B$, given by

$$p_A = \rho_A^* R_u T, \quad p_B = \rho_B^* R_u T \tag{7.50}$$

where $\rho_A^*$ and $\rho_B^*$ are the molar concentrations or molar densities of gases $A$ and $B$, respectively, and $R_u$ is the universal gas constant.

The mass diffusion rate is given by Fick's law, which states that the mass flux of a constituent, say $A$ in a binary mixture of species $A$ and $B$, per unit area is proportional to the concentration gradient of that species (that is, $\frac{\partial \rho_A}{\partial x}$). Thus,

$$\frac{\dot{m}_A}{A} = -D_{AB}\frac{\partial \rho_A}{\partial x} \tag{7.51}$$

where $D_{AB}$ is the proportionality constant, known as the diffusion coefficient, $\dot{m}_A$ is the mass flux per unit time and $\rho_A$ is the concentration per unit volume. An expression similar to Equation (7.51) could be written for constituent $A$ along $y$- and $z$-directions.

Note that Equation (7.51) is similar to the Fourier law of heat conduction

$$\left(\frac{\dot{q}}{A}\right)_A = -k\frac{\partial T}{\partial x}$$

and the equation for shear stress between fluid layers

$$\tau = \mu\frac{\partial u}{\partial y}$$

The heat conduction equation describes the *transport of energy*, the shear stress equation describes the *transport of momentum* across fluid layers, and the diffusion law describes the *transport of mass.*

For gases, Fick's law may be expressed in terms of partial pressure by making use of the perfect gas state equation

$$p = \rho R T \tag{7.52}$$

The gas constant $R$ for a particular gas is given by

$$R_A = \frac{R_u}{M_A} \tag{7.53}$$

where $R_u$ is the universal gas constant and is equal to 8314 J/(kmol K) and $M_A$ is the molecular weight of species $A$. Thus,

$$\rho_A = \frac{p_A M_A}{R_u T}$$

Therefore, for isothermal diffusion, Equation (7.51) becomes

$$\frac{\dot{m}_A}{A} = -D_{AB}\frac{M_A}{R_u T}\frac{dp_A}{dx} \tag{7.54}$$

Similarly,

$$\frac{\dot{m}_B}{A} = -D_{BA}\frac{M_B}{R_u T}\frac{dp_B}{dx} \tag{7.55}$$

The molar diffusion rates are given by

$$n_A = \frac{\dot{m}_A}{M_A} = -D_{AB}\frac{A}{R_u T}\frac{dp_A}{dx}$$

$$n_B = \frac{\dot{m}_B}{M_B} = -D_{BA}\frac{A}{R_u T}\frac{dp_B}{dx}$$

The total pressure $p$ of the system remains constant at steady state, so that

$$p = p_A + p_B$$

Differentiating with respect to $x$, we get

$$\frac{dp_A}{dx} + \frac{dp_B}{dx} = 0$$

or

$$\frac{dp_A}{dx} = -\frac{dp_B}{dx} \tag{7.56}$$

Since each molecule of $A$ is replacing a molecule of $B$, we may set the molar diffusion rates equal, thus,

$$-n_A = n_B$$

$$-D_{AB}\frac{A}{R_uT}\frac{dp_A}{dx} = D_{BA}\frac{A}{R_uT}\frac{dp_B}{dx}$$

$$= -D_{BA}\frac{A}{R_uT}\frac{dp_A}{dx}$$

The left-hand and right-hand sides are identical; therefore, we can express

$$D_{AB} = D_{BA} = D \tag{7.57}$$

Inserting into Equation (7.54), we get the mass flux of component $A$ as

$$\boxed{\frac{\dot{m}_A}{A} = -\left(\frac{DM_A}{R_uT}\right)\left(\frac{p_{A,2} - p_{A,1}}{\Delta x}\right)} \tag{7.58}$$

where $p_{A,1}$ is the partial pressure of species $A$ at location $x_1$ and $p_{A,2}$ is the partial pressure of species $A$ at location $x_2$, and $\Delta x = (x_2 - x_1)$.

Similarly, substituting $D_{AB} = D$, into Equation (7.55), we get the mass flux of component $B$ as

$$\boxed{\frac{\dot{m}_B}{A} = -\left(\frac{DM_B}{R_uT}\right)\left(\frac{p_{B,2} - p_{B,1}}{\Delta x}\right)} \tag{7.58a}$$

## Example 7.4

A large tank containing a uniform mixture of 90 mole percent of nitrogen gas and 10 mole percent of carbon dioxide gas is suddenly connected to another large tank containing a uniform mixture of 20 mole percent of nitrogen gas and 80 mole percent of carbon dioxide gas, by a duct of diameter 150 mm and length 1.5 m. If both the tanks were at pressure 1 atm and temperature 300 K and the mass diffusivity of nitrogen into carbon dioxide is $0.16 \times 10^{-4}$ m$^2$/s, determine the rate of transfer of nitrogen between the two tanks at steady state.

## Solution

Given, $d = 0.15$ m, $D = 0.16 \times 10^{-4}$ m$^2$/s, $p = 101325$ Pa, $T = 300$ K, $x_2 - x_1 = 1.5$ m.

Let subscripts $A$ and $B$ refer to nitrogen and carbon dioxide, respectively. The molecular weights (or molar mass) of $N_2$ and $CO_2$ are, $M_A = 28$ and $M_B = 44$.

The partial pressures of nitrogen in tanks 1 and 2 are

$$\begin{aligned} p_{A,1} &= 0.9 \times 101325 \\ &= 91192.5\,\text{Pa} \\ p_{A,2} &= 0.2 \times 101325 \\ &= 20265\,\text{Pa} \end{aligned}$$

By Equation (7.58), the mass diffusion of nitrogen is

$$\dot{m}_A = -\frac{ADM_A}{R_u T}\left(\frac{p_{A,2}-p_{A,1}}{x_2 - x_1}\right)$$

where $M_A = 28$, $R_u = 8314\ \text{m}^2/(\text{s}^2\,\text{K})$, $A = \frac{\pi d^2}{4}$. Therefore,

$$\begin{aligned} \dot{m}_A &= -\frac{\left(\frac{\pi\times 0.15^2}{4}\right)\times(0.16\times 10^{-4})\times 28}{8314\times 300}\times\left(\frac{20265-91192.5}{1.5}\right) \\ &= \boxed{1.5\times 10^{-7}\,\text{kg/s}} \end{aligned}$$

The mass diffusion of nitrogen obtained can be checked by verifying that *the total molar flux is zero.* In other words, $n_A^*$ should be equal to $-n_B^*$.

The partial pressures of carbon dioxide in tanks 1 and 2 are

$$\begin{aligned} p_{B,1} &= 0.1 \times 101325 = 10132.5\,\text{Pa} \\ p_{B,2} &= 0.8 \times 101325 = 81060\,\text{Pa} \end{aligned}$$

The mass diffusion of carbon dioxide is

$$\begin{aligned} \dot{m}_B &= -\frac{ADM_B}{R_u T}\left(\frac{p_{B,2}-p_{B,1}}{x_2-x_1}\right) \\ &= -\frac{\left(\frac{\pi\times 0.15^2}{4}\right)\times(0.16\times 10^{-4})\times 44}{8314\times 300}\times\left(\frac{81060-10132.5}{1.5}\right) \\ &= \boxed{-2.36\times 10^{-7}\,\text{kg/s}} \end{aligned}$$

The molar fractions of the species are

$$n_A^* = \frac{\dot{m}_A}{M_A}$$

$$= \frac{1.5 \times 10^{-7}}{28}$$

$$= 0.0538 \times 10^{-7}\,\text{kmol/s}$$

$$n_B^* = \frac{\dot{m}_B}{M_B}$$

$$= \frac{-2.36 \times 10^{-7}}{44}$$

$$= -0.0536 \times 10^{-7}\,\text{kmol/s}$$

Thus, $n_A^* \approx n_B^*$, and diffuse in the opposite directions.

## Example 7.5

Helium gas is stored in a spherical container of inner diameter 200 mm and wall thickness 2 mm. Initially the pressure and temperature of helium in the tank are 4 bars and 20°C. If the diffusion coefficient of helium through the container material is $0.4 \times 10^{-13}$ m$^2$/s, (a) determine the rate of diffusion of helium from the container to the surrounding air in a second from the beginning of diffusion and (b) the rate of decrease of pressure in the tank during that time. Assume the diffusion to be one-dimensional and the molar concentration of helium at the outer surface of the container to be negligible.

## Solution

Let $A$ and $B$ refer to helium and atmospheric air surrounding the container. Given $d = 0.2$ m, $x_2 - x_1 = 2 \times 10^{-3}$ m, $T = 20 + 273 = 293$ K, $p = 4 \times 10^5$ Pa, $D_{AB} = 0.4 \times 10^{-13}$ m$^2$/s.

The molecular weight of helium is $M_A = 4$; therefore, the gas constant becomes

$$R = \frac{R_u}{M_A} = \frac{8314}{4}$$

$$= 2078.5\,\text{m}^2/(\text{s}^2\ \text{K})$$

The density of helium in the container is

$$\rho = \frac{p}{RT} = \frac{4 \times 10^5}{2078.5 \times 293}$$

$$= 0.657\,\text{kg/m}^3$$

The partial pressure of helium at $x_1$ is $p_{A,1} = 4 \times 10^5$ Pa, and the partial pressure at $x_2$ is $p_{A,2} = 0$.

(a) By Equation (7.58), the mass diffusion rate of helium, per unit area is

$$\frac{\dot{m}_A}{A} = -\frac{D_{AB}M_A}{R_u T}\left(\frac{p_{A,2} - p_{A,1}}{x_2 - x_1}\right)$$

$$= -\frac{D_{AB}}{RT}\left(\frac{-p_{A,1}}{x_2 - x_1}\right)$$

$$= \frac{0.4 \times 10^{-13}}{2078.5 \times 293} \times \frac{4 \times 10^5}{2 \times 10^{-3}}$$

$$= 1.314 \times 10^{-11}\,\mathrm{kg/(m^2\ s)}$$

The area is the inner surface area of the container, thus

$$A = 4\pi r_i^2 = 4\pi \times 0.1^2$$

$$= 0.126\,\mathrm{m^2}$$

Hence,

$$\dot{m}_A = 1.314 \times 10^{-11} \times 0.126$$

$$= \boxed{0.166 \times 10^{-11}\,\mathrm{kg/s}}$$

(b) The mass can be expressed as $m = \rho \mathbb{V}$. Therefore,

$$\dot{m}_A = \frac{d}{dt}(\rho \mathbb{V})$$

$$= \frac{d}{dt}\left(\frac{p}{RT}\mathbb{V}\right)$$

Assuming $T$ to be a constant, we have

$$\dot{m}_A = \frac{\mathbb{V}}{RT}\frac{dp}{dt}$$

$$\frac{dp}{dt} = \frac{\dot{m}_A RT}{\mathbb{V}}$$

The volume is

$$\mathbb{V} = \frac{\pi d^3}{6} = \frac{\pi \times 0.2^3}{6}$$

$$= 4.12 \times 10^{-3}\,\mathrm{m^3}$$

Therefore, the pressure decrease in a second becomes

$$\frac{dp}{dt} = \frac{(0.166 \times 10^{-11}) \times 2078.5 \times 293}{4.12 \times 10^{-3}}$$

$$= 24537.4 \times 10^{-8}\,\mathrm{Pa/s}$$

$$= \boxed{245.37\,\mu\mathrm{Pa/s}}$$

## 7.11 Evaporation of Water

Let us examine the isothermal evaporation of water from the top surface of the water in a container and the subsequent diffusion of the water vapor through the stagnant air layer, as shown in Figure 7.9.

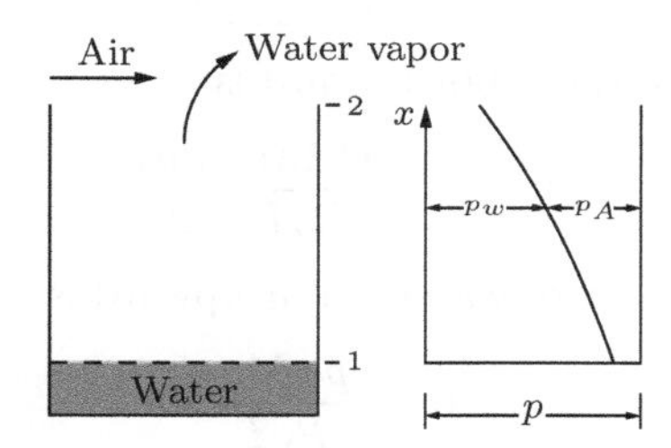


**Figure 7.9**
Diffusion of water vapor into air.

Let us assume that the evaporation is taking place at a steady state, and at constant pressure $p$ and constant temperature $T$. This steady-state evaporation at constant $p$ and $T$ requires that there be a slight air movement over the top of the container to remove the water vapor which diffused to that point. Let us also assume that the air movement required to accomplish this is without turbulence. Furthermore, let us assume that both air and water vapors behave as ideal gases.

The water vapor evaporating from the container diffuses upward through the air, and at steady state this upward movement must be balanced by a downward diffusion of air so that the concentration at any $x$ position will remain constant. But at the surface of water there can be no net movement of air downward. Consequently there must be a bulk mass movement upward with a velocity just large enough to balance the diffusion of air downward. This bulk mass movement then would induce an *additional* mass flux of water vapor upward.

The diffusion of air downward is given by

$$\dot{m}_A = -\frac{DAM_A}{R_u T}\frac{dp_A}{dx} \tag{7.59}$$

where $A$ is the cross-sectional area of the tank (that is, the area of the free-surface of the water). This must be balanced by the bulk-mass transfer upward so that

$$\rho_A A v = -\frac{p_A M_A}{R_u T} A v \tag{7.60}$$

where $v$ is the bulk mass velocity in the upward direction. The negative sign in this relation implies that the diffusion is from higher concentration to lower concentration.

From Equations (7.59) and (7.60), we get the bulk mass velocity as

$$\boxed{v = \frac{D}{p_A}\frac{dp_A}{dx}} \tag{7.61}$$

The mass diffusion of water vapor upward is

$$\dot{m}_w = -\frac{DAM_w}{R_u T}\frac{dp_w}{dx} \tag{7.62}$$

and the bulk-mass transfer of water vapor upward is

$$\rho_w A v = \frac{p_w M_w}{R_u T} A v \tag{7.63}$$

Adding Equations (7.62) and (7.63), we get the total mass transport as

$$\dot{m}_{w,total} = -\frac{DAM_w}{R_u T}\frac{dp_w}{dx} + \frac{p_w M_w}{R_u T} A v$$

Substituting the expression for $v$, this becomes

$$\dot{m}_{w,total} = \frac{DAM_w}{R_u T}\frac{dp_w}{dx} + \frac{p_w M_w}{R_u T} A \frac{D}{p_A}\frac{dp_A}{dx}$$

The partial pressures of the water vapor and air are related, by Dalton's law, as

$$p_A + p_w = p$$

But the total pressure $p$ is a constant; therefore,

$$\frac{dp_A}{dx} = -\frac{dp_w}{dx}$$

Thus the total mass flow of water vapor becomes

$$\begin{aligned}
\dot{m}_{w,total} &= -\frac{DAM_w}{R_u T}\left[\frac{dp_w}{dx}\left(1 + \frac{p_w}{p_A}\right)\right] \\
&= -\frac{DAM_w}{R_u T}\frac{p_A + p_w}{p_A}\frac{dp_w}{dx}
\end{aligned}$$

or

$$\boxed{\dot{m}_{w,total} = -\frac{DAM_w}{R_u T}\frac{p}{p - p_w}\frac{dp_w}{dx}} \quad (7.64)$$

This relation is called *Stefan's law.* Integrating this, we get

$$\dot{m}_{w,total} = \frac{DpAM_w}{R_u T\,(x_2 - x_1)} \ln\left(\frac{p - p_{w,2}}{p - p_{w,1}}\right)$$

or

$$\boxed{\dot{m}_{w,total} = \frac{DpAM_w}{R_u T\,(x_2 - x_1)} \ln\left(\frac{p_{A,2}}{p_{A,1}}\right)} \quad (7.65)$$

## Example 7.6

Estimate the diffusion rate of water from the surface of the water layer in a tank of diameter 300 mm and height 450 mm into dry atmospheric air at 300 K, taking the diffusion coefficient of water into air as 0.25 $cm^2/s$.

## Solution

Given, $d = 0.3$ m, $x_2 - x_1 = 0.45$ m, $p = 101325$ Pa, $T = 300$ K, $D = 0.25 \times 10^{-4}$ $m^2/s$.

Let subscripts $A$ and $w$ refer to air and water, respectively. Molecular weight of water $M_w = 18$ and the universal gas constant $R_u = 8314$ $m^2/(s^2\ K)$.

From properties table, the saturation pressure for water at 300 K is $p_{w,1} = 0.03531$ bar $= 0.03531 \times 10^5$ Pa. This is the partial pressure of water at the free surface of the water in the tank.

The vapor pressure of water $p_{w,2}$ at the top of the tank may be taken as zero, because it is diffusing into dry air.

Partial pressures of air at the bottom ($x_1$) and top ($x_2$) of the tank are

$$\begin{aligned} p_{A,1} &= p - p_{w,1} \\ &= 101325 - 3531 \\ &= 97794\,\text{Pa} \\ p_{A,2} &= p - p_{w,2} \\ &= 101325 - 0 \end{aligned}$$

$$= 101325\,\text{Pa}$$

The mass diffusion rate of water, given by Equation (7.65), is

$$\begin{aligned}
\dot{m}_w &= \frac{DpM_wA}{R_uT\,(x_2 - x_1)} \ln\left(\frac{p_{A,2}}{p_{A,1}}\right) \\
&= \frac{(0.25 \times 10^{-4}) \times 101325 \times 18 \times (\pi \times 0.15^2)}{8314 \times 300 \times 0.45} \ln\left(\frac{101325}{97794}\right) \\
&= \boxed{1.02 \times 10^{-7}\,\text{kg/s}}
\end{aligned}$$

## 7.12 Summary

Mass transfer can result from several different phenomena. Mass diffusion and turbulent fluctuations are the two primary causes of mass transfer.

The *concentration gradient is the driving potential* for mass transfer. The concentrations of species in a multi-component system may be expressed by their *mass density* $\rho_i$, defined as the mass of species $i$ per unit volume of the mixture and *molar density* $\rho_i^*$, defined as the number of moles of species $i$ per unit volume of the mixture.

The relation between the mass density and molar density is

$$\boxed{\rho_i^* = \frac{\rho_i}{M_i}}$$

where $M_i$ is the molecular weight of the species $i$.

The *mass fraction* $m_i$ is the ratio of the mass of species $i$ to the total mass of the mixture.

$$m_i = \frac{\rho_i}{\rho}$$

where $\rho_i$ is the mass density (mass/volume) of species $i$ and $\rho$ is the mass density of the mixture, given by

$$\rho = \Sigma\,\rho_i$$

The *mole fraction* $m_i^*$ is the ratio of the number of moles of species $i$ to the total number of moles of the mixture.

$$m_i^* = \frac{\rho_i^*}{\rho^*}$$

where $\rho_i^*$ is the molar density (molar mass/molar volume) of species $i$ and $\rho^*$ is the molar density of the mixture.

Molar mass $M$ is the mass of one mole of a substance. It is also called molecular weight.

From the definitions, we have

$$\boxed{\Sigma\, m_i = 1}$$

$$\boxed{\Sigma\, m_i^* = 1}$$

The *mean molecular weight* $M$ of the mixture can be defined as

$$M = \frac{\rho}{\rho^*}$$

The *mass average velocity* $V$ is

$$V = \boxed{\Sigma\, m_i V_i}$$

The *molar average velocity* $V^*$ is

$$V^* = \boxed{\Sigma m_i^* V_i^*}$$

The *mass diffusion velocity* is

$$\text{Mass diffusion velocity} = V_i - V$$

The *molar diffusion velocity* is

$$\text{Molar diffusion velocity} = V_i - V^*$$

The mass flux $n_i$, and the molar flux $n_i^*$, across a surface stationary surface in space can be expressed as

$$n_i = \rho_i V_i$$

and

$$n_i^* = \rho_i^* V_i^*$$

Across a surface moving with mass average velocity $V$, the mass diffusion flux $j_i$ becomes

$$j_i = \rho_i \left(V_i - V\right)$$

Across a surface moving with molar average velocity $V^*$, the molar diffusion flux $j_i^*$ becomes

$$j_i^* = \rho_i^* \left(V_i - V^*\right)$$

The mass fraction $m_i$ and and mass flux $n_i$ can be related as

$$\boxed{n_i = m_i \left(\Sigma\, n_i\right) + j_i}$$

Similarly, the molar flux $n_i^*$ becomes

$$\boxed{n_i^* = m_i^* \, (\Sigma \, n_i^*) + j_i^*}$$

For the diffusion of species $A$ into species $B$, the molecular flux relation, in the differential form, can be written as

$$\boxed{j_A^* = - \, \rho^* D_{AB} \frac{\partial m_A^*}{\partial y}}$$

where the constant $D_{AB}$ is called the *binary diffusivity* of $A$ in $B$. This relation is known as *Fick's first law of diffusion.*

The continuity equation for the mixture is

$$\boxed{\frac{\partial \rho}{\partial t} + \nabla \cdot (\rho \, V) = 0}$$

The continuity equation for a specific species is called the *species continuity equation.*

In dilute solutions of liquids or solids, the mass density $\rho$ of the mixture (also termed solution) can be approximately taken as a constant. For such a case of constant $\rho$, if the value for the binary diffusion coefficient $D_{AB}$ is also assumed to be a constant, we have

$$\boxed{\frac{\partial m_A}{\partial t} + \nabla . (m_A V) = D_{AB} \, \nabla^2 \, m_A + \frac{\dot{S}_{m,A}}{\rho}}$$

For gas mixtures of constant pressure and temperature, the molar density $\rho^*$ is constant. Thus, for constant pressure and temperature gas-diffusion (with constant value of $D_{AB}$), we have

$$\boxed{\frac{\partial m_A^*}{\partial t} + \nabla . (m_A^* V^*) = D_{AB} \, \nabla^2 \, m_A^* + \frac{\dot{S}_{m,A}^*}{\rho^*}}$$

The mass and molar conservations given by the above two equations are known as *Fick's second law.*

For steady-state condition, the species concentration equations reduce to

$$\nabla^2 m_A^* = 0$$

$$\nabla^2 m_A = 0$$

For one-dimensional diffusion of a single species $A$, the species concentration equation is

$$\nabla^2 m_A = 0$$

For this one-dimensional diffusion of single species $A$, the mass flux given by

$$\begin{aligned} n_A &= j_A = -\rho\, D_{AB}\, \frac{\partial m_A}{\partial x} \\ &= -\rho\, D_{AB}\, \frac{(m_{A2} - m_{A1})}{L} \end{aligned}$$

The mass flow current $\dot{Q}_{m,A}$ is defined as the rate of mass flow passing through per unit area and per unit time. In other words, the mass flow current is the product of mass flux and cross-sectional area of the passage. The mass flow current is given by

$$\boxed{\dot{Q}_{m,A} = \frac{\rho\, A\, D_{AB}}{L}\, (m_{A2} - m_{A1})}$$

The resistance to mass transfer is

$$\boxed{R_{m,A} = \frac{L}{\rho\, D_{AB}\, A}}$$

Following the same procedure as in convection heat transfer analysis in Chapter 5, the mass flow current or diffusion of fluid film on a solid surface can be expressed as

$$\dot{Q}_{m,A} = h_m A\, (m_{A2} - m_{A1})$$

or

$$\dot{Q}_{m,A} = \frac{(m_{A2} - m_{A1})}{\dfrac{1}{h_m A}}$$

where $h_m$ is the convection mass transfer coefficient. The resistance to the diffusion of fluid film on a solid surface becomes

$$\boxed{R_f = \frac{1}{h_m\, A}}$$

*The total molar flux across a surface moving with the molar average velocity is zero.*

The mass transfer associated with significant bulk velocity is termed *convective mass transfer*. That is, in a convective mass transfer the species involved in the mass diffusion will move with an appreciable speed.

The dimensionless parameter,

$$\mathrm{Gr}_m = -\frac{\beta_m\, g\, L^3\, (m_{A,w} - m_{A,f})}{\nu^2}$$

That is,

$$\mathrm{Gr}_m = \frac{g\, L^3\, (\rho_f - \rho_w)}{\nu^2\, \rho_f}$$

is the *Grashoff number* and the parameter

$$\text{Sc} = \frac{\nu}{D_{AB}}$$

is called *Schmidt number* (similar to Prandtl number in heat transfer), which is the ratio of momentum diffusivity to mass diffusivity and Re is the Reynolds number.

The convective mass transfer coefficient $h_m$ is

$$h_m\left(m_{A,w} - m_{A,f}\right) = -\rho D_{AB}\left.\frac{\partial m_A}{\partial y}\right|_{\text{wall}}$$

The *mass transfer Nusselt number* is

$$\text{Nu}_m = \frac{h_m}{\rho D_{AB}}$$

The mass transfer Nusselt number is also known as *Sherwood number* Sc.

In forced convection, $\text{Gr}_m$ is insignificant and thus $\text{Nu}_m$ depends only on Re and Sc. In free convection, Reynolds number is not important and hence $\text{Nu}_m$ depends only on Sc and $\text{Gr}_m$.

In a binary mixture of two gases, say $A$ and $B$, they will diffuse simultaneously in opposite direction to each other. That is, gas $A$ will diffuse through gas $B$, and vice versa. They would diffuse at the molar rate but in opposite directions.

In a steady-state equimolal counterdiffusion, the molar fluxes of gases $A$ and $B$ relative to stationary coordinates are equal, and in the opposite directions. That is,

$$n_A^* + n_B^* = 0 \text{ or } n_A^* = -n_B^*$$

The mass diffusion rate given by Fick's law, which states that the mass flux of a constituent per unit area is proportional to the concentration gradient. Thus,

$$\frac{\dot{m}_A}{A} = -D_{AB}\frac{\partial \rho_A}{\partial x}$$

The heat conduction equation describes the *transport of energy*, the shear stress equation describes the *transport of momentum* across fluid layers, and the diffusion law describes the *transport of mass*.

The mass flux of component $A$ is

$$\boxed{\frac{\dot{m}_A}{A} = -\left(\frac{DM_A}{R_u T}\right)\left(\frac{p_{A,2} - p_{A,1}}{\Delta x}\right)}$$

The total mass flow of water vapor into stagnant air is given by

$$\boxed{\dot{m}_{w,total} = -\frac{DAM_w}{R_u T}\,\frac{p}{p - p_w}\,\frac{dp_w}{dx}}$$

This relation is called *Stefan's law.* This can also be expressed as

$$\boxed{\dot{m}_{w,total} = \frac{DPAM_w}{R_u T\,(x_2 - x_1)} \ln\left(\frac{p_{A,2}}{p_{A,1}}\right)}$$

## 7.13 Exercise Problems

7.1 A mild steel component with a uniform carbon concentration of 0.15 percent is exposed to a carburizing environment and thereby the carbon concentration at the surface of the component is increased to 1.2 percent and maintained. If the diffusion coefficient of carbon in steel is $4.8 \times 10^{-10}$ m$^2$/s, determine the time required for the mass concentration of carbon at 0.5 mm below the surface to reach 1 percent.

[**Ans.** 4480 s]

7.2 A pipeline carrying ammonia gas at 5 kg/h and 25°C is connected to atmospheric air by a tube of diameter of 3 mm, extending to 20 m, in order to maintain the pressure of ammonia in the pipeline at 1 atm. If the atmospheric air is also at 25°C, (a) determine the rate at which ammonia diffuses to the atmosphere and (b) the mass fraction of air in the pipeline. Take $D_{NH_3} = 0.28 \times 10^{-4}$ m$^2$/s.

[**Ans.** (a) $0.0688 \times 10^{-10}$ kg/s, (b) $0.1008 \times 10^{-10}$ kg/s]

7.3 A vessel containing a layer of toluene at the bottom is connected by a long narrow tube to atmospheric air, to remove the toluene vapor from vent. The vent is vertical and the distance from the top of the toluene layer to the top of the vent tube is 1.5 m. The entire system is at pressure 1 atm and temperature 20°C. The pressure of the saturated vapor of toluene at the surface of the layer is 0.025 atm and the diffusivity of air-toluene vapor is $0.83 \times 10^{-5}$ m$^2$/s. Determine the rate at which the toluene is evaporating per unit surface area of the layer.

[**Ans.** $0.053 \times 10^{-5}$ kg/(m$^2$ s)]

7.4 A pipeline carries helium gas at the rate of 2 kg/s. The pressure in the pipeline is maintained at 1 atm and 25°C, by connecting a vertical venting tube of diameter 5 mm and length 15 m to the atmosphere which is at the same state. Determine (a) the mass flow rate of helium diffusing through the venting tube, into the atmospheric air and (b) the mass flow rate of air through the venting tube into the helium pipeline, if the partial pressure of helium at the top of the vent tube is 0.1 percent of the atmospheric pressure. Take $D_{He} = 7.2 \times 10^{-5}$ m$^2$/s.

[**Ans.** (a) $-1.54 \times 10^{-11}$ kg/(m$^2$ s), (b) $11.165 \times 10^{-11}$ kg/(m$^2$ s)]

7.5 Hydrogen gas is stored in a spherical tank of diameter 5 m and wall thickness 50 mm. The temperature and molar concentration of hydrogen in the tank are 360 K and 0.087 kmol/m$^3$, respectively. Hydrogen concentration at the outer surface of the tank is zero. If the diffusion coefficient of hydrogen through the tank wall is $1.2 \times 10^{-12}$ m$^2$/s, determine the rate at which hydrogen is diffusing from the tank to the surrounding, treating the diffusion as one-dimensional and the temperature in the tank as constant.

[**Ans.** $3.28 \times 10^{-10}$ kg/(m$^2$ s)]

# Chapter 8

# Boiling and Condensation

## 8.1 Introduction

Boiling and condensation are essentially convection heat transfer processes involving phase change. In many types of power and refrigeration cycles, phase changes from vapor to liquid or liquid to vapor are involved, depending on the particular part of the cycle considered. These are accomplished by boiling or condensation. From thermodynamics we know that when the temperature of a liquid at a given pressure is raised to the saturation temperature $T_{\text{sat}}$ corresponding to that pressure, the liquid will boil. Similarly, when the temperature of a vapor is lowered to $T_{\text{sat}}$, condensation occurs. Even though boiling and condensation are forms of convection heat transfer, they depend on the latent heat of vaporization $h_{fg}$ of the fluid and the surface tension $\sigma$ at the liquid-vapor interface, in addition to the properties of the fluid at each phase. Furthermore, since under equilibrium conditions the temperature remains constant during a phase-change process at a fixed pressure, large amount of heat can be transferred during boiling and condensation, essentially at constant temperature. However, in actual processes, it is necessary to maintain some difference between the surface temperature $T_s$ and saturation temperature $T_{\text{sat}}$ for an effective heat transfer. Heat transfer coefficient $h$ associated with boiling and condensation are several times higher than those associated with other forms of convection heat transfer processes that involve only a single phase.

## 8.2 Boiling Heat Transfer

Boiling is a liquid-to-vapor phase change process. In other words, boiling is a process in which a liquid phase is converted to a vapor phase. The energy for phase change is generally supplied by the surface on which boiling occurs. Even though it appears to be similar to evaporation, it differs significantly from evaporation. Evaporation occurs at the liquid-vapor interface, when the vapor pressure is less than the saturation pressure of the liquid at a given temperature. But boiling occurs at the solid-interface when a liquid is brought into contact with a solid surface maintained at a temperature $T$, which is sufficiently above the saturation temperature $T_{\text{sat}}$ of the liquid. Schematic sketches illustrating the evaporation and boiling are shown in Figure 8.1. It is seen that the evaporation involves no bubble formation, whereas the boiling involves formation and motion of bubbles.

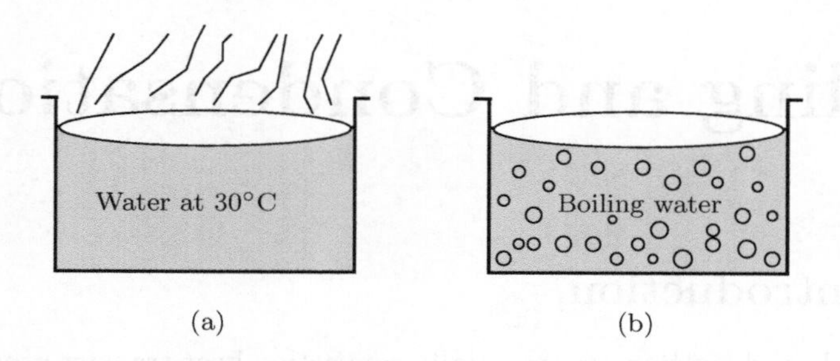


**Figure 8.1**
(a) Evaporation of water, (b) boiling of water.

The boiling process is characterized by the formation of vapor bubbles at the solid-liquid interface that detach from the surface when they reach a certain size and tend to rise to the free surface of the liquid. Boiling is a complicated phenomenon involving a large number of variables and a complex fluid motion caused by the formation and growth of the bubbles.

The heat transfer per unit time per unit surface area from a solid surface to the fluid is termed boiling heat flux. Being a form of convection heat transfer, the *boiling heat flux* can be expressed using Newton's law of cooling as

$$\boxed{\dot{q}_{\text{boiling}} = h\,(T_s - T_{\text{sat}}) = h\Delta T_{\text{excess}}} \tag{8.1}$$

where $T_s$ is the temperature at the surface of the solid, $T_{\text{sat}}$ is the saturation temperature of the liquid, and $\Delta T_{\text{excess}} = (T_s - T_{\text{sat}})$ is called the excess temperature.

For our discussions on free and forced convection heat transfer involving only a single phase of fluid, we need to consider the thermophysical properties, namely the density $\rho$, viscosity $\mu$, thermal conductivity $k$ and specific heat $c_p$ of the fluid, both at its liquid and vapor phases. In addition to these properties of the liquid and vapor, indicated by the subscripts $l$ and $v$, respectively, the latent heat of vaporization $h_{fg}$, and the surface tension $\sigma$, at the interface of liquid and vapor, are also necessary for the analysis of boiling heat transfer. Latent heat of vaporization $h_{fg}$ namely, the *energy required to vaporize unit mass of the liquid at a specified temperature or pressure* is the primary quantity of energy transferred during boiling heat transfer. The $h_{fg}$ values for water at different temperatures are given in Table A-9 in the Appendix.

Bubbles exist because of the surface tension ($\sigma$, N/m) at the liquid-vapor interface due to the attraction force on the molecules at the interface towards the liquid phase. This force is also referred to as *adhesive force.* The adhesive force becomes weaker when the temperature of the liquid rise to the saturation temperature level, at which the liquid begins to evaporate. The vapor thus formed gets liberated and moves towards the free surface of the liquid. In other words, when a liquid begins to evaporate, the cohesive force between the liquid molecules is not sufficient enough to hold them together in the liquid phase. The surface tension decreases with increase of temperature and becomes zero at the critical temperature. Therefore, no bubbles are formed when a liquid boils at supercritical pressure and temperature.

At this point, it will be useful to have a clear idea about the *supercritical* pressure and temperature. The temperature and pressure above the critical point is termed supercritical pressure and temperature, and the fluid at this state is called supercritical fluid. A supercritical fluid is any substance at a temperature and pressure above its critical point, where *distinct liquid and gas phases do not exist.* It can effuse through solids like a gas, and dissolve materials like a liquid. In addition, close to the critical point, small changes in pressure or temperature result in large changes in density, allowing many properties of a supercritical fluid to be "fine-tuned." Supercritical fluids are suitable as a substitute for organic solvents in a range of industrial and laboratory processes. Carbon dioxide and water are the most commonly used supercritical fluids, being used for decaffeination and power generation, respectively.

In the actual process of boiling, the temperature and pressure of the vapor in a bubble are usually different from those of the liquid. Because of this, the boiling process in practice does not occur under an equilibrium condition, and the bubbles are not in thermodynamic equilibrium with the surrounding liquid. During a boiling process,

- The pressure difference between the liquid and vapor is balanced by the surface tension $\sigma$ at the interface.
- The temperature difference between the vapor in a bubble and the surrounding liquid is the driving force for heat transfer between the two phases.

When the liquid is at a lower temperature than the bubble, heat will be transferred from the bubble to the liquid, causing some of the vapor inside the bubble to condense leading to collapse of the bubble. When the liquid temperature is higher than the temperature of the bubble, heat will be transferred from the liquid to the bubble, making the bubble grow in size and rise due to the buoyancy effect.

Boiling is called *pool boiling* if there is no bulk fluid motion and *flow boiling* or *forced convection boiling* in the presence of bulk fluid motion. In pool boiling, the fluid is stationary, and any motion of fluid is due to natural convection and motion of the bubbles is due to buoyancy. For example, the boiling of water in a container on a heater is pool boiling. In flow boiling, the fluid is forced to move in a heated pipe or on a heated surface by external means such as a pump. Further classification of boiling is based on the bulk liquid temperature. Boiling is termed *subcooled* or *local* when the temperature of the main body of the liquid is less than its saturation temperature $T_{\text{sat}}$. The boiling is called *saturated* or *bulk* when the temperature of the liquid is equal to $T_{\text{sat}}$.

### 8.2.1 Pool Boiling

In pool boiling, as we saw, the fluid is not forced to flow. Depending on the value of the excess temperature $\Delta T_{\text{excess}}$, the following four different boiling regimes can be observed in pool boiling:

- Natural convection boiling
- Nucleate boiling
- Transition boiling
- Film boiling

These four regimes of natural, nucleate, transition and film boiling are illustrated on the boiling curve for water shown in Figure 8.2.

From thermodynamics, we know that a pure substance at a specified pressure begins to boil when it is heated to its saturated temperature $T_{\text{sat}}$, at that pressure. But in an actual process, bubbles form on the heating surface and the liquid begins to boil only when the liquid is heated to a few degrees above $T_{\text{sat}}$. In Figure 8.2, up to point A, water is only slightly *superheated.* Therefore, the hot water, liberating from the bottom surface of the container in which it is heated and moving upwards, evaporates only when it reaches the free surface of the water layer. The liquid motion is governed by natural convection currents. The heat transfer from the heating surface to the fluid is by natural convection.

When $\Delta T_{\text{excess}}$ reaches about 5°C, bubbles begin to form (point A) on the heating surface. As the $\Delta T_{\text{excess}}$ increases from point A towards C, bubbles form at an increasing rate at an increasing number of nucleation (bubble formation) sites. In the region from A to B, isolated bubbles are formed at various nucleation sites on the heating surface. But these bubbles get diffused in the liquid soon after they liberate from the heating surface. From B to C bubbles form at such a large rate and at a large number of nucleation sites that they form numerous continuous columns of vapor in the liquid. These bubbles travel up to the free surface and break up and release their vapor

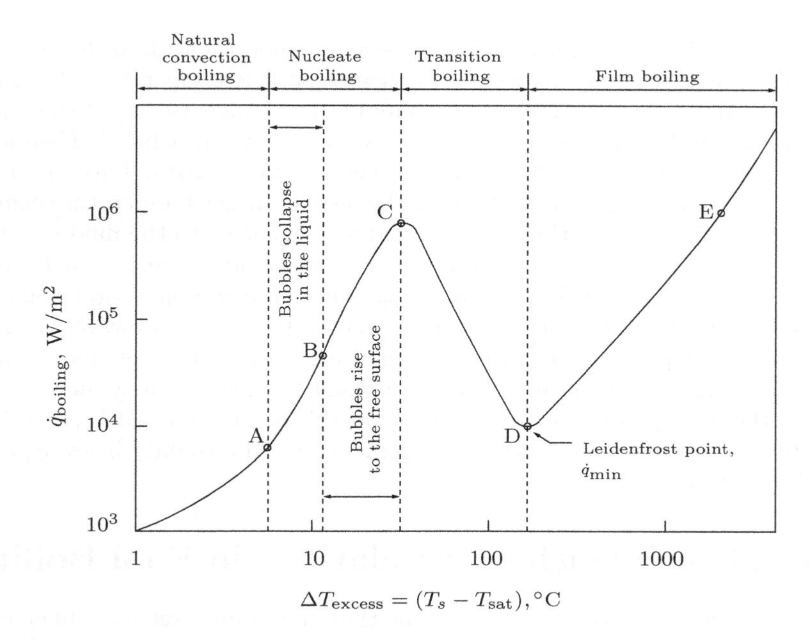


**Figure 8.2**
Boiling curve for water at 1 atm pressure.

content. The heat flux increases progressively with increasing $\Delta T_{excess}$, and reaches a maximum at point C. The heat flux at this point is called the *critical heat flux*, $\dot{q}_{max}$. For water $\dot{q}_{max} = 1$ MW/m$^2$. In the nucleate boiling regime very high heat transfer rates can be achieved with relatively small values of $\Delta T_{excess}$.

When $\Delta T_{excess}$ is increased beyond point C, the heat flux decreases as shown in Figure 8.2. This is because a larger fraction of the heater surface is covered by a vapor film. This cover acts as an insulation because of the low thermal conductivity of the vapor relative to that of the liquid. Both nucleate and film boiling co-exist in the transition boiling regime. The nucleate boiling which begins at point C is completely replaced by film boiling at point D. The transition boiling regime is also called the *unstable film boiling regime.*

In the film boiling regime the heater surface is completely covered by a continuous stable vapor film. The point D on the boiling curve where the heat flux reaches a minimum is called the *Leidenfrost point.* The presence of the vapor film between the heater surface and the liquid is responsible for the low heat transfer rates in the film boiling region. In the film boiling regime, the heat transfer rate increases with $\Delta T_{excess}$ increase as a result of heat transfer from the heated surface to the liquid through the vapor film by radiation, which becomes considerable at high temperatures.

Beyond point C, a typical boiling process will not follow. In order to move beyond point C where $\dot{q}_{\text{max}}$ occurs, the heater surface temperature $T_s$ has to be increased. To increase $T_s$, the heat flux has to be increased. But the liquid cannot receive this increased energy at a $\Delta T_{\text{excess}}$ beyond point C. Therefore, the heater surface absorbs the increased energy and $T_s$ rises. This continues till $T_s$ reaches a limiting value at which the heater surface temperature cannot increase any further and the heat supplied is transferred to the fluid steadily. This is point E on the boiling curve, which corresponds to very high $T_s$. Any further increase of heat flux beyond $\dot{q}_{\text{max}}$ will cause the operation point on the boiling curve to jump suddenly from point C to point E. However, $T_s$ that corresponds to point E is beyond the melting temperature of most heater materials, and *burnout* occurs. Therefore, point C on the boiling curve is also termed the boiling point and the heat flux at this point is termed *burnout heat flux*. In practice most heat transfer equipments operate slightly below $\dot{q}_{\text{max}}$ to avoid burnout.

## 8.3 Heat Transfer Correlations in Pool Boiling

From the above discussion, it is evident that the boiling regimes differ considerably in their character. Therefore, heat transfer relations for different boiling regimes are bound to be different. In the natural convection boiling regime, the heat transfer rates can be determined using the natural convection relations presented in Chapter 5. In the nucleate boiling regime, the rate of heat transfer depends on the number of active nucleate sites on the surface, the rate of bubble formation at each site, etc. Therefore, the heat transfer in the nucleate boiling regime is difficult to predict theoretically and we have to rely only on experimental data. The most widely used correlation for the rate of heat transfer in the nucleate boiling regime is the *Rohsenow correlation*,

$$\dot{q}_{\text{nucleate}} = \mu_l\, h_{fg} \left( \frac{g\,(\rho_l - \rho_v)}{\sigma} \right)^{1/2} \left( \frac{c_{pl}\,(T_s - T_{\text{sat}})}{C_{sf}\, h_{fg}\, \text{Pr}_l{}^n} \right)^3 \tag{8.2}$$

where $\dot{q}_{\text{nucleate}}$ is the nucleate boiling heat flux, $\mu_l$ is the viscosity of the liquid, $h_{fg}$ is the enthalpy of vaporization, $\rho_l$ is liquid density, $\rho_v$ is vapor density, $\sigma$ is the surface tension of liquid-vapor interface, $c_{pl}$ is the specific heat of the liquid, $T_s$ is the heater surface temperature, $T_{\text{sat}}$ is the saturation temperature of the fluid, $C_{sf}$ is a constant, $\text{Pr}_l$ is the Prandtl number of the liquid and $n$ is experimental constant. Equation (8.2) is called Rohsenow equation. The coefficient $C_{sf}$ and the index $n$ in Equation (8.2) depend on surface-fluid combination.

Note that the heat transfer rate or heat flux given by Equation (8.2) is per unit area. Therefore, the total heat transfer involved in the boiling process becomes

$$\dot{Q}_{\text{nucleate}} = A_s \dot{q}_{\text{nucleate}}$$

where $A_s$ is the surface area of the container base over which heat is supplied.

**Table 8.1** Surface tension of vapor-liquid interface for water

| $T$ °C | 0 | 20 | 40 | 60 | 80 | 100 |
|---|---|---|---|---|---|---|
| $\sigma$ N/m | 0.0757 | 0.0727 | 0.0696 | 0.0662 | 0.0627 | 0.0589 |

| 120 | 140 | 160 | 180 | 200 | 220 | 240 |
|---|---|---|---|---|---|---|
| 0.0550 | 0.0509 | 0.0466 | 0.0422 | 0.0377 | 0.0331 | 0.0284 |

| 260 | 280 | 300 | 320 | 340 | 360 | 374 |
|---|---|---|---|---|---|---|
| 0.0237 | 0.0190 | 0.0144 | 0.0099 | 0.0056 | 0.0019 | 0.0 |

**Table 8.2** Surface tension of some liquids

| Liquid ($T$ °C range) | $\sigma$ N/m |
|---|---|
| Ammonia (−75 to −40) | $0.0264 + 0.000223\,T$ |
| Benzene (10 to 80) | $0.0315 - 0.000129\,T$ |
| Butane (−70 to −20) | $0.0149 - 0.000121\,T$ |
| Carbon dioxide (−30 to −20) | $0.0043 - 0.000160\,T$ |
| Ethyl alcohol (10 to 70) | $0.0241 - 0.000083\,T$ |
| Mercury (5 to 200) | $0.4906 - 0.000205\,T$ |
| Methyl alcohol (10 to 60) | $0.0240 - 0.000077\,T$ |
| Pentane (10 to 30) | $0.0183 - 0.000110\,T$ |
| Propane (−90 to −10) | $0.0092 - 0.000087\,T$ |

The surface tension at the vapor-liquid interface of water and some selected liquids is given in Tables 8.1 and 8.2, respectively.

Experimentally found values of $C_{sf}$ are listed in Table 8.3, for some fluid-surface combinations. These values can be used for any geometry since it is found that the rate of heat transfer during nucleate boiling is independent of the geometry and orientation of the heated surface. The fluid properties in Equation (8.2) are to be evaluated at the saturation temperature $T_{\text{sat}}$.

The Rohsenow equation [Equation (8.2)] is applicable only to clean and smooth surfaces. It is essential to note that the error associated with the heat flux results obtained using Equation (8.2) can be as high as ± 100%.

**Table 8.3** $C_{sf}$ and $n$ for some fluid-surface combination

| Fluid-surface | $C_{sf}$ | $n$ |
|---|---|---|
| Water-copper (polished) | 0.0130 | 1.0 |
| Water-copper (scored) | 0.0068 | 1.0 |
| Water-stainless steel (mechanically polished) | 0.0130 | 1.0 |
| Water-stainless steel (ground and polished) | 0.0060 | 1.0 |
| Water-stainless steel (teflon pitted) | 0.0058 | 1.0 |
| Water-stainless steel (chemically etched) | 0.0130 | 1.0 |
| Water-brass | 0.0060 | 1.0 |
| Water-nickel | 0.0060 | 1.0 |
| Water-platinum | 0.0130 | 1.0 |
| n-Pentane-copper (polished) | 0.0154 | 1.7 |
| n-Pentane-chromium | 0.0150 | 1.7 |
| Benzene-chromium | 0.1010 | 1.7 |
| Ethyl alcohol-chromium | 0.0027 | 1.7 |
| Carbon tetrachloride-copper | 0.0130 | 1.7 |
| Isopropanol-copper | 0.0025 | 1.7 |

The uncertainty in the excess temperature, for a given heat transfer rate, calculated with Equation (8.2) can be of the order of ± 30%. Therefore, results obtained using this equation have to be interpreted with great care.

From thermodynamics, it is known that the enthalpy of vaporization $h_{fg}$ of a pure substance decreases with increase of pressure or temperature and reaches zero at the critical point. In Equation (8.2), $h_{fg}$ is in the denominator. Therefore, a significant rise in the rate of heat transfer will be experienced at high pressures during nucleate boiling. The rate of evaporation of a liquid is given by

$$\dot{m} = \frac{\dot{Q}_{\text{boiling}}}{h_{fg}}$$

where $\dot{m}$ is the mass of the fluid evaporating per second and $\dot{Q}$ rate at which heat is supplied.

## Example 8.1

Water at 1 atm is boiled in a container, using an electrical heater. If the heat supplied by the heater is 500 kW/m$^2$ and the temperature of the heater surface is 110°C, determine the coefficient $C_{sf}$ of the heater surface.

## Solution

Given, $\dot{q} = 500$ kW/m$^2$, $T_s = 110$°C.

At 1 atm pressure, the water will boil at 100°C. Thus, $T_{\text{sat}} = 100$°C.

For 100°C, from Table A-9,

$$\rho_l = 957.9\,\text{kg/m}^3,\ \rho_v = 0.5978\,\text{kg/m}^3,\ c_{pl} = 4217\,\text{J/(kg °C)},$$
$$\mu_l = 0.282 \times 10^{-3}\,\text{kg/(m s)},\ \text{Pr} = 1.75,\ h_{fg} = 2257\,\text{kW/kg}$$

At 100°C, from Table 8.1, the surface tension of the vapor-liquid interface for water is $\sigma = 0.0589$ N/m.

The heat flux for nucleate boiling, by Equation (8.2), is

$$\dot{q}_{\text{nucleate}} = \mu_l\, h_{fg} \left(\frac{g\,(\rho_l - \rho_v)}{\sigma}\right)^{1/2} \left(\frac{c_{pl}\,(T_s - T_{\text{sat}})}{C_{sf}\, h_{fg}\, \text{Pr}_l{}^n}\right)^3$$

For water-heater element combination, the exponent $n$ can be taken as 1. Therefore,

$$\begin{aligned} 500 \times 10^3 &= (0.282 \times 10^{-3})(2257 \times 10^3) \left(\frac{9.81 \times (957.9 - 0.5978)}{0.0589}\right)^{1/2} \\ &\quad \times \left(\frac{4217 \times (110 - 100)}{C_{sf} \times (2257 \times 10^3) \times 1.75}\right)^3 \end{aligned}$$

$$500 \times 10^3 = 636.474 \times \sqrt{159442} \times \left(\frac{0.010677}{C_{sf}}\right)^3$$

$$\left(\frac{0.010677}{C_{sf}}\right)^3 = 1.967$$

$$C_{sf} = \frac{0.010677}{(1.967)^{1/3}}$$

$$= \boxed{0.00852}$$

## 8.4 Peak Heat Flux

The peak or maximum heat flux in nucleate pool boiling was determined theoretically using different approaches and expressed as

$$\dot{q}_{\max} = C_{cr} h_{fg} \left(\sigma g \rho_v^2 \left(\rho_l - \rho_v\right)\right)^{1/4} \tag{8.3}$$

where $C_{cr}$ is a constant which depends on the heater geometry, $\sigma$ is the surface tension, $\rho_l$ and $\rho_v$ are the density of the liquid and vapor, respectively, and $g$ is the gravitational acceleration. Experimentally it has been found that, $C_{cr}$ is about 0.15. Values of $C_{cr}$ for some heater geometries, corresponding to characteristic parameter $L^*$, defined as

$$L^* = L\left[\frac{g\left(\rho_l - \rho_v\right)}{\sigma}\right]^{1/2} \tag{8.3a}$$

are listed in Table 8.4.

Note: The geometrical parameter $L$ in Equation (8.3a) should be taken as the length or width for a flat plate, and the radius for a tube or sphere.

The maximum heat flux is independent of the fluid-heating surface combination, the viscosity, thermal conductivity and specific heat of the liquid. A knowledge of the maximum heat flux is essential in the design of boiler heat transfer equipment to avoid burnout of the device.

### Example 8.2

Water at atmospheric pressure and saturation temperature is boiled in a stainless steel container, heated with an electrical heater. If the surface of the heating pan bottom is maintained at 110°C, calculate (a) the heat flux at the surface of the container bottom, and (b) the peak heat flux, assuming $C_{cr} = 0.15$.

**Table 8.4** $C_{cr}$ for some selected shapes

| Geometry | $C_{cr}$ | Characteristic parameter | $L^* = L\left(\sqrt{\frac{g(\rho_l - \rho_v)}{\sigma}}\right)$ |
|---|---|---|---|
| Large horizontal flat heater | 0.0149 | width or diameter | $L^* > 27$ |
| Small horizontal flat heater | $18.9\,a^{\#}$ | width or diameter | $9 < L^* < 20$ |
| Large horizontal cylinder | 0.12 | radius | $L^* > 1.2$ |
| Small horizontal cylinder | $0.12L^{*\,-0.25}$ | radius | $0.15 < L^* < 1.2$ |
| Large sphere | 0.11 | radius | $L^* > 4.26$ |
| Small sphere | $0.227L^{*\,-0.5}$ | radius | $0.15 < L^* < 4.26$ |

\# $a = \frac{\sigma}{g(\rho_l - \rho_v)A_{\text{heater}}}$

## Solution

Given, $T_s = 110°\text{C}$. At 1 atm pressure, $T_{\text{sat}} = 100°\text{C}$.

At 100°C, from Table A-9,

$$\rho_l = 957.9\,\text{kg/m}^3,\ \rho_v = 0.5978\,\text{kg/m}^3,\ h_{fg} = 2257\,\text{kJ/kg},\ \text{Pr}_l = 1.75,$$
$$c_{pl} = 4217\,\text{J/(kg °C)}, \mu_l = 0.282 \times 10^{-3}\,\text{kg/(m s)}$$

From Table 8.1, $\sigma = 0.0589$ N/m.

(a) By Equation (8.2), the heat flux at the surface of the container bottom is

$$\dot{q}_{\text{nucleate}} = \mu_l\, h_{fg} \left(\frac{g(\rho_l - \rho_v)}{\sigma}\right)^{1/2} \left(\frac{c_{pl}(T_s - T_{\text{sat}})}{c_{sf}\, h_{fg}\, \text{Pr}_l{}^n}\right)^3$$

From Table 8.3, for the given container, $C_{sf} = 0.013$ and $n = 1$. Therefore,

$$\begin{aligned}
\dot{q} &= (0.282 \times 10^{-3}) \times (2257 \times 10^3) \times \left(\frac{9.81 \times (957.9 - 0.5978)}{0.0589}\right)^{1/2} \\
&\quad \times \left(\frac{4217 \times (110 - 100)}{0.013 \times (2257 \times 10^3) \times 1.75}\right)^3 \\
&= 636.47 \times 399.30 \times 0.554 \\
&= 140794.93\,\text{W/m}^2 \\
&= \boxed{140.795\,\text{kW/m}^2}
\end{aligned}$$

(b) By Equation (8.3), the maximum heat flux is

$$\begin{aligned}\dot{q}_{\max} &= C_{cr} h_{fg} \left(\sigma g \rho_v^2 (\rho_f - \rho_v)\right)^{1/4} \\ &= 0.15 \times (2257 \times 10^3) \\ &\quad \times \left[0.0589 \times 9.81 \times 0.5978^2 \times (957.9 - 0.5978)\right]^{1/4} \\ &= \boxed{1269.43\,\text{kW/m}^2}\end{aligned}$$

## Example 8.3

A spherical electrical heating element of diameter 100 mm, made of copper is used to heat water at pressure 1 atm. If the surface of the heating element is maintained at 115°C, determine (a) the maximum heat flux and (b) the rate at which water evaporates.

## Solution

Given, $D = 0.1$ m, $T_{\text{sat}} = 100°\text{C}$, $T_s = 115°\text{C}$.

For 100°C, from Table A-9,

$$\rho_l = 957.9\,\text{kg/m}^3,\ \rho_v = 0.5978\,\text{kg/m}^3,\ h_{fg} = 2257\,\text{kJ/kg}$$

For 100°C, from Table 8.1, the surface tension is $\sigma = 0.0589$ N/m.

For the given sphere taking the radius as the length scale $L$, we get the characteristic parameter $L^*$ as

$$\begin{aligned}L^* &= L\left(\left[g(\rho_l - \rho_v)/\sigma\right]^{1/2}\right) \\ &= 0.05\left(\left[9.81\,(957.9 - 0.5978)/0.0589\right]^{1/2}\right) \\ &= 19.96\end{aligned}$$

For $L^* > 4.26$, from Table 8.4, $C_{cr} = 0.11$.

(a) The peak or maximum heat flux, by Equation (8.3) is

$$\begin{aligned}\dot{q}_{\max} &= C_{cr} h_{fg} \left(\sigma g \rho_v^2 (\rho_f - \rho_v)\right)^{1/4} \\ &= (0.11)(2257 \times 10^3)\left[0.0589 \times 9.81 \times 0.5978^2 (957.9 - 0.5978)\right]^{1/4} \\ &= \boxed{931\,\text{kW}}\end{aligned}$$

(b) The rate at which water mass evaporates is given by

$$\begin{aligned} \dot{m} &= \frac{\dot{Q}}{h_{fg}} = \frac{\dot{q}\,A_s}{h_{fg}} \\ &= \frac{(931 \times 10^3) \times (4\,\pi \times 0.05^2)}{2257 \times 10^3} \\ &= \boxed{0.013\,\text{kg/s}} \end{aligned}$$

## 8.5 Minimum Heat Flux

Minimum heat flux is the lower limit for the heat flux in the film boiling regime. Using the stability theory, the minimum heat flux for a large horizontal plate has been expressed as

$$\dot{q}_{\text{min}} = 0.09\rho_v h_{fg} \left( \frac{\sigma g\,(\rho_f - \rho_v)}{(\rho_f + \rho_v)^2} \right)^{1/4} \tag{8.4}$$

This is only an approximate relation and can be in error as high as 50%.

## 8.6 Film Boiling

The heat flux for film boiling on a horizontal cylinder or sphere of diameter $D$ is given by

$$\dot{q}_{\text{film}} = C_{\text{film}} \left( \frac{g k_v^3 \rho_v\,(\rho_l - \rho_v)\,[h_{fg} + 0.4 c_{pv}\,(T_s - T_{\text{sat}})]}{\mu_v D\,(T_s - T_{\text{sat}})} \right)^{1/4} (T_s - T_{\text{sat}}) \tag{8.5}$$

where $k_v$ is the thermal conductivity of the vapor and $C_{\text{film}}$ are 0.62 and 0.67, respectively, for horizontal cylinders and spheres. The vapor properties are to be evaluated at the film temperature $T_f = (T_s + T_{\text{sat}})\,/2$, which is the average temperature of the vapor film. The liquid properties and $h_{fg}$ are to be evaluated at the saturation temperature at the specified pressure.

At temperatures above 300°C, heat transfer across the vapor film by radiation becomes significant and needs to be accounted for. Treating the vapor film as a transparent medium sandwiched between two large parallel plates and approximating the liquid as a blackbody, radiation heat transfer can be determined from

$$\dot{q}_{\text{rad}} = \epsilon\sigma\left(T_s^4 - T_{\text{sat}}^4\right) \tag{8.6}$$

where $\epsilon$ is the emissivity of the heating surface and $\sigma = 5.67 \times 10^{-8}$ W/(m$^2$ K$^4$) is the Stefan-Boltzmann constant.

## Example 8.4

Water at 1 atm is boiled with a horizontal electrical heater wire of diameter 12 mm. The heater supplies 7.2 kW/m$^2$ of heat flux. If the surface temperature of the wire is 120°C, determine the boiling transfer coefficient $C_{\text{film}}$.

## Solution

Given, $p = 1$ atm, $D = 0.012$ m, $T_s = 120°\text{C}$, $\dot{q} = 7.2\ \text{kW/m}^2$.

At 1 atm, water will boil at 100°C. Therefore, the film temperature is

$$T_f = \frac{120 + 100}{2} = 110°\text{C}$$

At $T_f = 110°\text{C}$, from Table A-9, we have the vapor properties,

$$\rho_v = 0.8263\,\text{kg/m}^3,\ c_{pv} = 2071\,\text{J/(kg °C)},$$
$$k_v = 0.0262\,\text{W/(kg °C)},\ \mu = 1.261 \times 10^{-5}\,\text{kg/(m s)}$$

At $T = 100°\text{C}$, from Table A-9,

$$\rho_l = 957.9\,\text{kg/m}^3,\ h_{fg} = 2275\,\text{kJ/(kg °C)}$$

By Equation (8.5), the heat flux for film boiling the horizontal cylindrical heater is

$$\dot{q}_{\text{film}} = C_{\text{film}} \left( \frac{g k_v^3 \rho_v (\rho_l - \rho_v) \left[ h_{fg} + 0.4 c_{pv} (T_s - T_{\text{sat}}) \right]}{\mu_v D (T_s - T_{\text{sat}})} \right)^{1/4} (T_s - T_{\text{sat}})$$

Therefore,

$$\begin{aligned}
7200 &= C_{\text{film}} \Bigg( 9.81 \times 0.0262^3 \times 0.8263 \times (957.9 - 0.8263) \\
&\qquad \times \frac{2275 \times 10^3 + 0.4 \times 2071 \times (120 - 100)}{(1.261 \times 10^{-5}) \times 0.012 \times (120 - 100)} \Bigg)^{1/4} \times (120 - 100) \\
&= C_{\text{film}} \left( \frac{319733.96}{0.30264 \times 10^{-5}} \right)^{1/4} \times 20 \\
&= C_{\text{film}} \times 11402.4
\end{aligned}$$

$$\begin{aligned}
C_{\text{film}} &= \frac{7200}{11402.4} \\
&= \boxed{0.6314}
\end{aligned}$$

## 8.7 Flow Boiling

Flow boiling is that in which the fluid during its phase-change process is forced to move by an external source such as a pump. It is essentially a combination of natural and/or forced convection and pool boiling. When the fluid is forced to flow over a heated surface, it is termed *external flow boiling* and when the fluid is forced to flow inside a heated tube, it is called *internal flow boiling.*

External flow boiling over a plate or cylinder is similar to pool boiling (which involves a pool boiling of seemingly motionless liquid, with vapor bubbles rising to the top as a result of the buoyancy effect), but the added motion increases the nucleate boiling heat flux and the critical heat flux considerably, as illustrated in Figure 8.3.

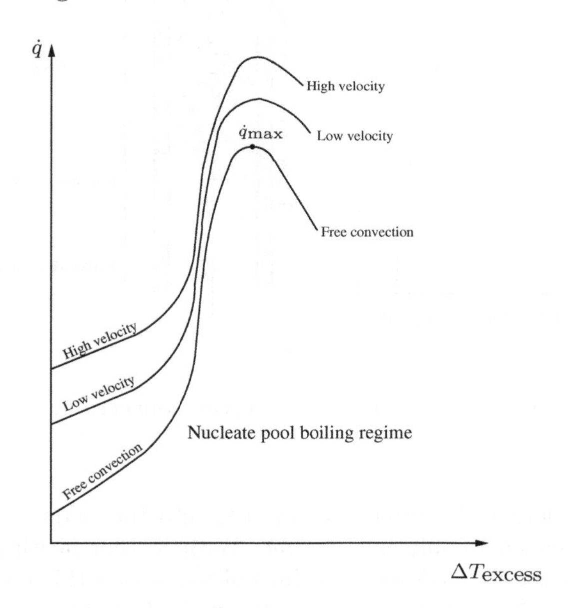


**Figure 8.3**
Forced convection effect on external flow boiling.

Internal flow boiling is more complicated than external flow boiling because there is no free surface for the vapor to escape, thus both the liquid and vapor are forced to flow together. The two-phase flow in a tube exhibits different flow boiling regimes, depending on the relative amounts of the liquid and the vapor phases.

Different stages and the associated heat transfer variation for flow boiling in a heated tube are shown in Figure 8.4.

The liquid is subcooled initially and the heat transfer to the liquid is by forced convection. When bubbles begin to form on the inner surface of

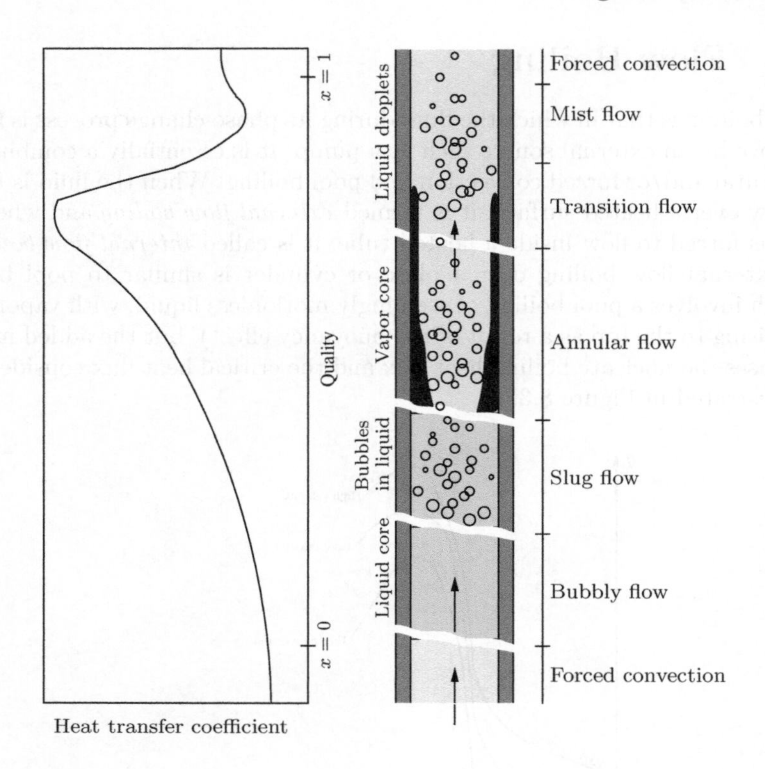


**Figure 8.4**
Regimes of flow boiling in a tube under forced convection.

the tube, the detached bubbles are drafted into the main flow stream and the regime is called bubbly flow regime. With further heating, the bubbles grow in size and fully coalesce into slugs of vapor and this regime is termed *slug-flow regime.* Soon after that, the core of the flow consists only of vapor, and the liquid layer is confined in the annular space between the vapor core and the tube wall. This regime is called *annular-flow regime.* With further heating, the annular liquid layer progressively gets thinner and finally dry spots begin to appear on the inner surface of the tube. This transition regime continues until the inner surface of the tube becomes completely dry. At this stage, any liquid present is in the form of droplets suspended in the vapor core and resembles a mist. This regime termed *mist-flow regime* continues until all the liquid droplets are vaporized. At the end of mist-flow regime, saturated vapor regime begins and the saturated vapor becomes superheated with further heating.

## 8.8 Condensation Heat Transfer

Condensation occurs when the temperature of a vapor is reduced below its saturation temperature $T_{\text{sat}}$, by bringing the vapor into contact with a solid surface whose temperature $T_s$ is less than $T_{\text{sat}}$ of the vapor. But condensation can also occur on the free surface of a liquid or even in a gas, when the temperature of the liquid or the gas to which the vapor is exposed is below the saturation temperature $T_{\text{sat}}$ of the vapor. In the second case, the liquid droplets suspended in the gas form a fog.

Condensation can occur in the following two distinct forms:

• *Film condensation*, in which the condensate wets the surface and forms a liquid film on the surface. The liquid film slides down under the influence of gravity. The thickness of the liquid film increases in the flow direction as more vapor condenses on the film. This kind of condensation is the one which is normally encountered in practice. The liquid layer between the solid surface and the vapor serves as a *resistance* to heat transfer. The heat of vaporization $h_{fg}$ released, as the vapor condenses, passes through the resistance and reaches the solid surface and then transferred to the medium on the other side of the solid surface.

• *Dropwise condensation*, in which the condensed vapor forms droplets on the surface instead of a continuous film, and the surface is covered with a large number of droplets of different diameters. The droplets slide down when they reach a certain size, clearing the surface and exposing it to vapor. There is no liquid film, offering resistance to heat transfer, as in the case of film condensation. Therefore, the heat transfer rates achieved in droplet condensation can be even more than 10 times larger than that with film condensation. However, the dropwise condensation achieved usually does not last long and gets converted to film condensation within a short period of time.

## 8.9 Film Condensation

Let us examine film condensation on a vertical plate illustrated in Figure 8.5. The liquid film flows down the plate because of gravity influence. The film thickness $\delta$ increases in the flow direction because of continued condensation at the liquid-vapor interface. Due to latent heat of vaporization, an amount of heat $h_{fg}$ is released during condensation. The heat released is transferred through the liquid film to the plate surface at temperature $T_s$. For condensation to take place, it is essential that the plate surface temperature is less than the saturation temperature of the vapor ($T_s < T_{\text{sat}}$).

The velocity of the condensate is zero at the wall, because of "no-slip" condition, and reaches a maximum at the liquid-vapor interface. The temperature of the condensate is $T_{\text{sat}}$ at the interface and decreases gradually to $T_s$ at the plate surface.

As in the case of forced convection, heat transfer in condensation also

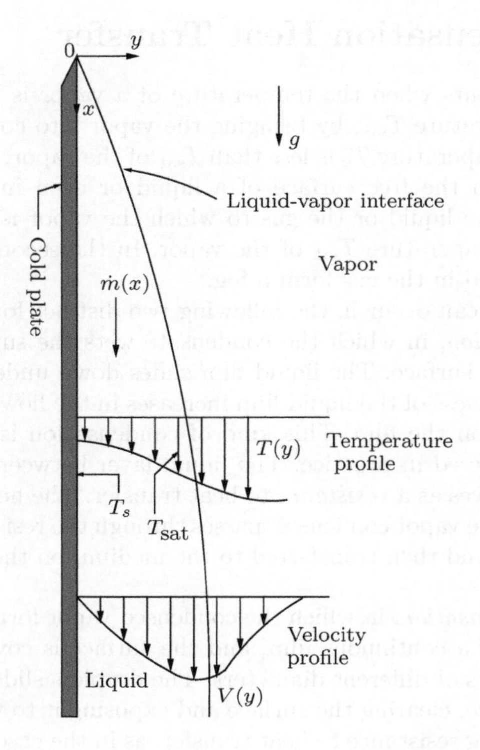


**Figure 8.5**
Film condensation on a vertical plate.

depends on the laminar or turbulent nature of the flow. The Reynolds number for the condensate flow is defined as

$$\mathrm{Re} = \frac{D_h \rho_l \overline{V}_l}{\mu_l} = \frac{4A_c \rho_l \overline{V}_l}{P\mu_l} = \frac{4\rho_l \overline{V}_l \delta}{\mu_l} \tag{8.7}$$

where $D_h = 4A_c/P = 4\delta$ = hydraulic diameter of the condensate flow, $P$ is wetted perimeter of the condensate, $A_c = P\delta$ is the cross-sectional area of the condensate flow at the lowest part of the flow, $\delta$ is film thickness, $\rho_l$ is density of the liquid, $\mu_l$ is viscosity of the liquid, $\overline{V}_l$ is the average velocity of the condensate at the lowest part of the flow and $\dot{m} = \rho_l A \overline{V}_l$ is the mass flow rate of the condensate at the lowest part.

The latent heat of vaporization $h_{fg}$ is the heat released as a unit mass of vapor condenses. It normally represents the heat transfer per unit mass of condensate formed during condensation. But the condensate in an actual condensation process is cooled down further to some average temperature between $T_{\rm sat}$ and $T_s$, releasing more heat in the process. Therefore, the actual heat transfer will be larger than the theoretical value. The cooling of the liquid below the saturation temperature can be accounted by replacing $h_{fg}$ by the modified latent heat of vaporization $h^*_{fg}$, defined as

$$h^*_{fg} = h_{fg} + 0.68\, c_{pl}\,(T_{\text{sat}} - T_s) \tag{8.8a}$$

where $c_{pl}$ is the specific heat of the liquid at the average film temperature.

For a vapor that enters the condenser as superheated at a temperature $T_v$ instead of as saturated, the vapor must be first cooled to $T_{\text{sat}}$ before it can condense, and the heat thus removed must also be transferred to the wall. The amount of heat released as a unit mass of superheated vapor at a temperature $T_v$ is cooled to $T_{\text{sat}}$ is $c_{pv}\,(T_v - T_{\text{sat}})$, where $c_{pv}$ is the specific heat of vapor at the average temperature of $(T_v + T_{\text{sat}})\,/2$. The modified latent heat of vaporization in this case becomes

$$h^*_{fg} = h_{fg} + 0.68\, c_{pl}\,(T_{\text{sat}} - T_s) + c_{pv}\,(T_v - T_{\text{sat}}) \tag{8.8b}$$

Now, the heat transfer can be expressed as

$$\dot{Q} = hA\,(T_{\text{sat}} - T_s) = \dot{m}\, h^*_{fg} \tag{8.9}$$

where $A$ is the surface area on which the condensate occurs.

By Equation (8.9), the mass flow becomes

$$\dot{m} = \frac{\dot{Q}}{h^*_{fg}}$$

But, $\dot{m}$ is also equal to $\rho_l A \overline{V}_l$. Therefore,

$$\begin{aligned} \rho_l A \overline{V}_l &= \frac{\dot{Q}}{h^*_{fg}} \\ \overline{V}_l &= \frac{\dot{Q}}{\rho_l A\, h^*_{fg}} \\ &= \frac{h\,A\,(T_{\text{sat}} - T_s)}{\rho_l A\, h^*_{fg}} \end{aligned}$$

Substituting this into Equation (8.7), we get

$$\begin{aligned} \text{Re} &= \frac{4\rho_l\,\delta}{\mu_l}\,\frac{h\,A\,(T_{\text{sat}} - T_s)}{\rho_l A\, h^*_{fg}} \\ &= \frac{4h\,A\,(T_{\text{sat}} - T_s)}{\mu_l\, h^*_{fg}}\,\frac{\delta}{A} \end{aligned}$$

But

$$\frac{\delta}{A} = \frac{1}{P}$$

Thus,

$$\mathrm{Re} = \frac{4\,\dot{Q}_{\text{conden}}}{P\mu_l\, h^*_{fg}} = \frac{4A\, h\,(T_{\text{sat}} - T_s)}{P\mu_l\, h^*_{fg}} \tag{8.10}$$

The temperature of the liquid film varies from $T_{\text{sat}}$ on the liquid-vapor interface of $T_s$ at the wall surface. Therefore, the properties of the liquid should be evaluated at the film temperature $T_f = (T_{\text{sat}} + T_s)/2$, which is approximately the average temperature of the liquid. But $h_{fg}$ should be evaluated only at $T_{\text{sat}}$ because it is not affected by the subcooling of the liquid.

### 8.9.1 Flow Regimes

The liquid film thickness $\delta$ on the plate increases in the flow direction. Thus, the Reynolds number of the condensate flow also increases in the flow direction. Because of this, the flow of liquid film exhibits different regimes, depending on the value of the Reynolds number. The outer surface of the liquid film will remain laminar and wave-free for Re $\leq$ 30, as shown in Figure 8.6.

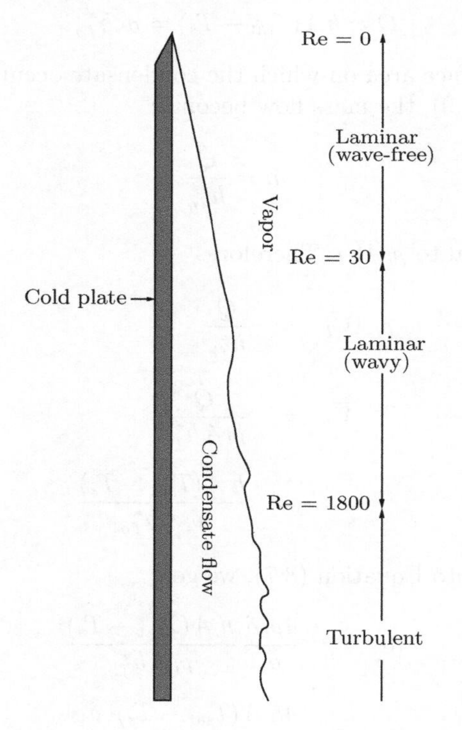


**Figure 8.6**
Flow regimes for film condensation over a vertical plate.

For Reynolds number more than 30, waves appear on the free surface of the condensate and the condensate flow becomes fully turbulent for Reynolds number about 1800. The condensate flow is a mixture of laminar and turbulent flows in the Reynolds number range from 450 to 1800.

## 8.9.2 Flow Condensation Heat Transfer Correlations

### Laminar Flow on Vertical Plates

Let us consider a vertical plate of length $L$ and width $b$, maintained at a constant temperature $T_s$, exposed to a vapor at $T_{\text{sat}}$, as shown in Figure 8.7.

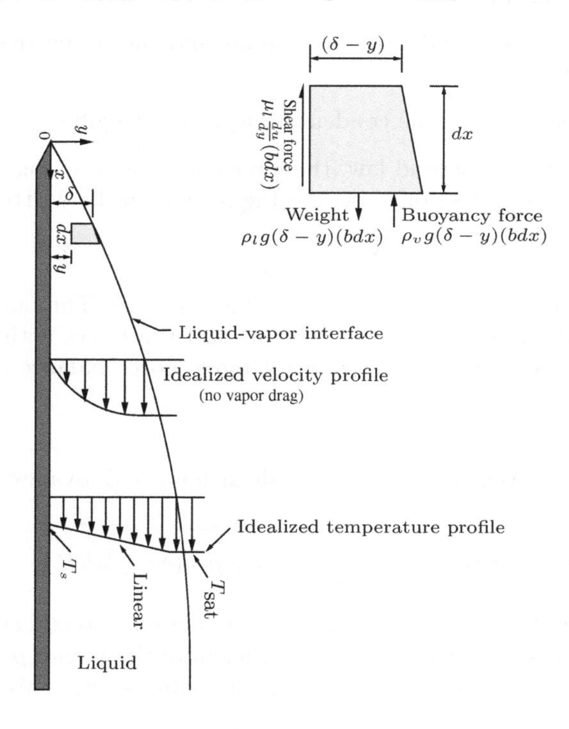


**Figure 8.7**
Laminar flow of condensate over a vertical plate.

The surface temperature $T_s$ of the plate is below the saturation temperature $T_{\text{sat}}$ of the vapor. Therefore, the vapor condenses on the plate surface. The liquid film flows downward under the influence of gravity. The film thickness $\delta$ and the mass flow rate of the condensate increases with $x$, due to continued condensation on the existing film. Therefore, the heat transfer from the vapor to the plate must occur through the liquid film at the plate surface, which offers resistance to heat transfer. Thus, the thicker the film the larger its thermal resistance and hence the lower the rate of heat transfer.

With the following simplifying assumptions, analytical relations for the heat transfer coefficient in film condensation on a vertical plate, illustrated in Figure 8.7, was developed by Nusselt.

- The plate and vapor are maintained at $T_s$ and $T_{\text{sat}}$, respectively, and the temperature across the film varies linearly.
- Heat transfer across the liquid film is purely by condensation.
- The velocity of the vapor is negligible so that it exerts no drag on the condensate, that is, there is no shear on the liquid-vapor interface.
- The flow of the condensate is laminar and the properties of the liquid are constant.
- The acceleration of the condensate layer is negligible.

Now, by Newton's second law, the force component acting on the volume element in the flow direction shown in Figure 8.7 can be written as

$$\sum F_x = m\,a_x = 0$$

since the acceleration of the condensate layer is zero. The only force acting downward is the weight of liquid element, and the forces acting upward are the viscous shear force and the buoyant force. Force balance on the volume element gives

$$\begin{aligned} \text{Weight} &= \text{Viscous shear force} + \text{Buoyancy force} \\ \rho_l g(\delta - y)(bdx) &= \mu_l \frac{du}{dy}(bdx) + \rho_v g(\delta - y)(bdx) \end{aligned}$$

where $\rho_l$ is the density of the liquid film, $g$ is gravitational acceleration, $\delta$ is the film thickness, $\mu_l$ is the viscosity coefficient of the liquid, $\rho_v$ is the density of the vapor and $b$ is the plate width (in the $z$-direction). This reduces to

$$\frac{du}{dy} = \frac{g\,(\rho_l - \rho_v)\,(\delta - y)}{\mu_l}$$

Integrating from $y = 0\,(u = 0)$ to $y = y\,(u = u(y))$, we get

$$\boxed{u(y) = \frac{g\,(\rho_l - \rho_v)}{\mu_l}\left(y\delta - \frac{y^2}{2}\right)} \tag{8.11}$$

The mass flow rate of the condensate at $x$, where the film thickness is $\delta$, is given by

$$\dot{m}(x) = \int_A \rho_l u(y) dA = \int_{y=0}^{\delta} \rho_l u(y)\, bdy \tag{8.12}$$

Substituting $u(y)$ from Equation (8.11), we get

$$\begin{aligned}
\dot{m}(x) &= \frac{gb\rho_l\,(\rho_l-\rho_v)}{\mu_l}\int_{y=0}^{\delta}\left(y\delta-\frac{y^2}{2}\right)dy \\
&= \frac{gb\rho_l\,(\rho_l-\rho_v)}{\mu_l}\left[\delta\frac{y^2}{2}-\frac{y^3}{6}\right]_0^{\delta} \\
&= \frac{gb\rho_l\,(\rho_l-\rho_v)}{\mu_l}\left[\frac{\delta^2}{2}-\frac{\delta^3}{6}\right] \\
&= \frac{gb\rho_l\,(\rho_l-\rho_v)}{\mu_l}\,\frac{\delta^3}{3}
\end{aligned}$$

That is,

$$\dot{m}(x)=\frac{gb\rho_l\,(\rho_l-\rho_v)\,\delta^3}{3\mu_l} \tag{8.13}$$

or

$$\boxed{\frac{d\dot{m}}{dx}=\frac{gb\rho_l\,(\rho_l-\rho_v)\,\delta^2}{\mu_l}\,\frac{d\delta}{dx}} \tag{8.14}$$

This represents the rate of condensation of vapor over a vertical length $dx$. The rate of heat transfer from the vapor to the plate, through the liquid film, is equal to the heat released by the vapor during condensation. Therefore, the heat transfer rate $\dot{Q}$ becomes

$$d\dot{Q}=h_{fg}\,d\dot{m}=k_l(bdx)\,\frac{T_{\text{sat}}-T_s}{\delta}$$

or

$$\frac{d\dot{m}}{dx}=\frac{k_l\,b}{h_{fg}}\,\frac{T_{\text{sat}}-T_s}{\delta} \tag{8.15}$$

where $k_l$ is the thermal conductivity of the liquid film.

From Equations (8.14) and (8.15), we get

$$\delta^3 d\delta=\frac{\mu_l\,k_l\,(T_{\text{sat}}-T_s)}{g\,\rho_l\,(\rho_l-\rho_v)\,h_{fg}}\,dx \tag{8.16}$$

Integrating from $x=0$ to $x=x$, the liquid film thickness at any location $x$ is determined to be

$$\boxed{\delta(x)=\left(\frac{4\,\mu_l\,k_l\,(T_{\text{sat}}-T_s)\,x}{g\,\rho_l\,(\rho_l-\rho_v)\,h_{fg}}\right)^{1/4}} \tag{8.17}$$

The heat transfer rate from the vapor to the plate, at a location $x$, can be expressed as

$$\dot{q}_x=h_x\,(T_{\text{sat}}-T_s)=k_l\frac{T_{\text{sat}}-T_s}{\delta}$$

This gives

$$h_x = \frac{k_l}{\delta\, x} \tag{8.18}$$

Substituting Equation (8.17), we get the local convection heat transfer as

$$h_x = \left( \frac{g\rho_l\,(\rho_l - \rho_v)\, h_{fg} k_l^3}{4\mu_l\,(T_{\text{sat}} - T_s)\, x} \right)^{1/4} \tag{8.19}$$

Integrating from $x = 0$ to $x = L$, we get the average heat transfer coefficient over the entire plate as

$$\begin{aligned}
h_{\text{ave}} &= \frac{1}{L}\int_0^L h_x dx \\
&= \frac{1}{L}\int_0^L \left( \frac{g\rho_l\,(\rho_l - \rho_v)\, h_{fg} k_l^3}{4\mu_l\,(T_{\text{sat}} - T_s)\, x} \right)^{1/4} dx \\
&= \frac{1}{L}\left( \frac{g\rho_l\,(\rho_l - \rho_v)\, h_{fg} k_l^3}{4\mu_l\,(T_{\text{sat}} - T_s)} \right)^{1/4} \int_0^L \frac{1}{x^{1/4}}\, dx \\
&= \frac{1}{L}\left( \frac{g\rho_l\,(\rho_l - \rho_v)\, h_{fg} k_l^3}{4\mu_l\,(T_{\text{sat}} - T_s)} \right)^{1/4} \left[\frac{4}{3}\, x^{3/4}\right]_0^L \\
&= \left(\frac{1}{4^{1/4}}\right)\left(\frac{4}{3}\right)\left(\frac{1}{L} \times L^{3/4}\right)\left( \frac{g\rho_l\,(\rho_l - \rho_v)\, h_{fg} k_l^3}{\mu_l\,(T_{\text{sat}} - T_s)} \right)^{1/4}
\end{aligned}$$

or

$$h_{\text{ave}} = 0.943 \left( \frac{g\,\rho_l\,(\rho_l - \rho_v)\, h_{fg} k_l^3}{\mu_l\,(T_{\text{sat}} - T_s)\, L} \right)^{1/4} \tag{8.20}$$

It is important to note that this expression will underestimate the heat transfer because it does not take into account the effects of the nonlinear temperature profile in the condensate film and the cooling of the liquid below $T_{\text{sat}}$. Both these effects can be accounted for by replacing $h_{fg}$ by $h_{fg}^*$. Thus, the modified relation for the average heat transfer coefficient for laminar film condensation over a vertical plate of length $L$ becomes

$$\boxed{h_{\text{vert}} = 0.943 \left( \frac{g\,\rho_l\,(\rho_l - \rho_v)\, h_{fg}^* k_l^3}{\mu_l\,(T_{\text{sat}} - T_s)\, L} \right)^{1/4}} \tag{8.21}$$

where $\rho_l$, $\rho_v$ are density of the liquid and vapor, respectively, $\mu_l$ is viscosity of the liquid and $k_l$ is thermal conductivity of the liquid. Equation (8.21) is valid for $0 < \text{Re} < 30$.

At a given temperature, $\rho_v \ll \rho_l$ and thus $(\rho_l - \rho_v) \approx \rho_l$ except near the critical point. The Reynolds number for the laminar film condensation over the vertical plate is

$$\mathrm{Re} \approx \frac{4g\rho_l\left(\rho_l - \rho_v\right)\delta^3}{3\mu_l^2} = \frac{4g\rho_l^2}{3\mu_l^2}\left(\frac{k_l}{h_{x=L}}\right)^3 = \frac{4g\rho_l^2}{3\mu_l^2}\left(\frac{k_l}{3h_{\mathrm{vert}}/4}\right)^3 \tag{8.22}$$

where $h_{\mathrm{vert}} = \dfrac{4}{3}h_{x=L}$, $\rho_v \ll \rho_l$ and $\delta_L = \dfrac{k_l}{h_{x=L}}$.

For $0 < \mathrm{Re} < 30$ and $\rho_v << \rho_l$ in terms of Re, the heat transfer coefficient becomes

$$h_{\mathrm{vert}} \approx 1.47\, k_l\, \mathrm{Re}^{-1/3}\left(\frac{g}{\nu_l^2}\right)^{1/3} \tag{8.23}$$

The results obtained with the above theoretical relations are in good agreement with the experimental results, when all properties, except $h_{fg}$, of the liquid are evaluated at the film temperature $T_f = \left(T_{\mathrm{sat}} + T_s\right)/2$. The $h_{fg}$ and $\rho_v$ are to be evaluated at $T_{\mathrm{sat}}$.

## Example 8.5

The copper bottom of a heater pan of 250 mm diameter, heating water, is maintained at 120°C by an electrical heater. Calculate the power required to (a) boil the water, (b) evaporation rate of water and (c) the peak heat flux.

## Solution

The heating pan with water and heating system is shown schematically in Figure E8.5.

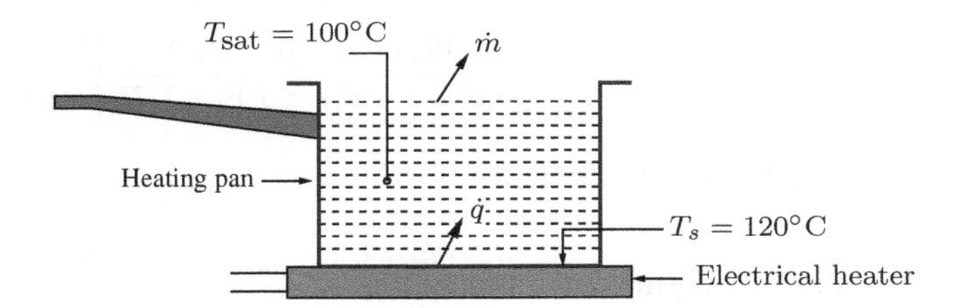


**Figure E8.5**
Heating pan with water on electrical heater.

The heating, boiling and evaporation process associated with this problem can be assumed to be steady. The heating pan bottom can be taken as polished copper and the heat loss from the heater can be regarded as negligible. This kind of appropriate assumptions are necessary for solving problems.

For liquid water at 100°C, from Table A-9, we have

$$\rho_l = 957.9 \text{ kg/m}^3, \; \rho_v = 0.5978 \text{ kg/m}^3, \; c_{pl} = 4217 \text{ J/(kg K)},$$

$$\mu_l = 0.282 \times 10^{-3} \text{ kg/(m s)}, \; \text{Pr}_l = 1.75, \; h_{fg} = 2257 \text{ kJ/kg}.$$

The surface tension of vapor-liquid interface for water at 100°C, from Table 8.1, is $\sigma = 0.0589$ N/m.

The excess temperature is

$$\Delta T_{\text{excess}} = T_s - T_{\text{sat}} = 120 - 100$$

$$= 20°\text{C}$$

(a) For this excess temperature of 20°C, from Figure 8.2 we see that the boiling process is nucleate boiling. Therefore, the rate of heat transfer given by Equation (8.2) is

$$\dot{q}_{\text{nucleate}} = \mu_l \, h_{fg} \left( \frac{g\,(\rho_l - \rho_v)}{\sigma} \right)^{1/2} \left( \frac{c_{pl}\,(T_s - T_{\text{sat}})}{C_{sf}\, h_{fg}\, \text{Pr}_l{}^n} \right)^3$$

For polished copper surface-water combination, from Table 8.3, we have,

$$C_{sf} = 0.013 \quad \text{and} \quad n = 1.0$$

Thus,

$$\dot{q}_{\text{nucleate}} = (0.282 \times 10^{-3}) \times (2257 \times 10^3) \times \left[ \frac{9.81(957.9 - 0.5978)}{0.0589} \right]^{1/2} \times \left[ \frac{4217 \times (120 - 100)}{0.013 \times 2257 \times 10^3 \times 1.75} \right]^3$$

$$= 636.474 \times 399.3 \times 4.43$$

$$= 1125.9 \, \text{kW/m}^2$$

Area of the pan bottom is

$$A_b = \frac{\pi\, D^2}{4} = \frac{\pi \times 0.25^2}{4}$$

$$= 0.0491 \, \text{m}^2$$

Hence, the heat transfer rate becomes

$$\dot{Q} = \dot{q}_{\text{nucleate}} A_b$$

$$= 1125.9 \times 0.0491$$

$$= \boxed{55.3\,\text{kW}}$$

(b) At steady state, the total heat supplied to the pan will result in water evaporation from the pan. Hence,

$$\dot{Q} = \dot{m}_w h_{fg}$$

where $\dot{m}_w$ is the rate at which water evaporates from the free surface of water. Therefore,

$$\begin{aligned} \dot{m}_w &= \frac{\dot{Q}}{h_{fg}} \\ &= \frac{55.3}{2257} \\ &= \boxed{0.0245\,\text{kg/s}} \end{aligned}$$

(c) The peak or maximum or critical heat flux in nucleate pool boiling, by Equation (8.3), is

$$\dot{q}_{\text{max}} = C_{cr} h_{fg} \left( \sigma g \rho_v^2 \left(\rho_l - \rho_v\right) \right)^{1/4}$$

The characteristic parameter for the given pan, by Equation (8.3a), is

$$\begin{aligned} L^* &= L \left( g \left(\rho_l - \rho_v\right) / \sigma \right)^{1/2} \\ &= 0.25 \left[ 9.81(957.9 - 0.5978)/0.0589 \right]^{1/2} \\ &= 99.82 \end{aligned}$$

For $L^* = 99.82$, from Table 8.4, $C_{cr} = 0.15$. Therefore,

$$\begin{aligned} \dot{q}_{\text{max}} &= 0.15 \times (2257 \times 10^3) \\ &\quad \times \left[ 0.0589 \times 9.81 \times 0.5978^2 \times (957.9 - 0.5978) \right]^{1/4} \\ &= 1269.43\,\text{kW/m}^2 \end{aligned}$$

For the given pan of surface area 0.0491 m$^2$, the maximum heat flux is

$$\begin{aligned}\dot{Q}_{\text{max}} &= \dot{q}_{\text{max}} A_b \\ &= 1269.43 \times 0.0491 \\ &= \boxed{62.33\,\text{kW}}\end{aligned}$$

## 8.10 Wavy Laminar Flow over a Vertical Plate

For condensate flow over a vertical plate, in the Reynolds number range from 30 to 1800, the interface at the liquid-vapor is wavy even though the flow in the liquid film is laminar. This flow is termed *wavy laminar*. The waves at the liquid-vapor interface increase the heat transfer. Also, because of the waves it becomes difficult to obtain analytical solutions for the heat transfer associated with this flow. Therefore, we have to solve these problems only experimentally. Due to the wavy nature the increase in heat transfer is usually about 20 percent, but it can exceed even 50 percent under certain conditions. The enhancement of heat transfer is strongly influenced by the Reynolds number. Based on experimental studies, the relation for the average heat transfer coefficient in wavy laminar condensate flow, for $\rho_v \ll \rho_l$ and $30 < \text{Re} < 1800$, is obtained as

$$h_{\text{vert,wavy}} = \frac{\text{Re}\, k_l}{1.08\,\text{Re}^{1.22} - 5.2} \left(\frac{g}{\nu_l^2}\right)^{1/3} \tag{8.24}$$

where $\nu_l = \mu_l / \rho_l$ is the kinematic viscosity of the fluid at liquid phase.

Based on his experimental studies, Kutateladze [1] proposed the following simpler relation, which relates the heat transfer coefficient in wavy laminar condensate flow to that in wave-free laminar flow

$$h_{\text{vert,wavy}} = 0.8\,\text{Re}^{0.11}\, h_{\text{vert,wave-free}} \tag{8.25}$$

Using Equations (8.24) and (8.10), a relation for the Reynolds number in the wavy laminar flow region can be expressed as follows.

By Equation (8.10), we have

$$\text{Re} = \frac{4Ah\,(T_{\text{sat}} - T_s)}{P\mu_l\, h_{fg}^*}$$

where $P$ is the perimeter equal to the length $L$ of the plate and $A$ is the surface area of the plate given by the product of width $w$ and length $L$ of the plate. Thus,

$$\text{Re} = \frac{4\,(w \times L)\, h\,(T_{\text{sat}} - T_s)}{L\mu_l\, h_{fg}^*}$$

$$= \frac{4\, w\, h\, (T_{\text{sat}} - T_s)}{\mu_l\, h^*_{fg}}$$

Note that the characteristic length for calculating the Reynolds number is the width $w$ of the plate and not the length $L$.

Substituting for $h$ from Equation (8.24),

$$\text{Re} = \frac{4\, w\ (T_{\text{sat}} - T_s)}{\mu_l\, h^*_{fg}} \frac{\text{Re}\, k_l}{1.08\, \text{Re}^{1.22} - 5.2} \left(\frac{g}{\nu_l^2}\right)^{1/3}$$

$$1.08\, \text{Re}^{1.22} - 5.2 = \frac{4\, w\ (T_{\text{sat}} - T_s)\ k_l}{\mu_l\, h^*_{fg}} \left(\frac{g}{\nu_l^2}\right)^{1/3}$$

$$1.08\, \text{Re}^{1.22} = 5.2 + \frac{4\, w\ (T_{\text{sat}} - T_s)\ k_l}{\mu_l\, h^*_{fg}} \left(\frac{g}{\nu_l^2}\right)^{1/3}$$

$$\text{Re}^{1.22} = \frac{5.2}{1.08} + \frac{4}{1.08} \frac{w\ (T_{\text{sat}} - T_s)\ k_l}{\mu_l\, h^*_{fg}} \left(\frac{g}{\nu_l^2}\right)^{1/3}$$

$$= \frac{5.2}{1.08} + \frac{4}{1.08} \frac{w\ (T_{\text{sat}} - T_s)\ k_l}{\mu_l\, h^*_{fg}} \left(\frac{g}{\nu_l^2}\right)^{1/3}$$

This simplifies to

$$\text{Re}_{\text{vert,wavy}} = \left(4.81 + \frac{3.7\, w\, k_l\, (T_{\text{sat}} - T_s)}{\mu_l\, h^*_{fg}} \left(\frac{g}{\nu_l^2}\right)^{1/3}\right)^{0.820} \tag{8.26}$$

where $h^*_{fg}$ is the modified latent heat of vaporization, given by Equation (8.8*a*).

## 8.11 Turbulent Flow on Vertical Plates

For Reynolds numbers above 1800, the condensate flow becomes turbulent. For this case, assuming $\rho_v \ll \rho_l$, for turbulent flow of condensate on vertical plates Labuntsov [2] proposed that,

$$h_{\text{vert,turb}} = \frac{\text{Re}\, k_l}{8750 + 58\, \text{Pr}^{-0.5} \left(\text{Re}^{0.75} - 253\right)} \left(\frac{g}{\nu_l^2}\right)^{1/3} \tag{8.27}$$

This is valid for turbulent flow of condensate with Reynolds number greater than 1800. The Reynolds number relation for this case can be obtained, by substituting Equation (8.27) into Equation (8.10), as

$$\mathrm{Re}_{\text{vert,turb}} = \left( \frac{0.0690L\, k_l\, \mathrm{Pr}^{0.5}\, (T_{\text{sat}} - T_s)}{\mu_l\, h^*_{fg}} \left( \frac{g}{\nu_l^2} \right)^{1/3} - 151\mathrm{Pr}^{0.5} + 253 \right)^{4/3} \tag{8.28}$$

The heat transfer coefficient (in dimensionless form) for wave-free laminar, wavy laminar, and turbulent flow of condensate over vertical plates are compared in Figure 8.8.

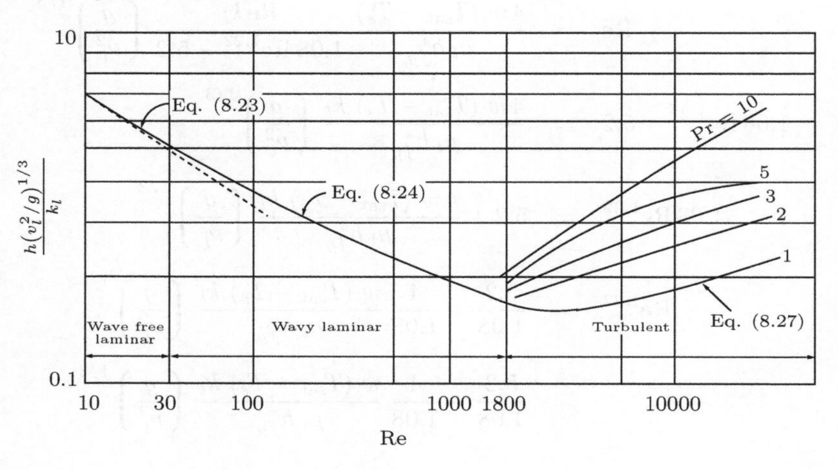


**Figure 8.8**
Dimensionless heat transfer coefficient for wave-free laminar, wavy laminar, and turbulent flow of condensate over a vertical plate.

At this stage, it is natural to have many doubts about the use of correlations proposed for the wave-free and wavy laminar and turbulent flow of condensate over vertical plates, because these correlations are valid only for the specified range of Reynolds number. Therefore, it is essential to have a thorough understanding of the calculation of the Reynolds number to decide on the choice of the average heat transfer coefficient correlation to be employed. The first step in this direction is that, for calculating the Reynolds number, the *width of the plate* should be taken as the characteristic length and not the length of the plate, as we use in fluid mechanics. To gain an understanding about the application aspects of the correlations, let us solve a problem of condensation of steam over a vertical plate, in the following example.

## Example 8.6

Saturated steam at atmospheric pressure condenses on a 3-m-wide and 2-m-high vertical plate, with its surface maintained at 80°C. Determine (a) the rate of heat transfer from steam to the plate, and (b) the rate at which steam condenses on the plate.

## Solution

(a) Given, $T_s = 80°\text{C}$, width $= L = 3$ m, height $= 2$ m.

At $p = 1$ atm, the saturation temperature $T_{\text{sat}} = 100°\text{C}$. Therefore, the film temperature is

$$\begin{aligned} T_f &= \frac{T_{\text{sat}} + T_s}{2} \\ &= \frac{100 + 80}{2} \\ &= 90°\text{C} \end{aligned}$$

For $T_f = 90°\text{C}$, from Table A-9,

$\rho_l = 965.3\,\text{kg/m}^3$, $\mu_l = 0.315 \times 10^{-3}\,\text{kg/(m s)}$, $k_l = 0.675\,\text{W/(m °C)}$

$c_{pl} = 4206\,\text{J/(kg °C)}, \text{Pr} = 1.96$

For $T_{\text{sat}} = 100°\text{C}$, from Table A-9,

$\rho_v = 0.5978\,\text{kg/m}^3$, $h_{fg} = 2257\,\text{kJ/kg}$

By Equation (8.8*a*), the modified latent heat of vaporization is

$$\begin{aligned} h_{fg}^* &= h_{fg} + 0.68\, c_{pl}\, (T_{\text{sat}} - T_s) \\ &= 2257 + 0.68 \times (4206 \times 10^{-3}) \times (100 - 80) \\ &= 2314\,\text{kJ/kg} \end{aligned}$$

Now, if the Reynolds number is in the range from 0 to 30, the average heat transfer coefficient $h$ can be calculated using Equation (8.20) or Equation (8.21). But for this the Reynolds number is given by Equation (8.22). Thus,

$$\text{Re} = \frac{4g\rho_l^2}{3\mu_l^2}\left(\frac{k_l}{3h_{\text{vert}}/4}\right)^3$$

The average heat transfer coefficient by Equation (8.20) is

$$h_{\text{ave}} = 0.943\left(\frac{g\,\rho_l\,(\rho_l - \rho_v)\,h_{fg}k_l^3}{\mu_l\,(T_{\text{sat}} - T_s)\,L}\right)^{1/4}$$

$$= \quad 0.943 \left( \frac{9.81 \times 965.3 \times (965.3 - 0.5978) \times (2257 \times 10^3) \times 0.675^3}{(0.315 \times 10^{-3}) \times (100 - 80) \times 3} \right)^{1/4}$$

$$= \quad 4036 \, \mathrm{W/(m^2 \, ^\circ C)}$$

Note that the width of the plate is taken as the characteristic length $L$ in the calculation of $h$.

Therefore,

$$\mathrm{Re} = \frac{4 \times 9.81 \times 965.3^2}{3 \times (0.315 \times 10^{-3})^2} \times \left( \frac{0.675}{(3/4) \times 4036} \right)^3$$

$$= 1362$$

Reynolds number is greater than 30; therefore, the average heat transfer coefficient calculated using Equation (8.20) is not appropriate. For 30 < Re < 1800, the condensate flow over the plate is wavy laminar flow. For this case the appropriate correlation for average heat transfer coefficient is Equation (8.24) and the corresponding expression for the Reynolds number is Equation (8.26). Thus,

$$\mathrm{Re} = \left[ 4.81 + \frac{3.7 \, w \, k_l \, (T_{\mathrm{sat}} - T_s)}{\mu_l \, h_{fg}^*} \left( \frac{g}{\nu_l^2} \right)^{1/3} \right]^{0.820}$$

$$= \left[ 4.81 + \frac{3.7 \times 3 \times 0.675 \times (100 - 80)}{(0.315 \times 10^{-3}) \times (2314.2 \times 10^3)} \times \left( \frac{9.81}{(3.263 \times 10^{-7})^2} \right)^{1/3} \right]^{0.820}$$

$$= \left( 4.81 + 0.2056 \times 45166 \right)^{0.820}$$

$$= (9291)^{0.820}$$

$$= 1794$$

Now, by Equation (8.24),

$$h_{\mathrm{vert,wavy}} = \frac{\mathrm{Re} \, k_l}{1.08 \, \mathrm{Re}^{1.22} - 5.2} \left( \frac{g}{\nu_l^2} \right)^{1/3}$$

$$= \frac{1794 \times 0.675}{1.08 \times 1794^{1.22} - 5.2} \times \left( \frac{9.81}{(3.263 \times 10^{-7})^2} \right)^{1/3}$$

$$= 0.1203 \times 45166$$

$$= 5433.5\,\mathrm{W/(m^2\,°C)}$$

The rate of heat transfer from steam to the plate is

$$\begin{aligned}\dot{Q} &= \dot{q}\,A_s \\ &= h_{\text{vert,wavy}}\,(T_{\text{sat}} - T_s)\,A_s \\ &= 5433.5 \times (100 - 80) \times (3 \times 2) \\ &= 652020\,\mathrm{W} \\ &= \boxed{652.02\,\mathrm{kW}}\end{aligned}$$

(b) The mass of steam condenses per second over the plate is

$$\begin{aligned}\dot{m} &= \frac{\dot{Q}}{h^*_{fg}} \\ &= \frac{652.02}{2314.2} \\ &= \boxed{0.282\,\mathrm{kg/s}}\end{aligned}$$

## 8.11.1 Inclined Plate

The velocity relation given by Equation (8.11) developed for vertical plate can also be used for laminar film condensation on the upper surfaces of plates that are inclined at an angle $\theta$, as shown in Figure 8.9, by replacing $g$ in the equation by $g \cos\theta$. That is,

$$h_{\text{inclined}} = h_{\text{vert}}\,(\cos\theta)^{1/4} \tag{8.29}$$

This approximation is valid for $\theta \leq 60°$. Further, this relation can also be used for wavy laminar flow as an approximate.

## 8.11.2 Vertical Tubes

The average heat transfer coefficient for vertical plate, given by Equation (8.21), can be applied for laminar condensation on the outer surfaces of vertical tubes when the diameter of the tube is large compared to the thickness of the

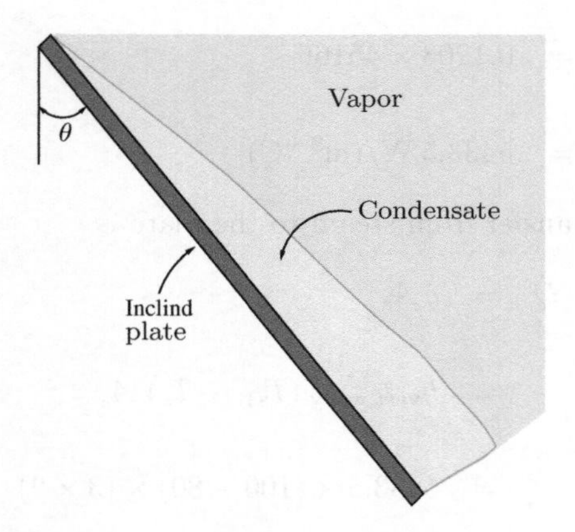


**Figure 8.9**
Film condensation on an inclined plate.

condensed liquid film on the tube surface. In other words, Equation (8.21), which is valid for wave-free flow of condensate over a vertical plate, in the range of Reynolds number from 0 to 30, can be applied for the entire range of wave-free and wavy laminar flow of condensate over a thin-walled vertical tube, that is, for Reynolds numbers from 0 to 1800.

## Example 8.7

A vertical tube of diameter 100 mm and length 1.2 m is submerged in a saturated steam at atmospheric pressure. If the outer surface of the tube is maintained at 60°C, by the flow of cool water through the tube, determine (a) the rate of heat transfer to the cooling water, and (b) the rate at which steam condenses at the tube surface.

## Solution

Given, $D = 0.1$ m, $L = 1.2$ m, $T_s = 60°\text{C}$.

At atmospheric pressure, the saturation temperature of steam is $T_{\text{sat}} = 100°\text{C}$.

From Table A-9, for saturated steam at $T_{\text{sat}} = 100°\text{C}$,

$$\rho_v = 0.5978 \text{ kg/m}^3,\ h_{fg} = 2257 \text{ kJ/kg}$$

The film temperature is

$$T_f = (T_{\text{sat}} + T_s)/2$$

$$= (100+60)/2$$

$$= 80°\text{C}$$

For saturated liquid at 80°C, from Table A-9,

$\rho_l = 971.8 \text{ kg/m}^3$, $\mu = 0.355 \times 10^{-3} \text{ kg/(m s)}$,

$k_l = 0.670 \text{ W/(m °C)}$, $c_{pl} = 4197 \text{ J/(kg °C)}$

(a) The heat transfer rate from steam to the cooling water can be determined from

$$\dot{q} = h_{\text{vert}}\,(T_{\text{sat}} - T_s)$$

Assuming the film condensation at the surface to be laminar (this assumption should be validated), we have the average heat transfer coefficient, from Equation (8.21), as

$$h_{\text{vert}} = 0.943\left(\frac{g\,\rho_l\,(\rho_l - \rho_v)\,h^*_{fg}k_l^3}{\mu_l\,(T_{\text{sat}} - T_s)\,L}\right)^{1/4}$$

where $h^*_{fg}$ is the modified latent of heat of vaporization, given by Equation (8.8$a$), as

$$\begin{aligned} h^*_{fg} &= h_{fg} + 0.68\,c_{pl}\,(T_{\text{sat}} - T_s) \\ &= 2257 + 0.68 \times 0.4197 \times (100 - 60) \\ &= 2268.42\,\text{kJ/kg} \end{aligned}$$

Therefore, the average heat transfer coefficient, with tube length as the characteristic dimension, is

$$\begin{aligned} h_{\text{ave}} &= 0.943\left(\frac{9.81 \times 971.8\,(971.8 - 0.5978)(2268.42 \times 10^3) \times 0.670^3}{0.355 \times 10^{-3} \times (100 - 60) \times 1.2}\right)^{1/4} \\ &= 4137.8\,\text{W/(m}^2\,°\text{C)} \end{aligned}$$

The heat transfer rate per unit area is

$$\begin{aligned} \dot{q} &= 4137.8 \times (100 - 60) \\ &= 165512\,\text{W/m}^2 \end{aligned}$$

The total heat transfer rate over the complete surface of the tube becomes

$$\begin{aligned}\dot{Q} &= \dot{q}\,A_s = \dot{q}\times(\pi d\,L)\\ &= 165512\times(\pi\times 0.1\times 1.2)\\ &= \boxed{62.4\,\text{kW}}\end{aligned}$$

The mass flow rate of steam condensation is

$$\begin{aligned}\dot{m} &= \frac{\dot{Q}}{h^*_{fg}} = \frac{62.4}{2268.42}\\ &= \boxed{0.0275\,\text{kg/s}}\end{aligned}$$

The Reynolds number for laminar condensation can be checked as follows.

$$\text{Re} = \frac{\rho V\delta}{\mu_l}$$

where $\delta$ is the thickness of the steam condensate. The Reynolds numbers can be expressed in terms of mass flow rate, as follows.

Taking the average thickness of the condensate as $\delta$, the mass flow rate of the condensate per meter depth becomes

$$\begin{aligned}\dot{m} &= \rho\times\text{the volume flow rate}\\ &= \rho\times(A\times V)\end{aligned}$$

But $A=\delta\times$ unity. Therefore,

$$\dot{m} = \rho\times(\delta\times V)\,\text{kg/(m s)}$$

In terms of mass flow rate, the Reynolds number becomes

$$\text{Re} = \frac{\dot{m}}{\mu_l}$$

Therefore,

$$\begin{aligned}\text{Re} &= \frac{0.0275}{0.355\times 10^{-3}}\\ &= 545\end{aligned}$$

The Reynolds number is less than 1800, thus the assumption made is valid.

### 8.11.3 Horizontal Tubes and Spheres

Following Nusselt's analysis of film condensation on vertical plates, the average heat transfer coefficient for film condensation on the outer surface of a horizontal tube can be expressed as [3]

$$h_{\text{horizontal}} = 0.729 \left( \frac{g\rho_l (\rho_l - \rho_v) h_{fg}^* k_l^3}{\mu_l (T_{\text{sat}} - T_s) D} \right)^{1/4} \tag{8.30}$$

where $D$ is the tube diameter. For spheres also Equation (8.30) can be used by replacing the constant 0.729 with 0.815.

Comparing the heat transfer relations for a vertical tube [Equation (8.21)] and horizontal tube [Equation (8.30)], we obtain

$$\frac{h_{\text{vert}}}{h_{\text{horizontal}}} = 1.294 \left( \frac{D}{L} \right)^{1/4} \tag{8.31}$$

From this relation, the following interesting feature can be inferred. When $h_{\text{vert}} = h_{\text{horizontal}}$, Equation (8.31) gives $L = 1.29^4 D = 2.77D$. This implies that, for a tube of diameter $D$ and length $2.77D$ the average heat transfer coefficient for laminar film condensation will be the same for vertical and horizontal orientations of the tube. When $L > 2.77D$, the heat transfer coefficient will be higher for the horizontal orientation. Because of this, in condensers, the tubes are kept horizontal with the length of the tubes several times its diameter to maximize the condensation heat transfer coefficient on the outer surface of the tubes.

### 8.11.4 Horizontal Tube Banks

Horizontal tubes arranged one on top of another, as illustrated in Figure 8.10, are commonly employed in condensers. Because of this arrangement, the average thickness of the liquid film at the lower tubes is larger as a result of condensed liquid falling on top of them from the tubes above them. Therefore, the average heat transfer coefficient at the lower tubes is smaller.

Assuming the drain of the condensate from the higher tubes to the lower tubes to be smooth, the average film condensation heat transfer coefficient for all tubes in a vertical tier can be expressed as

$$h_{\text{horizontal,tube bank}} = 0.729 \left( \frac{g\rho_l (\rho_l - \rho_v) h_{fg}^* k_l^3}{\mu_l (T_{\text{sat}} - T_s) ND} \right)^{1/4} \tag{8.32}$$

where $N$ is the number of horizontal tubes in the condenser. It is essential to note that this relation is valid only for horizontal tube bank with condensate draining smoothly and is not valid for any other arrangement of the tubes and also when the drain of the condensate is turbulent.

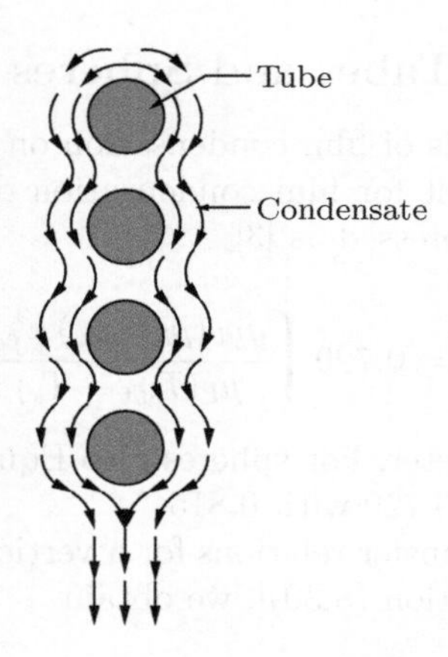


**Figure 8.10**
Vertical tier of horizontal tubes.

## Example 8.8

**Ten vertical tiers of 10 horizontal tubes in a tier, carrying a liquid, arranged in a square array are exposed to steam at atmospheric pressure. The temperature at the outer surface of the tubes is 90°C. If the diameter of the tubes is 12.5 mm, determine the mass flow rate of steam condensate per meter length of the tubes.**

## Solution

Given, $D = 0.0125$ m, $N = 10$, $T_s = 90$°C.

At 1 atm, the saturation temperature of steam is 100°C. Thus, the film temperature is

$$T_f = \frac{100 + 90}{2} = 95^\circ\text{C}$$

For $T_f = 95$°C, from Table A-9,

$$\rho_l = 961.5 \text{ kg/m}^3,\ k_l = 0.677 \text{ W/(m °C)},$$

$$\mu_l = 0.297 \times 10^{-3} \text{ kg/(m s)},\ c_{pl} = 4212 \text{ J/(kg °C)}$$

For $T_{\text{sat}} = 100$°C, from Table A-9,

$$h_{fg} = 2257 \text{ kJ/kg},\ \rho_v = 0.5978 \text{ kg/m}^3$$

The modified heat of vaporization $h^*_{fg}$, by Equation (8.8a), is

$$h^*_{fg} \quad = \quad h_{fg} + 0.68\, c_{pl}\ (T_{\text{sat}} - T_s)$$

$$= 2257 + 0.68 \times 4212 \times (100 - 90) \times 10^{-3}$$

$$= 2285.64\,\text{kJ/kg}$$

Now, assuming the flow of condensate to be laminar, the average film condensation heat transfer coefficient for the tubes in vertical tier, by Equation (8.32), is

$$h = 0.729 \left( \frac{g\rho_l (\rho_l - \rho_v) h^*_{fg} k_l^3}{\mu_l (T_{\text{sat}} - T_s) ND} \right)^{1/4}$$

$$= 0.729 \left( \frac{9.81 \times 961.5 \times (961.5 - 0.5978) \times (2285.64 \times 10^3) \times 0.677^3}{(0.297 \times 10^{-3}) \times (100 - 90) \times 10 \times 0.0125} \right)^{1/4}$$

$$= 8362.36\,\text{W/(m}^2\,^\circ\text{C)}$$

The surface area of 100 tubes, per meter length, is

$$A_s = (\pi DL) \times 100$$

$$= (\pi \times 0.0125 \times 1) \times 100$$

$$= 3.93\,\text{m}^2$$

$$\frac{A_s}{L} = 3.93\,\text{m}$$

The heat transfer from the steam to the tubes is

$$\dot{Q} = hA_s (T_{\text{sat}} - T_s)$$

$$= 8362.36 \times 3.93 \times (100 - 90)$$

$$= 328640.75\,\text{W/m}$$

The mass flow rate of the condensate is

$$\dot{m} = \frac{\dot{Q}}{h^*_{fg}}$$

$$= \frac{328640.75}{2285.64 \times 10^3}$$

$$= \boxed{0.144\,\text{kg/(m s)}}$$

The Reynolds number is

$$\begin{aligned} \mathrm{Re} &= \frac{\dot{m}}{\mu_l} \\ &= \frac{0.144}{0.297 \times 10^{-3}} \\ &= 485 \end{aligned}$$

This is within the range of 30 < Re < 1800 for wave-free laminar flow of condensate. Thus, the assumption made is valid.

## 8.12 Film Condensation inside Horizontal Tubes

Condensation processes encountered in devices such as refrigerators and air-conditioners are on the inner surfaces of horizontal or vertical tubes. The condensation inside a tube is strongly influenced by the vapor velocity and the rate of liquid accumulation on the walls of the tube, as illustrated in Figure 8.11.

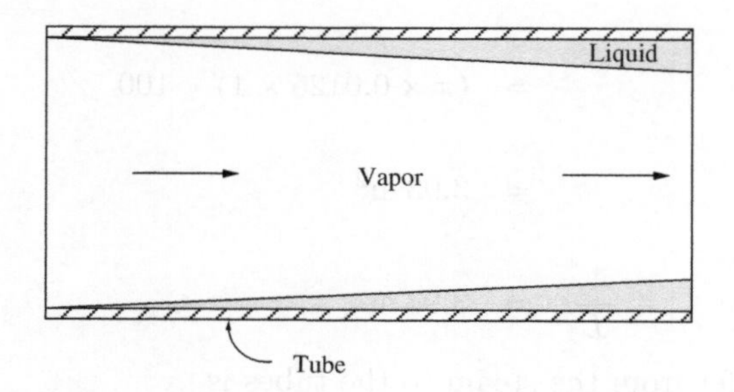


**Figure 8.11**
Condensate flow in a horizontal tube with large vapor velocities.

Because of the effects due to vapor velocity and the liquid accumulation on the walls, the analysis becomes difficult. For low vapor velocities, the following expression for condensation may be used to calculate the heat transfer coefficient $h$ [3].

$$h_{\text{internal}} = 0.555 \left[ \frac{g\rho_l (\rho_l - \rho_v) k_l^3}{\mu_l (T_{\text{sat}} - T_s)} \left( h_{fg} + \frac{3}{8} c_{pl} (T_{\text{sat}} - T_s) \right) \right]^{1/4} \quad (8.30a)$$

This relation is valid for

$$\mathrm{Re}_{\text{vapor}} = \left( \frac{\rho_v V_v D}{\mu_v} \right)_{\text{inlet}} < 35,000$$

where $D$ is the inner diameter of the tube, and the Reynolds number of the vapor is to be evaluated at tube inlet conditions.

## 8.13 Dropwise Condensation

Dropwise condensation is the process in which a large number of droplets of varying diameters are formed on the condensing surface instead of a continuous liquid film. This is one of the most effective mechanisms of heat transfer. Extremely large heat transfer coefficient can be achieved with dropwise condensation. In dropwise condensation, the small droplets that form at the nucleation sites on the surface grow as a result of continued condensation, and coalesce into large droplets. When a droplet reaches a certain size, it slides down clearing the surface and exposing it to vapor. There is no liquid film in this case to resist heat transfer. Because of this, the heat transfer coefficient achieved with dropwise condensation can be as high as 10 times of that associated with film condensation.

Sustaining the dropwise condensation for prolonged periods of time is not easy. This poses a challenge in designing a device for dropwise condensation. Dropwise condensation is achieved by adding a promotional chemical into the vapor, treating the surface with a promotional chemical, or coating the surface with a polymer such as teflon or a noble metal such as gold, platinum, silver, rhodium, or palladium. The promoters used are various waxes and fatty acids such as oleic, stearic, and ionic acids. However, the promoters lose their effectiveness because of fouling, oxidation, and the removal of their coated layer from the surface.

One of the widely used passages with dropwise condensation is copper tubes carrying steam. Some of the simple correlations for dropwise condensation of steam on a copper surface are the following:

For saturation temperature $T_{\text{sat}}$ in the range from 22°C to 100°C,

$$h_{\text{dropwise}} = 51140 + 2044\, T_{\text{sat}} \tag{8.33a}$$

For $T_{\text{sat}} > 100°\text{C}$

$$h_{\text{dropwise}} = 255310 \tag{8.33b}$$

It is essential to note that when the material of the condensing surface is not a good thermal conductor like copper, or when the thermal resistance on the other side of the surface is too large, the large heat transfer coefficient achievable with dropwise condensation is of no significance.

## 8.14 Summary

Boiling and condensation are essentially convection heat transfer processes involving phase change. When the temperature of a liquid at a given pressure is raised to the saturation temperature $T_{\text{sat}}$ corresponding to that pressure, the liquid will boil. Similarly, when the temperature of a vapor is lowered to $T_{sat}$, condensation occurs. Even though boiling and condensation are forms of convection heat transfer, they depend on the latent heat of vaporization $h_{fg}$ of the fluid and the surface tension $\sigma$ at the liquid-vapor interface, in addition to the properties of the fluid at each phase. Heat transfer coefficient $h$ associated with boiling and condensation are several times higher than those associated with other forms of convection heat transfer processes that involve only single phase.

Boiling is a process in which a liquid phase is converted to a vapor phase. The energy for phase change is generally supplied by the surface on which boiling occurs. Even though it appears to be similar to evaporation, it differs significantly from evaporation. Evaporation occurs at the liquid-vapor interface, when the vapor pressure is less than the saturation pressure of the liquid at a given temperature. But boiling occurs at the solid-interface when a liquid is brought into contact with a solid surface maintained at a temperature $T$, which is sufficiently above the saturation temperature $T_{sat}$ of the liquid.

The heat transfer per unit time per unit surface area from a solid surface to the fluid is termed boiling heat flux. It is given by

$$\boxed{\dot{q}_{boiling} = h\left(T_s - T_{sat}\right) = h\Delta T_{excess}}$$

where $T_s$ is the temperature at the surface of the solid, $T_{sat}$ is the saturation temperature of the liquid, and $\Delta T_{excess} = (T_s - T_{sat})$ is called the excess temperature.

During a boiling process,

- The pressure difference between the liquid and vapor is balanced by the surface tension $\sigma$ at the interface.
- The temperature difference between the vapor in a bubble and the surrounding liquid is the driving force for heat transfer between the two phases.

Boiling is called *pool boiling* if there is no bulk fluid motion and *flow boiling* or *forced convection boiling* in the presence of bulk fluid motion. In pool boiling, the fluid is stationary, and any motion of fluid is due to natural convection and motion of the bubbles is due to buoyancy.

Boiling is termed *subcooled* or *local* when the temperature of the main body of the liquid is less than its saturation temperature $T_{sat}$. The boiling is called *saturated* or *bulk* when the temperature of the liquid is equal to $T_{sat}$.

In pool boiling, the fluid is not forced to flow. The following four different boiling regimes can be observed in pool boiling:

- Natural convection boiling
- Nucleate boiling

- Transition boiling
- Film boiling

The most widely used correlation for the rate of heat transfer in the nucleate boiling regime is the *Rohsenow correlation*,

$$\dot{q}_{\text{nucleate}} = \mu_l \, h_{fg} \left( \frac{g\,(\rho_l - \rho_v)}{\sigma} \right)^{1/2} \left( \frac{c_{pl}\,(T_s - T_{\text{sat}})}{C_{sf}\, h_{fg}\, \text{Pr}_l{}^{n}} \right)^{3}$$

where $\dot{q}_{\text{nucleate}}$ is the nucleate boiling heat flux, $\mu_l$ is the viscosity of the liquid, $h_{fg}$ is the enthalpy of vaporization, $\rho_l$ is liquid density, $\rho_v$ is vapor density, $\sigma$ is the surface tension of liquid-vapor interface, $c_{pl}$ is the specific heat of the liquid, $T_s$ is the heater surface temperature, $T_{\text{sat}}$ is the saturation temperature of the fluid, $C_{sf}$ is a constant, $\text{Pr}_l$ is the Prandtl number of the liquid and $n$ is experimental constant.

The total heat transfer involved in the boiling process is

$$\dot{Q}_{\text{nucleate}} = A_s \dot{q}_{\text{nucleate}}$$

where $A_s$ is the surface area of the container base over which heat is supplied.

The peak or maximum heat flux in nucleate pool boiling was determined theoretically using different approaches and expressed as

$$\dot{q}_{\max} = C_{cr} h_{fg} \left( \sigma g \rho_v^2 \left(\rho_f - \rho_v\right) \right)^{1/4}$$

where $C_{cr}$ is a constant which depends on the heater geometry, $\sigma$ is the surface tension, $\rho_f$ and $\rho_v$ are the density of the liquid and vapor, respectively, and $g$ is the gravitational acceleration.

Minimum heat flux is the lower limit for the heat flux in the film boiling regime. Using the stability theory, the minimum heat flux for a large horizontal plate has been expressed as

$$\dot{q}_{\min} = 0.09 \rho_v h_{fg} \left( \frac{\sigma g \rho_v^2 \left(\rho_f - \rho_v\right)}{\left(\rho_f + \rho_v\right)^2} \right)^{1/4}$$

This is only an approximate relation and can be in error as high as 50%.

The heat flux for film boiling on a horizontal cylinder or sphere of diameter $D$ is given by

$$\dot{q}_{\text{film}} = C_{\text{film}} \left( \frac{g k_v^3 \rho_v \left(\rho_l - \rho_v\right) \left[h_{fg} + 0.4 c_{pv} \left(T_s - T_{\text{sat}}\right)\right]}{\mu_v D \left(T_s - T_{\text{sat}}\right)} \right)^{1/4} \left(T_s - T_{\text{sat}}\right)$$

where $k_v$ is the thermal conductivity of the vapor and $C_{\text{film}} = 0.63$ and $0.67$, respectively, for horizontal cylinders and spheres.

At temperatures above 300°C, heat transfer across the vapor film by radiation becomes significant and needs to be accounted for. Treating the vapor film as a transparent medium sandwiched between two large parallel plates and approximating the liquid as a blackbody, radiation heat transfer can be determined from

$$\dot{q}_{\text{rad}} = \epsilon\sigma\left(T_s^4 - T_{\text{sat}}^4\right)$$

where $\epsilon$ is the emissivity of the heating surface and $\sigma = 5.67 \times 10^{-8}$ W/(m$^2$ K$^4$) is the Stefan-Boltzmann constant.

Flow boiling is that in which the fluid during its phase-change process is forced to move by an external source such as a pump. It is essentially a combination of natural and/or forced convection and pool boiling. When the fluid is forced to flow over a heated surface, it is termed *external flow boiling* and when the fluid is forced to flow inside a heated tube, it is called *internal flow boiling.*

Condensation occurs when the temperature of a vapor is reduced below its saturation temperature $T_{\text{sat}}$, by bringing the vapor into contact with a solid surface whose temperature $T_s$ is less than $T_{\text{sat}}$ of the vapor. But condensation can also occur on the free surface of a liquid or even in a gas, when the temperature of the liquid or the gas to which the vapor is exposed is below the saturation temperature $T_{\text{sat}}$ of the vapor. In the second case, the liquid droplets suspended in the gas form a fog.

Condensation can occur in the following two distinct forms:

- *Film condensation*
- *Dropwise condensation*

The Reynolds number for the condensate flow is defined as

$$\text{Re} = \frac{D_h \rho_l \overline{V}_l}{\mu_l} = \frac{4A_c \rho_l \overline{V}_l}{P\mu_l} = \frac{4\rho_l \overline{V}_l \delta}{\mu_l}$$

where $D_h = 4A_c/P = 4\delta =$ hydraulic diameter of the condensate flow, $P$ is the wetted perimeter of the condensate, $A_c = P\delta$ is the cross-sectional area of the condensate flow at the lowest part of the flow, $\delta$ is film thickness, $\rho_l$ is density of the liquid, $\mu_l$ is viscosity of the liquid, $\overline{V}_l$ is the average velocity of the condensate at the lowest part of the flow and $\dot{m} = \rho_l A \overline{V}_l$ is the mass flow rate of the condensate at the lowest part.

The latent heat of vaporization $h_{fg}$ is the heat released as a unit mass of vapor condenses. It normally represents the heat transfer per unit mass of condensate formed during condensation. But the condensate in an actual condensation process is cooled further to some average temperature between $T_{\text{sat}}$ and $T_s$, releasing more heat in the process. Therefore, the actual heat transfer will be larger than the theoretical value. The cooling of the liquid below the saturation temperature can be accounted by replacing $h_{fg}$ by the modified latent heat of vaporization $h_{fg}^*$, defined as

$$h_{fg}^* = h_{fg} + 0.68\, c_{pl}\left(T_{\text{sat}} - T_s\right)$$

where $c_{pl}$ is the specific heat of the liquid at the average film temperature.

The heat transfer can be expressed as

$$\dot{Q} = hA\,(T_{\text{sat}} - T_s) = \dot{m}\,h^*_{fg}$$

where $A$ is the surface area on which the condensate occurs.

The mass flow becomes

$$\dot{m} = \frac{\dot{Q}}{h^*_{fg}}$$

The average heat transfer coefficient for laminar film condensation over a vertical plate is

$$h_{\text{ave}} = 0.943 \left(\frac{g\,\rho_l\,(\rho_l - \rho_v)\,h_{fg}k_l^3}{\mu_l\,(T_{\text{sat}} - T_s)\,L}\right)^{1/4}$$

The modified relation for the average heat transfer coefficient for laminar film condensation over a vertical plate of length $L$ is

$$\boxed{h_{\text{vert}} = 0.943 \left(\frac{g\,\rho_l\,(\rho_l - \rho_v)\,h^*_{fg}k_l^3}{\mu_l\,(T_{\text{sat}} - T_s)\,L}\right)^{1/4}}$$

This is valid for $0 < \text{Re} < 30$.

For $0 < \text{Re} < 30$ and $\rho_v << \rho_l$ in terms of Re, the heat transfer coefficient becomes

$$h_{\text{vert}} \approx 1.47\,k_l\,\text{Re}^{-1/3}\left(\frac{g}{\nu_l^2}\right)^{1/3}$$

For condensate flow over a vertical plate, in the Reynolds number range from 30 to 1800, the interface at the liquid-vapor is wavy even though the flow in the liquid film is laminar. This flow is termed *wavy laminar*.

The average heat transfer coefficient in wavy laminar condensate flow, for $\rho_v \ll \rho_l$ and $30 < \text{Re} < 1800$, is

$$h_{\text{vert}} = \frac{\text{Re}\,k_l}{1.08\,\text{Re}^{1.22} - 5.2}\left(\frac{g}{\nu_l^2}\right)^{1/3}$$

For Reynolds numbers above 1800 the condensate flow becomes turbulent. For this case, assuming $\rho_v \ll \rho_l$, for turbulent flow of condensate on vertical plates Labuntsov [2] proposed that,

$$h_{\text{vert,turb}} = \frac{\text{Re}\,k_l}{8750 + 58\,\text{Pr}^{-0.5}\,(\text{Re}^{0.75} - 253)}\left(\frac{g}{\nu_l^2}\right)^{1/3}$$

The velocity relation for vertical plate can also be used for laminar film condensation on the upper surfaces of plates that are inclined at an angle $\theta$ by replacing $g$ in the equation by $g\cos\theta$. That is,

$$h_{\text{inclined}} = h_{\text{vert}} \left(\cos\theta\right)^{1/4}$$

This approximation is valid for $\theta \le 60°$.

The average heat transfer coefficient for film condensation on the outer surface of a horizontal tube is

$$h_{\text{horizontal}} = 0.729 \left(\frac{g\rho_l\left(\rho_l - \rho_v\right) h_{fg}^* k_l^3}{\mu_l\left(T_{\text{sat}} - T_s\right) D}\right)^{1/4}$$

where $D$ is the tube diameter. For spheres Equation (8.30) can also be used by replacing the constant 0.729 with 0.815.

Assuming the drain of the condensate from the higher tubes to the lower tubes to be smooth, the average film condensation heat transfer coefficient for all tubes in a vertical tier can be expressed as

$$h_{\text{horizontal,tube bank}} = 0.729 \left(\frac{g\rho_l\left(\rho_l - \rho_v\right) h_{fg}^* k_l^3}{\mu_l\left(T_{\text{sat}} - T_s\right) ND}\right)^{1/4}$$

The condensation inside a tube is strongly influenced by the vapor velocity and the rate of liquid accumulation on the walls of the tube. For low vapor velocities, the following expression for condensation may be used to calculate the heat transfer coefficient $h$.

$$h_{\text{internal}} = 0.555 \left(\frac{g\rho_l\left(\rho_l - \rho_v\right) k_l^3}{\mu_l\left(T_{\text{sat}} - T_s\right)}\left(h_{fg} + \frac{3}{8}c_{pl}\left(T_{\text{sat}} - T_s\right)\right)\right)^{1/4}$$

This relation is valid for

$$\text{Re}_{\text{vapor}} = \left(\frac{\rho_v V_v D}{\mu_v}\right)_{\text{inlet}} < 35,000$$

where $D$ is the inner diameter of the tube, and the Reynolds number of the vapor is to be evaluated at the tube inlet conditions.

Dropwise condensation is the process in which a large number of droplets of varying diameters are formed on the condensing surface instead of a continuous liquid film. This is one of the most effective mechanisms of heat transfer. Extremely large heat transfer coefficient can be achieved with dropwise condensation. One of the widely used passages with dropwise condensation is copper tubes carrying steam. Some of the simple correlations for dropwise condensation of steam on copper surface are the following:

For saturation temperature $T_{\text{sat}}$ in the range from 22°C to 100°C,

$$h_{\text{dropwise}} = 51140 + 2044\,T_{\text{sat}}$$

For $T_{\text{sat}} > 100°\text{C}$

$$h_{\text{dropwise}} = 255310$$

## 8.15 Exercise Problems

8.1 At standard sea level, water is heated in a metallic pan of bottom diameter 250 mm and maintained at 120°C. Find the rate of heat transferred to the water and the evaporation of water.

[**Ans.** 55 kW, 0.024 kg/s]

8.2 At standard sea level, water is heated by a resistance heater of diameter 12 mm. Determine the maximum heat flux which can be transferred to the water by the heater.

[**Ans.** 1.017 MW/m$^2$]

8.3 At standard sea level, water is boiled using a copper wire heating element of diameter 9 mm. The emissivity of the heating element is 0.055. When the heater element surface temperature is maintained at 250°C, determine the heat flux per unit length of the heating element to the water.

[**Ans.** 64756.9 W/m$^2$]

8.4 Superheated steam at 1 atm and 150°C condenses on a vertical plate of height 100 mm and width 50 mm maintained at 70°C, at atmospheric pressure. Determine the rate of heat transfer to the plate by condensation. Also, calculate the mass flow rate of the condensed water dripping from the bottom of the plate.

[**Ans.** 4614.32 W, 0.002 kg/s]

8.5 The condenser of a power plant has 16 tubes, arranged as shown in Figure P8.5.

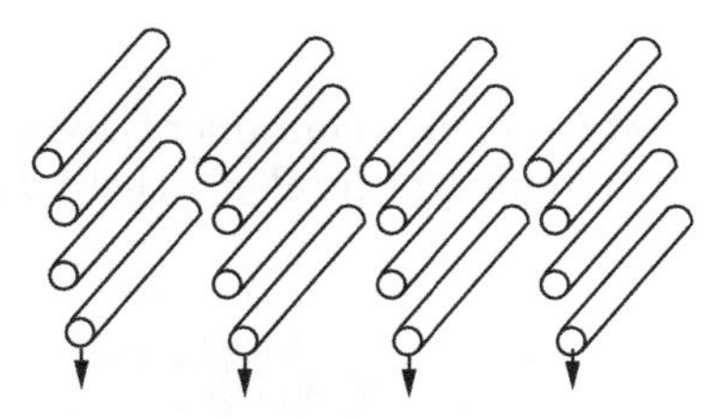

**Figure P8.5**
Array of condenser tubes.

Steam at a pressure of 6 MPa passing over the tubes condenses because of the cooling water flowing inside the tubes. If the tubes are of outer diameter 25 mm and the outer surface temperature is 28°C, determine (a) the rate of heat transferred to the cooling water, and (b) the rate of condensation of the steam per unit length of the tube.

[**Ans.** (a) 960.6 kW, (b) 0.31 kg/s]

8.6 Water is heated by nickel coated heating element of length 250 mm and 3 mm diameter at atmospheric pressure. (a) Determine the critical heat flux. (b) When the operating condition changes from nucleate boiling to film boiling, calculate the increase in the temperature of the heating wire. Assume the emissivity of the wire material to be 0.5.

[**Ans.** (a) 126.328 kW/m$^2$, (b) 631.55°C]

8.7 A heat exchanger has 12 tubes in each column. The outer diameter of the tubes is 12 mm. If saturated steam at 60°C condenses on the outer surfaces of the tubes, which are maintained at 22°C. Determine (a) the rate at which the steam condenses per every meter length of the tubes and (b) the average heat transfer coefficient.

[**Ans.** (a) 0.0325 kg/s, (b) 4663.45 W/(m$^2$ K)]

8.8 An insulated tube of diameter 30 mm with a surface temperature of 18°C passes through a room in which the air is at 30°C with a relative humidity of 70 percent. Determine the rate of condensation per unit length of the pipe, assuming (a) film condensation and (b) dropwise condensation.

[**Ans.** (a) $7.23 \times 10^{-4}$ kg/s, (b) 0.0251 kg/s]

8.9 A metallic heating rod of diameter 10 mm and emissivity 0.8 is immersed in water. Under steady-state condition, the surface of the heating element is 200°C. Determine the power consumption per unit length of the heating element. Assume

$$h_{\text{total}} = h_{\text{conv.}} + \frac{3}{2} h_{\text{rad.}}$$

when $h_{\text{conv.}} > h_{\text{rad.}}$.

[**Ans.** 1.194 kW]

8.10 Cold water at temperature $T_w$ in a container is exposed to saturated water vapor at the free-surface at temperature $T_{\text{sat}}$, where $T_w < T_{\text{sat}}$, as shown in the Figure P8.10.

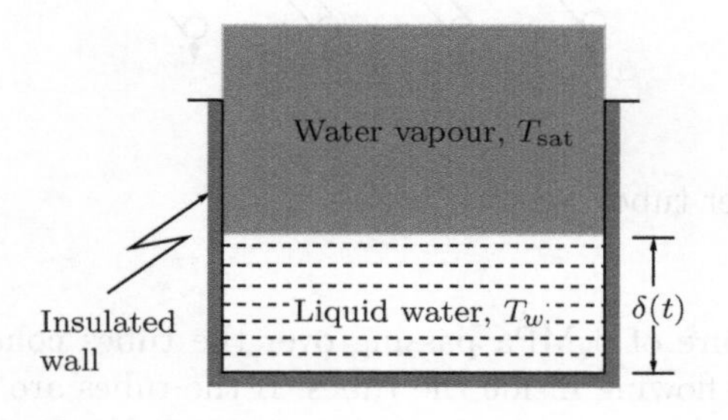


**Figure P8.10**

Assuming the linear temperature distribution in the liquid water, show that

$$\delta(t) = \left[\frac{2k_l\left(T_{\text{sat}} - T_w\right)}{\rho_l h_{fg}}\, t\right]^{\frac{1}{2}}$$

where $k_l$ is the thermal conductivity of the liquid, $\rho_l$ is the density of the liquid and $h_{fg}$ is the latent heat of condensation of water vapor.

8.11 Water is heated in a copper vessel with flat bottom of diameter 250 mm, at atmospheric pressure. If water evaporates at a rate of 2.5 kg/hour, determine the temperature of the vessel bottom.

[**Ans.** 106.09°C]

8.12 Saturated water at 100°C is boiled with a copper heating element of surface area 0.04 $m^2$, maintained at 115°C. Calculate the surface heat flux and rate of evaporation.

[**Ans.** 475.27 $kW/m^2$, 0.00842 kg/s]

8.13 Saturated steam at 7.38 kPa condenses on a horizontal tube bank made up of 4 rows of 3 tubes in a vertical tier. If the tubes are of diameter 30 mm and length 1 m, and their surface is maintained at 30°C by circulating water through them, determine (a) the rate of heat transfer to the cooling water circulating in the tubes, and (b) the rate of condensation of steam over the tubes.

[**Ans.** (a) 79.804 kW, (b) 0.03277 kg/s]

# References

1. Kutateladze, S. S., "A hydrodynamics theory of changes in boiling process under free convection," *Iz. Akad. Nauk SSSR, Otd, Tekh.Nauk*, (4): p. 524, 1951.

2. Labuntsov, D. A., "Heat transfer film condensation of pure steam on vertical surfaces and horizontal tubes," *Teploenergetika*, Vol. 4, pp. 72-80, 1957.

3. Cengel, Y., *Heat Transfer: A Practical Approach*, 2nd ed., McGraw-Hill, New York, 2002.

# Chapter 9

# Heat Exchangers

## 9.1 Introduction

Heat exchangers are devices that facilitate exchange of heat between two fluids that are at different temperatures, without making them mix with each other. Heat exchangers are widely used in numerous devices, such as automobile radiators, inter-coolers of high-pressure compressors, air-conditioning systems, large power plants, and so on. In a mixing chamber the fluids involved are allowed to mix, whereas in heat exchangers the fluids involved are not allowed to mix. For example, in a typical heat exchanger such as a car radiator, heat is transferred from the hot water flowing through the radiator tubes to the air flowing through the closely spaced thin plates attached to the tubes. Heat transfer in a heat exchanger usually involves convection in each fluid and conduction through the wall separating the two fluids. The temperature range, the phases of the fluids (liquid or gas), the quantity of thermal energy to be transferred, and the permissible pressure drops for the hot and cold fluids determine the heat exchanger configuration for a given application. In practice, the design and selection of heat exchangers often involve a trial-and-error procedure.

## 9.2 Types of Heat Exchangers

Depending on the application requirements, numerous types of heat exchangers are designed and used. In general, the classification is based on the following:

- The flow arrangement
- The transfer process
- Compactness

- Construction type
- Heat transfer mechanism

Thus, heat exchangers can be classified in many ways. One among them is based on the relative direction of the flow of the hot and cold fluids. The simplest type heat exchanger consists of two concentric tubes, as shown in Figure 9.1, and hence called *double-pipe heat exchanger*.

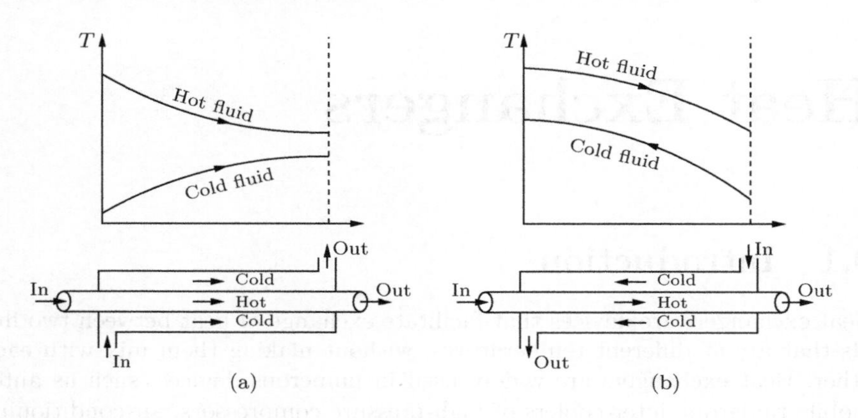


**Figure 9.1**
Double-pipe heat exchangers: (a) parallel flow, (b) counter flow.

When both the fluids flow in the same direction, as illustrated in Figure 9.1(a), it is termed *parallel flow* heat exchanger, when the fluids flow in parallel but in opposite directions, as shown in Figure 9.1(b), it is termed *counter flow* heat exchanger.

A type of heat exchanger which is designed to have a large heat transfer surface area per unit volume is the *compact* heat exchanger. The ratio of the heat transfer surface area of a heat exchanger to its volume is called the *area density* $\beta$. When $\beta > 700\ \text{m}^2/\text{m}^3$ a heat exchanger is classified as compact heat exchanger. For example, car radiators are compact heat exchangers with $\beta \approx 1000\ \text{m}^2/\text{m}^3$. Also, human lung ($\beta \approx 20,000\ \text{m}^2/\text{m}^3$) and regenerator of a Stirling engine ($\beta \approx 15,000\ \text{m}^2/\text{m}^3$) are compact heat exchangers. Compact heat exchangers can achieve high heat transfer rates between two fluids in a small volume. They are commonly used in applications with strict limitations on the volume and weight of the heat exchangers.

The large surface area in a compact heat exchanger is obtained by attaching closely spaced *thin plates* or *corrugated fins* to the walls separating the two fluids. In compact heat exchangers, hot and cold fluids usually move perpendicular to each other, and such a configuration is called *cross-flow*. The cross-flow is further classified into *mixed-flow* and *unmixed-flow*, depending on the flow configuration, as illustrated in Figures 9.2(a) and 9.2(b).

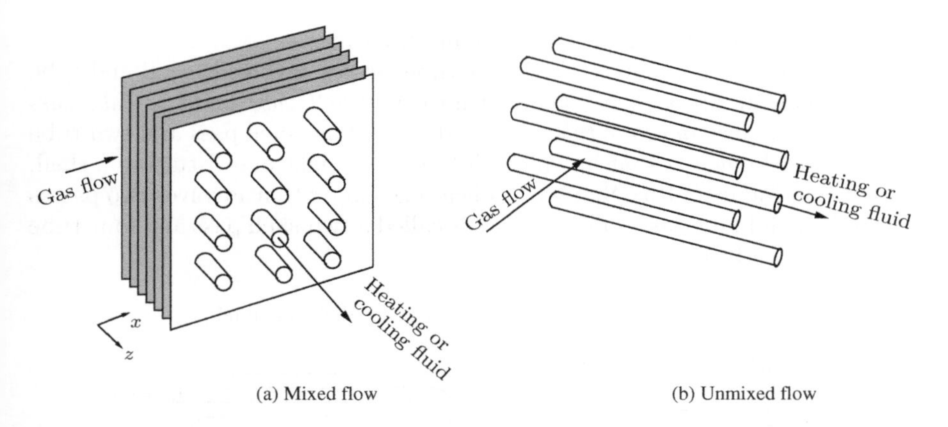


**Figure 9.2**
Cross-flow: (a) mixed and (b) unmixed flow configurations.

One of the most common types of heat exchanger for industrial applications is the *shell-in-tube* heat exchanger, shown in Figure 9.3. It has a large number of tubes packed in a shell with their axes parallel to the shell axis. Heat transfer takes place as one fluid flows inside the tubes, while the other fluid flows outside the tubes through the shell. Baffles are placed in the shell to force the shell-side flow to travel across the shell to enhance the heat transfer and to maintain uniform spacing between the tubes.

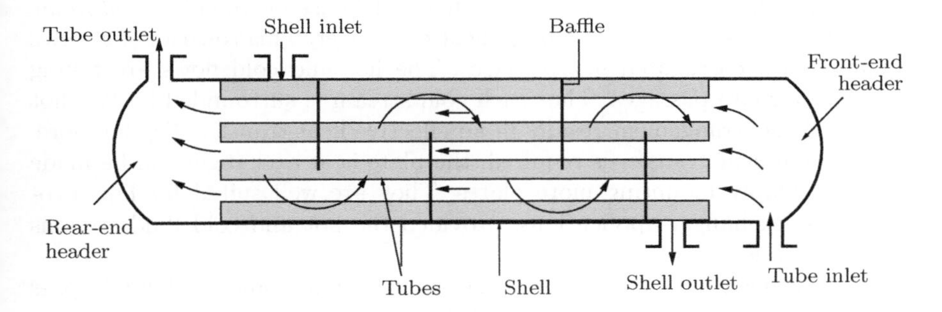


**Figure 9.3**
Schematic of shell-and-tube heat exchanger with one-shell and one-tube pass.

Even though widely used in industrial applications, because of their large size and heavy weight the shell-and-tube heat exchangers are not suitable for use in aircraft and marine applications. It is seen that the tubes in a shell-and-tube heat exchanger are open to some large areas called *headers* at both ends of the shell, as shown in Figure 9.3, where the tube-side fluid accumulates before entering the tubes and after leaving the tubes.

Based on the number of shell and tube passes involved, shell-and-tube heat exchangers are further classified into *one-shell pass and two-tube pass* and *two-shell pass and four-tube pass*, and so on. One-shell pass and two-tube pass heat exchanger is that in which all the tubes make one U-turn in a shell, as shown in Figure 9.4(a). Similarly, a heat exchanger that involves two passes in the shell and four passes in the tubes is called a two-shell pass and four-tube pass, as shown in Figure 9.4(b).

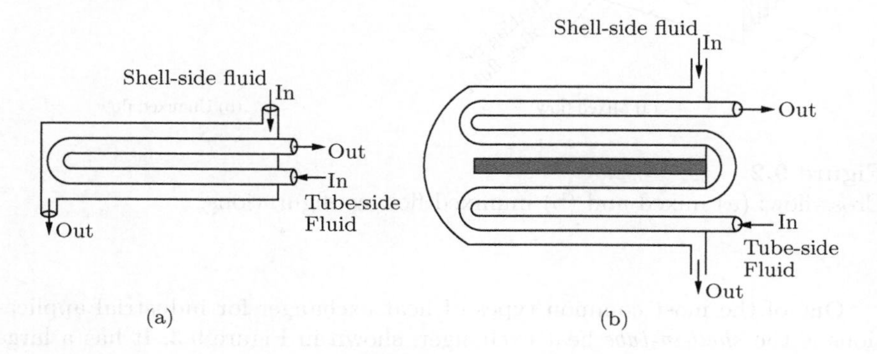


**Figure 9.4**
Shell-and-tube heat exchangers, (a) one-shell pass and two-tube pass, (b) two-shell pass and four-tube pass.

Another kind of heat exchanger which is widely used is the *plate and frame heat exchanger*, simply called a plate heat exchanger. This consists of a series of plates with corrugated flat passages. The hot and cold flows are taking place in alternate passages. Thus each cold stream is surrounded by two hot streams. This arrangement results in an effective heat transfer. Furthermore, if increase in heat transfer is required, the plate heat exchanger can be made larger by simply mounting more plates. They are well suited for liquid-to-liquid heat exchanger applications, provided the hot and cold fluid streams are at the same pressure.

Yet another kind of heat exchanger in which hot and cold fluids pass through alternate passages of identical flow area is the *regenerative* heat exchanger. The regenerative heat exchanger is further classified into *static-type* and *dynamic-type* regenerators.

• The static-type regenerator is a porous mass that has a large storage capacity, such as a ceramic wire mesh. Hot and cold fluids flow through the porous mass alternatively. Heat is transferred from the hot fluid to the material of the regenerator during the flow of the hot fluid, and from the material to the cold fluid during the flow of the cold fluid. Thus, the matrix serves as a temporary heat storage medium.

• The dynamic-type regenerator involves a rotating drum and continuous flow of hot and cold fluids through different portions of the drum. This enables a portion of the drum to *store* heat while passing through the hot stream and *reject* the stored heat while passing through the cold stream. Also, the drum serves as the medium to transport the heat from the hot stream to the cold fluid stream.

Usually heat exchangers are given names to reflect the specific application for which they are used. For example, a *boiler* is a heat exchanger in which one of the fluids absorbs heat and vaporizes. A *condenser* is a heat exchanger in which one of the fluids gives up heat and condenses as it flows through the heat exchanger.

## 9.3 The Overall Heat Transfer Coefficient

Usually a heat exchanger involves two fluids separated by a solid wall. Heat from the hot fluid is transferred by convection to the wall, the wall conducts the heat received by it and transports the same to the cold fluid again by convection. Radiation effect, if any, is usually included in the convection heat transfer coefficient.

The thermal resistance network associated with this heat transfer process involves two convection and one conduction resistances, as shown in Figure 9.5.

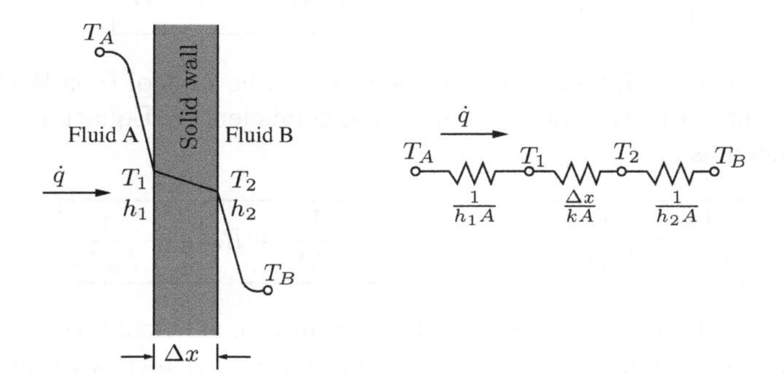


**Figure 9.5**
Thermal resistance network associated with heat transfer through a plane wall.

The heat transfer through the plane wall can be expressed as

$$\dot{q} = \frac{T_A - T_B}{\dfrac{1}{h_1 A} + \dfrac{\Delta x}{kA} + \dfrac{1}{h_2 A}} \tag{9.1}$$

where $k$ is the conductivity of the wall and $h_1$ and $h_2$ are the convection coefficients at the left and right surfaces of the wall.

From heat exchanger design point, the double pipe heat exchanger, shown in Figure 9.1, is more appropriate geometry than the plane wall shown in Figure 9.5. For the double pipe heat exchanger, the inside and outside areas of the inner tube are $A_i = \pi D_i L$ and $A_o = \pi D_o L$, respectively, where $D_i$ and $D_o$ are the inner and outer diameters of the tube, respectively, and $L$ is its length. The thermal resistance $R$ for the tube wall in this case is

$$R_{\text{wall}} = \frac{\ln(D_o/D_i)}{2\pi k L} \tag{9.2}$$

where $k$ is the thermal conductivity of the wall material and $L$ is the tube length. The total thermal resistance is the sum of the convective resistances at the inner and outer surfaces of the tube and the conduction resistance of the tube wall. Thus the total thermal resistance becomes

$$\boxed{R_{\text{total}} = R_i + R_{\text{wall}} + R_o = \frac{1}{h_i A_i} + \frac{\ln(D_o/D_i)}{2\pi k L} + \frac{1}{h_o A_o}} \tag{9.3}$$

It is convenient to combine all thermal resistances in the path of heat flow, from the hot fluid to the cold fluid, into a single resistance $R$ and express the rate of heat transfer between the two fluids as

$$\boxed{\dot{Q} = \frac{\Delta T}{R} = UA\Delta T = U_i A_i \Delta T = U_o A_o \Delta T} \tag{9.4}$$

where $U$ is the overall heat transfer coefficient. The unit of $U$ is W/(m$^2$ °C), which is identical to the unit of convection coefficient $h$. Equation (9.4) may be rearranged as

$$\boxed{\frac{1}{UA} = \frac{1}{U_i A_i} = \frac{1}{U_o A_o} = R = \frac{1}{h_i A_i} + R_{\text{wall}} + \frac{1}{h_o A_o}} \tag{9.5}$$

It is seen that, there are two overall heat transfer coefficients, $U_i$ and $U_o$, for a heat exchanger. This is because every heat exchanger has two heat transfer surface areas $A_i$ and $A_o$, in general $A_i \neq A_o$. However, $U_i A_i = U_o A_o$, implying $U_i \neq U_o$. Therefore, the overall heat transfer coefficient $U$ of a heat exchanger is meaningless unless the area on which it is based is specified. Especially this is the case when one side of the tube wall is finned and the other side is not, because the surface area of the finned side is several times that of the unfinned side.

When the thermal resistance of the tube $R_{\text{wall}}$ is negligibly small, due to the thin wall of the tube and high thermal conductivity of the tube material (which is the case usually), and the inner and outer surface areas of the tube are almost equal ($A_i \approx A_o \approx A$), the overall heat transfer coefficient [Equation

(9.5)] simplifies to

$$\frac{1}{U} \approx \frac{1}{h_i} + \frac{1}{h_o} \tag{9.6}$$

where $U \approx U_i \approx U_o$, and, $h_i$ and $h_o$ are the convection coefficients at the inner and outer surfaces of the tube.

As seen from Equation (9.6), $U$ is dictated by the smaller among $h_i$ and $h_o$. For example, when $h_i < h_o$, $U \approx h_i$. This situation is encountered when one of the fluids is a gas and the other is a liquid. In such a situation, fins can be used on the gas side to increase the magnitude of the product term $UA$ and thus the heat transfer on that side.

In heat exchanger applications, the heat transfer surface will foul with the accumulation of deposits. These deposits introduce an additional resistance in the path of the heat flow. The effect of fouling is generally introduced in the form of a *fouling factor* $R_f$ which has dimensions (m$^2$ °C)/W; and is discussed in Section 9.4.

Now, the total resistance $R_{\text{total}}$ in the path of heat flow across a tube, which is fouled due to deposit formation on both the inner and outer surfaces, is given by

$$R_{\text{total}} = R_i + R_{f,i} + R_{\text{wall}} + +R_{f,o} + R_o \tag{9.3a}$$

where $R_{f,i}$ and $R_{f,o}$ are the fouling factors (that is, unit fouling resistances) at the inner and outer surfaces of the tube, respectively, $R_i$ and $R_o$ are the thermal resistances at the inner and outer surfaces of the tube, respectively, and $R_{\text{wall}}$ is the thermal resistances across the tube wall thickness.

In heat transfer applications, the overall heat transfer coefficient is usually based on the outer surface of the tube. Therefore, Equation (9.3$a$) can be represented in terms of the overall heat transfer coefficient, based on the outer surface of the tube as

$$\boxed{U_o = \frac{1}{\dfrac{D_o}{D_i}\dfrac{1}{h_i} + \dfrac{D_o}{D_i}R_{f,i} + \dfrac{D_o}{2k}\ln\left(\dfrac{D_o}{D_i}\right) + R_{f,o} + \dfrac{1}{h_o}}} \tag{9.3b}$$

The overall heat transfer coefficient $U$ for some selected types of heat exchanger are given in Table 9.1.

It is seen that the magnitude of $U$ has a wide range from 10 W/(m$^2$ °C), for gas-to-gas heat exchangers, to about 8500 W/(m$^2$ °C), for heat exchangers involving phase changes. This high value of $U$ is because gases have very low thermal conductivities, and phase-change processes involve very high heat transfer coefficients.

**Table 9.1** Overall heat transfer coefficient for some heat exchangers

| Type of heat exchanger | $U$, W/(m$^2$ °C) |
|---|---|
| Water-to-water | 850–1700 |
| Water-to-oil | 100–350 |
| Water-to-kerosene | 300–1000 |
| Feed-water heaters | 1000–8500 |
| Steam-to-light fuel oil | 200–400 |
| Steam-to-heavy fuel oil | 50–200 |
| Steam condenser | 1000–6000 |
| Freon condenser (water cooled) | 300–1000 |
| Ammonia condenser (water cooled) | 800–1400 |
| Alcohol condenser (water cooled) | 250–700 |
| Gas-to-gas | 10–40 |
| Water-to-air in finned tubes (water in tubes) | 30–60* |
| | 400–850** |
| Steam-to-air in finned tubes (steam in tubes) | 30–60* |
| | 400–4000** |

* Based on air-side surface area, ** based on water-side or steam-side surface area

## 9.4 Fouling Factors

Over a period of operation, the heat-transferring surface of a heat exchanger may get coated with various deposits present in the flowing fluids. Also, the surface may get corroded due to the interactions of the fluids. The coating results in an additional resistance to the heat transfer and causes the rate of heat transfer to decrease. The effect of these accumulations on heat transfer is represented by a *fouling factor* $R_f$. Thus, fouling factor or fouling resistance is a measure of the thermal resistance introduced by fouling.

The most common type of fouling is the precipitation of solid deposits in a fluid on the heat transferring surface. For example, on the inner surfaces of a heater used for heating water, a layer of calcium-based deposits will form when the water is hard.

The fouling factor is zero for a new heat exchanger and increases with time as the solid deposits build up on the heat exchanger surface. The fouling factor depends on the operating temperature, the velocity of the fluids and the length of service. Fouling increases with increasing temperature and decreasing velocity.

Fouling factor is usually obtained experimentally by determining the values of the overall heat transfer coefficient $U$ for both clean and dirty conditions

**Table 9.2** Some selected fouling factors

| Fluid | $R_f$ (m$^2$ °C)/W |
|---|---|
| Sea water, river water boiler feed water: | |
| Below 50°C | 0.0001 |
| Above 50°C | 0.0002 |
| Fuel oil | 0.0009 |
| Steam (oil-free) | 0.0001 |
| Refrigerant (liquid) | 0.0002 |
| Refrigerant (vapor) | 0.0004 |
| Alcohol vapors | 0.0001 |
| Air | 0.0004 |

in the heat exchanger. Thus,

$$R_f = \frac{1}{U_{\text{dirty}}} - \frac{1}{U_{\text{clean}}}$$

Some representative values of fouling factor are given in Table 9.2.

## Example 9.1

Determine the overall heat transfer coefficient for a steel tube of diameter 25 mm and wall thickness 3 mm, based on the outside diameter. The convective heat transfer coefficients for the inner and outer surfaces are $h_i = 1850$ W/(m$^2$ °C) and $h_o = 1200$ W/(m$^2$ °C), and the fouling factors are $R_{f,i} = R_{f,o} = 0.0002$ (m$^2$ °C)/W. The conduction coefficient for the tube material is $k = 50$ W/(m$^2$ °C).

## Solution

Given $D_i = 25$ mm, $D_o = 31$ mm. By Equation (9.3b),

$$\begin{aligned} U_o &= \frac{1}{(D_o/D_i)(1/h_i) + (D_o/D_i)R_{f,i} + [D_o/(2k)]\ln(D_o/D_i) + R_{f,o} + 1/h_o} \\ &= \frac{1}{(31/25)(1/1850) + (31/25)(0.0002)} \end{aligned}$$

$$+ \frac{1}{[0.031/(2 \times 50)] \ln (31/25) + 0.0002 + 1/1200}$$

$$= \frac{1}{0.00067 + 0.000248 + 0.0000667 + 0.0002 + 0.000833}$$

$$= \frac{1}{2.0177 \times 10^{-3}}$$

$$= \boxed{495.61 \, \text{W}/(\text{m}^2 \, {}^\circ\text{C})}$$

## 9.5 Heat Exchanger Performance

Selection of a heat exchanger that will achieve the specified temperature change of a fluid stream of known mass flow rate, or will result in the required outlet temperature of the hot and cold fluid streams, is an important step in any heat exchanger application. To identify an appropriate heat exchanger to meet the requirement of a specific application, the performance analysis of the heat exchanger is essential. Usually the following two methods are used in the analysis of heat exchanger performance:

1. The *log mean temperature difference* (LMTD) method – this is well suited to identify a heat exchanger that will achieve the specified temperature change in a fluid stream of given mass flow rate.

2. The *effective NTU* (number of (heat) transfer unit) method – this is suitable to predict the outlet temperatures of the cold and hot fluid streams in a specified heat exchanger.

Heat exchangers can be modeled as steady-flow devices because usually they operate at the same operating conditions for long periods of time. Thus, the mass flow rate of each fluid practically remains constant, and the velocity and temperature at the inlet and outlet remain the same. Also, usually the fluid streams experience negligible or no change in their elevations. Thus, the kinetic and potential energy changes are negligible. Furthermore, the specific heat of the fluids flowing through a heat exchanger can be treated as constant, because in general, the specific heat changes only when there is a change in temperature. Axial heat conduction along the tube is usually insignificant and can be regarded negligible. Also, assuming the heat exchanger to be perfectly insulated, the heat loss from the heat exchanger to the surrounding medium can be neglected, hence any heat transfer in the system occurs between the two fluids only.

All the above simplifying assumptions closely agree with the practical situation. Under these assumptions, by first law of thermodynamics, we have:

The rate of heat transfer from the hot fluid = The rate of heat transfer to the cold fluid

that is,

$$\dot{Q} = \dot{m}_c c_{p_c} (T_{c,\text{out}} - T_{c,\text{in}}) \tag{9.7a}$$

and

$$\dot{Q} = \dot{m}_h c_{p_h} (T_{h,\text{in}} - T_{h,\text{out}}) \tag{9.7b}$$

where the subscripts $c$ and $h$ refer to the cold and hot fluids, respectively, $\dot{m}_c$, $\dot{m}_h$ are the mass flow rates, $c_{p_c}$, $c_{p_h}$ are the specific heats, $T_{c,\text{in}}$, $T_{h,\text{in}}$ are the inlet temperatures and $T_{c,\text{out}}$, $T_{h,\text{out}}$ are the outlet temperatures. In these relations $\dot{Q}$ is taken as positive with the convention that the heat transfer is from the hot fluid to the cold fluid in accordance with the second law of thermodynamics.

In this kind of analysis, it is convenient to use the *heat capacity rate* which is the product of the mass flow rate and the specific heat,

$$C = \dot{m} c_p \tag{9.8}$$

The heat capacity rate of a fluid stream is the rate of heat transfer required to change the temperature of the fluid stream by 1°C as it flows through a heat exchanger. From Equation (9.8), it is evident that in a heat exchanger a fluid with a large heat capacity rate will experience a small temperature change, and a fluid with a small heat capacity rate will experience a large temperature change. Therefore, doubling the mass flow rate of a fluid leaving the other parameters unchanged will halve the temperature change of that fluid.

Using the heat capacity rate, the heat transfer rate of a heat exchanger [Equations (9.7a) and (9.7b)] can be expressed as

$$\dot{Q} = C_c (T_{c,\text{out}} - T_{c,\text{in}}) \tag{9.9}$$

$$\dot{Q} = C_h (T_{h,\text{in}} - T_{h,\text{out}}) \tag{9.10}$$

That is, the heat transfer rate in a heat exchanger is equal to the heat capacity rate of the cold or hot fluid multiplied by the temperature change of that fluid. From Equations (9.9) and (9.10) it is seen that, when the heat capacity rate is equal to each other, the temperature rise of the cold fluid will be equal to the temperature drop of the hot fluid. In other words, in a well-insulated heat exchanger, two fluids of the same specific heat flowing with the same mass flow rate will experience the same temperature change as illustrated in Figure 9.6.

Two popular heat exchangers commonly used in practice are the *condensers* and *boilers*. In a condenser or a boiler, one of the fluids undergoes a phase-change process, and the rate of heat transfer is expressed as

$$\dot{Q} = \dot{m}\, h_{fg} \tag{9.11}$$

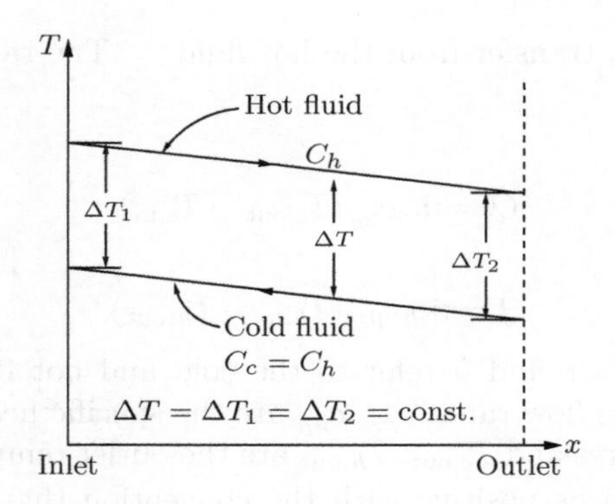


**Figure 9.6**
Heat exchange between two fluids of the same specific heat flowing with the same mass flow rate in a properly insulated heat exchanger.

where $\dot{m}$ is the rate of evaporation or condensation of the fluid and $h_{fg}$ is the enthalpy of vaporization of the fluid at the specified temperature or pressure.

A fluid usually absorbs or releases a large amount of heat essentially at constant temperature during a phase change process, as shown in Figure 9.7. The heat capacity rate of a fluid during a phase-change process would approach infinity because the temperature change is practically zero. That is, $C = \dot{m}c_p \to \infty$ when $\Delta T \to 0$. Thus, the heat transfer rate $\dot{Q} = \dot{m}c_p\Delta T$ is a finite quantity during this process. Therefore, in heat exchanger analysis, a condensing or boiling fluid can be conveniently modeled as a fluid whose heat capacity rate is infinity.

In an analogous manner to Newton's law of cooling, the rate of heat transfer in a heat exchanger can be expressed as

$$\dot{Q} = UA\Delta T_m \tag{9.12}$$

where $U$ is the overall heat transfer coefficient, $A$ is the heat transfer area, and $\Delta T_m$ is an appropriate average temperature difference between the two fluids. Here it is important to note that the surface area $A$ being a geometrical parameter can be precisely determined, whereas $U$ and $\Delta T_m$, in general are not constants and vary along the heat exchanger.

## 9.6 Log Mean Temperature Difference Method

We saw that the temperature difference between the hot and cold fluids varies along the heat exchanger. This situation can be handled if the *mean temperature difference* $\Delta T_m$ is used in the relation

$$\dot{Q} = UA\Delta T_m$$

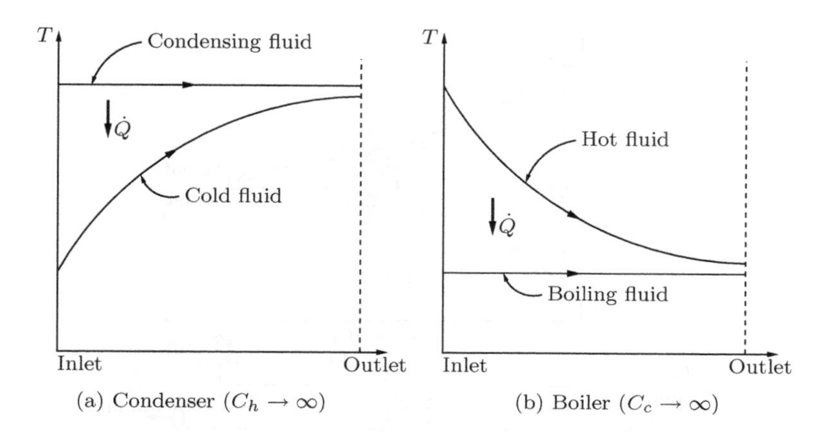


**Figure 9.7**
Temperature variation of a fluid during (a) condensing or (b) boiling, in a heat exchanger.

To develop a relation for the equivalent average temperature difference between the hot and cold fluids, consider the double-pipe parallel-flow heat exchanger shown in Figure 9.8.

It is seen that the mean temperature difference $\Delta T_m$ between the hot and cold fluids is the largest at the inlet of the heat exchanger and decreases exponentially towards the outlet and becomes minimum at the exit. It is obvious that the temperature of the hot fluid decreases and the cold fluid increases along the heat exchanger. Also, the temperature of the cold fluid can never exceed that of the hot fluid in a heat exchanger.

Let us assume that the outer surface of the heat exchanger is properly insulated so that the heat lost by the hot fluid is completely transferred to the cold fluid. Also, let us neglect any changes in the potential and kinetic energy. For this case, the energy balance on the hot and cold fluid mass in a differential section of the heat exchanger can be expressed as

$$\delta\dot{Q} = -\dot{m}_h c_{p_h} dT_h \tag{9.13}$$

and

$$\delta\dot{Q} = +\dot{m}_c c_{p_c} dT_c \tag{9.14}$$

where $\dot{Q}$ is the heat transfer rate, $c_p$ is the specific heat, $T$ is the temperature, and subscripts $h$ and $c$ refer to hot and cold fluids, respectively. The negative sign in Equation (9.13) implies that the hot fluid is losing heat and the positive sign in Equation (9.14) implies that the cold fluid is gaining the heat lost by the hot fluid. Solving these equations for $dT_h$ and $dT_c$, we get

$$dT_h = -\frac{\delta\dot{Q}}{\dot{m}_h c_{p_h}} \tag{9.15}$$

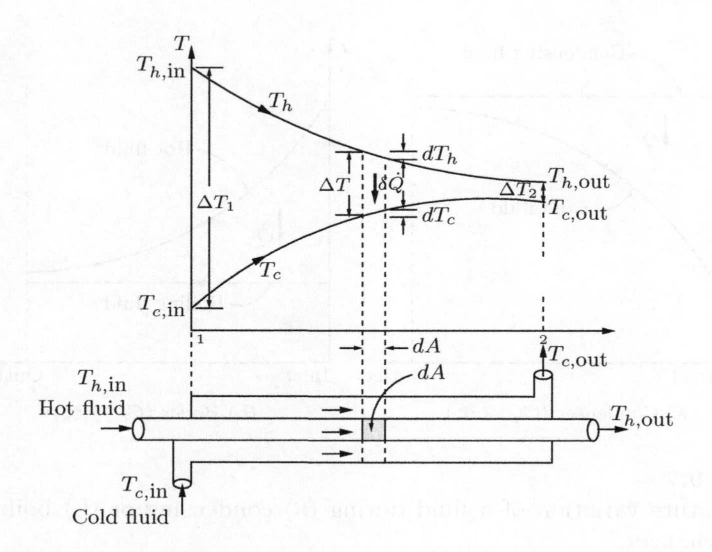


**Figure 9.8**
Temperature variation in a double-pipe parallel-flow heat exchanger.

and

$$dT_c = \frac{\delta \dot{Q}}{\dot{m}_c \, c_{p_c}} \tag{9.16}$$

Therefore,

$$dT_h - dT_c = d\left(T_h - T_c\right) = -\delta \dot{Q} \left( \frac{1}{\dot{m}_h c_{p_h}} + \frac{1}{\dot{m}_c \, c_{p_c}} \right) \tag{9.17}$$

Thus, the rate of heat transfer in the differential section of the heat exchanger can be expressed as

$$\delta \dot{Q} = U\left(T_h - T_c\right) dA \tag{9.18}$$

where $U$ is the overall heat transfer coefficient and $dA$ is the area over which heat transfer is taking place.

Substituting for $\delta \dot{Q}$ from Equation (9.17), we obtain

$$\frac{d\left(T_h - T_c\right)}{T_h - T_c} = -UdA \left( \frac{1}{\dot{m}_h c_{p_h}} + \frac{1}{\dot{m}_c c_{p_c}} \right) \tag{9.19}$$

Integration of this equation from the inlet to the outlet of the heat exchanger yields

$$\ln \left( \frac{T_{h,\text{out}} - T_{c,\text{out}}}{T_{h,\text{in}} - T_{c,\text{in}}} \right) = -UA \left( \frac{1}{\dot{m}_h c_{p_h}} + \frac{1}{\dot{m}_c c_{p_c}} \right) \tag{9.20}$$

where $A = (A_{\text{out}} - A_{\text{in}})$, the difference between the outlet and inlet area.

Now, solving Equations (9.9) and (9.10) for $\dot{m}_c c_{p_c}$ and $\dot{m}_h c_{p_h}$, we get

$$\dot{m}_c c_{p_c} = \frac{\dot{Q}}{T_{c,\text{out}} - T_{c,\text{in}}}$$

$$\dot{m}_h c_{p_h} = \frac{\dot{Q}}{T_{h,\text{in}} - T_{h,\text{out}}}$$

Substituting these into Equation (9.20), we obtain

$$\begin{aligned}
\ln\left(\frac{T_{h,\text{out}} - T_{c,\text{out}}}{T_{h,\text{in}} - T_{c,\text{in}}}\right) &= -\frac{UA}{\dot{Q}}\left(T_{h,\text{in}} - T_{h,\text{out}} + T_{c,\text{out}} - T_{c,\text{in}}\right) \\
&= -\frac{UA}{\dot{Q}}\Big((T_{h,\text{in}} - T_{c,\text{in}}) - (T_{h,\text{out}} - T_{c,\text{out}})\Big) \\
\ln\left(\frac{\Delta T_2}{\Delta T_1}\right) &= -\frac{UA}{\dot{Q}}\left(\Delta T_1 - \Delta T_2\right)
\end{aligned}$$

where $\Delta T_1 = (T_{h,\text{in}} - T_{c,\text{in}})$ is the difference between the temperatures of the hot and cold fluids at the inlet and $\Delta T_2 = (T_{h,\text{out}} - T_{c,\text{out}})$ is that at the outlet. Thus,

$$\begin{aligned}
\dot{Q} &= \frac{UA\,(\Delta T_1 - \Delta T_2)}{-\ln\,(\Delta T_2/\Delta T_1)} \\
&= \frac{UA\,(\Delta T_1 - \Delta T_2)}{\ln\,(\Delta T_1/\Delta T_2)}
\end{aligned}$$

or

$$\boxed{\dot{Q} = UA\Delta T_{\text{lm}}} \tag{9.21}$$

where

$$\Delta T_{\text{lm}} = \frac{\Delta T_1 - \Delta T_2}{\ln\,(\Delta T_1/\Delta T_2)} \tag{9.22}$$

is the log mean temperature difference and $\Delta T_1$ and $\Delta T_2$ refer to the temperature difference between the two fluids at the two ends (inlet and outlet) of the heat exchanger. It is interesting to note that in this analysis [Equation (9.22)] which end is designated as the inlet or outlet is of no consequence, as illustrated in Figure 9.9.

The temperature difference between the hot and cold fluids decreases from the inlet to outlet. Thus, it appears that the arithmetic mean temperature

$$\Delta T_{\text{am}} = \frac{\Delta T_1 + \Delta T_2}{2}$$

may be used as the average temperature difference in the analysis.

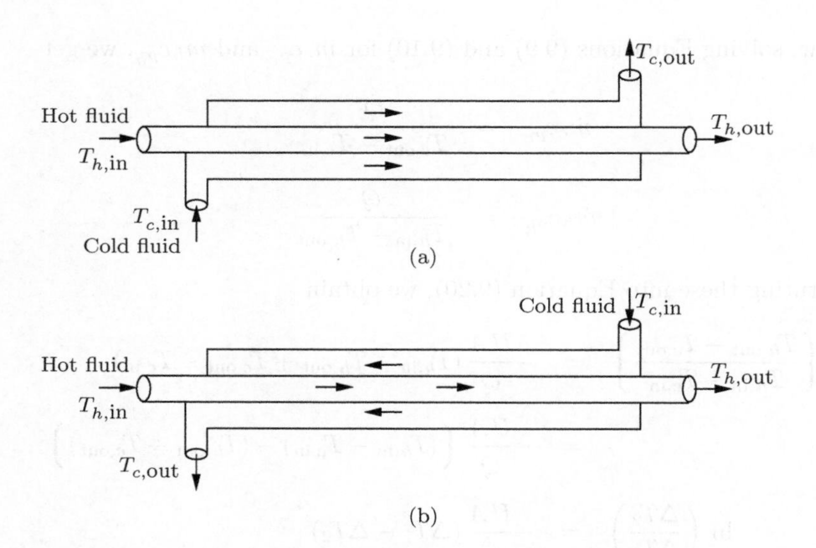


**Figure 9.9**
(a) Parallel-flow and (b) counter-flow heat exchangers.

It is important to note that $\Delta T_{\text{am}}$ is the mean based on the difference between the temperatures of the hot and cold fluids just at the inlet and outlet, and not based on the actual temperature variation all along the length of the heat exchanger. Therefore, the heat transfer determined with $\Delta T_{\text{am}}$ cannot be accurate. But the logarithmic mean temperature difference $\Delta T_{\text{lm}}$ is obtained considering the actual temperature profile of the fluid along the heat exchanger. Thus, $\Delta T_{\text{lm}}$ is the exact representation of the average temperature difference between the cold and hot fluids. It essentially reflects the exponential decay of the local temperature difference. Further, $\Delta T_{\text{lm}}$ is always less than $\Delta T_{\text{am}}$. Therefore, use of $\Delta T_{\text{am}}$ in place of $\Delta T_{\text{lm}}$ will overestimate the rate of heat transfer in a heat exchanger.

## 9.6.1 Counter-Flow Heat Exchanger

In a counter-flow heat exchanger, the hot and cold fluids flow in opposite directions, as shown in Figure 9.10. The hot and cold fluids enter the heat exchanger from opposite ends. The outlet temperature of the cold fluid in this device may exceed the outlet temperature of the hot fluid. In the limiting case, the cold fluid will be heated to the level of the inlet temperature of the hot fluid. But in reality this limiting situation is impossible since this would violate increase of entropy principle.

As shown in Figure 9.10, the temperature difference between the hot and cold fluids at the inlet is $\Delta T_1 = (T_{h\text{in}} - T_{c\text{in}})$ and the temperature difference

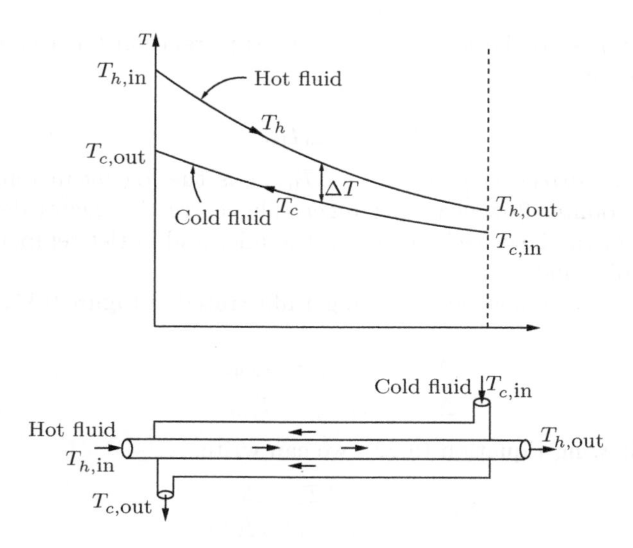


**Figure 9.10**
Schematic of counter-flow double-pipe heat exchanger and the temperature variation of hot and cold fluids.

between the fluids at the outlet is $\Delta T_2 = (T_{h_{\mathrm{out}}} - T_{c_{\mathrm{out}}})$. In terms of $\Delta T_1$ and $\Delta T_2$, the log mean temperature difference relation Equation (9.20) expressed using parallel-flow heat exchanger can be used for a counter-flow heat exchanger. For specified inlet and outlet temperatures of the hot and cold fluids, the log mean temperature difference for a counter-flow heat exchanger is always greater than that for a parallel-flow heat exchanger. Thus, relatively a smaller surface area (that is, smaller heat exchanger) is needed to achieve a specified heat transfer rate in a counter-flow heat exchanger.

When the heat capacity rates of the hot and cold fluids are equal, that is, when $C_h = C_c$ or $\dot{m}_h c_{p_h} = \dot{m}_c c_{p_c}$, the temperature difference between the two fluids will remain constant along the heat exchanger. In other words, $\Delta T_1 = \Delta T_2$ and $\Delta T_{\mathrm{lm}} = 0/0$, which is indeterminate. Using l'Hospital's rule, it can be shown that, for this case $\Delta T_{\mathrm{lm}} = \Delta T_1 = \Delta T_2$, as expected.

### 9.6.2 Multipass and Cross-Flow Heat Exchanger

As in the case of parallel-flow and counter-flow heat exchangers, the log mean temperature difference $\Delta T_{\mathrm{lm}}$ relation can be developed for cross-flow and multipass shell-and-tube heat exchangers too. But the resulting expressions would be complicated owing to the complex nature of the flow conditions. Therefore, for this kind of heat exchangers it is convenient to relate the equivalent temperature difference to the log mean temperature relation for a counter-flow heat exchanger as

$$\Delta T_{\text{lm}} = F\,\Delta T_{\text{lm},cf} \tag{9.23}$$

where $F$ is the *correction factor* and $\Delta T_{\text{lm},cf}$ is the log mean temperature difference for counter-flow heat exchanger. This correction factor depends on the geometry of the heat exchanger and the inlet and outlet temperatures of the hot and cold fluids.

Examine the cross-flow heat exchanger illustrated in Figure 9.11. It is seen that,

$$\Delta T_1 = T_{h,\text{in}} - T_{c,\text{out}}$$
$$\Delta T_2 = T_{h,\text{out}} - T_{c,\text{in}}$$

Therefore, following Equation (9.22), we can write,

$$\Delta T_{\text{lm},cf} = \frac{\Delta T_1 - \Delta T_2}{\ln\left(\Delta T_1/\Delta T_2\right)}$$

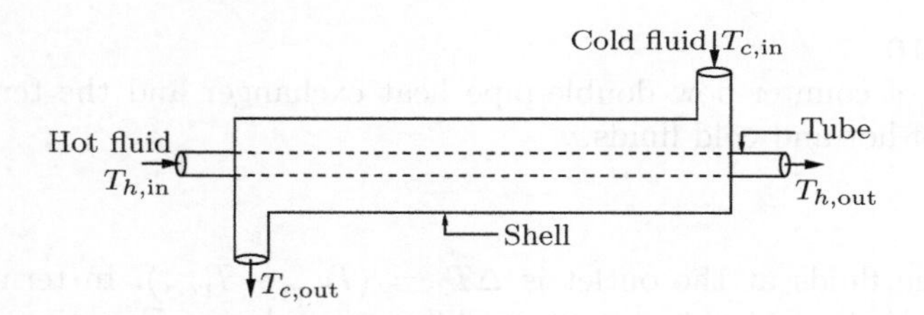


**Figure 9.11**
Cross-flow or multipass shell-and-tube heat exchanger.

Using the $\Delta T_{\text{lm},cf}$, the heat transfer rate can be expressed as

$$\dot{Q} = UAF\,\Delta T_{\text{lm},cf}$$

The correction factor $F$ is less than unity for a cross-flow or multipass shell-and-tube heat exchanger. The limiting value of $F = 1$ corresponds to the counter-flow heat exchanger. Thus $F$ for a heat exchanger is a measure of the deviation of the $\Delta T_{\text{lm}}$ from the corresponding values for the counter-flow case.

The correction factor $F$ for common cross-flow and shell-and-tube heat exchanger configuration are given in Figures 9.12a to 9.12d, as a function of two temperature ratios $P$ and $R$, defined as

$$P = \frac{t_2 - t_1}{T_1 - t_1} \tag{9.24}$$

$$R = \frac{T_1 - T_2}{t_2 - t_1} = \frac{(\dot{m}c_p)_{\text{tube side}}}{(\dot{m}c_p)_{\text{shell side}}} \tag{9.25}$$

where the subscripts 1 and 2 refer to the inlet and outlet, respectively. From Equation (9.25), it is seen that both the inlet and outlet temperatures should be known for determining the correction factor $F$. Note that, in Equations (9.24) and (9.25), $T_1$, $T_2$, $t_1$ and $t_2$ represent temperatures. $T_1$ and $T_2$ are the temperatures of the hot fluid at the inlet and outlet, respectively, and $t_1$ and $t_2$ are the temperatures of the cold fluid at the inlet and outlet, respectively.

The value of $P$ ranges from 0 to 1. But the value of $R$ ranges from 0 to infinity, $R = 0$ corresponds to the phase-change (condensation or boiling) on the shell side and $R \to \infty$ corresponds to the phase-change on the tube side. The correction for both these limiting cases is unity ($F = 1$). Therefore, the correction factor for a condensation or boiling is $F = 1$, irrespective of the configuration of the heat exchanger.

## 9.7 The Effective-$NTU$ Method

The log mean temperature difference (LMTD) approach to heat exchanger analysis is useful when the inlet and outlet temperatures of the hot and cold fluids are known or can easily be determined. Once $\Delta T_{\text{lm}}$, which is the difference between the temperatures of the hot and cold fluids, the mass flow rates of the hot and cold fluids, and the overall heat transfer coefficient $U$ are available, the heat transfer surface area $A$ of the heat exchanger can be determined from

$$\dot{Q} = UA\,\Delta T_{\text{lm}}$$

The LMTD method is apt for determining the size of a heat exchanger to realize the prescribed outlet temperatures when the mass flow rates and the inlet and outlet temperatures of the hot and cold fluids are specified. In other words, the primary task here is to select the heat exchanger that will meet the prescribed performance requirements. For this we have to follow the steps given below.

- Select the inlet and outlet temperatures of heat exchanger suitable for the specified application.
- Find the $\Delta T_{\text{lm}}$ and the correction factor $F$.
- Calculate the overall heat transfer coefficient $U$.
- Calculate the heat transfer surface area $A$.

Once the heat exchanger that has a heat transfer surface area equal to or larger than area $A$ is selected, this task is over. Once the type and size of the heat exchanger is chosen, the next step in the analysis is to determine the heat transfer rate and the outlet temperatures of the hot and cold fluids for the prescribed mass flow rates and inlet temperatures. In this case, the

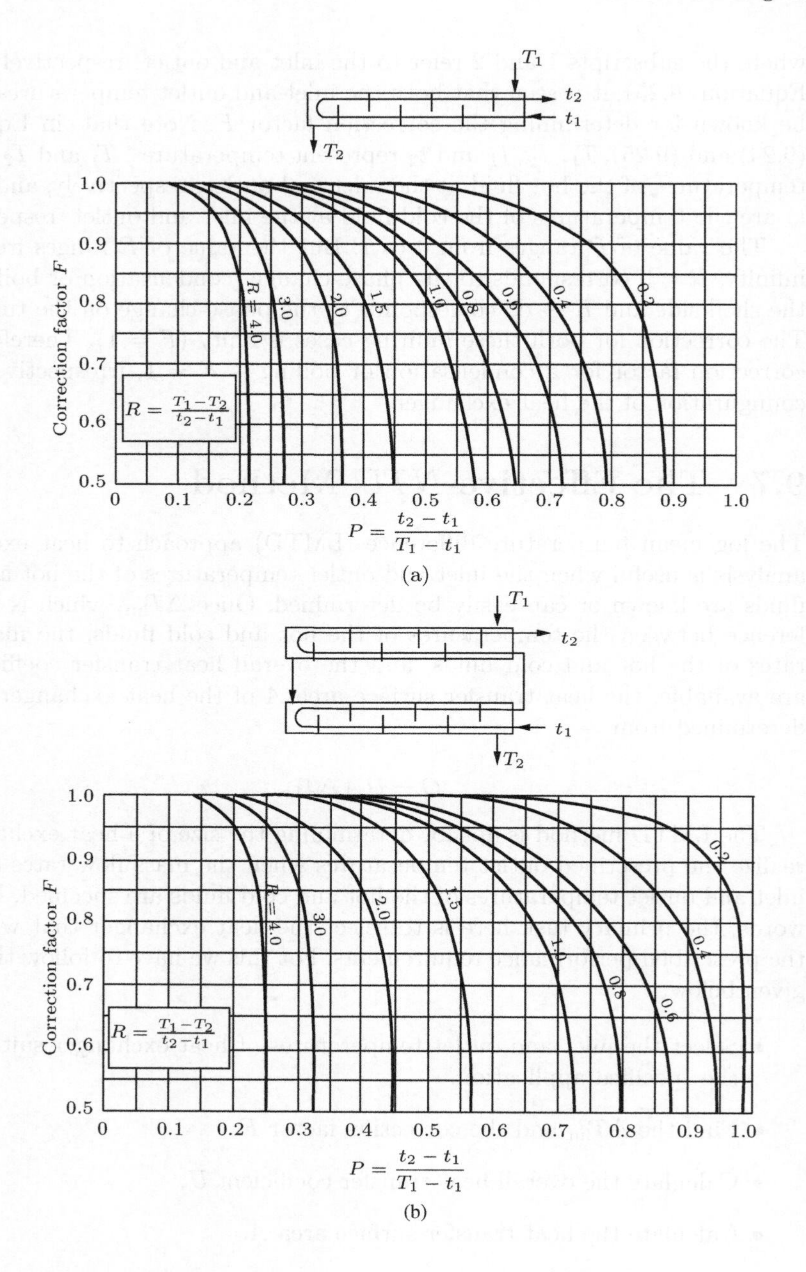


**Figure 9.12**
(a) One-shell pass and 2, 4, 6, etc. (multiple of 2) tube passes, (b) two-shell passes and 4, 8, 12, etc. (multiple of 4) tube passes, (c) single-pass cross-flow with both fluids unmixed, and (d) single-pass cross-flow with one fluid mixed and the other unmixed [1].

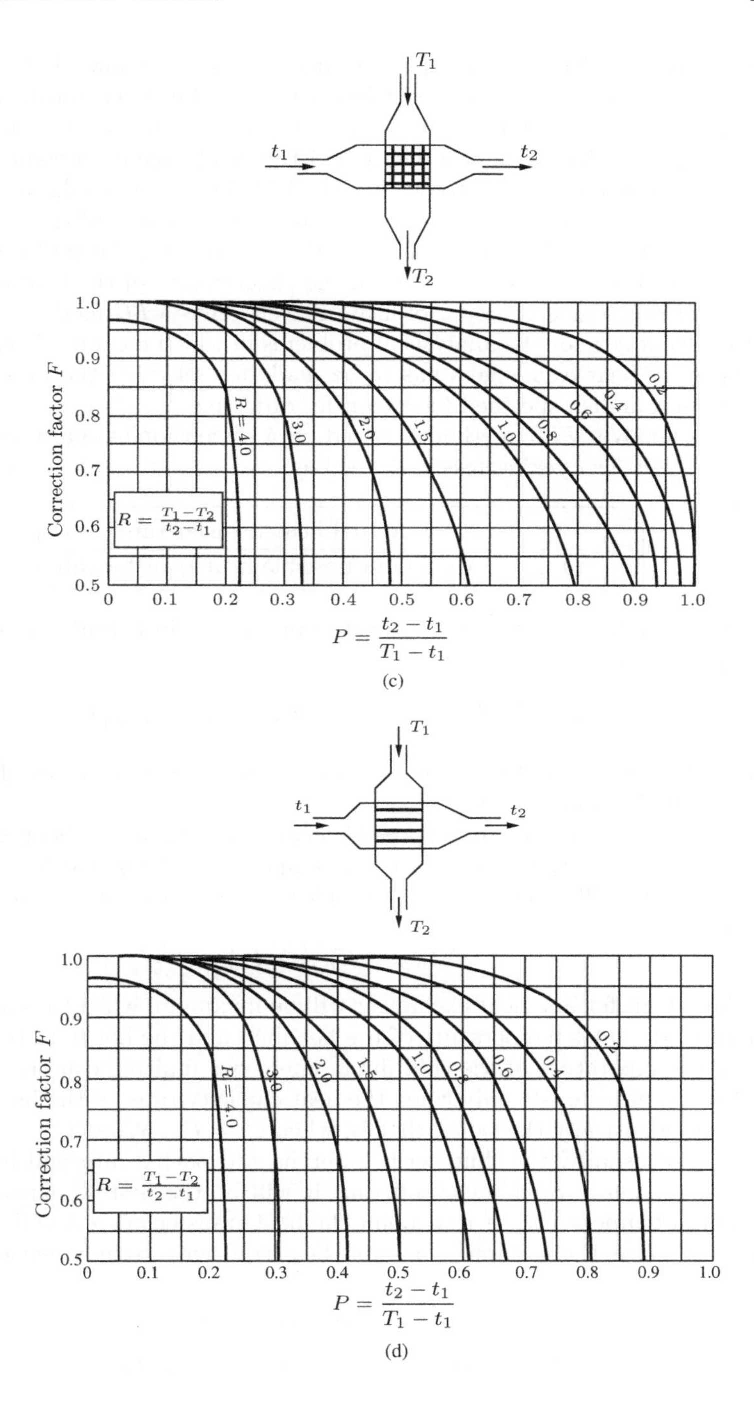


**Figure 9.12**
(Continued.)

heat transfer surface area $A$ of the heat exchanger is known but the outlet temperatures of the hot and cold fluids have to be determined. Thus, the task here is to determine the heat transfer performance of a specified heat exchanger or to determine whether an existing heat exchanger could meet the performance requirements. To do this, the LMTD method could be used. But the calculations would require iterative procedure because of the logarithmic function of the LMTD. In these cases the analysis can be performed more easily by utilizing a method based on the effectiveness of the heat exchanger in transferring a given amount of heat. The effective-$NTU$ method also offers many advantages for the analysis of problems in which a comparison between the types of heat exchangers has to be made for selecting the most suitable type to accomplish specified heat transfer objective.

The effective-$NTU$ method is based on a dimensionless parameter called the *heat transfer effectiveness* $\varepsilon$, defined as

$$\boxed{\varepsilon = \frac{\dot{Q}}{\dot{Q}_{\text{max}}} = \frac{\text{Actual heat transfer rate}}{\text{Maximum possible heat transfer rate}}} \tag{9.26}$$

Using energy balance, the actual heat transfer rate in a heat exchanger can be expressed as

$$\dot{Q} = C_c\left(T_{c,\text{out}} - T_{c,\text{in}}\right) = C_h\left(T_{h,\text{in}} - T_{h,\text{out}}\right) \tag{9.27}$$

where $C_c = \dot{m}_c c_{pc}$ and $C_h = \dot{m}_h c_{ph}$ are the heat capacity rates of the cold and hot fluids, respectively.

To determine the maximum possible heat transfer rate in a heat exchanger, we just recognize that the maximum temperature difference in a heat exchanger is the difference between the inlet temperatures of the hot and cold fluids. That is,

$$\Delta T_{\text{max}} = T_{h,\text{in}} - T_{c,\text{in}} \tag{9.28}$$

The heat transfer in a heat exchanger will be maximum when the cold fluid is heated to the inlet temperature of the hot fluid and the hot fluid is cooled to the inlet temperature of the cold fluid. These two limiting conditions will be reached simultaneously only when the heat capacity rates of the hot and cold fluids are identically the same, that is, when $C_c = C_h$. When $C_c \neq C_h$, which is the case usually, the fluid with lower heat capacity rate will experience a larger temperature change, and thus it will be the first to experience the maximum temperature, at that point the heat transfer process will come to a halt. Therefore, the maximum possible heat transfer rate in a heat exchanger is

$$\dot{Q}_{\text{max}} = C_{\text{min}}\left(T_{h,\text{in}} - T_{c,\text{in}}\right) \tag{9.29}$$

where $C_{\text{min}} = \left(\dot{m}c_p\right)_{\text{min}}$ is the smaller of $\dot{m}_h c_{ph}$ and $\dot{m}_c c_{pc}$.

## Example 9.2

Determine the maximum rate of heat transfer in the heat exchanger illustrated in Figure E9.2. Assume the specific heat of cold fluid as $c_{pc} = 4.18$ kJ/(kg °C) and that of the hot fluid as $c_{ph} = 2.3$ kJ/(kg °C).

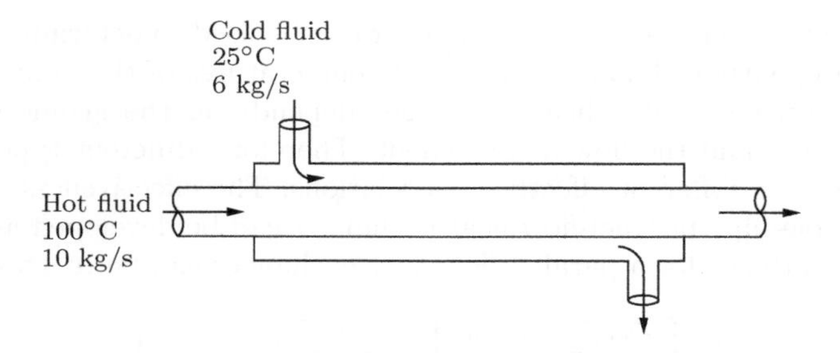


**Figure E9.2**
Heat exchanger.

## Solution

Given, $\dot{m}_c = 6$ kg/s, $\dot{m}_h = 10$ kg/s, $T_{c,\text{in}} = 25$°C, $T_{h,\text{in}} = 100$°C.

Therefore,

$$C_{\min} = \dot{m}c_{ph} = 10 \times 2.3$$

$$= 23\,\text{kW/°C}$$

$$\Delta T_{\max} = T_{h,\text{in}} - T_{c,\text{in}} = 100 - 25$$

$$= 75\,°\text{C}$$

Thus,

$$\dot{Q}_{\max} = C_{\min}\left(T_{h,\text{in}} - T_{c,\text{in}}\right)$$

$$= 23 \times 75$$

$$= \boxed{1725\,\text{kW}}$$

To determine the maximum possible heat transfer $\dot{Q}_{\max}$ in a heat exchanger, the inlet and outlet temperatures of the hot and cold fluids and their mass flow rates should be known. Then, if the effectiveness of the heat exchanger is also known, the actual heat transfer rate $\dot{Q}$ can be determined from the following equation.

$$\dot{Q} = \varepsilon \dot{Q}_{\text{max}} = \varepsilon C_{\text{min}} \left(T_{h,\text{in}} - T_{c,\text{in}}\right) \tag{9.30}$$

Thus, the effectiveness of a heat exchanger can be used to determine the heat transfer rate, without knowing the outlet temperatures of the fluids.

The effectiveness of a heat exchanger depends on the geometry of the heat exchanger and the flow arrangement. Therefore, different types of heat exchangers have different effectiveness relations. The effectiveness $\varepsilon$ relation for the double-pipe parallel-flow heat exchanger can be developed as follows.

Equation (9.20) for a parallel-flow heat exchanger can be rearranged as

$$\ln\left(\frac{T_{h,\text{out}} - T_{c,\text{out}}}{T_{h,\text{in}} - T_{c,\text{in}}}\right) = -\frac{UA}{C_c}\left(1 + \frac{C_c}{C_h}\right) \tag{9.31}$$

Solving Equation (9.27) for $T_{h,\text{out}}$, we get

$$T_{h,\text{out}} = T_{h,\text{in}} - \frac{C_c}{C_h}\left(T_{c,\text{out}} - T_{c,\text{in}}\right) \tag{9.32}$$

Substituting this into Equation (9.31), we have

$$\ln\left(\frac{T_{h,\text{in}} - \frac{C_c}{C_h}\left(T_{c,\text{out}} - T_{c,\text{in}}\right) - T_{c,\text{out}}}{T_{h,\text{in}} - T_{c,\text{in}}}\right) = -\frac{UA}{C_c}\left(1 + \frac{C_c}{C_h}\right)$$

Now, adding and subtracting $T_{c,\text{in}}$ in the numerator of left-hand side, we get

$$\ln\left(\frac{T_{h,\text{in}} + T_{c,\text{in}} - T_{c,\text{in}} - \frac{C_c}{C_h}\left(T_{c,\text{out}} - T_{c,\text{in}}\right) - T_{c,\text{out}}}{T_{h,\text{in}} - T_{c,\text{in}}}\right) = -\frac{UA}{C_c}\left(1 + \frac{C_c}{C_h}\right)$$

$$\ln\left(\frac{\left(T_{h,\text{in}} - T_{c,\text{in}}\right) - \left(T_{c,\text{out}} - T_{c,\text{in}}\right) - \frac{C_c}{C_h}\left(T_{c,\text{out}} - T_{c,\text{in}}\right)}{T_{h,\text{in}} - T_{c,\text{in}}}\right) = -\frac{UA}{C_c}\left(1 + \frac{C_c}{C_h}\right)$$

This simplifies to

$$\ln\left(1 - \left(1 + \frac{C_c}{C_h}\right)\frac{T_{c,\text{out}} - T_{c,\text{in}}}{T_{h,\text{in}} - T_{h,\text{out}}}\right) = -\frac{UA}{C_c}\left(1 + \frac{C_c}{C_h}\right) \tag{9.33}$$

By definition of effectiveness, we have

$$\varepsilon = \frac{\dot{Q}}{\dot{Q}_{\text{max}}}$$

Substituting for $\dot{Q}$ and $\dot{Q}_{\text{max}}$ from Equations (9.27) and (9.29), we get

$$\varepsilon = \frac{C_c\left(T_{c,\text{out}} - T_{c,\text{in}}\right)}{C_{\text{min}}\left(T_{h,\text{in}} - T_{c,in}\right)}$$

Thus,

$$\frac{T_{c,\text{out}} - T_{c,\text{in}}}{T_{h,\text{in}} - T_{c,\text{in}}} = \varepsilon \frac{C_{\min}}{C_c}$$

Substituting this into Equation (9.33) and solving for $\varepsilon$, we get the effectiveness of the parallel-flow heat exchanger as

$$\varepsilon_{\text{parallel-flow}} = \frac{1 - \exp\left[-\dfrac{UA}{C_c}\left(1 + \dfrac{C_c}{C_h}\right)\right]}{\left(1 + \dfrac{C_c}{C_h}\right)\dfrac{C_{\min}}{C_c}} \tag{9.34}$$

Taking the smaller of $C_h$ and $C_c$ as $C_{\min}$, Equation (9.34) simplifies to

$$\boxed{\varepsilon_{\text{parallel-flow}} = \frac{1 - \exp\left[-\dfrac{UA}{C_{\min}}\left(1 + \dfrac{C_{\min}}{C_{\max}}\right)\right]}{\left(1 + \dfrac{C_{\min}}{C_{\max}}\right)}} \tag{9.35}$$

The dimensionless group $UA/C_{\min}$ in the effectiveness relation Equation (9.35) is called the *number of transfer units NTU*, and is expressed as

$$\boxed{NTU = \frac{UA}{C_{\min}} = \frac{UA}{(\dot{m}c_p)_{\min}}} \tag{9.36}$$

where $U$ is the overall heat transfer coefficient and $A$ is the heat transfer surface area of the heat exchanger. It is seen in this relation that $NTU$ is proportional to $A$. Therefore, for specified values of $U$ and $C_{\min}$, the value of $NTU$ is a measure of the heat transfer surface area $A$. Thus, larger the $NTU$, the larger would be the heat exchanger area.

It is convenient to define another dimensionless quantity termed *capacity ratio* $C$ as

$$C = \frac{C_{\min}}{C_{\max}} \tag{9.37}$$

It can be shown that, the effectiveness of a heat exchanger is a function of the $NTU$ and $C$. That is,

$$\varepsilon = f(UA/C_{\min}, C_{\min}/C_{\max}) = f(NTU, C)$$

Effectiveness relations have been developed for a large number of heat exchangers by Kays and London [2]. Some important relations among them are given in Table 9.3. The effectiveness of some common types of heat exchangers are plotted in Figures 9.13(a)–9.13(f). In Figure 9.13(f), the solid curves correspond to $C_{\min}$ mixed and $C_{\max}$ unmixed and the dashed curves correspond to $C_{\min}$ unmixed and $C_{\max}$ mixed. From Table 9.3 and plots 9.13(a)–(f), it can be inferred that,

**Table 9.3** Effectiveness relations for heat exchangers [2]

| Heat exchanger type | Effectiveness relation |
|---|---|
| *Double pipe*: | |
| Parallel-flow | $\varepsilon = \dfrac{1 - \exp\left[-NTU(1+C)\right]}{1+C}$ |
| Counter-flow | $\varepsilon = \dfrac{1 - \exp\left[-NTU(1-C)\right]}{1 - C\exp\left[-NTU(1-C)\right]}$ |
| *Shell and tube*: One-shell pass | |
| (2,4,....tube passes) | $\varepsilon = 2\left\{1 + C + \sqrt{1+C^2}\,\dfrac{1+\exp\left[-NTU\sqrt{1+C^2}\right]}{1-\exp\left[-NTU\sqrt{1+C^2}\right]}\right\}^{-1}$ |
| *Cross-flow (single pass)*: | |
| Both fluids unmixed | $\varepsilon = 1 - \exp\left(\frac{NTU^{0.22}}{C}\left[\exp\left(-C \times NTU^{0.78}\right) - 1\right]\right)$ |
| $C_{\max}$ mixed $C_{\min}$ unmixed | $\varepsilon = \frac{1}{C}\left(1 - \exp\left(1 - C\left[1 - \exp(-NTU)\right]\right)\right)$ |
| $C_{\min}$ mixed $C_{\max}$ unmixed | $\varepsilon = 1 - \exp\left(-\frac{1}{C}\left[1 - \exp\left(-C \times NTU\right)\right]\right)$ |
| *All heat exchangers* with $C = 0$ | $\varepsilon = 1 - \exp\left(-NTU\right)$ |

- For $C = 0$, all heat exchangers have the same effectiveness $\varepsilon$.
- For $NTU \leq 0.25$, all heat exchangers have approximately the same $\varepsilon$.
- For $C > 0$ and $NTU \geq 0.25$, the counter-flow heat exchanger is the most effective.
- For $\varepsilon < 40\%$, the capacity ratio $C$ does not have any significant effect on the effectiveness.

From the effectiveness relations charts (Figures 9.13(a) to 9.13(f)) for heat exchangers, we can infer the following:

• The value of effectiveness ranges from 0 to 1. It increases rapidly with $NTU$ up to about $NTU = 1.5$ and slowly for lower values of $NTU$. This implies that, large size heat exchangers with $NTU$ larger than 3 will not be economical.

• For a given capacity ratio $C$ and $NTU$, the counter-flow heat exchanger has the highest effectiveness followed by the cross-flow heat exchangers with fluids unmixed. As we can expect, the lowest effectiveness values are encountered in parallel-flow heat exchangers. The effectiveness of counter-flow, cross-flow and parallel-flow heat exchangers for a given $NTU$ and $C$ are compared in Figure 9.14.
• The effectiveness of a heat exchanger is independent of the capacity ratio $C$, for $NTU$ less than 0.3.

• The capacity ratio $C$ ranges from 0 to 1. For a given $NTU$, the effectiveness $\varepsilon$ becomes a maximum for $C = 0$ and a minimum for $C = 1$. When $C \to 0$, $C_{\text{max}} \to \infty$. This corresponds to a phase-change process in a condenser or boiler. All effectiveness relations in this case reduce to

$$\varepsilon = \varepsilon_{\text{max}} = 1 - \exp(-NTU) \tag{9.38}$$

The variation of effectiveness with $NTU$ is illustrated in Figure 9.15.
For the case of $C = 0$, the temperatures of the condensing or boiling fluids remain constant. The effectiveness is the lowest in the other limiting case of $C = 1$, which is experienced when $C$ of the two (cold and hot) fluids are equal.

### 9.7.1 Use of $\epsilon$-$NTU$ Relations for Rating and Sizing of Heat Exchangers

The $\epsilon$-$NTU$ relations can be used to address the rating and sizing requirements of heat exchangers. Let us consider the *rating problem* first.

Suppose $T_{c,\text{in}}$, $T_{h,\text{in}}$ are the inlet temperatures of the cold and hot fluids and $\dot{m}_c$ and $\dot{m}_h$ are the mass flow rates of cold and hot fluids, respectively. Also, let us assume that the overall heat transfer coefficient $U$, the total heat transfer area $A$, and the type and flow arrangement of the heat exchanger are known. Now the problem is, for the above given parameters, find the total

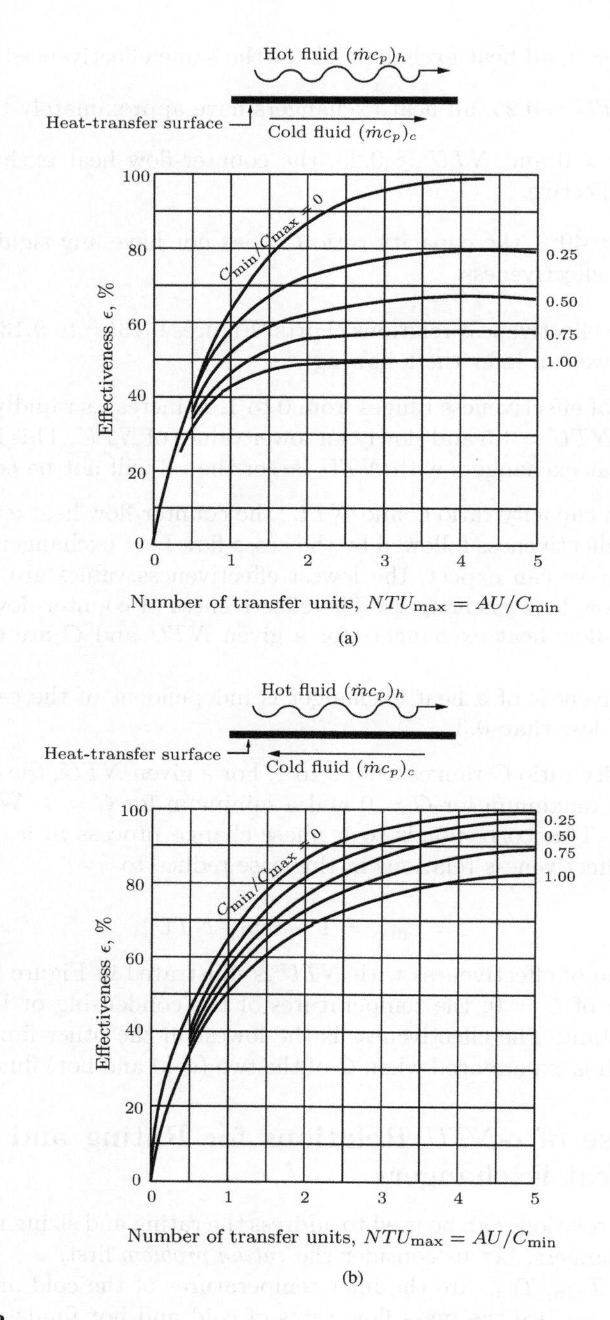


**Figure 9.13**
(a) Parallel-flow, (b) counter-flow, (c) one-shell pass and 2, 4, 6 tubes, (d) two-shell pass and 4, 8, 12 tube passes, (e) cross-flow with both fluids unmixed, (f) cross-flow with one fluid mixed and the other unmixed [2].

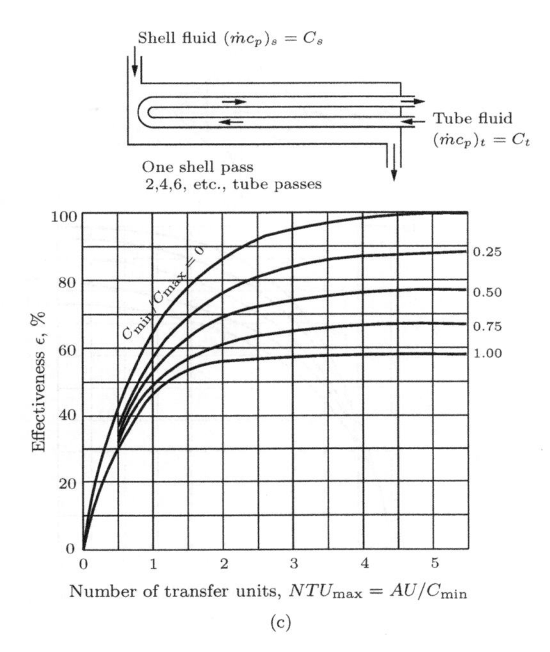


(c)

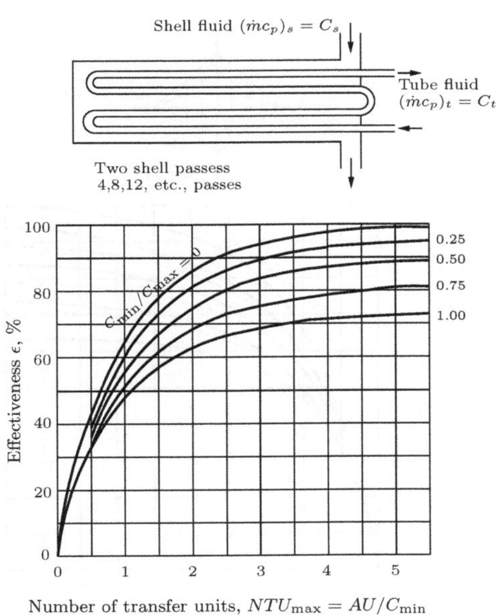


(d)

**Figure 9.13**
(Continued.)

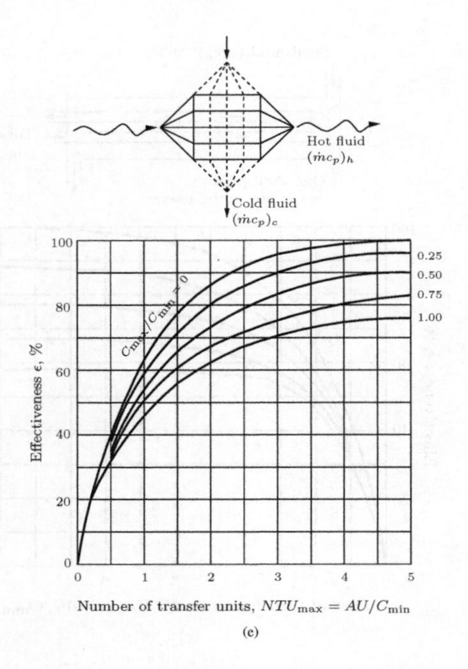

(e)

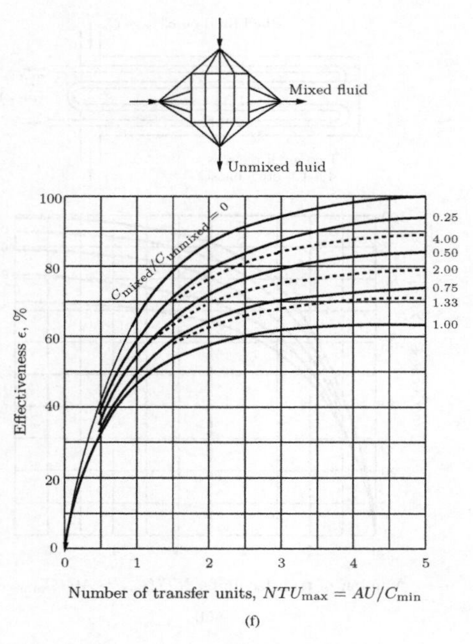

(f)

**Figure 9.13**
(Continued.)

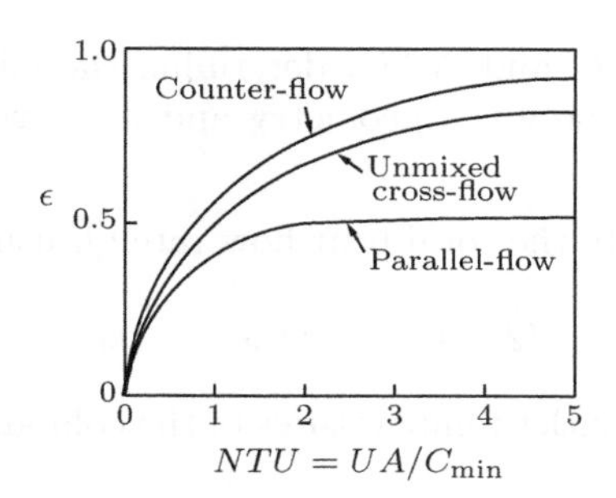


**Figure 9.14**
Effectiveness of counter-flow, cross-flow and parallel-flow heat exchangers.

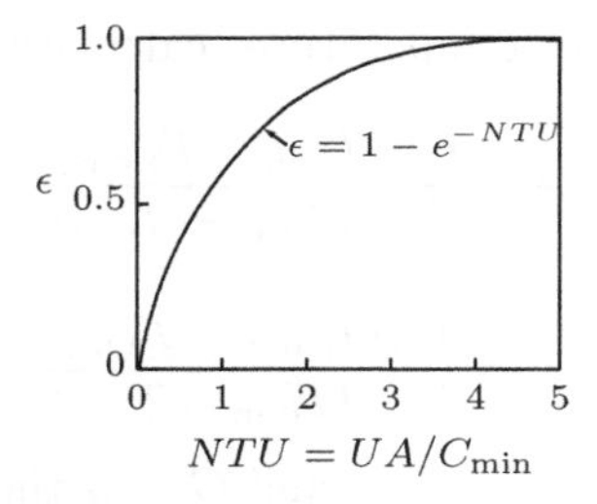


**Figure 9.15**
The effectiveness variation with $NTU$ when $C = 0$.

heat flow rate $\dot{Q}$ and the outlet temperatures $T_{h,\text{out}}$, $T_{c,\text{out}}$ of the hot and cold fluids to be determined. This problem can be solved as follows:

• Calculate the capacity ratio $C = C_{\min}/C_{\max}$ and $NTU = UA/C_{\min}$, for the given data.

• For these values of $C$ and $NTU$, determine the effectiveness $\epsilon$ from the chart or equation for the given geometry and flow arrangement of the heat exchanger.

• Using this $\epsilon$, calculate the total heat flow rate $\dot{Q}$, using the relation

$$\dot{Q} = \epsilon\, C_{\min}\,(T_{h,\text{in}} - T_{c,\text{in}})$$

Now calculate the outlet temperatures of the cold and hot fluids [Equation (9.270)], as

$$T_{h,\text{out}} = T_{h,\text{in}} - \frac{\dot{Q}}{C_h}$$

$$T_{c,\text{out}} = T_{c,\text{in}} + \frac{\dot{Q}}{C_c}$$

The above procedure clearly shows that the rating problem where the outlet temperatures are not given can be addressed with the $\epsilon$-*NTU* method. Note that a tedious iteration procedure would be required to solve this problem with LMTD method.

For addressing the *sizing problem*, let $T_{c,\text{in}}$, $T_{c,\text{out}}$, $T_{h,\text{in}}$, $T_{h,\text{out}}$, $\dot{m}_c$, $\dot{m}_h$, $U$, and the type and flow arrangement of the heat exchanger are given and the total heat transfer area $A$ has to be determined. This can be done as follows:

• Using the inlet and outlet temperatures calculate $\epsilon$, using the relation

$$\epsilon = \frac{C_h\,(T_{h,\text{in}} - T_{h,\text{out}})}{C_{\min}\,(T_{h,\text{in}} - T_{c,\text{in}})}$$

or

$$\epsilon = \frac{C_c\,(T_{c,\text{out}} - T_{c,\text{in}})}{C_{\min}\,(T_{h,\text{in}} - T_{c,\text{in}})}$$

where $C_h = \dot{m}_h c_{p_h}$ and $C_c = \dot{m}_c c_{p_c}$, and $C_{\min}$ is the smaller of $C_h$ and $C_c$.

• Find $C = C_{\min}/C_{\max}$.

• For these values of $\epsilon$ and $C$, determine $NTU$ from the appropriate $\epsilon$-*NTU* chart.

• For this $NTU$ calculate the heat transfer surface $A$, from Equation (9.36),

$$A = \frac{NTU\, C_{\min}}{UA}$$

For designing compact heat exchangers for automotive, aircraft, air conditioners, and other industrial applications, where the inlet temperatures of the hot and cold fluids are specified and the heat transfer rates are to be determined, the $\epsilon$-$NTU$ method is usually preferred. For designing large heat exchangers for process, power, and petrochemical industries, where both inlet and outlet temperatures of the hot and cold fluids are specified, the LMTD method is generally used.

## 9.8 Compact Heat Exchangers

Compact heat exchangers are devices with very large surface area per unit volume. The surface area of compact heat exchangers do not fall into the categories discussed so far. Compact heat exchangers are used for applications where gas flow and low values of convection heat transfer coefficient $h$ are encountered. Some typical configurations of compact heat exchangers are shown in Figure 9.16.

A finned-tube heat exchanger is illustrated in Figure 9.16(a) and a circular finned-tube array in a different configuration is shown in Figure 9.16(b). Ways to achieve very high surface areas on both sides of the exchanger is shown in Figures 9.16(c) and 9.16(d). The configurations in 9.16(c) and 9.16(d) are used when gas-to-gas heat transfer is involved.

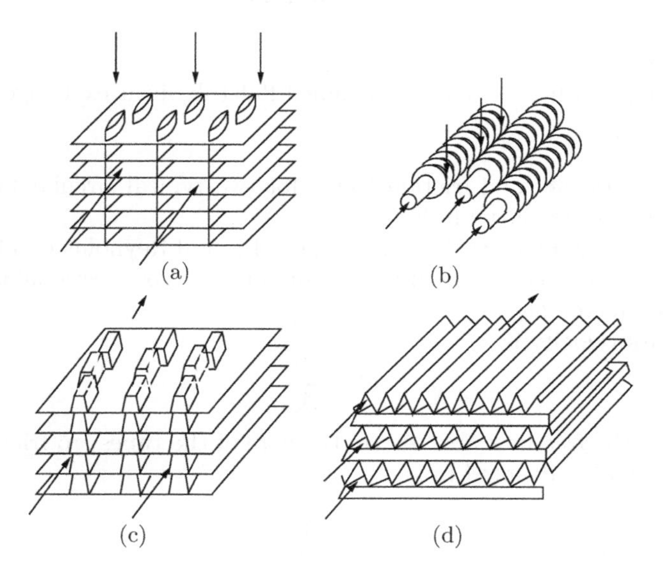


**Figure 9.16**
Some compact heat exchanger configurations.

The heat transfer and friction factor variation with Reynolds number, for two typical flat-tube heat exchangers, are shown in Figure 9.17. Note: A tube with its width very large compared to the height, as shown in Figure 9.17, is termed flat-tube.

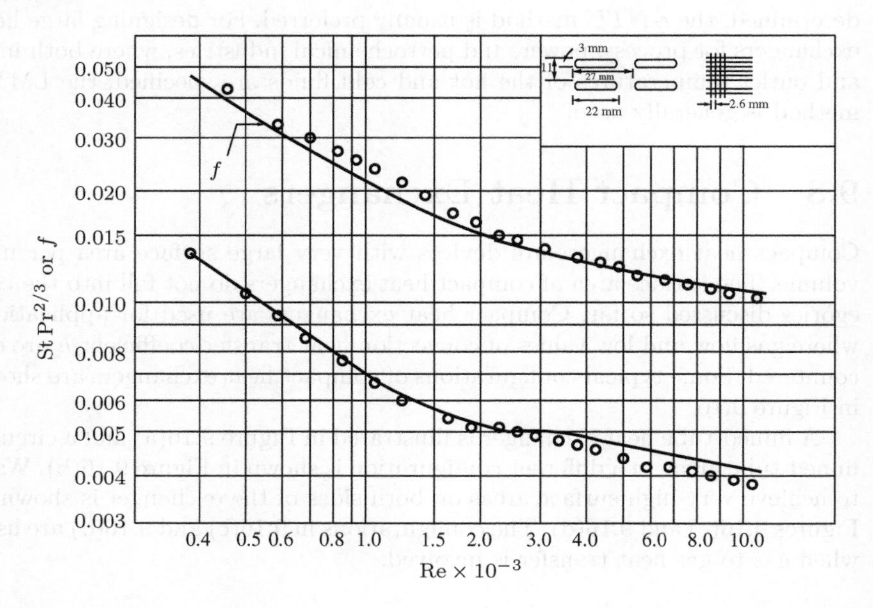


**Figure 9.17**
Heat transfer and friction factor for finned flat-tube heat exchanger.

The heat transfer and friction factor for two typical circular finned-tube heat exchangers are shown in Figure 9.18.

The Stanton number St, Prandtl number Pr, and Reynolds number Re are based on the mass velocity $G$ in the minimum flow cross-sectional area and a hydraulic diameter $D_h$.

The mass velocity is

$$G = \frac{\dot{m}}{A_c} \tag{9.39}$$

The ratio of the free-flow area $A_c$ (in the core of the tubes) to frictional area $A$ (surface area) is denoted by $\sigma$

$$\sigma = \frac{A_c}{A}$$

Thus, the Stanton number and Reynolds number in terms of $G$ become

$$\text{St} = \frac{h}{G\,c_p}$$

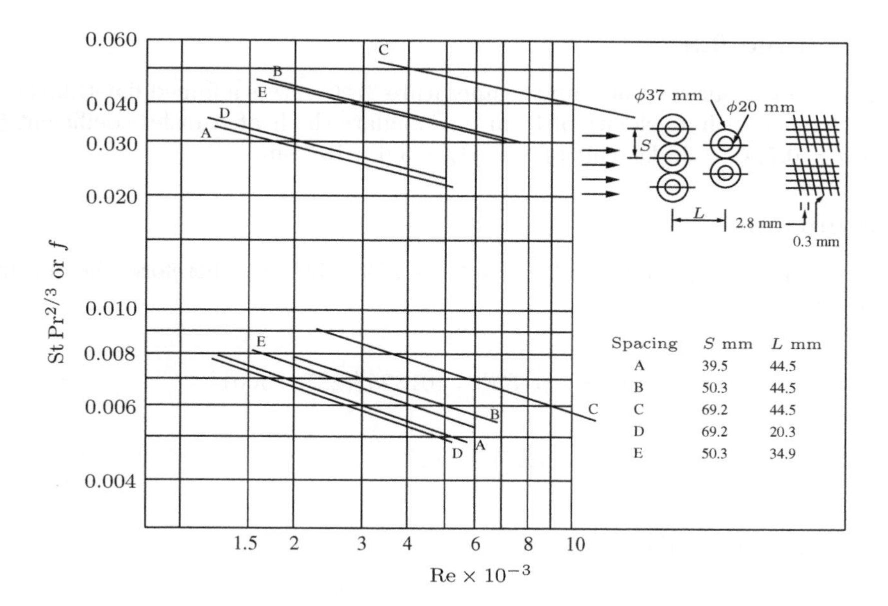


**Figure 9.18**
Heat transfer and friction factor for finned circular-tube heat exchanger.

$$\mathrm{Re} = \frac{D_h G}{\mu}$$

Fluid properties in these expressions of Stanton and Reynolds numbers are evaluated at the average bulk temperature. Heat transfer and fluid friction inside the tubes are evaluated with the hydraulic diameter, defined as

$$D_h = \frac{4A}{P}$$

where $A$ is the cross-sectional area of the flow and $P$ is the wetted perimeter.

The pressure drop $\Delta p$ for flow across fined-tube banks may be calculated using the relation

$$\Delta p = \frac{v_1 G^2}{2}\left[\left(1+\sigma^2\right)\left(\frac{v_2}{v_1}-1\right) + f\frac{A}{A_c}\frac{v_m}{v_1}\right] \tag{9.40}$$

where $v_1$ and $v_2$ are the specific volumes at the inlet and exit, respectively, $v_m$ is the mean specific volume in the heat exchanger, defined as $v_m = (v_1+v_2)/2$ and $f$ is the friction factor.

Note: Design of compact heat exchanger is an involved task. For an elaborate detail about the design aspects, one may refer to Reference 2.

## Example 9.3

Air at a pressure of 1 atm and temperature 27°C enters a finned flat-tube heat exchanger with a velocity of 15 m/s. Calculate the heat transfer coefficient $h$, if $\sigma = 0.7$, $G = 38$ kg/(m$^2$ s) and $D_h = 3.6 \times 10^{-3}$ m.

## Solution

Given, $p = 1$ atm and $T = 27°\text{C} = 27 + 273 = 300$ K. Therefore, the density becomes

$$\begin{aligned} \rho &= p/RT = 101325/(287 \times 300) \\ &= 1.177\,\text{kg/m}^2 \end{aligned}$$

The viscosity at $T = 300$ K is

$$\begin{aligned} \mu &= 1.46 \times 10^{-6}\,\frac{300^{3/2}}{300 + 111} \\ &= 1.85 \times 10^{-5}\,\text{kg/(m s)} \end{aligned}$$

For air at 300 K, from Table A-4, $c_p = 1.005$ kJ/(kg K) and Pr = 0.7.

The Reynolds number becomes

$$\begin{aligned} \text{Re} &= \frac{D_h G}{\mu} \\ &= \frac{3.6 \times 10^{-3} \times 38}{1.85 \times 10^{-5}} \\ &= 7395 \end{aligned}$$

For Re = 7395, from Figure 9.17, we have

$$\text{St}\,\text{Pr}^{2/3} \approx 0.004$$

That is,

$$\frac{h}{G\,c_p}\,\text{Pr}^{2/3} = 0.004$$

Thus,

$$\begin{aligned} h &= (0.004)(38)(1005)(0.7)^{-2/3} \\ &= \boxed{193.77\,\text{W/(m}^2\,°\text{C)}} \end{aligned}$$

## 9.9 Analysis for Variable Properties

The convection heat transfer coefficient to be used in the analysis of heat exchanger performance depends on the fluid being considered. Therefore, the overall heat transfer coefficient for a heat exchanger may vary significantly through the heat exchanger, if the fluids used are such that their properties are influenced by temperature. In such a situation the best way is analyzing the flow process through the heat exchanger by a numerical method.

As an example, let us analyze the flow through a parallel-flow double-pipe heat exchanger, using a finite difference method. Let us divide the heat exchanger surface area over which heat transfer takes place into $n$ incremental surface areas of area $\Delta A_j$ each, where $j = 1, n$. For surface area $\Delta A_j$, let the temperatures of the hot and cold fluids be $T_{hj}$ and $T_{cj}$, respectively. Let the overall heat transfer coefficient be expressed as

$$U_j = U_j\left(T_{hj} - T_{cj}\right)$$

The incremental heat transfer in $\Delta A_j$ is

$$\Delta\dot{q}_j = -\left(\dot{m}_h c_h\right)_j\left(T_{hj+1} - T_{hj}\right) = \left(\dot{m}_c c_c\right)_j\left(T_{cj+1} - T_{cj}\right) \tag{9.40}$$

Also,

$$\Delta\dot{q}_j = U_j \Delta A_j\left(T_h - T_c\right)_j \tag{9.41}$$

The finite-difference equation analogous to Equation (9.19) is

$$\frac{\left(T_h - T_c\right)_{j+1} - \left(T_h - T_c\right)_j}{\left(T_h - T_c\right)_j} = -U_j\left(\frac{1}{\left(m_h c_h\right)_j} + \frac{1}{\left(m_c c_c\right)_j}\right)\Delta A_j$$

or

$$\frac{\left(T_h - T_c\right)_{j+1} - \left(T_h - T_c\right)_j}{\left(T_h - T_c\right)_j} = -k_j\left(T_h, T_c\right)\Delta A_j \tag{9.42}$$

where

$$k_j\left(T_h, T_c\right) = U_j\left(\frac{1}{\left(m_h c_h\right)_j} + \frac{1}{\left(m_c c_c\right)_j}\right)$$

Simplifying Equation (9.42), we get

$$\frac{\left(T_h - T_c\right)_{j+1}}{\left(T_h - T_c\right)_j} = 1 - k_j \Delta A_j \tag{9.43}$$

When the inlet temperatures are known, the numerical analysis can be carried out through the following steps:

- Choose a convenient value of $\Delta A_j$.
- Calculate the overall heat transfer coefficient $U$ for the inlet conditions and through the initial increment area $\Delta A$.

- Calculate the value of $\dot{q}$ for this increment, using Equation (9.41).
- Calculate $T_h$, $T_c$ and $(T_h - T_c)$ for the next increment of area, using Equations (9.40) and (9.43).
- Repeat the above steps until all the increments in $\Delta A$ are covered.

Now, the total heat-transfer rate becomes

$$\dot{q}_{\text{total}} = \sum_{j=1}^{n} \Delta \dot{q}_j$$

where $n$ is the number of increments in $\Delta A$.

From the above discussions it is evident that the analysis of heat exchangers with unknown outlet temperatures is simpler with the effective-$NTU$ method, but requires tedious interactive procedure with LMTD method.

Now, we know that when all the inlet and outlet temperatures are specified, the size of the heat exchanger can easily be determined using the LMTD method. The size can also be determined using effective-$NTU$ method, by evaluating the effectiveness $\varepsilon$, and then the $NTU$ from the approximate $NTU$ relations listed in Table 9.4. It is essential to note that the relations in Table 9.4 are equivalent to those in Table 9.3. The relations in Table 9.3 give the effectiveness directly when $NTU$ is known, and the relations in Table 9.4 give the $NTU$ directly when the effectiveness $\varepsilon$ is known.

## 9.10 Boilers and Condensers

We saw that, in a boiling or condensation process, the fluid temperature essentially remains constant, or the fluid acts as if it had infinite specific heat. In these cases, $C_{\min}/C_{\max} \to 0$ and the heat exchanger effectiveness relation simplifies to

$$\varepsilon = 1 - e^{-NTU}$$

This relation is listed as the last item in Tables 9.3 and 9.4.

## 9.11 Selection of Heat Exchangers

Selection of a heat exchanger for the process and power industries is made on the basis of cost and specifications furnished by the manufacturer. This is because for these applications heat exchangers are purchased as off-the-shelf items. But for specified applications in high-tech fields, such as aerospace and electronic industries, a particular design to precisely meet the heat transfer requirements is called for. Whether the heat exchanger is selected either as an off-the-shelf item or specifically designed for a specified application, the following factors are usually considered during the selection:

**Table 9.4** *NTU* relations for heat exchangers [2]

| Heat exchanger type | *NTU* relation |
|---|---|
| *Double pipe*: | |
| Parallel-flow | $NTU = -\frac{\ln[1-\varepsilon(1+C)]}{1+C}$ |
| Counter-flow | $NTU = \frac{1}{C-1}\ln\left(\frac{\varepsilon-1}{\varepsilon C-1}\right)$ |
| *Shell and tube*: | |
| One-shell pass (2,4,....tube passes) | $NTU = -\frac{1}{\sqrt{1+C^2}}\ln\left(\frac{2/\varepsilon-1-C-\sqrt{1+C^2}}{2/\varepsilon-1-C+\sqrt{1+C^2}}\right)$ |
| *Cross-flow (single pass)* | |
| Both fluids unmixed | $NTU = -\ln\left(1+\frac{\ln(1-\varepsilon C)}{C}\right)$ |
| $C_{\text{max}}$ mixed $C_{\text{min}}$ unmixed | $NTU = -\ln\left(1+\frac{\ln(1-\varepsilon C}{C}\right)$ |
| $C_{\text{min}}$ mixed $C_{\text{max}}$ unmixed | $NTU = -\frac{\ln[C\ln(1-\varepsilon)+1]}{C}$ |
| *All heat exchangers* | $NTU = -\ln(1-\varepsilon)$ |

- Heat transfer rate
- Cost
- Size and weight
- Pumping power
- Materials

### 9.11.1 Heat Transfer Rate

The heat transfer rate is the most important parameter in the selection of a heat exchanger. A heat exchanger should have the capability to transfer heat at the specified rate to achieve the required temperature change of the fluid, at the specified mass flow rate.

### 9.11.2 Cost

An off-the-shelf heat exchanger usually has cost advantage over those made specially for an application. However, in situations such as when the heat exchanger is an integral part of the device to be manufactured, it often becomes essential to design and make a heat exchanger to suit the specific needs. The operation and maintenance costs of the heat exchanger are also important factors in assessing the overall cost.

### 9.11.3 Size and Weight

For many practical applications, small and light heat exchangers are the requirement. For example, in aerospace and automotive industries, the size and weight requirements are very stringent. In some cases the space available for the heat exchanger limits the length of the tube that can be used.

### 9.11.4 Pumping Power

In a heat exchanger, the cold and hot fluids are usually forced to flow by pumps or fans operated by electric power. Therefore, minimizing the pressure drop and the mass flow rate of the fluids will minimize the power requirement, leading to the reduction of operating cost of the heat exchanger. But this will maximize the size of the heat exchanger and thus the initial cost. For instance, doubling the mass flow rate will reduce the initial cost by, say, about half but will increase the pumping power required by a factor of about eight. Fluid velocities encountered in heat exchangers range between 0.7 to 7 m/s for liquids, and between 3 to 30 m/s for gases.

### 9.11.5 Materials

The material used for making the heat exchanger is also an important consideration in the selection of heat exchangers. For example, when the pressures and temperatures are of the order of 10 atm and 100°C, the thermal and structural stress effects are of important consideration and limit the acceptable materials for heat exchanger construction.

## 9.12 Summary

Heat exchangers are devices that facilitate exchange of heat between two fluids that are at different temperatures, without making them mix with each other. Heat transfer in a heat exchanger usually involves convection in each fluid and conduction through the wall separating the two fluids.

In general, the classification is based on the following:

- The flow arrangement
- The transfer process
- Compactness
- Construction type
- Heat transfer mechanism

When both the fluids flow in the same direction, it is termed *parallel flow* heat exchanger, when the fluids flow in parallel but in opposite directions, it is termed *counter flow* heat exchanger.

A type of heat exchanger which is designed to have a large heat transfer surface area per unit volume is the *compact* heat exchanger. The large surface area in a compact heat exchanger is obtained by attaching closely spaced *thin plates* or *corrugated fins* to the walls separating the two fluids. In compact heat exchangers, hot and cold fluids usually move perpendicular to each other, and such a configuration is called *cross-flow.* The cross-flow is further classified in to *mixed-flow* and *unmixed-flow*, depending on the flow configuration.

One of the most common types of heat exchanger for industrial applications is the *shell-and-tube* heat exchanger. It has a large number of tubes packed in a shell with their axes parallel to the shell axis.

Based on the number of shell and tube passes involved, shell-and-tube heat exchangers are further classified into *one-shell pass and two-tube pass* and *two-shell pass and four-tube pass*, and so on.

Another kind of heat exchanger which is widely used is the *plate and frame heat exchanger*, or simply called a plate heat exchanger. This consists of a series of plates with corrugated flat passages. The hot and cold flows are taking place in alternate passages. Thus each cold stream is surrounded by two hot streams. This arrangement results in an effective heat transfer.

Yet another kind of heat exchanger in which hot and cold fluids pass through alternate passages of identical flow area is the *regenerative* heat exchanger. The regenerative heat exchanger is further classified into *static-type* and *dynamic-type* regenerators.

Usually heat exchangers are given names to reflect the specific application for which they are used. For example, a *boiler* is a heat exchanger in which one of the fluids absorbs heat and vaporizes. A *condenser* is a heat exchanger in which one of the fluids gives up heat and condenses as it flows through the heat exchanger.

Usually a heat exchanger involves two fluids separated by a solid wall. Heat from the hot fluid is transferred by convection to the wall, the wall conducts the heat received by it to the cold fluid again by convection. Radiation effect, if any, is usually included in the convection heat transfer coefficient.

Over a period of operation, the heat-transferring surface of a heat exchanger may get coated with various deposits present in the flowing fluids. Also, the surface may get corroded due to the interactions of the fluids. The coating results in an additional resistance to the heat transfer and causes the rate of heat transfer to decrease. The effect of these accumulations on heat transfer is represented by a *fouling factor* $R_f$. Thus, fouling factor or fouling resistance is a measure of the thermal resistance introduced by fouling.

The most common type of fouling is the precipitation of solid deposits in a fluid on the heat transferring surface.

Fouling factor is usually obtained experimentally by determining the values of the overall heat transfer coefficient $U$ for both clean and dirty conditions in the heat exchanger. Thus,

$$R_f = \frac{1}{U_{\text{dirty}}} - \frac{1}{U_{\text{clean}}}$$

Selection of a heat exchanger that will achieve the specified temperature change of a fluid stream of known mass flow rate, or will result in the required outlet temperature of the hot and cold fluid streams, is an important step in any heat exchanger application. Usually the following two methods are used in the analysis of heat exchanger performance.

1. The *log mean temperature difference* (LMTD) method – this is well suited to identify a heat exchanger that will achieve the specified temperature change in a fluid stream of given mass flow rate.

2. The *effective NTU* (number of (heat) transfer unit) method – this is suitable to predict the outlet temperatures of the cold and hot fluid streams in a specified heat exchanger.

Two popular heat exchangers commonly used in practice are the *condensers* and *boilers*. In a condenser or a boiler, one of the fluids undergoes a phase-change process, and the rate of heat transfer is expressed as

$$\dot{Q} = \dot{m}\, h_{fg}$$

where $\dot{m}$ is the rate of evaporation or condensation of the fluid and $h_{fg}$ is the enthalpy of vaporization of the fluid at the specified temperature or pressure.

In an analogous manner to Newton's law of cooling, the rate of heat transfer in a heat exchanger can be expressed as

$$\dot{Q} = UA\Delta T_m$$

where $U$ is the overall heat transfer coefficient, $A$ is the heat transfer area, and $\Delta T_m$ is an appropriate average temperature difference between the two fluids. Here it is important to note that the surface area $A$ being a geometrical parameter can be precisely determined, whereas $U$ and $\Delta T_m$, in general are not constant and vary along the heat exchanger.

The temperature difference between the hot and cold fluids varies along the heat exchanger. This situation can be handled if the *mean temperature difference* $\Delta T_m$ is used in the relation

$$\dot{Q} = UA\Delta T_m$$

where

$$\Delta T_{\text{lm}} = \frac{\Delta T_1 - \Delta T_2}{\ln(\Delta T_1/\Delta T_2)}$$

is the log mean temperature difference and $\Delta T_1$ and $\Delta T_2$ refer to the temperature difference between the two fluids at the two ends (inlet and outlet) of the heat exchanger.

In a counter-flow heat exchanger, the hot and cold fluids flow in opposite directions. The hot and cold fluids enter the heat exchanger from opposite ends. The outlet temperature of the cold fluid in this device may exceed the outlet temperature of the hot fluid. In the limiting case, the cold fluid will be heated to the level of the inlet temperature of the hot fluid. But in reality this limiting situation is impossible since this would violate increase of entropy principle.

The log mean temperature difference (LMTD) approach to heat exchanger analysis is useful when the inlet and outlet temperatures of the hot and cold fluids are known or can easily be determined. Once $\Delta T_{\text{lm}}$, which is the difference between the temperatures of the hot and cold fluids, the mass flow rates of the hot and cold fluids, and the overall heat transfer coefficient $U$ are available, the heat transfer surface area $A$ of the heat exchanger can be determined from

$$\dot{Q} = UA\,\Delta T_{\text{lm}}$$

The LMTD method is apt for determining the size of a heat exchanger to realize the prescribed outlet temperatures when the mass flow rates and the inlet and outlet temperatures of the hot and cold fluids are specified.

The effective-$NTU$ method is based on a dimensionless parameter called the *heat transfer effectiveness* $\varepsilon$, defined as

$$\boxed{\varepsilon = \frac{\dot{Q}}{\dot{Q}_{\max}} = \frac{\text{Actual heat transfer rate}}{\text{Maximum possible heat transfer rate}}}$$

The heat transfer in a heat exchanger will be maximum when the cold fluid is heated to the inlet temperature of the hot fluid and the hot fluid is cooled to the inlet temperature of the cold fluid. These two limiting conditions will be reached simultaneously only when the heat capacity rates of the hot and cold fluids are identically the same, that is, when $C_c = C_h$. When $C_c \neq C_h$, which is the case usually, the fluid with lower heat capacity rate will experience a larger temperature change, and thus it will be the first to experience the maximum temperature, at that point the heat transfer process will come to a halt. Therefore, the maximum possible heat transfer rate in a heat exchanger is

$$\dot{Q}_{\max} = C_{\min}\left(T_{h,\text{in}} - T_{c,\text{in}}\right)$$

where $C_{\min} = \left(\dot{m}c_p\right)_{\min}$ is the smaller of $\dot{m}_h c_{ph}$ and $\dot{m}_c c_{pc}$.

The $\epsilon$-$NTU$ relations can be used to address the rating and sizing requirements of heat exchangers.

For designing compact heat exchangers for automotive, aircraft, air conditioners, and other industrial applications, where the inlet temperatures of the hot and cold fluids are specified and the heat transfer rates are to be determined, the $\epsilon$-$NTU$ method is usually preferred. For designing large heat exchangers for process, power, and petrochemical industries, where both inlet and outlet temperatures of the hot and cold fluids are specified, the LMTD method is generally used.

Compact heat exchangers are devices with a very large surface area per unit volume. The surface area of compact heat exchangers does not fall into the categories discussed so far. Compact heat exchangers are used for applications where gas flow and low values of convection heat transfer coefficient $h$ are encountered.

The convection heat transfer coefficient to be used in the analysis of heat exchanger performance depends on the fluid being considered. Therefore, the overall heat transfer coefficient for a heat exchanger may vary significantly through the heat exchanger, if the fluids used are such that their properties are influenced by temperature. In such a situation the best way is analyzing the flow process through the heat exchanger by a numerical method.

Selection of a heat exchanger for the process and power industries is made on the basis of cost and specifications furnished by the manufacturer. This is because for these applications heat exchangers are purchased as off-the-shelf items. But for specified applications in high-tech fields, such as aerospace and electronic industries, a particular design to precisely meet the heat transfer requirements is called for. The following factors are usually considered during the selection.

- Heat transfer rate
- Cost
- Size and weight
- Pumping power
- Materials

## 9.13 Exercise Problems

9.1 A shell-and-tube counter-flow heat exchanger has to be designed to heat 4 kg/s of water from 20°C to 80°C, by passing it through 10 thin-walled tubes of diameter 20 mm running through the shell, with hot oil at 180°C. The convection coefficient of the oil is 400 W/(m$^2$ K). All the tubes carrying water make 10 passes through the shell. Determine (a) the flow rate of oil and (b) the tube length required for the water to reach 80°C, if the oil leaves the heat exchanger at 90°C. For the oil, $c_p$ = 2300 J/(kg K) and for water, $c_p$ = 4180 J/(kg K), $\mu = 550 \times 10^{-6}$ kg/(m s), $k$ = 0.640 W/(m K) and the Prandtl number Pr = 3.6. Assume $h_i$ to be 3000 W/(m$^2$ K).

[**Ans.** (a) 4.846 kg/s, (b) 63.30 m]

9.2 In a counter-flow heat exchanger, hot gas is used to heat water from 30°C to 116°C. The water flow rate is 1.3 kg/s, and the hot gas enters the heat exchanger at 295°C and leaves at 90°C. Assuming the specific heat of the hot gas as 1005 J/(kg K) and the overall heat transfer coefficient $U_h$, based on the gas-side surface area, as 100 W/(m$^2$ K), determine the gas-side surface area required, using NTU method.

[**Ans.** 50.30 m$^2$]

9.3 In a counter-flow heat exchanger, water enters at 35°C and at a rate of 1 kg/s. The hot gas, which heats the water, enters with temperature 250°C at a rate of 1.5 kg/s. The overall heat transfer coefficients of the gas and surface area are 110 W/m$^2$ and 43 m$^2$, respectively. Determine the rate of heat transfer taking place in the heat exchanger, and the outlet temperatures of the gas and water. Assume $c_p$ for water as 4195 J/(kg K) and $c_p$ for gas as 1105 J/(kg K).

[**Ans.** 291701.25 W, 56.5°C, 104.5°C]

9.4 A large shell-and-tube heat exchanger consisting of a single shell and 10,000 tubes, each containing 2 passes, is to be designed for use in a large steam power plant to condense steam to liquid water. The tubes to be used are thin-walled ones with 25 mm diameter. The steam condenses on the outer surface of the tubes with an associated convection coefficient of $h_o$ = 11000 W/(m$^2$ K). The heat transfer rate associated with the heat exchanger is 2000 MW and the overall heat transfer coefficient is 4500 W/(m$^2$ K). The cooling water flow rate through the tubes is 25000 kg/s. The water enters at 20°C and the steam condenses at 50°C. Determine the temperature of the cooling water at the outlet of the condenser, the convection coefficient $h_i$, based on the tube diameter, and the length of the tube for each pass.

[**Ans.** 39.1°C, 7616.15 W/(m$^2$ K), 15 m]

9.5 In a counter-flow heat exchanger, oil at 110°C enters at 2.2 kg/s and the cooling water at 22°C enters at 0.5 kg/s. For the oil, $c_p$ = 2000 J/(kg K) and for water, $c_p$ = 4170 J/(kg K). If the heat transfer area is 12 m$^2$ and the $NTU$ = 2.34, calculate the exit temperature of water, the overall heat transfer coefficient and the total heat transfer rate.

[**Ans.** 92.4°C, 406.6 W/(m$^2$ K), 146.78 kW]

9.6 A cross-flow heat exchanger, as in Figure 9.13(e), has to cool compressed air at 95°C entering at 0.5 kg/s to 30°C, with water. If the water at 20°C enters at 1 kg/s, determine the exit temperature of water, the effectiveness of the heat exchanger and the total heat transfer rate, when the overall heat transfer coefficient is 250 W/(m$^2$ °C) and the heat transfer area is 6 m$^2$. Take $c_p$ for water and air as 4180 J/(kg °C) and 1005 J/(kg °C), respectively.

[**Ans.** 27.8°C, 0.92, 32662.5 W]

9.7 A 2-shell pass and 8-tube pass heat exchanger, of arrangement as shown in Figure 9.13(d), has to cool compressed air from 80°C to 30°C, with water entering at 20°C. The mass flow rates of air and water are 0.3 kg/s and 1.2 kg/s, respectively, and their $c_p$ are 1005 J/(kg °C) and 4180 J/(kg °C), respectively. If the heat transfer area is 4 m$^2$, determine the overall heat transfer coefficient.

[**Ans.** 120.6 W/(m$^2$ °C]

9.8 In a steam condenser, cooling water at 20°C enters at the rate of 0.65 kg/s per tube and leaves at 40°C. The condenser has single-pass tubes of diameter 25 mm and the steam condenses at 55°C. If the overall heat transfer coefficient based on the outer surface of the tube is 3500 W/(m$^2$ °C), calculate the tube length and the heat transfer rate, taking the $c_p$ for water as 4180 J/(kg °C).

[**Ans.** 8.36 m, 54204.15 W]

# References

1. Bowman, R. A., Mueller, A. C. and Nagle, W. M., "Mean temperature difference in design," *Trans. ASME*, Vol. 62, pp. 283-294, 1940.

2. Kays, W. M. and London, A. L., *Compact Heat Exchangers*, 2nd ed., McGraw-Hill Book Co., New York, 1964.

# Appendix

**Table A.1** Properties of some metals at 20°C

| *Metal* | $\rho$ kg/m$^3$ | $c$ kJ/(kg °C) | $k$ W/(m °C) | $\alpha \times 10^5$ m$^2$/s |
|---|---|---|---|---|
| Aluminum (pure) | 2707 | 0.896 | 204 | 8.418 |
| Al-Cu (Duralumin), 94-96% Al, 3-5% Cu, trace Mg | 2787 | 0.883 | 164 | 6.676 |
| Copper (pure) | 8954 | 0.3831 | 386 | 11.234 |
| Bronze 75% Cu, 25% Sn | 8666 | 0.343 | 26 | 0.859 |
| Brass 70% Cu, 30% Zn | 8522 | 0.385 | 111 | 3.412 |
| Iron (pure) | 7897 | 0.452 | 73 | 2.304 |
| Wrought iron 5% C | 7849 | 0.46 | 59 | 1.626 |
| Carbon steel C $\approx$ 0.5% | 7833 | 0.465 | 54 | 1.474 |
| C $\approx$ 1.0% | 7801 | 0.473 | 43 | 1.172 |
| C $\approx$ 1.5% | 7753 | 0.486 | 36 | 0.970 |
| Chrome steel Cr, 0% | 7897 | 0.452 | 73 | 2.026 |
| Cr, 1% | 7865 | 0.46 | 61 | 1.665 |
| Cr, 5% | 7833 | 0.46 | 40 | 1.110 |
| Cr, 20% | 7689 | 0.46 | 22 | 0.635 |
| Invar 36% Ni | 8137 | 0.46 | 10.7 | 0.286 |
| Magnesium (pure) | 1746 | 1.013 | 171 | 9.708 |
| Molybdenum | 10220 | 0.251 | 123 | 4.790 |
| Nickel steel Ni $\approx$ 0% | 7897 | 0.452 | 73 | 2.026 |
| Ni $\approx$ 20% | 7933 | 0.46 | 19 | 0.526 |
| Ni $\approx$ 40% | 8169 | 0.46 | 10 | 0.279 |
| Ni $\approx$ 80% | 8618 | 0.46 | 35 | 0.872 |
| Silver (100% pure) | 10524 | 0.2340 | 419 | 17.004 |
| (99.9% pure) | 10525 | 0.2340 | 407 | 16.563 |
| Tin (pure) | 7304 | 0.2265 | 64 | 3.884 |
| Tungsten | 19350 | 0.1344 | 163 | 6.271 |
| Zinc (pure) | 7144 | 0.3843 | 112.2 | 4.106 |

**Table A.2** Properties of some nonmetals

| *Substance* | $T$ °C | $\rho$ kg/m$^3$ | $c$ kJ/(kg °C) | $k$ W/(m °C) | $\alpha \times 10^5$ m$^2$/s |
|---|---|---|---|---|---|
| Asbestos (loosely packed) | -45 | 0.149 | | | |
| | 0 | 0.154 | 470-570 | 0.816 | 3.3-4 |
| | 100 | 0.161 | | | |
| Asbestos-cement | 20 | 0.74 | | | |
| boards and sheets | 51 | 0.166 | | | |
| Asphalt | 20-55 | 0.74-0.76 | | | |
| Brick | | | | | |
| Building brick | 20 | 0.69 | 1600 | 0.84 | 5.2 |
| Carborundum brick | 600 | 18.5 | | | |
| | 1400 | 11.1 | | | |
| Chrome brick | 200 | 2.32 | 3000 | 0.84 | 9.2 |
| Fireclay brick, | | | | | |
| burnt at 1330°C | 500 | 1.04 | 2000 | 0.96 | 5.4 |
| | 800 | 1.07 | | | |
| | 1100 | 1.09 | | | |
| burnt at 1450°C | 500 | 1.28 | 2300 | 0.96 | 5.8 |
| | 800 | 1.37 | | | |
| | 1100 | 1.4 | | | |
| Cement (portland) | | 0.29 | 1500 | | |
| Concrete (cinder) | 23 | 0.76 | | | |
| Cork board ($\approx$ 40 kg/m$^3$) | 30 | 0.043 | 160 | | |
| Cork (re-granulated) | 32 | 0.045 | 45-120 | 1.88 | 2-5.3 |
| Felt, hair | 30 | 0.036 | 130-200 | | |
| Fiber, insulating board | 20 | 0.048 | 240 | | |
| Glass (window) | 20 | 0.78 (ave) | 2700 | 0.84 | 3.4 |
| Glass wool | 23 | 0.038 | 24 | 0.7 | 22.6 |

**Table A.2** Continued.

| *Substance* | $T$ °C | $\rho$ kg/m$^3$ | $c$ kJ/(kg °C) | $k$ W/(m °C) | $\alpha \times 10^5$ m$^2$/s |
|---|---|---|---|---|---|
| Insulex, dry | 32 | 0.064 | | | |
| Plaster (gypsum) | 20 | 0.48 | 1440 | 0.84 | 4.0 |
| Metal lath | 20 | 0.47 | | | |
| Wood lath | 20 | 0.28 | | | |
| Sawdust | 23 | 0.059 | | | |
| Silica aerogel | 32 | 0.024 | 140 | | |
| Wood (across the grain) | | | | | |
| Balsa | 30 | 0.055 | 140 | | |
| Cypress | 30 | 0.097 | 460 | | |
| Fir | 23 | 0.11 | 420 | 2.72 | 0.96 |
| Maple or oak | 30 | 0.0166 | 540 | 2.4 | 1.28 |
| Yellow pine | 23 | 0.147 | 640 | 2.8 | 0.82 |
| White pine | 30 | 0.112 | 430 | | |
| Wood shavings | 23 | 0.059 | | | |
| Wool | 30 | 0.036 | 130-200 | | |

**Table A.3** Properties of some saturated liquids

| $T$ °C | $\rho$ kg/m$^3$ | $c_p$ kJ/(kg °C) | $\nu \times 10^6$ m$^2$/s | $k$ W/(m °C) | $\alpha \times 10^7$ m$^2$/s | Pr | $\beta \times 10^3$ K$^{-1}$ |
|---|---|---|---|---|---|---|---|
| | | | Ammonia | | | | |
| −50 | 703.69 | 4.463 | 0.435 | 0.547 | 1.742 | 2.61 | |
| −40 | 691.68 | 4.467 | 0.406 | 0.547 | 1.775 | 2.28 | |
| −30 | 679.34 | 4.476 | 0.387 | 0.549 | 1.801 | 2.15 | |
| −20 | 666.69 | 4.509 | 0.381 | 0.547 | 1.819 | 2.09 | |
| −10 | 653.55 | 4.564 | 0.378 | 0.543 | 1.825 | 2.07 | |
| 0 | 640.10 | 4.635 | 0.373 | 0.540 | 1.819 | 2.05 | |
| 10 | 626.16 | 4.714 | 0.368 | 0.531 | 1.801 | 2.04 | |
| 20 | 611.75 | 4.798 | 0.359 | 0.521 | 1.775 | 2.02 | 2.45 |
| 30 | 596.37 | 4.890 | 0.349 | 0.507 | 1.742 | 2.01 | |
| 40 | 580.99 | 4.999 | 0.340 | 0.493 | 1.701 | 2.00 | |
| 50 | 564.33 | 5.116 | 0.330 | 0.476 | 1.654 | 1.99 | |
| | | | Carbon dioxide | | | | |
| −50 | 1156.34 | 1.84 | 0.119 | 0.855 | 0.4021 | 2.96 | |
| −40 | 117.77 | 1.88 | 0.118 | 0.1011 | 0.4810 | 2.46 | |
| −30 | 1076.76 | 1.97 | 0.117 | 0.1116 | 0.5272 | 2.22 | |
| −20 | 1032.39 | 2.05 | 0.115 | 0.1151 | 0.5445 | 2.12 | |
| −10 | 938.38 | 2.18 | 0.113 | 0.1099 | 0.5133 | 2.20 | |
| 0 | 926.99 | 2.47 | 0.108 | 0.1045 | 0.4578 | 2.38 | |
| 10 | 860.03 | 3.14 | 0.101 | 0.0971 | 0.3608 | 2.80 | |
| 20 | 772.57 | 5.00 | 0.091 | 0.0872 | 0.2219 | 4.10 | 14.00 |
| 30 | 597.81 | 36.4 | 0.080 | 0.0703 | 0.0279 | 28.7 | |

**Table A.3** Continued.

| $T$ °C | $\rho$ kg/m$^3$ | $c_p$ kJ/(kg °C) | $\nu \times 10^6$ m$^2$/s | $k$ W/(m °C) | $\alpha \times 10^7$ m$^2$/s | Pr | $\beta \times 10^3$ K$^{-1}$ |
|---|---|---|---|---|---|---|---|
| | | | Freon | | | | |
| −50 | 1546.75 | 0.8750 | 0.310 | 0.067 | 0.501 | 6.2 | 2.63 |
| −40 | 1518.71 | 0.8847 | 0.279 | 0.069 | 0.514 | 5.4 | |
| −30 | 1489.56 | 0.8956 | 0.253 | 0.069 | 0.526 | 4.8 | |
| −20 | 160.57 | 0.9073 | 0.235 | 0.071 | 0.539 | 4.4 | |
| −10 | 1429.49 | 0.9203 | 0.221 | 0.073 | 0.550 | 4.0 | |
| 0 | 1397.45 | 0.9345 | 0.214 | 0.073 | 0.557 | 3.8 | |
| 10 | 1364.30 | 0.9496 | 0.203 | 0.073 | 0.560 | 3.6 | |
| 20 | 1330.18 | 0.9659 | 0.198 | 0.073 | 0.560 | 3.5 | |
| 30 | 1295.10 | 0.9835 | 0.194 | 0.071 | 0.560 | 3.5 | |
| 40 | 1257.13 | 1.0019 | 0.191 | 0.069 | 0.555 | 3.5 | |
| 50 | 1215.96 | 1.0216 | 0.190 | 0.067 | 0.545 | 3.5 | |
| | | | Glycerin | | | | |
| 0 | 1276.03 | 2.261 | 0.00831 | 0.282 | 0.983 | 84.7 | |
| 10 | 1270.11 | 2.319 | 0.00300 | 0.284 | 0.965 | 31.0 | |
| 20 | 1264.02 | 2.386 | 0.00118 | 0.286 | 0.947 | 12.5 | 0.50 |
| 30 | 1258.09 | 2.445 | 0.00050 | 0.286 | 0.929 | 5.38 | |
| 40 | 1252.01 | 2.512 | 0.00022 | 0.286 | 0.914 | 2.45 | |
| 50 | 1244.96 | 0.2.583 | 0.00015 | 0.287 | 0.893 | 1.63 | |
| | | | Ethylene glycol | | | | |
| 0 | 1130.75 | 2.294 | 57.53 | 0.242 | 0.934 | 615 | |
| 20 | 1116.65 | 2.382 | 19.18 | 0.249 | 0.939 | 204 | 0.65 |
| 40 | 1103.43 | 2.474 | 8.69 | 0.256 | 0.939 | 93 | |
| 60 | 1087.66 | 2.562 | 4.75 | 0.260 | 0.932 | 51 | |
| 80 | 1077.56 | 2.650 | 2.98 | 0.261 | 0.921 | 32.4 | |
| 100 | 1058.50 | 0.2.742 | 2.03 | 0.263 | 0.908 | 22.4 | |

**Table A.3** Continued.

| $T$ °C | $\rho$ kg/m$^3$ | $c_p$ kJ/(kg °C) | $\nu$ m$^2$/s | $k$ W/(m °C) | $\alpha$ m$^2$/s | Pr | $\beta \times 10^3$ K$^{-1}$ |
|---|---|---|---|---|---|---|---|
| | | | Engine oil | | | | |
| 0 | 899.12 | 1.796 | 0.00428 | 0.147 | $0.911 \times 10^{-7}$ | 47100 | |
| 20 | 888.23 | 1.880 | 0.00090 | 0.145 | $0.872 \times 10^{-7}$ | 10400 | 0.70 |
| 40 | 876.05 | 1.964 | 0.00024 | 0.144 | $0.834 \times 10^{-7}$ | 2870 | |
| 60 | 864.04 | 2.047 | 0.0000839 | 0.140 | $0.800 \times 10^{-7}$ | 1050 | |
| 80 | 852.02 | 2.131 | 0.0000375 | 0.138 | $0.769 \times 10^{-7}$ | 490 | |
| 100 | 840.01 | 2.219 | 0.0000203 | 0.137 | $0.738 \times 10^{-7}$ | 276 | |
| 120 | 828.96 | 2.307 | 0.124 | 0.135 | $0.710 \times 10^{-7}$ | 175 | |
| 140 | 816.94 | 2.395 | 0.0000080 | 0.133 | $0.686 \times 10^{-7}$ | 116 | |
| 160 | 805.89 | 2.483 | 0.0000056 | 0.132 | $0.663 \times 10^{-7}$ | 84 | |
| | | | Mercury | | | | |
| 0 | 13628.22 | 0.1403 | $0.124 \times 10^{-6}$ | 8.20 | $42.99 \times 10^{7}$ | 0.0288 | |
| 20 | 13579.04 | 0.1394 | $0.144 \times 10^{-6}$ | 8.69 | $46.06 \times 10^{7}$ | 0.0249 | 0.182 |
| 50 | 13505.84 | 0.1386 | $0.104 \times 10^{-6}$ | 9.40 | $50.22 \times 10^{7}$ | 0.0207 | |
| 100 | 13384.58 | 0.1373 | $0.0928 \times 10^{-6}$ | 10.51 | $57.16 \times 10^{7}$ | 0.0162 | |
| 150 | 13264.28 | 0.1365 | $0.0853 \times 10^{-6}$ | 11.49 | $63.54 \times 10^{7}$ | 0.0134 | |
| 200 | 13144.94 | 0.1570 | $0.0802 \times 10^{-6}$ | 12.34 | $69.08 \times 10^{7}$ | 0.0116 | |
| 250 | 13025.60 | 0.1357 | $0.0765 \times 10^{-6}$ | 13.07 | $74.06 \times 10^{7}$ | 0.0103 | |
| 315.5 | 12847 | 0.134 | $0.0673 \times 10^{-6}$ | 14.02 | $81.50 \times 10^{7}$ | 0.0083 | |

**Table A.4** Properties of air at atmospheric pressure $\mu$, $k$ $c_p$ and Pr only weakly depend on pressure and may be used over a fairly wide range of pressure.

| $T$ K | $\rho$ kg/m$^3$ | $c_p$ kJ/(kg °C) | $\mu \times 10^5$ kg/(m s) | $k$ W/(m °C) | $\alpha \times 10^4$ m$^2$/s | $\nu \times 10^6$ m$^2$/s | Pr |
|---|---|---|---|---|---|---|---|
| 100 | 3.6010 | 1.0266 | 0.6924 | 0.009246 | 0.02501 | 1.923 | 0.770 |
| 150 | 2.3635 | 1.0099 | 1.0283 | 0.013735 | 0.05745 | 4.343 | 0.753 |
| 200 | 1.7684 | 1.0061 | 1.3289 | 0.018090 | 0.10165 | 7.490 | 0.739 |
| 250 | 1.4128 | 1.0053 | 1.5990 | 0.022270 | 0.15675 | 11.31 | 0.722 |
| 300 | 1.1774 | 1.0057 | 1.8462 | 0.026240 | 0.22160 | 15.69 | 0.708 |
| 350 | 0.9980 | 1.0090 | 2.0750 | 0.030030 | 0.29830 | 20.76 | 0.697 |
| 400 | 0.8826 | 1.0140 | 2.2860 | 0.033650 | 0.37600 | 25.90 | 0.689 |
| 450 | 0.7833 | 1.0207 | 2.4840 | 0.037070 | 0.42220 | 31.71 | 0.683 |
| 500 | 0.7048 | 1.0295 | 2.6710 | 0.040380 | 0.55640 | 37.90 | 0.680 |
| 550 | 0.6423 | 1.0392 | 2.8480 | 0.043600 | 0.65320 | 44.34 | 0.680 |
| 600 | 0.5879 | 1.0551 | 3.0180 | 0.046590 | 0.75120 | 51.34 | 0.680 |
| 650 | 0.5430 | 1.0635 | 3.1770 | 0.049530 | 0.85780 | 58.51 | 0.682 |
| 700 | 0.5030 | 1.0752 | 3.3320 | 0.052300 | 0.96720 | 66.25 | 0.684 |
| 750 | 0.4703 | 1.0856 | 3.4810 | 0.055090 | 1.07740 | 73.91 | 0.686 |
| 800 | 0.4405 | 1.0978 | 3.6250 | 0.057790 | 1.19510 | 82.29 | 0.689 |
| 850 | 0.4149 | 1.1095 | 3.7650 | 0.060280 | 1.30970 | 90.75 | 0.692 |
| 900 | 0.3925 | 1.1212 | 3.8990 | 0.062790 | 1.42710 | 99.30 | 0.696 |
| 950 | 0.3716 | 1.1321 | 4.0230 | 0.065250 | 1.55100 | 108.2 | 0.699 |

**Table A.4** Continued.

| $T$ K | $\rho$ kg/m$^3$ | $c_p$ kJ/(kg °C) | $\mu \times 10^5$ kg/(m s) | $k$ W/(m °C) | $\alpha \times 10^4$ m$^2$/s | $\nu \times 10^6$ m$^2$/s | Pr |
|---|---|---|---|---|---|---|---|
| 1000 | 0.3524 | 1.1417 | 4.1520 | 0.067250 | 1.67790 | 117.8 | 0.702 |
| 1100 | 0.3204 | 1.1600 | 4.4400 | 0.073200 | 1.96900 | 138.6 | 0.704 |
| 1200 | 0.2947 | 1.1790 | 4.6900 | 0.078200 | 2.25100 | 159.1 | 0.707 |
| 1300 | 0.2707 | 1.1970 | 4.9300 | 0.083700 | 2.58300 | 182.1 | 0.705 |
| 1400 | 0.2515 | 1.2140 | 5.1700 | 0.089100 | 2.92000 | 205.5 | 0.705 |
| 1500 | 0.2355 | 1.2300 | 5.4000 | 0.094600 | 3.26200 | 229.1 | 0.705 |
| 1600 | 0.2211 | 1.2480 | 5.6300 | 0.100000 | 3.60900 | 254.5 | 0.705 |
| 1700 | 0.2082 | 1.2670 | 5.8500 | 0.105000 | 3.97700 | 280.5 | 0.705 |
| 1800 | 0.1970 | 1.2870 | 6.0700 | 0.111000 | 4.37900 | 308.1 | 0.704 |
| 1900 | 0.1858 | 1.3090 | 6.2900 | 0.117000 | 4.81100 | 338.5 | 0.704 |
| 2000 | 0.1762 | 1.3380 | 6.5000 | 0.124000 | 5.26000 | 369.0 | 0.702 |
| 2100 | 0.1682 | 1.3720 | 6.7200 | 0.131000 | 5.71500 | 399.6 | 0.700 |
| 2200 | 0.1602 | 1.4190 | 6.9300 | 0.139000 | 6.12000 | 432.6 | 0.707 |
| 2300 | 0.1538 | 1.4820 | 7.1400 | 0.149000 | 6.54000 | 464.0 | 0.710 |
| 2400 | 0.1458 | 1.5740 | 7.3500 | 0.161000 | 7.02000 | 504.0 | 0.718 |
| 2500 | 0.1394 | 1.6880 | 7.5700 | 0.175000 | 7.44100 | 543.5 | 0.730 |

**Table A.5** Properties of some gases at atmospheric pressure $\mu$, $k$ $c_p$ and Pr only weakly depend on pressure and may be used over a fairly wide range of pressure.

| $T$ K | $\rho$ $kg/m^3$ | $c_p$ kJ/(kg °C) | $\mu \times 10^5$ kg/(m s) | $k$ W/(m °C) | $\alpha \times 10^4$ $m^2/s$ | Pr |
|---|---|---|---|---|---|---|
| | | | Helium | | | |
| 144 | 0.33790 | 5.200 | 125.5 | 0.0928 | 0.5275 | 0.70 |
| 200 | 0.24350 | 5.200 | 156.6 | 0.1177 | 0.9288 | 0.694 |
| 255 | 0.19060 | 5.200 | 181.7 | 0.1357 | 1.3675 | 0.70 |
| 366 | 0.13280 | 5.200 | 230.5 | 0.1691 | 2.4490 | 0.71 |
| 477 | 0.10204 | 5.200 | 275.0 | 0.1970 | 3.7160 | 0.72 |
| 589 | 0.08282 | 5.200 | 311.3 | 0.2250 | 5.2150 | 0.72 |
| 700 | 0.07032 | 5.200 | 347.5 | 0.2510 | 6.6610 | 0.72 |
| 800 | 0.06023 | 5.200 | 381.7 | 0.2750 | 8.7740 | 0.72 |
| | | | Hydrogen | | | |
| 150 | 0.16371 | 12.602 | 5.595 | 0.0981 | 0.475 | 0.718 |
| 200 | 0.12270 | 13.540 | 6.813 | 0.1282 | 0.772 | 0.719 |
| 250 | 0.09819 | 14.059 | 7.919 | 0.1561 | 1.130 | 0.713 |
| 300 | 0.08185 | 14.314 | 8.693 | 0.1820 | 1.554 | 0.706 |
| 350 | 0.07016 | 14.346 | 9.954 | 0.2060 | 2.031 | 0.697 |
| 400 | 0.06135 | 14.491 | 10.864 | 0.2280 | 2.568 | 0.690 |
| 450 | 0.05462 | 14.499 | 11.799 | 0.2510 | 3.164 | 0.682 |
| 500 | 0.04918 | 14.507 | 12.636 | 0.2720 | 3.817 | 0.675 |
| 550 | 0.04469 | 14.532 | 13.475 | 0.2920 | 4.561 | 0.668 |
| 600 | 0.04085 | 14.537 | 14.285 | 0.3150 | 5.306 | 0.664 |
| 700 | 0.03492 | 14.574 | 15.890 | 0.3510 | 6.903 | 0.660 |
| 800 | 0.03060 | 14.675 | 17.400 | 0.3840 | 8.563 | 0.665 |
| 900 | 0.02723 | 14.821 | 18.780 | 0.4120 | 10.217 | 0.676 |

**Table A.5** Continued.

| $T$ K | $\rho$ kg/m$^3$ | $c_p$ kJ/(kg °C) | $\mu \times 10^5$ kg/(m s) | $k$ W/(m °C) | $\alpha \times 10^4$ m$^2$/s | Pr |
|---|---|---|---|---|---|---|
| | | | Oxygen | | | |
| 150 | 2.6190 | 0.9178 | 11.49 | 0.01367 | 0.05688 | 0.773 |
| 200 | 1.9559 | 0.9131 | 14.85 | 0.01824 | 0.10214 | 0.745 |
| 250 | 1.5618 | 0.9157 | 17.87 | 0.02259 | 0.15794 | 0.725 |
| 300 | 1.3007 | 0.9203 | 20.63 | 0.02676 | 0.22353 | 0.709 |
| 350 | 1.1133 | 0.9291 | 23.16 | 0.03070 | 0.29680 | 0.702 |
| 400 | 0.9755 | 0.9420 | 25.54 | 0.03461 | 0.37680 | 0.695 |
| 450 | 0.8682 | 0.9567 | 27.77 | 0.03828 | 0.46090 | 0.694 |
| 500 | 0.7801 | 0.9722 | 29.91 | 0.04173 | 0.55020 | 0.697 |
| 550 | 0.7096 | 0.9881 | 31.97 | 0.04517 | 0.64100 | 0.700 |
| | | | Nitrogen | | | |
| 200 | 1.7108 | 1.0429 | 12.947 | 0.01824 | o.10224 | 0.747 |
| 300 | 1.1421 | 1.0408 | 17.840 | 0.02620 | 0.22044 | 0.713 |
| 400 | 0.8538 | 1.0459 | 21.980 | 0.03335 | 0.37340 | 0.691 |
| 500 | 0.6824 | 1.0555 | 25.700 | 0.03984 | 0.55300 | 0.684 |
| 600 | 0.5687 | 1.0756 | 29.110 | 0.04580 | 0.74860 | 0.686 |
| 700 | 0.4934 | 1.0969 | 32.130 | 0.05123 | 0.94660 | 0.691 |
| 800 | 0.4277 | 1.1225 | 34.840 | 0.05609 | 1.16850 | 0.700 |
| 900 | 0.3796 | 1.1464 | 37.490 | 0.06070 | 1.39460 | 0.711 |
| 1000 | 0.3412 | 1.1677 | 40.000 | 0.06475 | 1.62500 | 0.724 |
| 1100 | 0.3108 | 1.1857 | 42.280 | 0.06850 | 1.85910 | 0.736 |
| 1200 | 0.2851 | 1.2037 | 44.500 | 0.07184 | 2.09320 | 0.748 |

**Table A.5** Continued.

| $T$ K | $\rho$ kg/m$^3$ | $c_p$ kJ/(kg °C) | $\mu \times 10^5$ kg/(m s) | $k$ W/(m °C) | $\alpha \times 10^4$ m$^2$/s | Pr |
|---|---|---|---|---|---|---|
| | | | Carbon dioxide | | | |
| 220 | 2.4733 | 0.783 | 11.105 | 0.010805 | 0.05920 | 0.818 |
| 250 | 2.1657 | 0.804 | 12.590 | 0.012884 | 0.07401 | 0.793 |
| 300 | 1.7973 | 0.871 | 14.958 | 0.016572 | 0.10588 | 0.770 |
| 350 | 1.5362 | 0.900 | 17.205 | 0.020470 | 0.14808 | 0.755 |
| 400 | 1.3424 | 0.942 | 19.320 | 0.024610 | 0.19463 | 0.738 |
| 450 | 1.1918 | 0.980 | 21.340 | 0.028970 | 0.24813 | 0.721 |
| 500 | 1.0732 | 1.013 | 23.260 | 0.033520 | 0.30840 | 0.702 |
| 550 | 0.9739 | 1.047 | 25.080 | 0.038210 | 0.37500 | 0.685 |
| 600 | 0.8938 | 1.076 | 26.830 | 0.043110 | 0.44830 | 0.668 |
| | | | Ammonia, $NH_3$ | | | |
| 273 | 0.7929 | 2.177 | 9.353 | 0.0220 | 0.1308 | 0.90 |
| 323 | 0.6487 | 2.177 | 11.035 | 0.0270 | 0.1920 | 0.88 |
| 373 | 0.5590 | 2.236 | 12.886 | 0.0327 | 0.2619 | 0.87 |
| 423 | 0.4934 | 2.315 | 14.672 | 0.0391 | 0.3432 | 0.87 |
| 473 | 0.4405 | 2.395 | 16.490 | 0.0467 | 0.4421 | 0.84 |
| | | | Water vapor | | | |
| 380 | 0.5863 | 2.060 | 12.71 | 0.0246 | 0.2036 | 1.060 |
| 400 | 0.5542 | 2.014 | 13.44 | 0.0261 | 0.2338 | 1.040 |
| 450 | 0.4902 | 1.980 | 15.25 | 0.0299 | 0.3070 | 1.010 |
| 500 | 0.4405 | 1.985 | 17.04 | 0.0339 | 0.3870 | 0.996 |
| 550 | 0.4005 | 1.997 | 18.84 | 0.0379 | 0.4750 | 0.991 |
| 600 | 0.3652 | 2.026 | 20.67 | 0.0422 | 0.5730 | 0.986 |
| 650 | 0.3380 | 2.056 | 22.47 | 0.0464 | 0.6660 | 0.995 |
| 700 | 0.3140 | 2.085 | 24.26 | 0.0505 | 0.7720 | 1.000 |
| 750 | 0.2931 | 2.119 | 26.04 | 0.0549 | 0.8830 | 1.005 |
| 800 | 0.2739 | 2.152 | 27.86 | 0.0592 | 1.0010 | 1.010 |
| 850 | 0.2579 | 2.186 | 29.69 | 0.0637 | 1.1310 | 1.019 |

**Table A.6** Properties of some low-melting metals

| Metal | Melting point °C | Boiling point °C | $\rho \times 10^{-3}$ kg/m$^3$ | $\mu \times 10^3$ kg/(m s) kJ/(kg °C) | Heat capacity W/(m °C) | $k$ | Pr |
|---|---|---|---|---|---|---|---|
| Bismuth | 271 | 1477 | 10.01 | 1.62 | 0.144 | 16.4 | 0.014 |
| Lead | 327 | 1737 | 10.5 | 2.40 | 0.159 | 16.1 | 0.024 |
| Lithium | 179 | 1317 | 0.51 | 0.60 | 4.19 | 38.1 | 0.065 |
| Mercury | −39 | 357 | 13.6 | 1.59 | 0.138 | 8.1 | 0.027 |
| Potassium | 63.8 | 760 | 0.81 | 0.37 | 0.796 | 45.0 | 0.0066 |
| Sodium | 97.8 | 883 | 0.90 | 0.43 | 1.34 | 80.3 | 0.0072 |

**Table A.7** Diffusion coefficients of some gases and vapors at 25°C and 1 atm

| Substance | $D$ cm$^2$/s | Sc $= \nu/D$ | Substance | $D$ cm$^2$/s | Sc $= \nu/D$ |
|---|---|---|---|---|---|
| Ammonia | 0.28 | 0.78 | Formic acid | 0.159 | 0.97 |
| Carbon dioxide | 0.164 | 0.94 | Acetic Acid | 0.133 | 1.16 |
| Hydrogen | 0.410 | 0.22 | Aniline | 0.073 | 2.14 |
| Oxygen | 0.206 | 0.75 | Benzene | 0.088 | 1.76 |
| Water | 0.256 | 0.60 | Toluene | 0.084 | 1.84 |
| Ethyl ether | 0.093 | 1.66 | Ethyl benzene | 0.077 | 2.01 |
| Methanol | 0.159 | 0.97 | Propyl benzene | 0.059 | 2.62 |
| Ethyl alcohol | 0.119 | 1.30 | | | |

**Table A.8** Properties of water (saturated liquid)

| $T$, °C | $c_p$, kJ/(kg °C) | $\rho$, kg/m$^3$ | $\mu$, kg/(m s) | $k$ W/(m °C) | Pr |
|---|---|---|---|---|---|
| 0 | 4.225 | 999.8 | $1.79 \times 10^{-3}$ | 0.566 | 13.25 |
| 4.44 | 4.208 | 999.8 | $1.55 \times 10^{-3}$ | 0.575 | 11.35 |
| 10 | 4.195 | 999.2 | $1.31 \times 10^{-3}$ | 0.585 | 9.40 |
| 15.56 | 4.186 | 998.6 | $1.12 \times 10^{-3}$ | 0.595 | 7.88 |
| 21.11 | 4.179 | 997.4 | $9.8 \times 10^{-4}$ | 0.604 | 6.78 |
| 26.67 | 4.179 | 995.8 | $8.6 \times 10^{-4}$ | 0.614 | 5.85 |
| 32.22 | 4.174 | 994.9 | $7.65 \times 10^{-4}$ | 0.623 | 5.12 |
| 37.78 | 4.174 | 993.0 | $6.82 \times 10^{-4}$ | 0.630 | 4.53 |
| 43.33 | 4.174 | 990.6 | $6.16 \times 10^{-4}$ | 0.637 | 4.04 |
| 48.89 | 4.174 | 988.8 | $5.62 \times 10^{-4}$ | 0.644 | 3.64 |
| 54.44 | 4.179 | 985.7 | $5.13 \times 10^{-4}$ | 0.649 | 3.30 |
| 60 | 4.179 | 983.3 | $4.71 \times 10^{-4}$ | 0.654 | 3.01 |
| 65.55 | 4.186 | 980.3 | $4.30 \times 10^{-4}$ | 0.659 | 2.73 |
| 71.11 | 4.186 | 977.3 | $4.01 \times 10^{-4}$ | 0.665 | 2.53 |
| 76.67 | 4.191 | 973.7 | $3.72 \times 10^{-4}$ | 0.668 | 2.33 |
| 82.22 | 4.195 | 970.2 | $3.47 \times 10^{-4}$ | 0.673 | 2.16 |
| 87.78 | 4.199 | 966.7 | $3.27 \times 10^{-4}$ | 0.675 | 2.03 |
| 93.33 | 4.204 | 963.2 | $3.06 \times 10^{-4}$ | 0.678 | 1.90 |
| 104.4 | 4.229 | 955.1 | $2.67 \times 10^{-4}$ | 0.684 | 1.66 |
| 115.6 | 4.229 | 946.7 | $2.44 \times 10^{-4}$ | 0.685 | 1.51 |
| 126.7 | 4.250 | 937.2 | $2.19 \times 10^{-4}$ | 0.685 | 1.36 |
| 137.8 | 4.271 | 928.1 | $1.98 \times 10^{-4}$ | 0.685 | 1.24 |
| 148.9 | 4.296 | 918.0 | $1.86 \times 10^{-4}$ | 0.684 | 1.17 |
| 176.7 | 4.467 | 890.4 | $1.57 \times 10^{-4}$ | 0.677 | 1.02 |
| 204.4 | 4.467 | 859.4 | $1.36 \times 10^{-4}$ | 0.665 | 1.00 |
| 232.2 | 4.585 | 825.7 | $1.20 \times 10^{-4}$ | 0.646 | 0.85 |
| 260 | 4.731 | 785.2 | $1.07 \times 10^{-4}$ | 0.616 | 0.83 |
| 287.7 | 5.024 | 735.5 | $9.51 \times 10^{-5}$ | | |
| 315.6 | 5.703 | 678.7 | $8.68 \times 10^{-5}$ | | |

**Table A.9** Properties of saturated water

| $T$ | $p_{sat}$ | $\rho$ | | $h_{fg}$ | $c_p$ | | $k$ | | $\mu$ | | Pr | | $\beta \times 10^3$ |
|---|---|---|---|---|---|---|---|---|---|---|---|---|---|
| °C | kPa | kg/m$^3$ | | kJ/kg | J/(kg °C) | | W/(m °C) | | kg/(m s) | | | | 1/K |
| | | liq | vap | | liq | vap | liq | vap | liq | vap | liq | vap | |
| 0.01 | 0.6113 | 999.8 | 0.0048 | 2501 | 4217 | 1854 | 0.561 | 0.0171 | $1.792 \times 10^{-3}$ | $0.922 \times 10^{-5}$ | 13.5 | 1 | −0.068 |
| 5 | 0.8721 | 999.9 | 0.0068 | 2490 | 4205 | 1857 | 0.571 | 0.0173 | $1.519 \times 10^{-3}$ | $0.934 \times 10^{-5}$ | 11.2 | 1 | 0.015 |
| 10 | 1.2276 | 999.7 | 0.0094 | 2478 | 4194 | 1863 | 0.580 | 0.0176 | $1.307 \times 10^{-3}$ | $0.946 \times 10^{-5}$ | 9.45 | 1 | 0.733 |
| 15 | 1.7051 | 999.1 | 0.0128 | 2466 | 4186 | 1863 | 0.589 | 0.0179 | $1.138 \times 10^{-3}$ | $0.959 \times 10^{-5}$ | 8.09 | 1 | 0.138 |
| 20 | 2.339 | 998.0 | 0.0173 | 2454 | 4182 | 1867 | 0.598 | 0.0182 | $1.002 \times 10^{-3}$ | $0.973 \times 10^{-5}$ | 7.01 | 1 | 0.195 |
| 25 | 3.169 | 997.0 | 0.0231 | 2442 | 4180 | 1870 | 0.607 | 0.0186 | $0.891 \times 10^{-3}$ | $0.987 \times 10^{-5}$ | 6.14 | 1 | 0.247 |
| 30 | 4.246 | 996.0 | 0.0304 | 2431 | 4178 | 1875 | 0.615 | 0.0189 | $0.798 \times 10^{-3}$ | $1.001 \times 10^{-5}$ | 5.42 | 1 | 0.294 |
| 35 | 5.628 | 994.0 | 0.0397 | 2419 | 4178 | 1880 | 0.623 | 0.0192 | $0.720 \times 10^{-3}$ | $1.016 \times 10^{-5}$ | 4.83 | 1 | 0.337 |
| 40 | 7.384 | 992.1 | 0.0512 | 2407 | 4179 | 1885 | 0.631 | 0.0196 | $0.653 \times 10^{-3}$ | $1.031 \times 10^{-5}$ | 4.32 | 1 | 0.377 |
| 45 | 9.593 | 990.1 | 0.0655 | 2395 | 4180 | 1892 | 0.637 | 0.0200 | $0.596 \times 10^{-3}$ | $1.046 \times 10^{-5}$ | 3.91 | 1 | 0.415 |
| 50 | 12.35 | 988.0 | 0.0831 | 2383 | 4181 | 1900 | 0.644 | 0.0204 | $0.547 \times 10^{-3}$ | $1.062 \times 10^{-5}$ | 3.55 | 1 | 0.451 |

**Table A.9** Continued.

| $T$ | $p_{sat}$ | $\rho$ | | $h_{fg}$ | $c_p$ | | $k$ | | $\mu$ | | Pr | | $\beta \times 10^3$ |
|---|---|---|---|---|---|---|---|---|---|---|---|---|---|
| °C | kPa | kg/m$^3$ | | kJ/kg | J/(kg °C) | | W/(m °C) | | kg/(m s) | | | | 1/K |
| | | liq | vap | | liq | vap | liq | vap | liq | vap | liq | vap | |
| 55 | 15.76 | 985.2 | 0.0145 | 2371 | 4183 | 1908 | 0.649 | 0.0208 | $0.504 \times 10^{-3}$ | $1.077 \times 10^{-5}$ | 3.25 | 1 | 0.484 |
| 60 | 19.74 | 983.3 | 0.1304 | 2359 | 4185 | 1916 | 0.654 | 0.0212 | $0.467 \times 10^{-3}$ | $1.093 \times 10^{-5}$ | 2.99 | 1 | 0.517 |
| 65 | 25.03 | 980.4 | 0.1614 | 2346 | 4187 | 1926 | 0.659 | 0.0216 | $0.433 \times 10^{-3}$ | $1.110 \times 10^{-5}$ | 2.75 | 1 | 0.548 |
| 70 | 31.19 | 977.5 | 0.1983 | 2334 | 4190 | 1936 | 0.663 | 0.0221 | $0.404 \times 10^{-3}$ | $1.126 \times 10^{-5}$ | 2.55 | 1 | 0.578 |
| 75 | 38.58 | 974.7 | 0.2421 | 2321 | 4193 | 1948 | 0.667 | 0.0225 | $0.378 \times 10^{-3}$ | $1.142 \times 10^{-5}$ | 2.38 | 1 | 0.607 |
| 80 | 47.39 | 971.8 | 0.2935 | 2309 | 4197 | 1962 | 0.670 | 0.0230 | $0.355 \times 10^{-3}$ | $1.159 \times 10^{-5}$ | 2.22 | 1 | 0.653 |
| 85 | 57.83 | 968.1 | 0.3536 | 2296 | 4201 | 1977 | 0.673 | 0.0235 | $0.333 \times 10^{-3}$ | $1.176 \times 10^{-5}$ | 2.08 | 1 | 0.670 |
| 90 | 70.14 | 965.3 | 0.4235 | 2283 | 4206 | 1993 | 0.675 | 0.0240 | $0.315 \times 10^{-3}$ | $1.193 \times 10^{-5}$ | 1.96 | 1 | 0.702 |
| 95 | 84.55 | 961.5 | 0.5045 | 2270 | 4212 | 2010 | 0.677 | 0.0246 | $0.297 \times 10^{-3}$ | $1.210 \times 10^{-5}$ | 1.85 | 1 | 0.716 |
| 100 | 101.33 | 957.9 | 0.5978 | 2257 | 4217 | 2029 | 0.679 | 0.0251 | $0.282 \times 10^{-3}$ | $1.227 \times 10^{-5}$ | 1.75 | 1 | 0.750 |

**Table A.9** Continued.

| $T$ | $p_{\text{sat}}$ | $\rho$ | | $h_{fg}$ | $c_p$ | | $k$ | | $\mu$ | | Pr | | $\beta \times 10^3$ |
|---|---|---|---|---|---|---|---|---|---|---|---|---|---|
| °C | kPa | kg/m$^3$ | | kJ/kg | J/(kg °C) | | W/(m °C) | | kg/(m s) | | | | 1/K |
| | | liq | vap | | liq | vap | liq | vap | liq | vap | liq | vap | |
| 110 | 143.27 | 950.6 | 0.8263 | 2230 | 4229 | 2071 | 0.682 | 0.0262 | $0.255 \times 10^{-3}$ | $1.261 \times 10^{-5}$ | 1.58 | 1 | 0.798 |
| 120 | 198.53 | 943.4 | 1.121 | 2203 | 4244 | 2120 | 0.683 | 0.0275 | $0.232 \times 10^{-3}$ | $1.296 \times 10^{-5}$ | 1.44 | 1 | 0.858 |
| 130 | 270.1 | 934.6 | 1.496 | 2174 | 4263 | 2177 | 0.684 | 0.0288 | $0.213 \times 10^{-3}$ | $1.330 \times 10^{-5}$ | 1.33 | 1.01 | 0.913 |
| 140 | 361.3 | 921.7 | 1.965 | 2145 | 4286 | 2244 | 0.683 | 0.0301 | $0.197 \times 10^{-3}$ | $1.365 \times 10^{-5}$ | 1.24 | 1.02 | 0.970 |
| 150 | 475.8 | 916.6 | 2.546 | 2114 | 4311 | 2314 | 0.682 | 0.0316 | $0.183 \times 10^{-3}$ | $1.399 \times 10^{-5}$ | 1.16 | 1.02 | 1.025 |
| 160 | 617.8 | 907.4 | 3.256 | 2083 | 4340 | 2420 | 0.680 | 0.0331 | $0.170 \times 10^{-3}$ | $1.434 \times 10^{-5}$ | 1.09 | 1.05 | 1.145 |
| 170 | 791.7 | 897.7 | 4.119 | 2050 | 4370 | 2490 | 0.677 | 0.0347 | $0.160 \times 10^{-3}$ | $1.468 \times 10^{-5}$ | 1.03 | 1.05 | 1.178 |
| 180 | 1002.1 | 887.3 | 5.153 | 2015 | 4410 | 2590 | 0.673 | 0.0364 | $0.150 \times 10^{-3}$ | $1.502 \times 10^{-5}$ | 0.983 | 1.07 | 1.210 |
| 190 | 1254.4 | 876.4 | 6.338 | 1979 | 4460 | 2710 | 0.669 | 0.0382 | $0.142 \times 10^{-3}$ | $1.537 \times 10^{-5}$ | 0.947 | 1.09 | 1.280 |
| 200 | 1553.8 | 864.3 | 7.852 | 1941 | 4500 | 2840 | 0.663 | 0.0401 | $0.134 \times 10^{-3}$ | $1.571 \times 10^{-5}$ | 0.910 | 1.11 | 1.350 |

**Table A.9** Continued.

| $T$ | $p_{sat}$ | $\rho$ | | $h_{fg}$ | $c_p$ | | $k$ | | $\mu$ | | Pr | | $\beta \times 10^3$ |
|---|---|---|---|---|---|---|---|---|---|---|---|---|---|
| °C | kPa | kg/m$^3$ | | kJ/kg | J/(kg °C) | | W/(m °C) | | kg/(m s) | | | | 1/K |
| | | liq | vap | | liq | vap | liq | vap | liq | vap | liq | vap | |
| 220 | 2318 | 840.3 | 11.60 | 1859 | 4610 | 3110 | 0.650 | 0.0442 | $0.122 \times 10^{-3}$ | $1.641 \times 10^{-5}$ | 0.865 | 1.15 | 1.520 |
| 240 | 3344 | 813.7 | 16.73 | 1767 | 4760 | 3520 | 0.632 | 0.0487 | $0.111 \times 10^{-3}$ | $1.712 \times 10^{-5}$ | 0.836 | 1.24 | 1.720 |
| 260 | 4688 | 783.7 | 23.69 | 1663 | 4970 | 4070 | 0.609 | 0.0540 | $0.102 \times 10^{-3}$ | $1.788 \times 10^{-5}$ | 0.832 | 1.35 | 2.000 |
| 280 | 6412 | 750.8 | 33.15 | 1544 | 5280 | 4835 | 0.581 | 0.0605 | $0.094 \times 10^{-3}$ | $1.870 \times 10^{-5}$ | 0.854 | 1.49 | 2.380 |
| 300 | 8581 | 713.8 | 46.15 | 1405 | 5750 | 5980 | 0.548 | 0.0695 | $0.086 \times 10^{-3}$ | $1.965 \times 10^{-5}$ | 0.902 | 1.69 | 2.950 |
| 320 | 11,274 | 667.1 | 64.57 | 1239 | 6540 | 7900 | 0.509 | 0.0836 | $0.078 \times 10^{-3}$ | $2.084 \times 10^{-5}$ | 1.00 | 1.97 | - |
| 340 | 14,586 | 610.5 | 92.62 | 1028 | 8240 | 11870 | 0.469 | 0.110 | $0.070 \times 10^{-3}$ | $2.255 \times 10^{-5}$ | 1.23 | 2.43 | - |
| 360 | 18.651 | 528.3 | 144.0 | 720 | 14,690 | 25,800 | 0.427 | 0.178 | $0.060 \times 10^{-3}$ | $2.571 \times 10^{-5}$ | 2.06 | 3.73 | - |
| 374.14 | 22,090 | 317.0 | 317.0 | 0 | $\infty$ | $\infty$ | $\infty$ | $\infty$ | $0.043 \times 10^{-3}$ | $4.313 \times 10^{-5}$ | - | - | - |

# Index

For Product Safety Concerns and Information please contact our EU representative GPSR@taylorandfrancis.com Taylor & Francis Verlag GmbH, Kaufingerstraße 24, 80331 München, Germany